Prostaglandins and their Inhibitors
in Clinical Obstetrics and Gynaecology

Prostaglandins and their Inhibitors in Clinical Obstetrics and Gynaecology

Edited by

Marc Bygdeman, MD

Gary S. Berger, MD, MSPH

Louis G. Keith, MD

MTP PRESS LIMITED
a member of the KLUWER ACADEMIC PUBLISHERS GROUP
LANCASTER / BOSTON / THE HAGUE / DORDRECHT

Published in the UK and Europe by
MTP Press Limited
Falcon House
Lancaster, England

British Library Cataloguing in Publication Data

Prostaglandins and their inhibitors in clinical obstetrics and gynaecology.
  1. Prostaglandins—Therapeutic use
  2. Obstetrical pharmacology
  I. Bygdeman, M.   II. Berger, Gary S.
  III. Keith, Louis G.
  618   RG129.P7

ISBN-13: 978-94-011-6736-9   e-ISBN-13: 978-94-011-6734-5
DOI: 10.1007/978-94-011-6734-5

Published in the USA by
MTP Press
A division of Kluwer Boston Inc
190 Old Derby Street
Hingham, MA 02043, USA

Library of Congress Cataloging in Publication Data

Prostaglandins and their inhibitors in clinical obstetrics and gynaecology.

  Includes bibliographies and index.
  1. Prostaglandins—Therapeutic use. 2. Prostaglandins—Antagonists—Therapeutic
use. 3. Generative organs, Female—Effect of drugs on. I. Bygdeman, Marc.
II. Berger, Gary S. III. Keith, Louis G. [DNLM: 1. Genital diseases, Female—drug
therapy. 2. Genitalia, Female—drug effects. 3. Pregnancy—drug effects.
4. Prostaglandin Antagonists—physiology. 5. Prostaglandins—physiology.
WQ 200 P9663]
RG129.P7P77   1986   618.1'061   86–119
ISBN-13: 978-94-011-6736-9

Contents

List of Contributors

J.-J. AMY
Department of Gynecology,
  Andrology and Obstetrics
Free University of Brussels
Academic Hospital
Laarbeeklaan 101
B-1090 Brussels
Belgium

G. S. BERGER
Chapel Hill Fertility Services
109 Conner Drive, Suite 2104
Chapel Hill, NC 27514
USA

P. W. BUDOFF
Women's Medical Center
4300 Hempstead Turnpike
Bethpage, New York 11714
USA

M. BYGDEMAN
Department of Obstetrics and
  Gynecology
Karolinska Hospital
S-104 01 Stockholm
Sweden

A. A. CALDER
University Department of
  Obstetrics and Gynaecology
Royal Infirmary
10 Alexandra Parade
Glasgow G31 2ER
United Kingdom

K. GREEN
Department of Clinical
  Chemistry
Karolinska Hospital
S-104 01 Stockholm
Sweden

L. HAMBERGER
Department of Obstetrics and
  Gynecology
University of Gothenburg
S-413 45 Gothenburg
Sweden

P. O. JANSSON
Department of Obstetrics and
  Gynecology
University of Gothenburg
S-413 45 Gothenburg
Sweden

L. G. KEITH
Department of Obstetrics and
  Gynecology
Northwestern University Medical
  School
333 E. Superior Street
Chicago, IL 60611
USA

F. A. KIMBALL
The Upjohn Company
Kalamazoo MI 49001
USA

K. KIRTON
Unit 7253
The Upjohn Company
Building 126 - 3rd Floor
7000 Portage Avenue
Kalamazoo, MI 49001
USA

A. P. LANGE
Department of Obstetrics and
    Gynecology
Glostrup University Hospital
DK 2600 Copenhagen
Denmark

N. LAUERSEN
Department of Obstetrics and
    Gynecology
New York Medical College
784 Park Avenue
New York, NY 10021
USA

B. LINDBLOM
Department of Obstetrics and
    Gynecology
University of Gothenburg
S-413 45 Gothenburg
Sweden

V. LUNDSTRÖM
Department of Obstetrics and
    Gynecology
Karolinska Hospital
S-104 01 Stockholm
Sweden

E. L. MARUT
Department of Obstetrics and
    Gynecology
Michael Reese Hospital and
    Medical Center
Lake Shore Drive at 31st Street
Chicago, IL 60616
USA

L. NILSSON
Department of Obstetrics and
    Gynecology
University of Gothenburg
S-413 45 Gothenburg
Sweden

O. M. OWENS
Department of Obstetrics and
    Gynecology
University of Cincinnati
College of Medicine
Cincinnati, OH 45167
USA

F. B. PIPKIN
Department of Obstetrics and
    Gynaecology
University Hospital
Queen's Medical Centre
Nottingham NG7 2UH
United Kingdom

N. L. POYSER
Department of Pharmacology
University of Edinburgh Medical
    School
1 George Square
Edinburgh EH10 64A
United Kingdom

R. N. V. PRASAD
Department of Obstetrics and
    Gynaecology
National University Hospital
Kent Ridge
Singapore 0511

S. S. RATNAM
Department of Obstetrics and
    Gynaecology
National University Hospital
Kent Ridge
Singapore 0511

P. T. RUSSELL
Department of Obstetrics and
  Gynecology
University of Cincinnati
College of Medicine
Eden and Bethesda
Cincinnati, OH 45167
USA

E. M. SYMONDS
Department of Obstetrics and
  Gynaecology
University Hospital
Queen's Medical Centre
Nottingham NG7 2UH
United Kingdom

M. THIERY
Verloskundige Kliniek
Academisch Ziekenhuis
De Pintelaan 185
B-9000 Gent
Belgium

M. TOPPOZADA
Department of Obstetrics and
  Gynecology
University of Alexandria
Shatby Hospital for Women
Alexandria
Egypt

U. ULMSTEN
Department of Obstetrics and
  Gynecology
Karolinska Institutet
S-182 88 Stockholm
Sweden

Preface

It is clear today that several prostaglandins play an important role in the regulation of many of the physiological events of the reproductive organs in the human. Both naturally occurring prostaglandins and their analogues are used routinely in many countries to ripen the cervix and induce labour at term as well as to dilate the cervix and to terminate pregnancy. Prostaglandin biosynthesis inhibitors are widely used in the treatment of primary dysmenorrhoea.

The editors have aimed at an accurate, thorough, yet easily understandable review of the status in 1986 of medical knowledge regarding both the physiological importance and the clinical use of prostaglandins and their inhibitors in obstetrics and gynaecology.

I believe this book will be of value for all clinicians concerned with reproductive health. The list of authors guarantees an authoritative and up-to-date review of this active field.

SUNE BERGSTRÖM
Karolinska Institutet, Stockholm, Sweden

Table Examples of clinical use of natural prostaglandins and some prostaglandin analogues

Indications	Prostaglandin	Route of administration	Manufacturer
Dilatation of the cervix prior to vacuum aspiration	15-methyl-PGF$_{2\alpha}$ 16-phenoxy-PGE$_2$ methyl sulphonylamide 16,16-dimethyl-PGE$_1$ methyl ester	Intramuscular Intramuscular Vaginal	Upjohn, USA Schering AG, West Germany ONO, Japan; May & Baker, UK
Second trimester abortion	PGF$_{2\alpha}$ PGE$_2$ 15-methyl-PGF$_{2\alpha}$ Intra-amniotic 16-phenoxy-PGE$_2$ methyl sulphonylamide 16,16-dimethyl-PGE$_1$ methyl ester	Intra-amniotic Vaginal Intra-amniotic, intramuscular Intramuscular Vaginal	Upjohn, USA; ONO, Japan Upjohn, USA Upjohn, USA Schering AG, West Germany ONO, Japan; May & Baker, UK
Abnormal pregnancy (missed abortion, molar pregnancy)	15-methyl-PGF$_{2\alpha}$ 16-phenoxy-PGE$_2$ methyl sulphonylamide PGE$_2$	Intramuscular Intramuscular Vaginal	Upjohn, USA Schering AG, West Germany Upjohn, USA
Ripening of the cervix at or near term	PGE$_2$	Vaginal, intracervical or extra-amniotic	Upjohn, USA
Labour induction	PGE$_2$	Intravenous Oral, vaginal	Upjohn, USA; ONO, Japan Upjohn, USA
Post-partum haemorrhage	15-methyl-PGF$_{2\alpha}$	Intramuscular	Upjohn, USA

1
Prologue

M. BYGDEMAN, G. S. BERGER and L. G. KEITH

The present book regarding the clinical aspects of prostaglandins and their inhibitors is being published exactly 50 years after von Euler named the smooth muscle stimulatory agent of seminal and prostate extracts *prostaglandin*. The introduction of the term *prostaglandin* followed the discovery 5 years earlier by Krusrok and Leib that semen introduced into the vagina during artificial insemination could stimulate intense uterine activity. Following von Euler's recommendations, Bergström at the Karolinska Institute in Stockholm initiated two decades of work culminating in the isolation of the first two prostaglandins (PGE_1 and $PGF_{1\alpha}$) and subsequently described the chemical structures of many more prostaglandins by the late 1960s. In the two decades immediately past, tens of thousands of scientific articles have been reported in the medical literature characterizing the nature and function of these compounds. This enormous effort has uncovered the tip of an iceberg about which much remains enigmatic. While much remains unknown about the prostaglandins and their role in health and disease, certain features are known to have definite clinical implications for physicians concerned with reproductive health. It is to these physicians that this book is primarily directed.

Prostaglandins have been described as locally active hormone-like substances which are made by nearly every tissue in the body. The natural prostaglandins are unsaturated fatty acids which are derived from phospholipids, which at first have been converted to arachidonic acid, then subsequently to an intermediate form called endoperoxide and finally converted into prostaglandins of which numerous individual subtypes have been categorized. The multiple steps are called the arachidonic acid cascade.

The determination of which cells produce which type of prostaglandins is genetically determined by the type of enzymes produced in each cell. The quantity of prostaglandins produced is dependent upon the availability of precursors as well as substances which are capable of blocking one or more steps in the enzymatic conversion of the precursors described in the arachidonic acid cascade. As more has become known about the prostaglandins, the terminology and nomenclature has become increasingly complex and often is confusing to the non-biochemist. For example, additional groups of compounds, termed *thromboxanes* and *prostacyclines*, have also been described; these are prostaglandins but merit different names because of their sites of production (in the thrombocytes in the case of thromboxane and in blood vessel walls in the case of prostacyclines). Also confusing are the facts that a specific prostaglandin may exert (1) multiple and different effects, based upon which organ of the body is affected; and (2) different effects on the same organ at varying concentrations or even in similar concentrations at different times, for example, in the menstrual cycle. Thus, the function of these hormone-like substances is not unique for each chemical entity but depends upon a variety of other factors and regulatory mechanisms which may or may not be operative at a given time. For example, prostaglandin E_2 (PGE_2) affects the smooth muscle of the uterus by causing contraction, the smooth muscle of the cervix by affecting relaxation, and also results in the contraction of the vascular bed of the lung while at the same time causing relaxation of the vascular bed in other organs. The explanation for these variable effects in different tissues is not known at present, but apparently is related to the type and availability of receptors in different tissues.

The prostaglandins are produced intracellularly and transported locally where they exert their effects on other cells of the same organ or tissue. One situation where prostaglandins act as classical hormones in exerting an effect at a distant target organ is in animals, where prostaglandin $F_{2\alpha}$ produced by the endometrium has been demonstrated to be transported to the ovary by the uterine vein, from which it passes into the ovarian artery by a counter-current mechanism, and then results in dissolution of the corpus luteum. In most situations, however, the prostaglandins act within the cell in which they are produced or in immediately neighbouring cells, reaching them through the intercellular space.

Prostaglandins have been described as *modulators* of responses to exogenous and endogenous stimulators or inhibitors such as hormones or nervous stimulation. For example, the effect of LH stimulation to the ovary at a given time in the reproductive cycle is influenced by the production of prostaglandin $F_{2\alpha}$ which has a regulatory effect on the availability of LH receptors. The local production of prostaglandins under the influence of hormones thus results in a direct effect on the end organ. The process of cervical ripening, for example, is modulated by the increase of prostaglandin production in response to changing ratios of oestrogen and progesterone, and it is the increased prostaglandin concentration which results in increased collagenase activity which leads to disintegration of the collagen fibrils.

While prostaglandins are hormone-like substances, they differ from steroid hormones in their chemical structure, their synthetic pathways, and in their mechanism of action. Moreover, they serve as intermediates to the steroid hormones in their functional effects. In contrast to the classical hormones, the prostaglandins are synthesized immediately before release, are not stored, and exist in their natural form for only a matter of seconds.

Given the history of their discovery in prostatic secretions and seminal fluid which contain huge quantities of prostaglandins in comparison with other body sites, it seems ironic that almost nothing is known about prostaglandin functions with respect to their effects on male reproductive physiology. This deficiency in our knowledge is underscored by a review of the existing prostaglandin literature which shows a ratio of over 100 investigations regarding the female reproductive physiology to each investigation regarding male reproductive physiology.

With this background in mind, the present book was conceived in hopes of having the most knowledgeable experts contribute individual chapters which would condense available knowledge with respect to practical clinical applications for the use of prostaglandins and their inhibitors in obstetrics and gynaecology and provide a rational basis for these clinical applications. In order to achieve this goal, the basic science of prostaglandins, including their structure, biosynthesis, metabolism, bioregulation and modes of action are described in Chapters 2 and 3. The following four chapters review the anatomy and physiology of the female reproductive system, subdivided into its separate organs (the cervix, the uterus, the Fallopian tube and the ovary). This is followed in Chapter 8 by a discussion of the process of implantation and in Chapter 9 by a summary of the meagre knowledge currently available regarding the relationship between prostaglandins and male fertility. The majority of Chapters 2-9 regarding basic science and reproductive physiology of the prostaglandins have been contributed by Swedish colleagues, reflecting the fact that most of the original discoveries regarding these topics have been contributed by the Swedish investigators.

Chapters 2-9 provide the basis for the clinical chapters (Chapters 10-16). Although we have requested that each author present clinically useful information in as simple a format as possible, such as with the use of summary flow diagrams to emphasize the key elements of each chapter, this volume is not a simple cookbook approach but a scholarly treatise which describes the rational basis on which clinical applications for the use of prostaglandins and their inhibitors in obstetrics and gynaecology exists.

Our efforts as the editors have concentrated on providing a uniformity of style to each chapter in the hope of making the presentation of the information as easy to digest as possible. We thank the authors for adhering to our recommendations and are grateful that we have been able to see this book come to fruition. We believe that it represents an accurate, thorough, yet easily understandable résumé of the status in 1986 of medical knowledge regarding the clinical use of prostaglandins and their inhibitors in obstetrics and gynaecology.

ACKNOWLEDGEMENTS

We are very grateful to Professor Sune Bergström, the 'father' of prostaglandins, for writing the preface to this book. Despite his tremendous achievements in research in prostaglandin biochemistry, we believe no one has devoted so much interest and effort in facilitating the progress of the clinical use of this family of compounds.

One person, Astrid Häggblad, deserves our special gratitude. She has skilfully and thoroughly redesigned, checked and retyped all manuscripts, an enormous piece of work, which has been the prerequisite for the uniformity of the book in its final appearance.

The Center for the Advancement of Reproductive Health, Chapel Hill, North Carolina, contributed to the support of this endeavour.

Last, but by no means least, it has been a pleasure to work with Mr D. Bloomer and his staff at MTP Press, in particular their editor Mr P. Johnstone.

SECTION I
BASIC SCIENCE

2
Bioregulation and mode of action

F. A. KIMBALL and K. T. KIRTON

INTRODUCTION

The natural prostaglandins are intimately associated with a formidable array of physiological processes[1]. Not unexpectedly, the administration of these compounds usually elicits a complex spectrum of physiological and pharmacological responses. Some of the naturally occurring prostaglandins such as PGE_1, E_2, $F_{2\alpha}$ have been synthesized and are currently available for clinical use[1,2]; however, the routine use of natural prostaglandins as therapeutic agents is limited by their chemical instability, lack of target tissue specificity, and rapid metabolic inactivation[3]. For ideal use, orally administered prostaglandins should remain available in therapeutic levels for several hours after ingestion, act on specific selected tissues, produce a minimum of undesirable effects, and have acceptable pharmaceutical stability. These challenges have been met with less than uniform degrees of success. The purpose of this chapter is to review the bioregulation, modes of action and the development of the analogues of prostaglandins in the field of obstetrics and gynaecology.

BIOREGULATION OF PROSTAGLANDIN SYNTHESIS

Nearly all mammalian tissues produce prostaglandins locally under the control of several enzymes. These enzymes are known collectively as prostaglandin synthetases, and are located in the microsomes of the cells in which the prostaglandins are produced. The biosynthesis of prostaglandins and related compounds (i.e. thromboxane, lipoxins, leukotrienes) proceeds via a series of reactions known as the 'arachidonic acid cascade'. In this series of reactions phospholipids, triglycerides and cholesterol esters are converted to free fatty acids (such as arachidonic) in the initial step of the cascade by acylhydrolase enzymes such as phospholipase (Figure 2.1). Only three free unsaturated fatty acids can serve as precursors for prostaglandin synthesis. It is the degree of unsaturation of the esterified fatty acid that dictates which

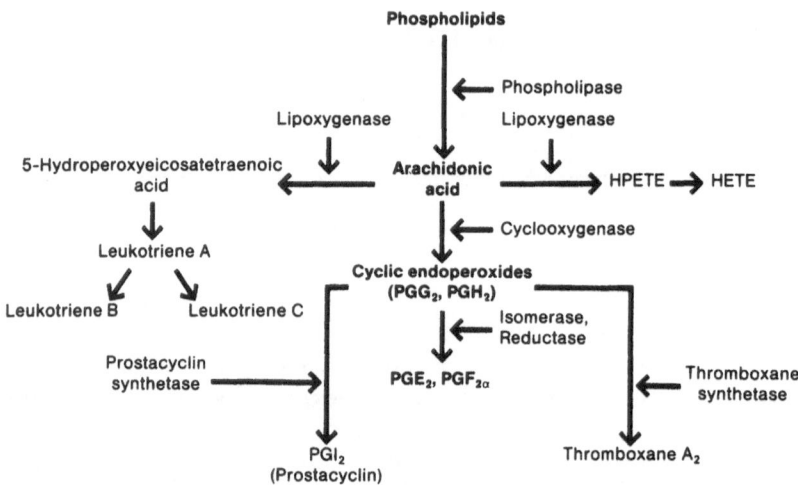

Figure 2.1 Pathway for the biosynthesis of prostaglandins and related compounds via th arachidonic acid cascade. (Adapted with permission from Dawood, M.Y. (1981). Hormones, prostaglandins, and dysmenorrhea. In Dawood, M.Y. (ed.) *Dysmenorrhea*. (Baltimore: Williams & Wilkins))

prostaglandin series, i.e. the E_1, E_2, or E_3, is synthesized. High concentrations of phospholipids are located in the plasma membrane of several mammalian tissues. Triglycerides are stored in nearly all adipose cells and cholesterol is concentrated especially in secretory or endometrial cells.

Arachidonic acid, a 20-carbon fatty acid with four unsaturated centres, is the precursor of the predominant group of (dienoic) prostaglandin, i.e. those with two double bonds (see Figure 2.1). It is formed from the essential fatty acid linoleate (18:2) by elongation and desaturation. The relative abundance of different classes of prostaglandins is controlled by a number of factors, such as the oxidation reduction state of the tissue, the presence of cofactors or the availability of metallic ions[4].

Phospholipids are hydrolysed to arachidonic acid by phospholipase, a lysosomal enzyme (see Figure 2.1). When phospholipase is released and activated, it converts phospholipids to arachidonic acid or other precursors, which then are converted into fatty acids called cyclic endoperoxides. Cyclic endoperoxide intermediates such as prostaglandin G_2 and H_2 have very short half-lives *in vivo*. They are, however, potent vasoconstrictors and uterotonic agents. Prostaglandin G_2 can be converted to PGE_2 by two pathways. In one, the reduction of the parent PGG_2 by the peroxidase enzyme is followed by an enzymatic isomerization step; in the other, the isomerization of PGG_2 to form 15-hydroperoxy-PGE_2 is followed by a reduction involving peroxidase. The dominance of one pathway or the other appears to depend on the tissue of origin. Other controlling factors are not known at present. Other prostaglandin compounds, such as those of the F category, are similarly synthesized from the same pool of cyclic endoperoxides by a similar enzymatic mechanism[5].

In most tissues, prostaglandin compounds are synthesized immediately before cellular release and are not stored for long periods of time. The factors that regulate prostaglandin biosynthesis in different tissues are not completely understood. However, the availability of precursors such as arachidonic acid appears to control the rate of synthesis in most instances. The availability of precursors in turn is controlled by the activity of the liberating enzyme. The following series of examples illustrates this point. Cyclic AMP triggers prostaglandin synthesis by causing lipases to liberate precursor acids. Therefore, the production of prostaglandins can be indirectly stimulated by cAMP-generating hormones, such as peptides, steroids and catecholamines[6].

Steroids active in the female endocrine system (such as oestrogens or progestins) are only one of the multitude of factors controlling the availability and activity of prostaglandins in reproductive tissues. For example, after oestrogen stimulation, prostaglandin production in endometrial tissue is correlated with oestradiol concentrations in uterine vein blood. Moreover, oestradiol increases the *in vitro* production of prostaglandins in endometrial tissue of several species.

Prostaglandin synthesis is stimulated in some instances by progesterone as well. For example, PGF_1 and $PGF_{2\alpha}$ concentrations in menstrual fluids are higher during ovulatory cycles than during non-ovulatory cycles. In addition, endometrial concentrations of PGE_2 and $PGF_{2\alpha}$ are higher during the secretory phase of the menstrual cycle than during the proliferative phase. High progesterone concentrations stabilize lysosomal membranes; stabilization may cause regression of the corpus luteum in several mammalian species[7].

In damaged or traumatized tissue, lysosomes are involved in the biosynthesis of prostaglandins. Cellular injury lowers intracellular pH thereby inducing the release of lysosomal enzymes. Once synthesized, the naturally occurring prostaglandins are rapidly metabolized to biologically inactive products. This metabolism is especially effective in peripheral circulation through the lungs or other normal metabolically active tissues[3].

MODES OF ACTION

The initial steps in the release of arachidonic acid require calcium to activate the esterases which remove arachidonic acid from phospholipids. This mechanism establishes a relationship between receptors that work through calcium and the release of arachidonic acid. It has been hypothesized that various arachidonic acid metabolites provide feedback which influences the calcium signalling system positively as well as negatively. Hydrolysis of phospholipids may function as a general mechanism to initiate a series of intracellular signals through receptors[8]. In support of this hypothesis, the concentration of phospholipids drops significantly during stimulation of smooth muscle, the injury of platelets and the activation of other arachidonic cascade-related phenomena. Such a decline in phospholipids may starve controlling systems of the essential precursor(s) necessary to translate external signals into internal secondary messengers. Inactivation systems have been described for some agonists, for example those dependent on calcium channels for the production of prostaglandins and for the generation of cyclic GMP. One hypothesis has proposed that the hydrolysis of phospholipids is part of a multifunctional mechanism responsible for generating several diverse signals. This would account for the basic similarity of all of these phenomena in which inactivation occurs when the cells run short of phospholipid[8].

Several prostaglandin-related reproductive processes involve an interaction with smooth muscle cells. This interaction is either direct, for example through stimulation of myometrial tissue, or indirect by mechanisms such as altering blood flow by constricting smooth muscle cells in the vascular tree[9]. All the details of the interaction of prostaglandins with smooth muscle have not yet been determined. It is believed that changes in the length of smooth muscle occur as a result of sliding of the thick and thin filaments. Many of the gross molecular components and molecular properties of striated muscle resemble those of smooth muscle. Therefore, the results of the more extensive studies with skeletal muscle may be extrapolated to smooth muscle mechanisms. Regulation of smooth muscle in various invertebrates is linked to calcium binding by myosin light chains. This myosin length regulation has been detected in smooth muscle of several tissues.

Early studies demonstrated that components other than actinomycin alone were required to activate ATPase. However, the ATPase system continues to be classified as myosin-linked, since these additional components may modify the myosin molecule. Pure actin and myosin form an inactive complex. The function of the regulatory components is to activate Mg-ATPase activity, but only in the presence of calcium ion. That activation constitutes the fundamental requirement for the regulatory mechanism of smooth muscle. The nature of activation and whether or not the system is myosin linked remains to be determined. A correlation between different degrees of phosphorylation and the degree of actin-activated ATPase must be established. In addition, the role of myosin light chain phosphatase in the regulatory process cannot be evaluated without more detailed information. However, available evidence strongly implicates these mechanisms in the activation of smooth muscle contractility[10,11].

In addition to acting on smooth muscle, the prostaglandins appear to interact with cells that synthesize and secrete hormones. Some tissues, such as the anterior pituitary and corpus luteum, respond directly to prostaglandin-induced alterations in their function. Moreover, the mechanisms underlying these actions have not yet been determined. It is known that prostaglandins inhibit LH-activated adenylcyclase[7]. In addition, recent studies have demonstrated that prostaglandins enter luteal cells and alter membrane fluidity[12]. This could be responsible for their negating the LH stimulation of progesterone secretion. The original hypothesis of Pharriss[13], which suggested a vascular mechanism for the regulation of menstrual function, may still be viable; however, prostaglandins do not appear to act on the large vessels draining the ovarian blood flow as originally hypothesized. The prostaglandin endoperoxides not only stimulate smooth muscle but also increase the sensitivity of neuroterminals to pain by lowering the threshold to pain-producing chemicals and mechanical stimuli. Such a mechanism could increase the sensitivity of a responsive tissue to external stimuli. This would allow an increase in response without a corresponding amplification of the signal. However, the exact mechanism of these activities has not yet been determined.

DEVELOPMENT OF ANALOGUES

The primary purpose of developing analogues for the prostaglandins was to produce compounds with higher therapeutic indices, greater specificity, longer half-lives, and more chemical and metabolic stability than those available with the natural compounds[3]. These shortcomings were logical targets for initial efforts. In this regard, the prostaglandins were somewhat unusual in that scientists very quickly determined the details of their metabolic degradation (Figure 2.2)[14,15]. This information then helped to stimulate and direct the synthesis of analogues. Initial

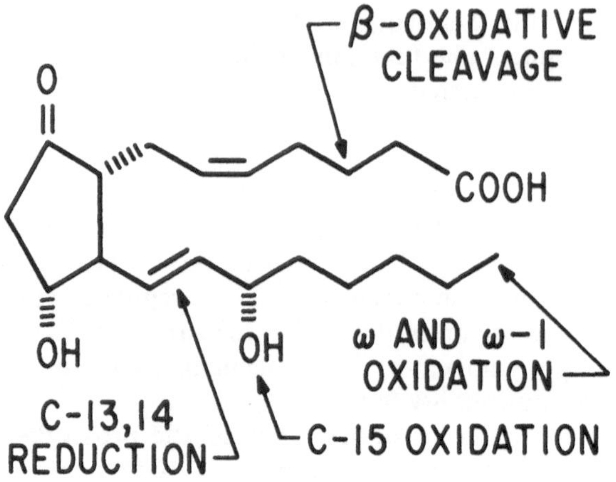

β-OXIDATIVE CLEAVAGE

O

COOH

OH   OH   ω AND ω-I OXIDATION

C-13,14 REDUCTION

C-15 OXIDATION

Figure 2.2   Sites of enzymatic inactivation of the prostaglandin molecule

efforts attempted to block the natural degradation by chemical modification of the naturally occurring molecules in the vicinity of carbon 15 (Figure 2.3). Several modifications were made near the 15-hydroxyl group[16], because the first metabolic degradation occurred at this location in most cases via the action of 15-hydroxyprostaglandin dehydrogenase. The resulting 15-methyl, 16,16-dimethyl, and 16,16-difluoro analogues are totally resistant to 15-hydroxy-prostaglandin dehydrogenase and have longer half-lives *in vitro* than naturally occurring prostaglandins. Modifications of the ω-chain (ω, Figure 2.2) near carbon 15 also confers resistance to the dehydrogenase enzyme as does

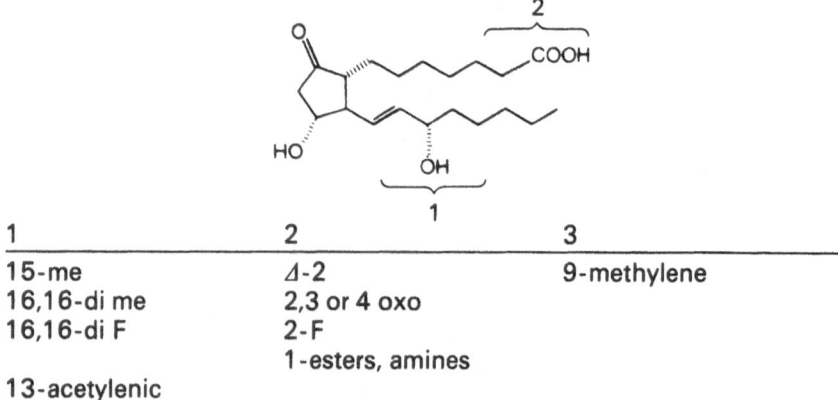

1	2	3
15-me	Δ-2	9-methylene
16,16-di me	2,3 or 4 oxo	
16,16-di F	2-F	
	1-esters, amines	
13-acetylenic		
16-phenoxy		

Figure 2.3   Some selected sites of chemical modification of the prostaglandin molecule

replacement of the 13-*trans*-double bond by an acetylenic unit and mono substitutions at C_{16}. In summary, alterations of the electronic or steric configurations of the molecule reduce the ease of oxidation of the 15-hydroxy atom, thereby increasing *in vivo* activity. In addition, substitution with aromatic groups at C_{17} provides analogues with greater biological activity and resistance to ω-oxidation. Some of these bulky substitutions have also inhibited 15-dehydrogenase activity[1,17].

When ω-chain degradation has been accomplished, β-oxidation of the α-chain becomes the primary means of inactivation. Substitution of alkyl groups or hetero atoms in the α-chain blocks this degradation. Unfortunately, with many of these substitutions, biological activity also decreases. On the other hand, substitution of aromatic rings for carbon atoms at positions 4–6, shifting the double bond to the 2 or 4 position, or introducing fluorine atoms at position 2, all block β-oxidation, usually without decreasing biological activity[1].

Improving chemical stability and increasing the likelihood of precipitation into the crystalline form have also been important targets for synthetic analogue programmes. An example is the 9-methylene series of E-type prostaglandins[18]. Derivatives, such as esters or amides, particularly attached at C_1, sometimes have had beneficial effects on crystalline precipitation and analogue stability. Molecular modifications such as these have achieved many of the original objectives of analogue design. These include increased stability, ease of formulation and prolonged *in vivo* activity.

Improving tissue specificity is more difficult and has been less successful in terms of analogues. On the other hand, efforts to separate gastrointestinal from reproductive system effects have been more successful. Synthetic molecules have been designed which retain fertility control efficacy with reduced gastrointestinal side-effects. Examples are the 9-methylene analogues (less gastrointestinal side-effects), and 17-phenoxy analogues (increased corpus luteum inhibiting activity).

INHIBITION OF SYNTHESIS

Glucocorticoids and non-steroidal anti-inflammatory agents (NSAIDs), such as indomethacin and aspirin, inhibit prostaglandin biosynthesis[19,20]. The precise modes of their action are still not known. However, the therapeutic and some of the toxic effects of these compounds (glucocorticoids and NSAIDs) are mediated through inhibition of prostaglandin biosynthesis. The sites of action of these compounds are illustrated in Figure 2.4. One group of NSAIDs suppress prostaglandin biosynthesis by inhibiting the cyclo-oxygenase enzyme required to convert arachidonic acid to the cyclic endoperoxides[21]. Consequently,

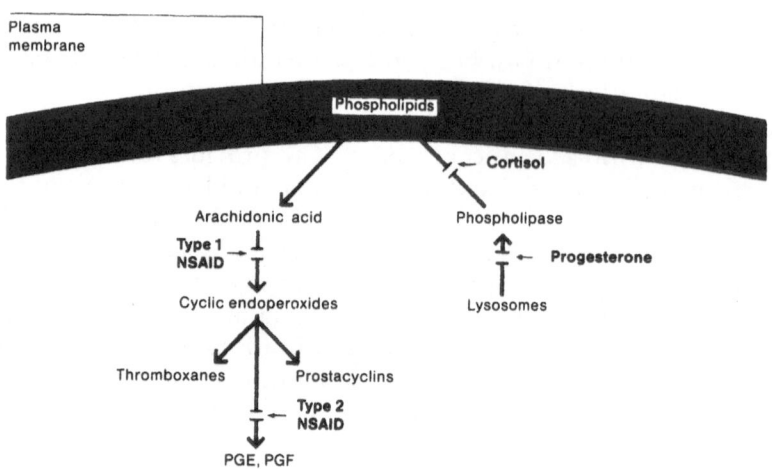

Figure 2.4 Effect of steroid hormones and prostaglandin synthetase inhibitors on biosynthesis of prostaglandins. (Adapted with permission from Dawood, M. Y. (1981). Hormones, prostaglandins, and dysmenorrhea. In Dawood, M. Y. (ed.) *Dysmenorrhea*. (Baltimore: Williams & Wilkins))

these agents reduce the concentration of all arachidonic acid biotransformation products but apparently do not significantly change the activity of peroxidase and endoperoxide isomerase enzymes. This group of agents includes aspirin, indomethacin and the fenamates.

The other group of NSAIDs includes, for example, phenylbutazone; these drugs inhibit endoperoxide isomerase and endoperoxide reductase. Therefore, their action follows the formation of the cyclic endoperoxides in the arachidonic acid cascade. These reagents are more selective and possess a potential therapeutic advantage as they do not inhibit or suppress the biosynthesis of prostacyclin or thromboxanes.

Glucocorticoids suppress prostaglandin release from some types of inflamed tissue. The fact that this suppression can be overcome by adding arachidonic acid to the cAMP generating system suggests that suppression occurs before arachidonic acid has been produced in the reaction sequence. When glucocorticoids are present and when the cyclo-oxygenase and lipoxygenase pathways are blocked, stimulated lung tissue does not release free arachidonic acid from phospholipids. Moreover, glucocorticoids do not interfere with phospholipase activity in cell-free tissue homogenates. These findings demonstrate that corticosteroids stabilize plasma membranes and protect arachidonic acid from phospholipase action. It has been suggested that the stabilizing effect of corticosteroids on plasma membranes inhibits inflammation and prevents the digestion of pathogenic micro-organisms[22]. Additional studies aimed at more clearly defining the cellular and subcellular location of the cyclo-oxygenase enzyme system will aid in evaluating the mechanisms by which corticosteroids control the bio-synthesis of prostanoids.

Prostaglandins have specific binding sites in several tissues including the corpus luteum and uterus[6]. Agents which directly inhibit the action of prostaglandins at such receptor sites are termed prostaglandin antagonists[23]. Examples include the phosphorylated polymers, 7-oxy-13-prostynoic acid and dibenzoxazapine derivatives. Because these compounds do not specifically inhibit binding to prostaglandin-binding sites, studies of the mechanisms of action of these agents have been difficult to carry out. Nevertheless, preliminary pharmacological studies with these inhibitors have assisted our understanding of the physiological functions of prostaglandins. To date, these studies have demonstrated a prostaglandin association with dysmenorrhoea, the ovulatory process in the dominant follicle, and the transport of gametes in the male and female reproductive tract.

References

1 Muchowski, J. M. (1984). Synthesic prostanoids. *Chemical Rubber Co. Handbook* (in press)
2 Kirton, K. T. and Kimball, F. A. (1984). Potential for prostaglandin use in controlling human reproduction. In Toppozada, M. and Bygdeman, M., Hafez, E. S. E. (eds.) *Prostaglandins and Fertility Regulation.* pp. 3-10 (Lancaster: MTP Press)
3 Bergstrom, S., Carlson, L. A. and Weeks, J. R. (1968). The prostaglandins: a family of biologically active lipids. *Pharmacologic Rev.,* **20,** 1-48
4 Lands, W. E. M. (1979). The biosynthesis and metabolism of prostaglandins. *Ann. Rev. Physiol.,* **41,** 633-52
5 Samuelsson, B. and Hamberg, M. (1974). Role of endoperoxides in the biosynthesis and action of prostaglandins. In Robinson, H. J. and Vane, J. R. (eds.) *Prostaglandin Synthetase Inhibitors.* pp. 107-19. (New York: Raven Press)
6 Gorman, R. R. (1978). Prostaglandins, thromboxanes and prostacyclin. In Rikenberg, H. V. (ed.) *International Review of Biochemistry, Biochemistry and Mode of Action of Hormones II.* pp. 81-107. (Baltimore: Univ. Park Press)
7 Behrman, H. R. (1979). Prostaglandins in hypothalamo-pituitary and ovarian function. *Ann. Rev. Physiol.,* **41,** 685-700
8 Berridge, M. J. (1984). A novel cellular signaling system based on the integration of phospholipid and calcium metabolism. In Cheung, W. Y. (ed.) *Calcium and Cell Function VIII.* pp. 1-32. (New York: Academic Press)
9 Kimball, F. A. and Kirton, K. T. (1977). Prostaglandins as antifertility

agents. In Goldberg, N. E. (ed.) *Pharmacological and Biochemical Properties of Drug Substances*. pp. 373–85. (Washington, DC: Am. Pharmaceutical Assn., Acad. of Pharmaceutical Sci.)

10 Adelstein, R. S. and Kee, C. B. (1980). Calcium and cell function. In Cheung, W. Y. (ed.) *Calmodulin*. Vol. 1, pp. 150–82. (New York: Academic Press)

11 Hartshorne, D. J. and Siemankowski, R. F. (1981). Regulation of smooth muscle actomyosin. In Edelman, I. S. (ed.) *Ann. Rev. Physiol.*, **43**, 519–30

12 Chegini, N. and Rao Ch. V. (1984). A quantitative electron microscope autoradiographic study of ^{3}H-PGE$_1$ and ^{3}H-PGF$_{2\alpha}$ internalization in bovine luteal slices. In Ryan R. J. (ed.) *Fifth Ovarian Workshop*, Laramie, Wyoming, July 21–23

13 Pharriss, B. B. (1970). The possible vascular regulation of luteal function. *Perspect. Biol. Med.*, **13**, 434–44

14 Anggard, E. and Samuelsson, B. (1966). Purification and properties of a 15-hydroxyprostaglandin dehydrogenase from swine lung. *Arkiv. Kem.*, **25**, 293–300

15 Bergstrom, S. (1967). Prostaglandins: Members of a new hormonal system. *Science*, **157**, 382–91

16 Bundy, G., Lincoln, F., Nelson, N., Pike, J. and Schneider, W. (1971). Novel prostaglandin syntheses. *Ann. NY Acad. Sci.*, **180**, 76–90

17 Nelson, N., Kelly, R. C. and Johnson, R. A. (1982). Prostaglandins and the arachidonic acid cascade. *Chem. Eng. News*, **60**, 1–15

18 Kimball, F. A., Bundy, G. L., Robert, A. and Weeks, J. R. (1979). Synthesis and biological properties of 9-deoxo-9-methylene-16,16-dimethyl-90-methylene-PGE$_2$. *Prostaglandins*, **17**, 657

19 Miyamoto, T., Yimamoto, S. and Hayaiski, O. (1974). Prostaglandin synthetase system – resolution into oxygenase and isomerase components. *Proc. Nat. Acad. Sci.*, **71**, 3645–8

20 Sannes, J. H. (1974). Substances that inhibit the actions of prostaglandins. *Arch. Intern. Med.*, **133**, 133–46

21 Vane, J. R. (1976). The mode of action of aspirin and similar compounds. *J. Allergy Clin. Immunol.*, **58**, 691–712

22 Schieren, H., Weissmann, G. and Seligman, M. (1978). Interaction of immunoglobulins with liposomes: an ESR study demonstrating protection by hydrocortisone. *Biochem. Biophys. Res. Commun.*, **82**, 1160–7

23 Samuelsson, B. (1973). Quantitative aspects of prostaglandin synthesis in man. In Bergstrom, S. and Bernhard, S. (eds.) *Advances in the Biosciences*. Vol. 9, pp. 7–14. (New York: Pergamon Press)

3
Structure, biosynthesis and metabolism

K. GRÉEN

INTRODUCTION

The purpose of this chapter is to review in more detail the chemical structure, biosynthesis and metabolism of the prostaglandins and related compounds.

CHEMICAL STRUCTURE

The chemical structures of the 'classical' or naturally occurring prostaglandins were elucidated during the late 1950s and early 1960s by S. Bergström and colleagues (see Ref. 1). They are all oxygenated unsaturated hydroxy fatty acids with 20 carbon atoms. The characteristic structure is the prostanoic acid molecule, a cyclopentane ring with two side-chains (Figure 3.1, top). Depending on the number of substituents on different carbon atoms and the structure in the five-membered ring, the different groups of prostaglandins have been designated with letters from A to I (Figure 3.1, bottom). All prostaglandins have one or more double bonds. The number of double bonds characterizes the members of each group and is indicated by numerical subscripts that follow the corresponding letter referring to the parent structure, e.g. E_1, E_2, and E_3 (Figure 3.2). Prostaglandins G_2 and H_2 have been characterized as short-lived intermediates in the reactions leading to formation of the classical prostaglandins as well as prostacyclin (PGI_2) and thromboxane A_2 (TxA_2), the latter two with different ring structures (Figure 3.3).

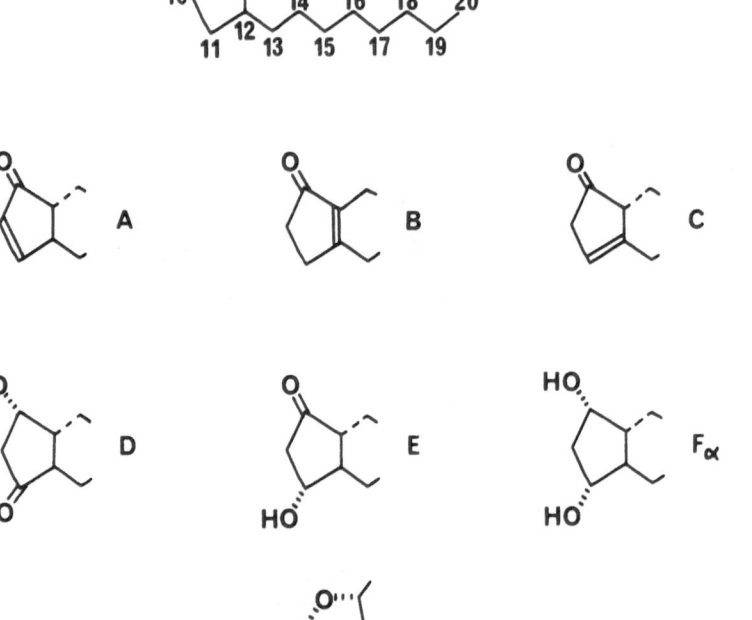

Figure 3.1   Basic structures of the different groups of prostaglandins

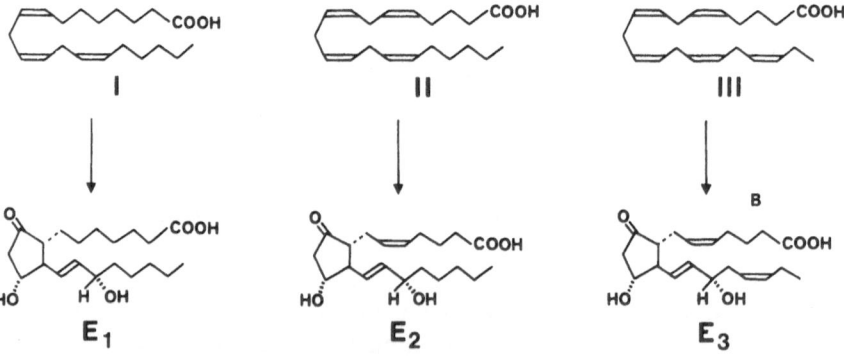

Figure 3.2   Structures of the essential precursor fatty acids and of PGE₁, PGE₂ and PGE₃. Fatty acid I: 8,11,14-all-*cis*-eicosatrienoic acid; fatty acid II: 5,8,11,14-all-*cis*-eicosatetraenoic acid (arachidonic acid); fatty acid III: 5,8,11,14,17-all-*cis*-eicosapentaenoic acid

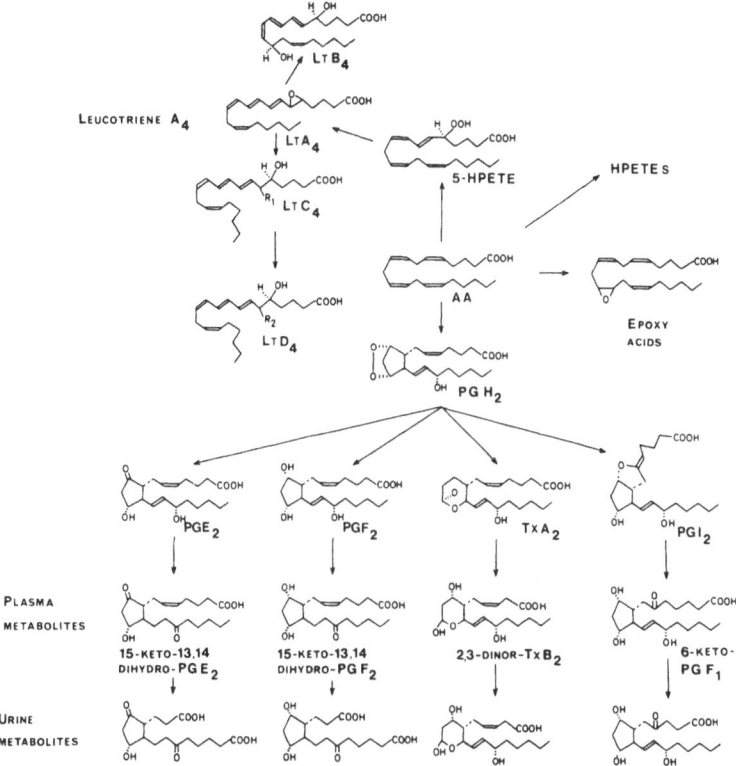

Figure 3.3 Summary of the bioconversion of arachidonic acid (AA) into prostaglandins and metabolism thereof and into epoxy acids, hydroperoxy acids and leukotrienes

BIOSYNTHESIS

All naturally occurring prostaglandins and thromboxanes are biosynthesized from three essential unsaturated fatty acids containing 20 carbon atoms. Of these, arachidonic acid (5,8,11,14-all-*cis*-eicosatetraenoic acid) is quantitatively the most important (Figure 3.3). The other two are 8,11,14-all-*cis*-eicosatrienoic and 5,8,11,14,17-all-*cis*-eicosapentaenoic acid and these give rise to prostaglandins of the 1 and 3 series, respectively.

Some of the presently known routes in the bioconversion of arachidonic acid are illustrated in Figure 3.3. The first step in the route leading to biosynthesis of PGE_2 and $PGF_{2\alpha}$, PGI_2 and TxA_2 is the formation of the cyclic endoperoxides PGG_2 and PGH_2 (only PGH_2 is shown in Figure 3.3). The other two precursor acids also undergo similar reactions resulting in the formation of prostaglandins with one double-bond (e.g. PGE_1) or three double bonds (e.g. PGE_3), respectively. The cyclo-oxygenase catalysed reactions are inhibited by several non-steroidal anti-inflammatory drugs (NSAIDs) such as aspirin and indomethacin. The endoperoxides can be converted non-enzymatically into PGE_2 or enzymatically into PGE_2, $PGF_{2\alpha}$, PGI_2, TxA_2 and other products depending on the type of cells in which the biosynthesis occurs[1,2].

Bioconversion of arachidonic acid through the action of lipoxygenases may lead to formation of various hydroperoxy-tetraenoic acids (HPETEs), which can be converted to hydroxy-tetraenoic acids (HETEs)[3] (see Figure 3.3, top). In addition, hydroperoxy acids of various structures can be formed non-enzymatically through auto-oxidation of unsaturated fatty acids.

Other reactions are also initiated by lipoxygenase activity; they begin with 5-HPETE as a common intermediate and lead to the formation of various leukotrienes (Figure 3.3, top left). LtA_4 is a biologically active epoxide that can be further transformed into LtB_4 or LtC_4 (Figure 3.3). The reactions leading to formation of LtC_4 involve coupling of glutathione to carbon 6 of LtA_4. Cleavage of one peptide bond in LtC_4 yields LtD_4 and cleavage of the second peptide bond yields LtE_4 (not shown in Figure 3.3)[3].

The biosynthesis of leukotrienes occurs in white blood cells especially upon exposure to ionophores. Like the classical prostaglandins, leukotrienes can also be formed from 8,11,14-all-*cis*-eicosatrienoic acid and 5,8,11,14,17-all-*cis*-eicosapentaenoic acid. These reactions yield leukotrienes of the 3 and 5 series (number of double bonds; e.g. LtC_3, LtC_5, etc.). Leukotrienes C and D are extremely potent bronchoconstrictors, up to 20 000 times more so than histamine, and their action is to increase the permeability of the microvessels. LtB_4, on the other hand, has a very strong chemotactic effect on leukocytes. It is not presently known whether leukotrienes, HPETEs, HETEs or epoxy acids are involved in the reproductive processes of humans, although some recent data indicate that such compounds may be involved in the immunology of human pregnancy[4].

In summary, the bioconversion of arachidonic acid can result in formation of a wide variety of biologically potent compounds. These compounds are collectively designated as 'products of the arachidonic acid cascade'. The arachidonic acid cascade in reality represents a very complex phenomenon. The multitude of compounds, some of which are probably formed simultaneously and possibly interact with each other's formation as well as biological effects, is evidence of this complexity. In this regard it is worth noting that non-steroidal anti-inflammatory drugs (NSAIDs) inhibit formation of PGG_2 and PGH_2, whereas anti-inflammatory steroids seem to inhibit the release of arachidonic acid from phospholipids and thus inhibit the formation of all products of the arachidonic acid cascade[5].

The formation of the endoperoxides, PGG_2 and PGH_2, as well as the compounds derived therefrom, appears to take place in all mammalian tissues. This process, as well as the further conversion of the endoperoxides into prostaglandins or thromboxane, is exceedingly rapid. Once biosynthesis is initiated, it is almost always completed within a few minutes in *in vitro* systems[6,7].

As a result of this rapid synthesis, 'tissue levels' of a given compound measured in biopsies do not reflect the level *in vivo* prior to taking the biopsy. Because platelets are also capable of forming PGE_2, $PGF_{2\alpha}$ and TxA_2, and because they accomplish this rapidly upon stimulation[6], the levels of PGE_2, $PGF_{2\alpha}$ and TxB_2 found in serum or plasma do not accurately reflect the corresponding levels in the circulation but are more representative of what is artifactually formed during collection of serum or plasma.

The biosynthesis of TxA_2 was first demonstrated in thrombocytes in which this compound apparently is the major product formed from arachidonic acid. Although small amounts of TxB_2 (a chemical degradation product of TxA_2) can be found in many tissues, this does not necessarily mean that the tissue cells *per se* are capable of synthesizing TxA_2, since thrombocytes are present in all vascularized tissues. The same is true for PGI_2 which seems to be synthesized primarily in vessel walls.

The enzyme responsible for conversion of arachidonic acid into PGH_2 (Figure 3.3) (variously referred to as 'fatty acid cyclo-oxygenase', 'PG endoperoxide synthetase' or 'PGH synthase') has been isolated from different tissue sources and studied in purified form[8]. This enzyme shows the following peculiarities *in vitro*: (1) It requires haeme as a cofactor, but in the presence of haeme enzymatic activity is lost relatively rapidly; (2) The rate of this 'self-catalysed destruction' is dependent on the haeme/enzyme ratio, but the enzyme is protected from destruction by the presence of other compounds such as tryptophan and epinephrine. It is important to keep these features in mind when interpreting data from incubation experiments in which the 'capacity' of tissue homogenates to convert arachidonic acid into prostaglandins is studied. The results of such experiments will be dependent not only on the amount of enzyme present, but also on the amount of haeme and the presence of protecting compounds.

INHIBITION OF BIOSYNTHESIS

Several groups of compounds are capable of inhibiting the biosynthesis of prostaglandins[9]. Among these are structural analogues of arachidonic acid, such as arachidynoic acid and the acetylenic 20-carbon acids, both of which have been used in *in vitro* systems for inhibition of prostaglandin synthesis. In addition, anti-inflammatory steroids suppress prostaglandin formation by reducing the liberation of precursor acids, e.g. arachidonic acid, from phospholipids. More important, however, are the non-steroidal anti-inflammatory drugs (NSAIDs). Most studies on the inhibitory effects of those compounds have been performed in various *in vitro* systems. In general, the formation of only one of all possible products (e.g. $PGF_{2\alpha}$) has been measured and used as indicator of the inhibition. The conversion of radiolabelled arachidonic acid into a prostaglandin has frequently been used to study the effects of NSAIDs. In these experiments the amount of radioactivity appearing as a prostaglandin has been interpreted as representing the formation of this prostaglandin. However, endogenous and exogenous arachidonic acid do not mix freely in *in vitro* systems, and therefore this type of methodology may lead to erroneous interpretations[10].

The effects of several different NSAIDs have been studied extensively in purified enzyme preparations and in more complex *in vitro* systems. The inhibitory effect seems to be exerted mainly through an interaction with the initial reaction in the biosynthesis of prostaglandins, e.g. the cyclo-oxygenase which converts arachidonic acid to PGH_2. It appears that the mechanism of action of various NSAIDs is different. Some seem to cause an irreversible inhibition (indomethacin, aspirin), while others such as diflunesal seem to act as reversible inhibitors[9]. An irreversible inhibitor should theoretically destroy the enzyme, and the subsequent recovery of prostaglandin biosynthesis would then depend on synthesis of new enzymes. For example, platelet cyclo-oxygenase is irreversibly inhibited through acetylation by aspirin[11]. Since circulating platelets are not capable of synthesizing proteins, aspirin ingestion leads to a prolonged inhibition (10–14 days) of platelet cyclo-oxygenase activity; the recovery of this activity is probably due to the appearance of newly formed platelets with unacetylated cyclo-oxygenase in the circulation.

This aspirin-induced inhibition and the subsequent recovery of cyclo-oxygenase activity is also paralleled by the *in vivo* formation of thromboxane[12]. Although indomethacin has been considered an irreversible inhibitor, enzyme preparation studies show that the effect of the indomethacin on the synthesis of thromboxane *in vivo* is different from that of aspirin and that the duration of the inhibitory effect is only 24 h. It is therefore obvious that the results from studies on the effect of NSAIDs on enzyme preparations and other *in vitro* systems cannot be directly extrapolated to an *in vivo* situation.

Another way to establish the inhibitory effect of various NSAIDs on the formation of prostaglandins in the human is to monitor plasma or urinary metabolites. Such an approach gives information on the total body production, metabolism and excretion of the prostaglandin under study. Few such studies have been published to date, however. It has been shown in a few cases that the excretion of the major urinary metabolite of PGE_2 is reduced following indomethacin, aspirin and sodium salicylate intake and that there is a considerable variation in the effect of these drugs in different individuals[13]. More information exists on the *in vivo* drug effects on the biosynthesis of thromboxane and prostacyclin. For example, *in vivo* production of thromboxane and prostacyclin are reduced by aspirin[12,14-16]. However, a single dose of aspirin results in vastly different effects on the biosynthesis of these two compounds. Whereas 500 mg of aspirin causes a pronounced (80–90%) and long lasting (10–14 days) inhibition of thromboxane synthesis, the inhibition of prostacyclin synthesis is less pronounced (40–50%) and of much shorter duration (2–3 h). Similarly, the inhibitory effect of indomethacin is different; a 50 mg dose causes about a 50% reduction of the *in vivo* synthesis of prostacyclin in contrast to an 85% reduction of the thromboxane synthesis in the same individual on the same occasion[12,16]. Although it seems logical that the penetration of NSAIDs into different cells will vary and thus result in varying degrees of inhibition of the arachidonic acid cascade, this is rarely considered when data on the effect of NSAIDs have been interpreted in the literature.

METABOLISM OF PROSTAGLANDINS

The metabolism of the classical prostaglandins has been studied extensively in tissue homogenates as well as *in vivo* in animals and man[1]. The major route for metabolic degradation is initiated by an oxidation at the site of carbon 15. This process yields 15-keto-prostaglandins which are biologically inactive compounds when tested in various systems. This reaction is followed by a reduction of the Δ-13-double bond. The enzymes responsible for these initial steps are particularly abundant in lung, liver and kidney[17]. The resulting metabolites, 15-keto-13,14-dihydro-prostaglandins, are formed quickly following an intravenous injection of the parent prostaglandin into humans[1]. Thus, within 1·5 min after intravenous bolus injection of tritium-labelled PGE_2, less than 5% is still present as intact PGE_2, while approximately 50% of the administered compound has been metabolized to 15-keto-

13,14-dihydro-PGE_2 in the circulation[18]. The estimated half-life of PGE_2 in the circulation is less than 15 sec, whereas that of its metabolite is approximately 8 min. Thus, the metabolite remains longer and in much greater amounts in the circulation than the parent prostaglandin. Using the most specific methods of assay presently available, i.e. gas chromatography–mass spectrometry (GC–MS) with deuterated internal standards/carriers, the levels of the metabolites of PGE_2 and $PGF_{2\alpha}$ in plasma have been shown[19] to be in the range 15–100 pg/ml. When comparing the half-lives of PGE_2 and its major plasma metabolite, it is clear that the maximal levels of the parent prostaglandin in the circulation are considerably lower, at most a few pg/ml, than those of the metabolite. Even so, primary prostaglandin levels ranging from several hundred to several thousand pg/ml have been reported in the literature. Apart from the utilization of non-specific analytical variation can be explained by the fact that platelets are quite active in synthesizing PGE_2 and $PGF_{2\alpha}$[1]. Since platelets are easily and invariably activated when plasma is collected, even the use of highly specific analytical methods cannot overcome the artifactual formation of primary prostaglandins during collection of plasma. Kinetic, biochemical and analytical data demonstrate beyond doubt that the analysis of primary prostaglandins such as PGE and PGF in plasma is meaningless.

Further metabolic degradation of primary prostaglandins occurs via β- and ω-oxidation and yields metabolites with 18, 16 and 14 carbons[1]. The ω-oxidations lead to ω-1- or ω-2-hydroxylated products as well as dicarboxylic acids. Combinations of such reactions lead to formation of the major metabolites of PGE_2 and $PGF_{2\alpha}$ found in human urine (Figure 3.3, bottom)[1,18]. Quantitative analyses of these metabolites open other avenues of monitoring prostaglandin biosynthesis *in vivo* in humans. Since the half-life times of plasma metabolites are about 8 min, they are often used for monitoring changes in prostaglandin production over minutes to hours. On the other hand, measurement of the urinary metabolites is more suitable for monitoring changes over hours to days[18].

Knowledge about the kinetics of the metabolism of prostaglandins and data from quantitative determination of the urinary excretion of these metabolites (about 2–50 μg/24 h) have allowed calculations of 'total daily production rates' of $PGE_1 + PGE_2$ and $PGF_{1\alpha} + PGF_{2\alpha}$[23]. These figures range from 15 to 300 μg/24 h for PGE and 50–370 μg/24 h for PGF.

It has not been possible to study the metabolism of thromboxane A_2 because of its instability and strong platelet aggregatory and vasoconstrictory effects. However, TxA_2 undergoes rapid hydrolysis to TxB_2 in aqueous solutions and the metabolism of TxB_2 has been more easily studied[20]. A large number of urinary metabolites of TxB_2 have been identified, but the major metabolite, 2,3-dinor-TxB_2, is formed by only one step of β-oxidation (Figure 3.3). This metabolite is found in plasma. The findings that TxA_2 can react in platelet-rich plasma to form protein-bound derivatives raises the question of how much of the TxA_2 formed *in vivo* is converted into the major urinary metabolite[21,22] of TxB_2. Quantitative data on the excretion of urinary 2,3-dinor-TxB_2 presently indicate that the *in vivo* metabolism of TxA_2 proceeds at least in part via TxB_2.

Quantitative determination of the *in vivo* synthesis of thromboxane is associated with the same type of problems as described above for PGE and PGF. 'Plasma levels' of TxB_2 found after analysis of a blood sample certainly do not represent the level in the circulation but rather the *ex vivo* formation of TxB_2. Studies on the metabolism of TxB_2 demonstrate that about 10% of intravenously injected TxB_2 occurs in urine as the major metabolite[20]. Quantitative determination of this metabolite with GC–MS demonstrates excretions between 200 and 750 ng/24 h in normal healthy individuals[12,24]. Using the same reasoning described above for prostaglandins, this would correspond to a 'daily production rate' of TxB_2 of about 2–7.5 μg/24 h.

Since prostacyclin (PGI_2) can be administered intravenously, it has been possible to study the metabolism of this compound in greater detail[25-27]. A large number of metabolites have been identified[25,27]. Prostacyclin seems to be initially converted into 6-keto-$PGF_{1\alpha}$ (Figure 3.3) and its 15-keto-13,14-dihydro analogue. The oxidation at carbon 15 and reduction of the Δ-13 double bond seem to be of much less quantitative importance in the metabolism of prostacyclin than of the primary prostaglandins. Further metabolic degradation occurs via β- and ω-oxidation leading to dinor-, tetranor-, ω-1 and ω-2 hydroxylated products as well as dicarboxylic acids. The major urinary metabolite is 2,3-dinor-$PGF_{1\alpha}$ which once again is formed through only one step of β-oxidation of 6-keto-$PGF_{1\alpha}$ (Figure 3.3). This latter compound also occurs in the urine but in smaller amounts.

To date the metabolic and analytical data obtained have been used to calculate the estimated rate of entry of endogenous PGI_2 into the circulation[26]. The figures arrived at are approximately 0.1 ng kg^{-1} min^{-1}. This corresponds to a daily production of prostacyclin of about 10 μg/24 h. These metabolic, kinetic and analytical data also demonstrate that the plasma level of 6-keto-$PGF_{1\alpha}$ should be in the very low pg/ml range. Based on these considerations, the highly variable and high concentrations of this compound in peripheral plasma as reported in the literature, sometimes several thousands pg/ml, must be erroneous and may depend on artifactual formation during the collection of plasma and/or measurements with suboptimal techniques.

PROSTAGLANDIN ANALOGUES

The rapid metabolic inactivation of classical prostaglandins means that those compounds have to be administered continuously through intravenous infusion or slow release formulations if a continuous pharmacological effect is desired. Intravenous infusions have been used for induction of labour and abortion. Studies of the metabolic reactions causing the initial inactivation provided the information necessary for development of prostaglandin analogues that were resistant to

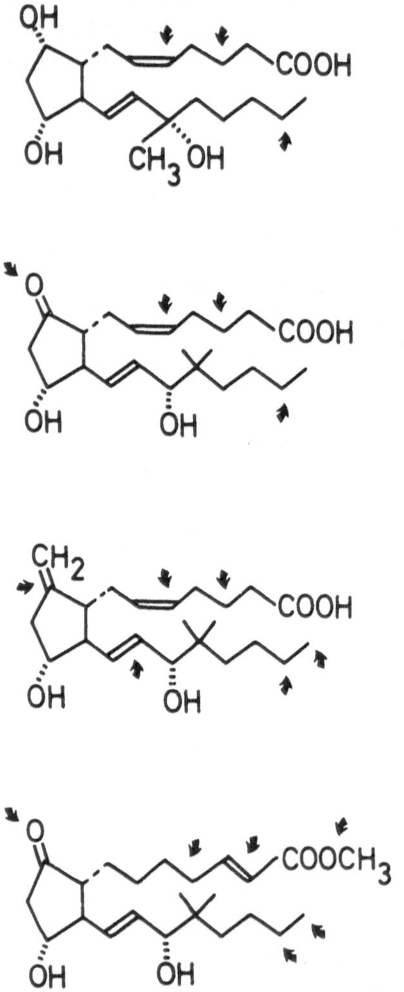

15-METHYL-PG F$_{2\alpha}$

16,16-DIMETHYL-PG E$_2$

POINTS OF ENZYMATIC ATTACK

9-DEOXO-16,16-DIMETHYL-9-METHYLENE PG E$_2$

16,16-DIMETHYL-TRANS-DELTA-2 PG E$_1$

Figure 3.4   Structures of prostaglandin analogues and points of enzymatic attack demonstrated in humans

inactivation while retaining their biological activity. The first such compound successfully tested in humans for induction of uterine contractility was 15-methyl-PGF$_{2\alpha}$ (Figure 3.4, first formula). In this compound the hydrogen at carbon 15 in the PGF$_{2\alpha}$ molecule has been replaced by a methyl group. This modification prevents the initial enzymatic reaction that leads to the inactivation of the classical primary prostaglandins and results in a prolonged half-life in the circulation[28]. Another successful approach to inhibit the rapid inactivation was the introduction of two methyl groups at carbon 16 (16,16-dimethyl prostaglandin analogues). Two such analogues, 16,16-dimethyl-PGE$_2$ (Figure 3.4, second formula) and 16,16-dimethyl-trans-delta-2-PGE$_1$ (Figure 3.4, fourth formula), have been successfully used in clinical trials for induction of uterine activity leading to abortion. However, the ring structure of PGE compounds undergoes dehydration relatively easily and results in inactive PGAs and PGBs. To overcome this problem, 9-deoxo-16,16-dimethyl-9-methylene PGE$_2$ was synthesized (Figure 3.4, third formula). This is a chemically stable analogue with PGE-like properties. The metabolism of these analogues has been studied extensively[28-31]. In all of them the oxidation at carbon 15 (responsible for the rapid inactivation of primary prostaglandin) is completely blocked. Instead, the analogues are inactivated by β- and ω-oxidations as indicated by arrows shown in all formulas in Figure 3.4.

The resistance to metabolic inactivation caused by the modifications illustrated in Figure 3.4 renders the compounds useful as pharmacological agents. The half-lives of these analogues in the human circulation are approximately 10 min and their metabolism in man has been investigated. As a part of their early development as drugs, the 'therapeutic levels' of these compounds have also been determined and the characteristics of different routes of administration and formulations have been demonstrated[31-40].

References

1 Samuelsson, B., Granström, E., Green, K., Hamberg, M. and Hammarström, S. (1975). Prostaglandins. *Ann. Rev. Biochem.*, **44**, 669-95

2 Samuelsson, B., Boldyne, M., Granström, E., Hamberg, M., Hammarström, S. and Malmsten, C. (1978). Prostaglandins and thromboxanes. *Ann. Rev. Biochem.*, **47**, 997-1029

3 Hansson, G., Malmsten, C. and Rådmark, O. (1983). The leucotrienes and other lipoxygenase products. In Pace-Asciak and Granstrom (eds.) *Prostaglandins and Related Substances*. pp. 127-69. (Amsterdam: Elsevier Science Publishers)

4 Johnsen, S-A., Olding, L. and Green, K. (1983). Conversion of arachidonic

acid in human maternal and neonatal mononuclear leucocytes. *Immunol. Letters*, **6**, 213–18

5 Hirata, F., Schiffman, E., Venkatasubramanian, K., Salomon, D. and Axelrod, J. A phospholipase A₂ inhibitory protein in rabbit neutrophils induced by glucocorticoids. (1980). *Proc. Natl. Acad. Sci. USA*, **77**, 2533–6

6 Hamberg, M., Svensson, J. and Samuelsson, B. (1974). Prostaglandin endoperoxides. A new concept concerning the mode of action and release of prostaglandins. *Proc. Nat. Acad. Sci.*, **71**, 3824–8

7 Christensen, N. J. and Green, K. (1983) Bioconversion of arachidonic acid in human pregnant reproductive tissues. *Biochem. Med.*, **30**, 162–80

8 Yamamoto, S. (1983). Enzymes in the arachidonic acid cascade. In Pace-Asciak, C. and Granstrom, E. (eds.) *Prostaglandins and Related Substances.* pp. 171–202. (Amsterdam: Elsevier Science Publishers)

9 Lands, W. E. M. and Hanel, A. M. (1983). Inhibitors and activators of prostaglandin biosynthesis. In Pace-Asciak, C. and Granstrom, E. (eds.) *Prostaglandins and Related Substances.* pp. 203–23. (Amsterdam: Elsevier Science Publishers)

10 Dimov, V., Christensen, N. and Green, K. (1983). Analyses of prostaglandins formed from endogenous and exogenous arachidonic acid in homogenates of human reproductive tissues. *Biochim. Biophys. Acta*, **754**, 38–43

11 Burch, J. W., Stanford, N. and Majerus, P. W. (1978). Inhibition of platelet prostaglandin synthetase by oral aspirin. *J. Clin. Invest.*, **61**, 314–19

12 Vesterqvist, O. and Green, K. (1984). Urinary excretion of 2,3-dinor-thromboxane B₂ in man under normal conditions, following drugs and during some pathological conditions. *Prostaglandins*, **27**, 627–44

13 Hamberg, M. (1972). Inhibition of prostaglandin synthesis in man. *Biochem. Biophys. Res. Commun.*, **49**, 720–6

14 Fitzgerald, G. A., Oates, J. A., Hawiger, J., Maas, R. L., Jackson-Roberts, L., Lawson, J. A. and Brash, A. R. (1983). Endogenous biosynthesis of prostacyclin and thromboxane and platelet function during chronic administration of aspirin in man. *J. Clin. Invest.*, **71**, 676–88

15 Vesterqvist, O. (1985). Rapid recovery of *in vivo* prostacyclin formation after inhibition by aspirin. Evidence from measurements of the major urinary metabolite of prostacyclin by GC–MS. To be published

16 Vesterqvist, O. and Green, K. (1984). Development of a GC–MS method for quantitation of 2,3-dinor-6-keto-PGF₁ and determination of the urinary excretion rates in healthy humans under normal conditions and following drugs. *Prostaglandins*, **28**, 139–54

17 Änggård, E., Larsson, C. and Samuelsson, B. (1971). The distribution of 15-hydroxy-prostaglandin-dehydrogenase and prostaglandin Δ-13-reductase in different tissues of the swine. *Acta Phys. Scand.*, **81**, 396–404

18 Hamberg, M. and Samuelsson, B. (1971). On the metabolism of prostaglandins E₁ and E₂ in man. *J. Biol. Chem.*, **246**, 6713–21

19 Green, K. and Samuelsson, B. (1974). Endogenous levels of 15-keto-dihydro-prostaglandins in human plasma. Parameters for monitoring prostaglandin synthesis. *Biochem. Med.*, **11**, 298–303

20 Jackson-Roberts, L., Sweetman, B. J. and Oates, J. A. (1981). Metabolism of thromboxane B_2 in man. Identification of twenty urinary metabolites. *J. Biol. Chem.*, **256**, 8384–93

21 Fitzpatrick, F. A. and Gorman, R. R. (1977) Platelet-rich plasma transforms exogenous prostaglandin endoperoxide H_2 into thromboxane A_2. *Prostaglandins*, **14**, 881–9

22 Maclouf, J., Kindahl, H., Granström, E. and Samuelsson, B. (1980). Interactions of prostaglandin endoperoxide H_2 and thromboxane A_2 with human serum albumin. *Eur. J. Biochem.*, **109**, 561–6

23 Samuelsson, B. (1973). Quantitative aspects on prostaglandin synthesis in man. In Bergström, S. and Bernard, S. (eds.) *Advances in the Biosciences*. Vol. 9, pp. 7–14. (Vieweg: Pergamon Press)

24 Maas, R. L., Jackson-Roberts, L., Taber, D. F. and Oates, J. A. (1980). Urinary dinor thromboxane B_2: levels in normal males and in cardiovascular disease. *Clin. Res.*, **28**, 319

25 Rosenkranz, B., Fischer, C., Weimer, K. E. and Frölich, J. C. (1980). Metabolism of prostacyclin and 6-keto-prostaglandin F_1 in man. *J. Biol. Chem.*, **255**, 10194–8

26 Fitzgerald, G. A., Brash, A. R., Falardeau, P. and Oates, J. A. (1981). Estimated rate of prostacyclin secretion into the circulation of normal man. *J. Clin. Invest.*, **68**, 1272–6

27 Brash, A. R., Jackson, E. K., Saggese, C. A., Lawson, J. A., Oates, J. A. and Fitzgerald, G. A. (1983). Metabolic disposition of prostacyclin in humans. *J. Pharm. Exp. Ther.*, **226**, 78–87

28 Hansson, G. and Granström, E. (1977). Metabolism of 15-methyl-prostaglandin $F_{2\alpha}$ in the cynomolgus monkey and the human female. *Biochem. Med.*, **18**, 420–39

29 Steffenrud, S. (1980). Metabolism of 16,16-dimethyl-prostaglandin E_2 in the human female. *Biochem. Med.*, **24**, 274–92

30 Steffenrud, S. (1983). Metabolism of 9-deoxo-16,16-dimethyl-9-methylene prostaglandin E_2 in humans. *Drug Metab. Dispos.*, **2**, 255–65

31 Green, K. (1984). Metabolism and pharmacokinetics of prostaglandin analogs in man. In Toppozada, M., Bygdeman, M. and Hafez, E. S. E. (eds.) *Prostaglandins and Fertility Regulation*. pp. 11–19. (Lancaster: MTP Press)

32 Green, K., Granström, E. Bygdeman, M. and Wiqvist, N. (1976). Kinetic and metabolic studies on 15-methyl-$PGF_{2\alpha}$ administered intra-amniotically for induction of abortion. *Prostaglandins*, **11**, 699–711

33 Green, K. and Bygdeman, M. (1976). Plasma levels of the methyl ester of 15-methyl-$PGF_{2\alpha}$ in connection with intravenous and vaginal administration to the human. *Prostaglandins*, **11**, 879–92

34 Bergström, S., Green, K. and Bygdeman, M. (1976). Metabolism and pharmacokinetics of 15-methyl-$PGF_{2\alpha}$ and its ester after administration via various routes. *Prostaglandins Suppl.*, **12**, 17–26

35 Green, K. and Bygdeman, M. (1977). Plasma levels of 15(S)15-methyl-$PGF_{2\alpha}$ following administration via various routes for induction of abortion. *Prostaglandins*, **14**, 1013–23

36 Green, K., Bygdeman, M. and Bremme, K. (1978). Interruption of early

first trimester pregnancy by single vaginal administration of 15-methyl-PGF$_{2\alpha}$-methylester. *Contraception*, **18**, 551–60

37 Steffenrud, S. and Lincoln, F. H. (1979). Method for quantitative analysis of 16,16-dimethyl-prostaglandin E$_2$ from plasma using deuterated carrier and gas-chromatography–mass spectrometry. *Anal. Biochem.*, **100**, 109–17

38 Green, K., Vesterqvist, O., Bygdeman, M., Christensen, N. J. and Bergström, S. (1982). Plasma levels of 9-deoxo-16,16-dimethyl-9-methylene-PGE$_2$ in connection with its development as an abortifacient. *Prostaglandins*, **24**, 451–66

39 Dimov, V., Green, K., Bygdeman, M., Konishi, Y., Imaki, K. and Hayshi, M. (1983). Gas chromatographic mass spectrometric quantitation of 16,16-dimethyl-*trans*-delta-2-PGE$_1$. *Prostaglandins*, **25**, 225–35

40 Bygdeman, M., Christensen, N. J., Dimov, V. and Green, K. 1984). Termination of pregnancy by a slow release device containing 16,16-dimethyl-*trans*-Δ-2-PGE$_1$ methyl ester. *Asia-Oceania J. Obstet. Gynecol.*, **10**, 359–65

SECTION II
REPRODUCTIVE PHYSIOLOGY

4
The cervix

U. ULMSTEN

INTRODUCTION

The normal non-pregnant human cervix has a length of about 3 cm, a diameter of 2–3 cm and in general a thickness of 1 cm. According to its position above or below the reflection of the vaginal epithelium, the cervix can be divided into supravaginal and vaginal portions. Although movable, the cervix and uterus are stabilized by several ligaments. The most important of these is the transverse cervical ligament which encircles the cervix like a collar just above the reflection of the vaginal mucosa. This ligament is often referred to as the cardinal ligament based on its function and Mackinrodt's ligament in eponymic terminology.

Functionally, the human cervix is a biological valve controlling the passage of a variety of fluids and biological matter into and out of the uterine cavity. Since the corpus and cervix uteri form a functional unit in which both parts possess several common regulating mechanisms, i.e. hormones, enzymes and neuroregulators, cervical function cannot truly be separated from that of the corpus. Nonetheless, recent investigations have focused on the cervix as an organ with its own activity which is at least partly independent of the uterus. Hence, the previously held opinion that the cervix was a purely passive and inferior part of the corpus uteri should be abandoned or at least modified.

The cervical control of the uterine portal in the non-pregnant state involves facilitation of sperm transport, prevention of invasion by unwanted micro-organisms and facilitation of the outflow of menstrual blood as well as uterine secretions. During pregnancy the cervix acts to secure preservation of the growing conceptus within

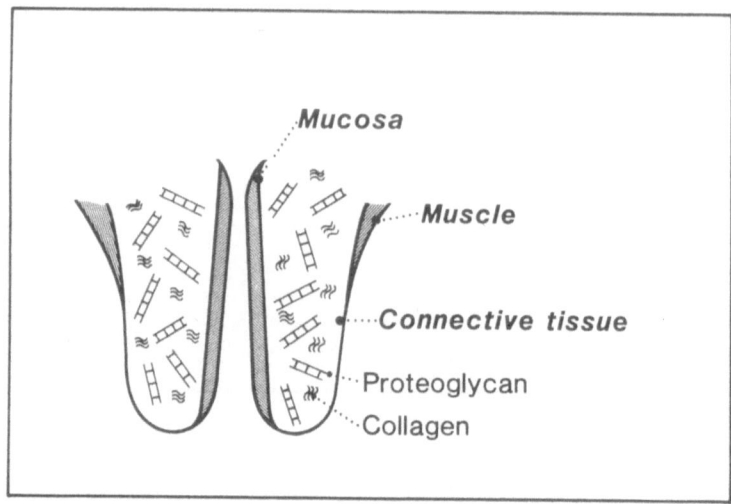

Figure 4.1   Schematic outline of the morphology of the human uterine cervix. The dominating structure is the fibrous connective tissue in the middle of the cervical body. The main components of this tissue are huge proteoglycan molecules and bundles of collagens. The inner part (towards the cervical canal) is lined with a complex mucosa whereas the outer part of the organ is covered by a thin smooth muscle coat

the uterine cavity. At term, on the other hand, it dilates rapidly to allow easy passage of the fetus. In the case of pregnancy, the inability of the cervix to fulfil all of these conflicting demands may result in at least one or more of the following: infection during the first or second trimester, preterm delivery, or prolonged labour-delivery times.

The human cervix possesses a unique construction in order to fulfil its multiple functions (Figure 4.1). Its centre possesses a narrow canal lined by columnar epithelium. Within and immediately beneath this epithelium are hundreds of secretory units arranged in crypts and gland-like structures; these secretory units produce the cervical mucus. The main part of the cervical body consists of a stroma made up primarily of connective tissue whose main constituents are bundles of collagen separated or surrounded by ground substance, i.e. proteoglycans. This construction provides the cervix with its palpable firmness, a characteristic finding in the non-pregnant state and in early pregnancy. Outside the stroma and facing the vagina, somewhat above the level of the external os, the cervix is covered by a thin layer of smooth muscle. For a long time this muscular coating was considered as an important component of cervical function, acting as a real sphincter. However, Danforth and co-workers demonstrated that the connective tissue was the most important component of the cervical closure mechanism[1-3]. More recent investigations indicate that the smooth muscle layer should not be completely discounted in this regard[4-6]. The outside of the cervix or the *portio vaginalis* is covered by stratified squamous epithelium.

CERVICAL PHYSIOLOGY

The cervical canal

In the non-pregnant uterus, the endocervical canal has a fusiform shape, a length of about 30 mm and an average diameter of approximately 5 mm. The upper part, i.e. where the cervix meets the corpus, is strictly defined as the fibromuscular junction; however, in clinical terms this site is normally referred to as the internal os. The cervical canal is lined with columnar epithelium except for its most distal lower part, including the ectocervix and the external os, where squamous epithelium predominates.

Non-ciliated secretory cells in and beneath the columnar epithelium produce an abundant sticky liquid, the cervical mucus. The mucus production varies from a high of 600 mg/day during midcycle to a low of about 50 mg/day during other times of the cycle. The columnar epithelium also contains kinociliated cells, i.e. cells with cilia that are capable of motion. The kinocilia of these cells appear to beat rhythmically towards the external os. It is assumed, although not proven, that these ciliated cells are involved in some sort of mucociliary clearance of micromolecules produced by the secretory cells. The kinociliated cells direct the flow of mucus towards the vagina. Since the strongest flow occurs in the periphery of the canal close to the epithelium, it has been suggested that this arrangement, according to physical fluid theories, encourages rapid movement of the spermatozoa swimming in the centre of the canal and delays defective sperm in the periphery. It has been suggested that prostaglandins may influence both kinociliary activity and mucus production but this hypothesis remains unproven[7,8].

Great interest has been directed towards determining the composition of the cervical mucus, and several investigations have elicited the biophysical and biochemical properties of this unique 'liquid'[9,10]. The viscosity of cervical mucus undergoes cyclical changes, being minimal at time of ovulation and thereby facilitating sperm penetration. The changes in the viscosity of the mucus are measured and expressed in terms of spinnbarkeit which refers to the capacity of the mucus to be drawn into threads. A progressive increase in spinnbarkeit occurs immediately prior to ovulation, after which it decreases. The variations in spinnbarkeit form the basis of the symptothermal method of periodic abstinence for contraceptive practice. The stickiness of the mucus, also referred to as tack, is most pronounced during pregnancy. This characteristic had been considered previously of little clinical importance; however, the growing attraction for intracervical drug deposition has focused scientific interest on this phenomenon once again.

Ferning refers to the arborization pattern which can be seen at microscopic examination of dried cervical mucus. Ferning is most pronounced during the late part of the proliferative phase and is directly related to spermatozoal receptivity.

The molecular arrangement of the cervical mucus has been the subject of several investigations using different techniques, e.g. nuclear magnetic resonance (NMR)[11], electron microscopy (EM)[12] and laser spectroscopy[8]. Different results have been obtained from these examinations and no agreement presently exists on the unique rheological properties of the mucus which form the basis for its function[8,9].

The biochemical characteristics of cervical mucus have been investigated, but our knowledge of this substance remains obscure due to difficulties in obtaining it without sampling artifacts. However, despite incomplete knowledge it appears that both the cervical epithelium and its mucus individually and collectively exert important functions. By their action, sperm transport is facilitated at time of ovulation and impeded at other times. Moreover, they also protect the uterine cavity from invasion by unwanted micro-organisms. Progesterone and oestrogen both affect these properties, but it has also been suggested that relaxin and prostaglandins are involved as well. The observations made in this regard, however, are few and do not allow clinical applications at present, although it appears that prostaglandin treatment induces changes in the composition of the mucus of the unripe pregnant cervix similar to that seen in spontaneous ripening[10].

The stroma

Located beneath the epithelium is the stroma which forms the most important part of the cervical body. It is mainly composed of connective tissue including ground substance (proteoglycans), collagen and elastin, which constitute more than 70% of its dry weight. The predominant structural elements are collagen fibres which lie in different directions, separated or surrounded by huge proteoglycan molecules (Figure 4.1). Although this firm connective tissue traditionally has been viewed as the inert substance of the cervix, it recently has been found to possess dynamic properties. Significant changes in important biochemical and biophysical properties of this connective tissue occur during spontaneous and pharmacologically induced cervical ripening in pregnant women at term[13]. Data also indicate that cyclical changes occur in the stroma of the non-pregnant human cervix[14]. Following these observations, Scott[15] has postulated a possible interaction between the proteoglycans and the collagen fibres in connective tissue. In the following section, the histological, biochemical and biophysical characteristics of the cervical stroma will be discussed, as this background is essential for an understanding of human cervical function.

HISTOLOGY OF THE CERVIX

In the non-pregnant uterus and during the first trimester of pregnancy, the cervical connective tissue can be subdivided into a superficial loose zone and a deeper and dense stromal zone. The deep zone is composed of dense but irregular connective tissue. It is relatively poor in cells except for mast cells, but eosinophils, neutrophils, plasma cells, lymphocytes and macrophages may also be found. The collagen fibrils are arranged in parallel bundles forming fibres with a magnitude of 1–6 nm in diameter. They run in all directions (Figure 4.2). The matrix or ground substance between the collagen consists of huge proteoglycan molecules (Figure 4.3). This network is infiltrated by cells;

Figure 4.2 Electron micrograph of fibrous connective tissue from an unripe human cervix (enlarged 16 800). The dominating structure is numerous collagen bundles running in different directions

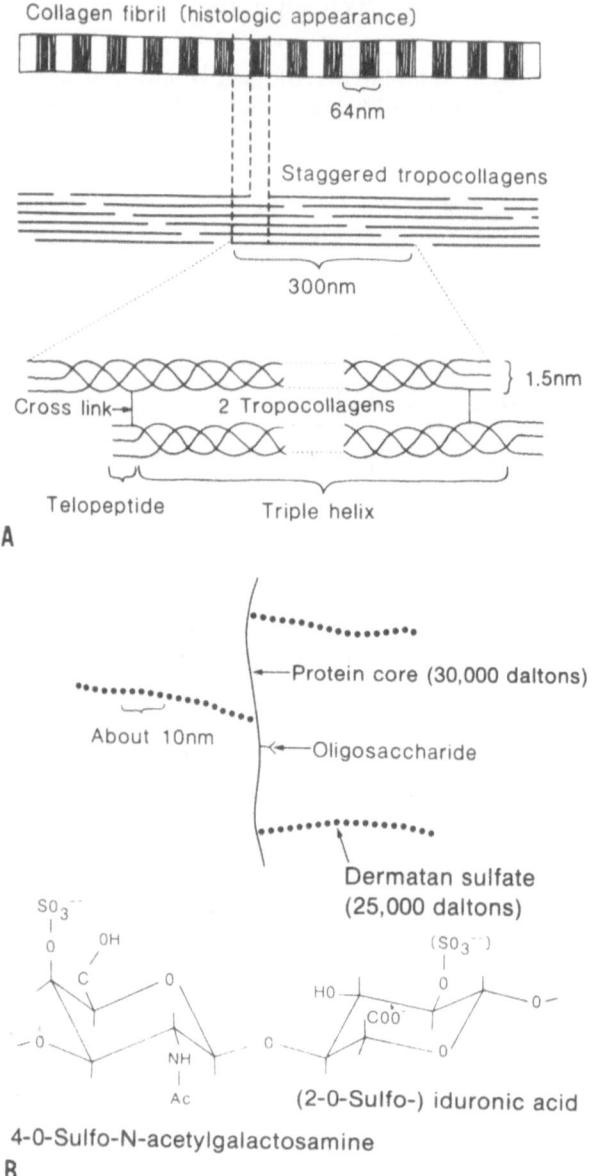

Collagen fibril (histologic appearance)

64nm

Staggered tropocollagens

300nm

1.5nm

Cross link → 2 Tropocollagens

Telopeptide   Triple helix

A

Protein core (30,000 daltons)

About 10nm

Oligosaccharide

Dermatan sulfate
(25,000 daltons)

SO_3^{--}

OH

(SO_3^{--})

(2-0-Sulfo-) iduronic acid

4-0-Sulfo-N-acetylgalactosamine

B

Figure 4.3 **A:** The cross-striated collagen fibril is composed of tropocollagen molecules staggered in such a way that they give rise to the dark and light bands. Each tropocollagen contains three polypeptides wrapped around each other in a triple helix with relatively short non-helical telopeptides at either end. The physical strength of the fibril is determined by the cross-links between the triple helices and the telopeptides. **B:** The dominating proteoglycan in the human uterine cervix is a dermatan sulphate proteoglycan with a molecular weight of 100 000 daltons. It contains a protein core with two or three dermatan sulphate chains. Small amounts of oligosaccharides are also present. (From Uldbjerg, Umsten and Ekman, Ref. 13, with permission)

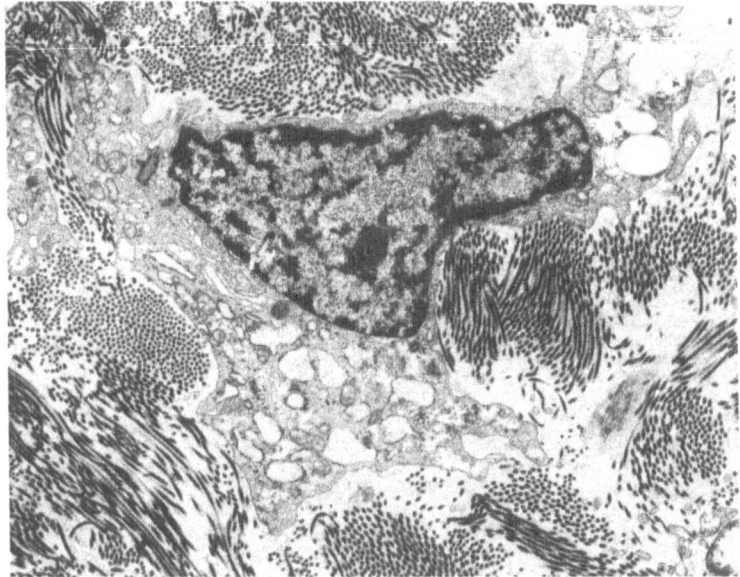

Figure 4.4 Electron micrograph of non-pregnant cervical tissue. The dominating structure is a fibroblast surrounded by bundles of collagen (× 16 000)

fibroblasts at various levels of activity predominate (Figure 4.4). A
striking feature of the cervical fibroblasts is the presence of numerous
long cytoplasmatic processes which radiate from one cell body to
another. It is possible that these cervical 'cell junctions' resemble those
of the myometrial gap junctions.

Typical smooth muscle cells also penetrate deeply into the stroma,
but these cells occur less frequently than other fibroblasts. Blood and
lymphatic vessels are also present.

As pregnancy proceeds the microscopic features of the cervical connective tissue change. In the ripened and softened cervix at term, an
increased vascularity is noted, along with dilated vessels and 'activated'
fibroblasts producing secretions. In addition, white cells and macrophages migrate out of the vessel walls into the tissues. Moreover, an
almost 20-fold increase of haemoglobin has been found in pregnant
term compared to non-pregnant cervices. Changes such as these are
closely related to biochemical and biophysical changes which occur in
the cervix during pregnancy. The microscopic appearance is comparable to that seen in inflammatory reactions and wound healing[16]. As
a result of the increased cellular activity, the previously intact collagen
bundles, tightly surrounded by proteoglycans, are broken up (Figure
4.5).

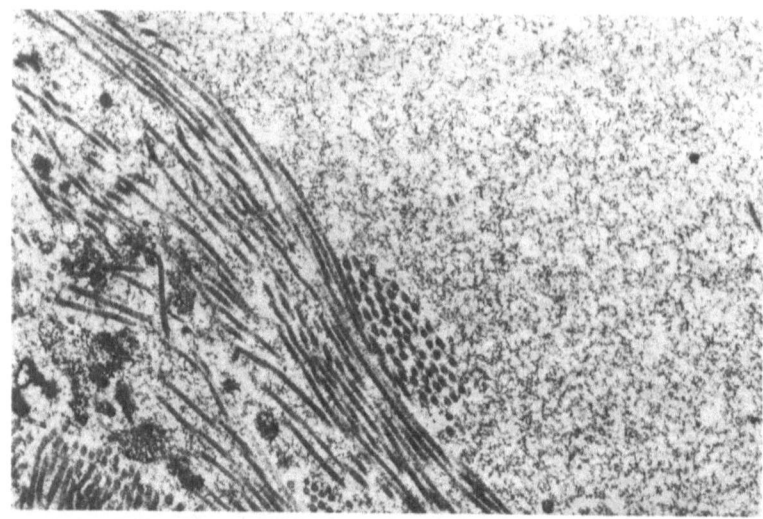

Figure 4.5 Electron micrograph of a ripened human uterine cervix. The collagen bundles are partly scattered and dissolved (× 17 000). For comparison see also Figure 4.2

Biochemical characteristics (for review, see References 13 and 17)

Figure 4.6 outlines the most important biochemical characteristics of the cervix in the non-pregnant, pregnant and postpartum states. The identification of these molecules has been incomplete, partly as a result of analytical problems and partly because of problems in obtaining accurate tissue specimens from humans. Many postulations on the biochemical composition of the human cervix are extrapolated from animal studies, although it is well known that variations between species may invalidate reliable conclusions. The following sections will describe the basic composition of the human cervix in biochemical terms and note how these parameters change in normal and pathological pregnancy. The results of pharmacological intervention with prostaglandins will also be described briefly.

The collagen fibril is the main constituent in the human cervix. The basic molecule, tropocollagen, has a molecular weight of about 300 000 daltons. It has a stiff rod-like shape with unusual dimensions (300 nm long × 1·5 nm in diameter). The native collagen fibril is made up of tropocollagen molecules that are orientated in parallel and staggered in such a way that they create the typical light and dark bands as seen on electron microscopy (Figure 4.3). The fibril is held together by cross-links. Newly synthesized collagen without cross-links can be extracted from cervical tissue samples with acetic acid. The number of

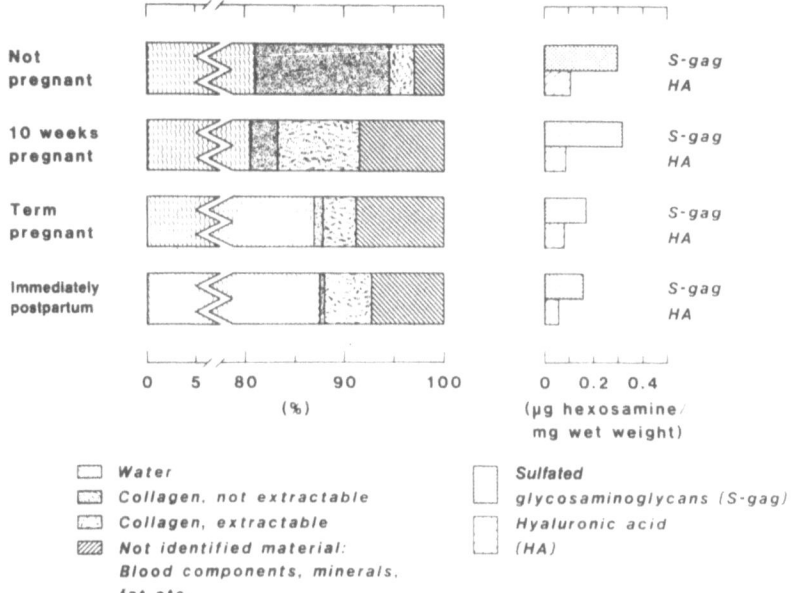

Figure 4.6 The chemical composition of the lower part of the human uterine cervix and changes herein during pregnancy. The collagen concentration decreases by 70% and the solubility increases during pregnancy but to a less extent during labour. The concentration of sulphated glycosaminoglycans follows that of collagen. Significant amounts of unidentified materials are present. (From Uldbjerg, Ulmsten and Ekman, Ref. 13, with permission)

cross-links as well as their acid stability is higher in older collagen but seems to differ also from one tissue to another.

The determination of hydroxyproline content is generally used for quantitation of the collagen[13,17].

Although the collagen fibril is a stable structure, it can be broken down by enzymes (proteinases), the most well-known of which is collagenase. The activity of this latter substance is considered essential for the regulation of collagen degradation; however, a satisfactory model explaining this regulation has not yet been developed. The rate of synthesis and secretion of collagenase or procollagenase, the concentrations of inhibitors (anticollagenase and α_2-macroglobulin) as well as the concentration and activity of collagenase activators all appear to be important factors. Prostaglandin and PG inhibitors are involved in all these regulatory processes. Several types of cells in the cervix, the most important of which are fibroblasts and polymorphonuclear leukocytes, are capable of producing collagenase. It is possible to follow the dynamic changes in cervical collagen by determination of the amount of hydroxyproline and the activity of collagenase.

Elastin is also present in human cervical connective tissue along with collagen and proteoglycans. Elastin has characteristic elastic fibres which can be stretched to several times their length and then rapidly return to their original size and shape, once tension has been released. In parallel with the action of collagenase on collagen, leukocyte-elastase is important for the catabolism of elastin[13,17].

Other important molecules in the human cervix include the proteoglycans mentioned above. These huge molecules are specific to connective tissues. They are made up of a number of glycosaminoglycans (acid mucopolysaccharides) connected to a protein core (Figure 4.3). A great number of sulphate groups are present in the glycosaminoglycans and these give the molecules a distinct appearance with highly hydrophilic properties. The proteoglycans are the largest molecules in the human body. They can be as large as 3×10^6 daltons. The predominant

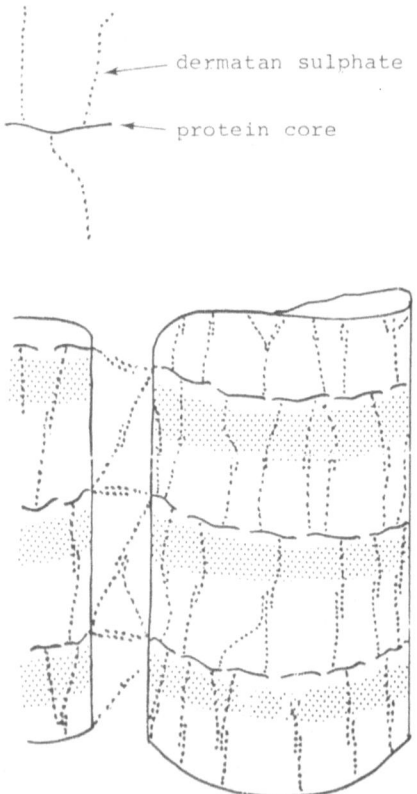

Figure 4.7 Proteoglycans lingering around the collagen bundles (outlined as dotted horizontal bands). This arrangement may have a stabilizing effect on the collagen fibril

proteoglycan molecule in the human cervix has recently been identified[17] to be a dermatan sulphate proteoglycan with a molecular weight
just below 100 000 daltons (Figure 4.3). Interestingly, it resembles the
predominant proteoglycan in the cornea of the eye.

Recently it has been postulated[15] that an interaction exists between
dermatan sulphate proteoglycans and collagen molecules with the proteoglycans forming a coat or network around the collagen bands
(Figure 4.7). The physiological significance of such an interaction may
be to stabilize the collagen[18]. It has also been shown that the orienta-
tion and physical properties of the collagen can be modified by glycosaminoglycans. One action of prostaglandins may be to attack the
interplay between collagen and proteoglycan resulting in an increased
cervical compliance. In addition to the molecules described above,
water is an important cervical constituent and the amount of water
increases in pregnancy close to term. An increasing amount of so-
called unidentified material is also found in the term pregnant cervix
(Figure 4.6).

Parts of this material most certainly consist of blood components,
but minerals and fat may also be present. We have studied structural
changes in cervical tissue in a long series of investigations on patients
with normal pregnancies and on those complicated by failure of cervical ripening at term[13,17,19,20]. Biochemical analyses were made on cervical biopsies obtained by a special rotating needle instrument which
facilitated the tissue sampling. The *in vitro* analyses have been closely
related to the clinical state of the cervix, the course of the pregnancy,
and to physical characteristics of the cervix measured by different
techniques such as intrauterine and intracervical pressure recordings of
the tissues[21].

Figures 4.6 and 4.8 summarize the biochemical composition of the
human cervix and its changes during pregnancy. As can be seen from
these illustrations, there is a pronounced decrease in the amount of
collagen from that of the non-pregnant to that of the normally ripened
term pregnant cervix. Simultaneously, collagenase activity and elastase
activity are increased considerably during pregnancy. Concerning the
proteoglycans, the amount of sulphated glycosaminoglycans (hexos-
amine) is markedly decreased in the normal term pregnant cervix compared to that of the non-pregnant cervix. These biochemical changes
are all important for cervical ripening at term.

Biochemical composition differs significantly between the ripe and
the unripe cervix at term. Patients with unripe cervices and prolonged
labour–delivery times have significantly higher amounts of cervical
collagen compared to those with favourable (ripe) cervices and normal
delivery times. Interestingly, this occurs despite a high collagenase
activity[22,23].

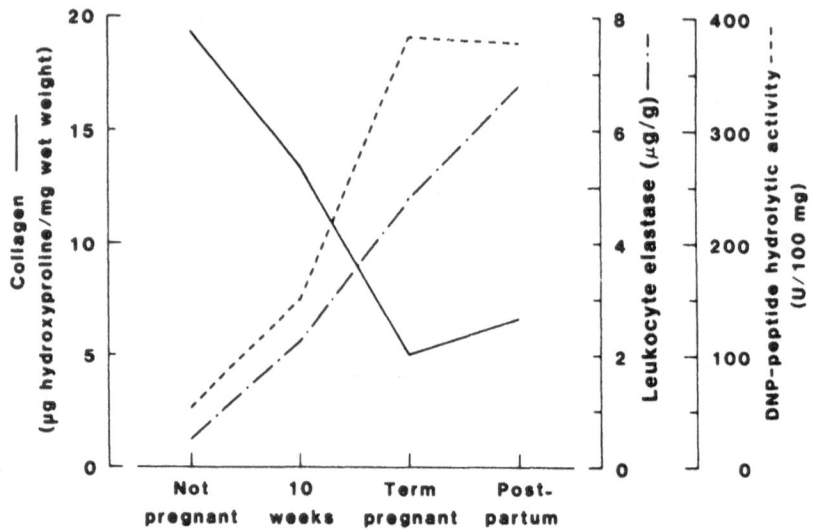

Figure 4.8   Collagen (hydroxyproline) collagenase (DNP-peptide hydrolytic activity) and elastase in the lower part of the human uterine cervix in non-pregnant women, in pregnant women and immediately postpartum. (From Uldbjerg, Ulmsten and Ekman, Ref. 13)

The biochemical changes described in the cervix are also reflected in the uterus. The myometrium in patients with unripe cervices and concomitant dystocia contains a higher amount of collagen and proteoglycans (sulphated glycosaminoglycans) despite an increased collagenase activity compared to patients with spontaneously ripened cervices and normal labour–delivery times[22,23]. Based on the amount of collagen in the cervix one can predict the labour–delivery times[17,22].

Biophysical characteristics

The biochemical changes of the cervix are reflected in changes of the biophysical characteristics such as changes in elasticity, distensibility or compliance. Most well-known clinically and easily demonstrated is the palpable change from the firm rigid structure of the non-pregnant cervix to the oedematous and softened cervix at term. Although the biophysical characteristics of the human cervix or its mechanical properties have generally been expressed in terms of stromal elasticity, distensibility or extensibility, occasionally the bioengineering term 'compliance' is used to describe the overall capacity of the cervix to resist dilatation. The higher the compliance the easier dilatation and vice versa.

The mechanical properties of the cervix can also be illustrated on a

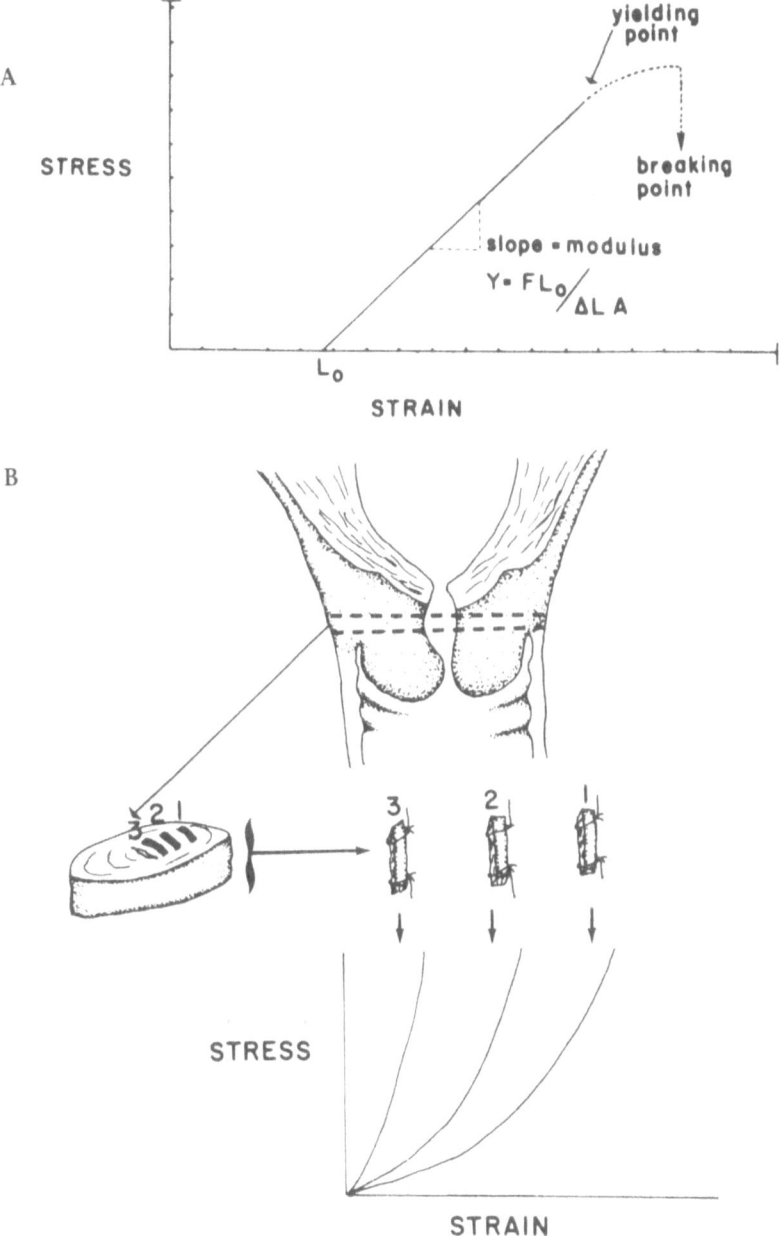

Figure 4.9 **A:** Idealized stress–strain curve used to calculate cervical compliance. L_o represents the resting length. The formula for the calculation of Young's modulus is given. (From Conrad and Ueland, Ref. 21, with permission). **B:** Stress–strain curves from different portions of the human uterine cervix of non-pregnant patients. The curves indicate that the tissue is least extensible in an area surrounding the central canal and progressively more extensible toward the periphery (from Conrad and Ueland, Ref. 21, with permission)

graph as tension v. length (Figure 4.9). In their important work on the physical characteristics of the cervix, Conrad and Ueland have used similar strain–stretch curves to illustrate their results[21]. Cervical biopsies taken from non-pregnant patients at different gestational stages show that the strain–stretch curve flattens out as the cervix becomes softer (Figure 4.9). Consistent with biochemical studies are the findings that PGE_2 treated cervices have a strain–stretch curve similar to that of spontaneously ripened cervices[21]. The distensibility of the non-pregnant cervix also increases at the time of menstruation.

Prostaglandins administered systemically or locally are able to prime the cervix and induce labour in patients with unfavourable cervices. Biochemical analyses of cervical stroma from these patients show that

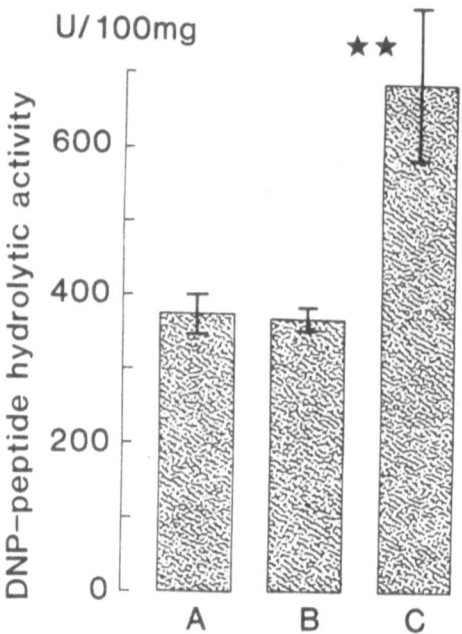

A : Control
B : Moderate clinical response
C : Prompt clinical response

Figure 4.10 Collagenase activity in human cervix of three categories of patients delivered vaginally. A: spontaneously delivered; B: oxytocin-induced; C: patients induced with local application of PGE_2 intracervically. As seen, a significantly higher collagenase activity is present in the PGE_2-induced patients

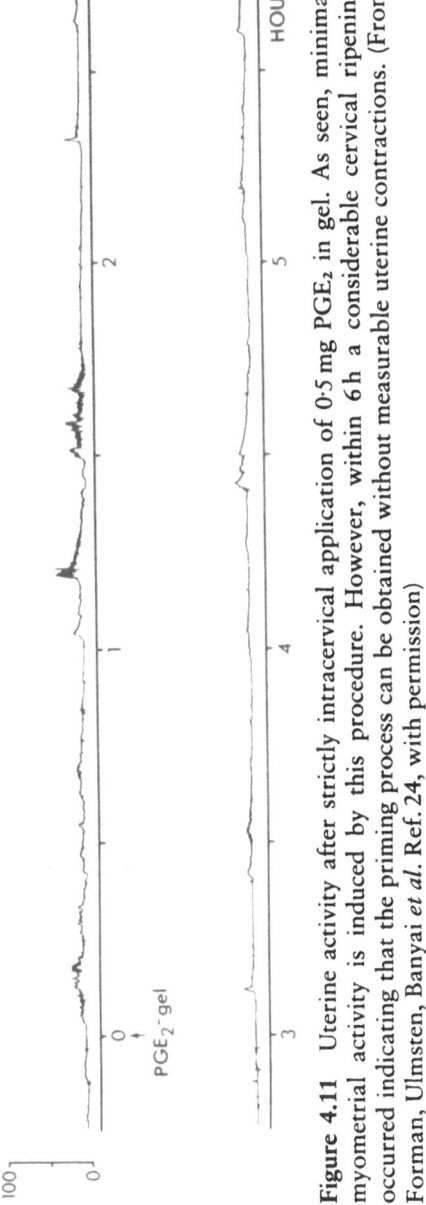

Figure 4.11 Uterine activity after strictly intracervical application of 0·5 mg PGE$_2$ in gel. As seen, minimal myometrial activity is induced by this procedure. However, within 6h a considerable cervical ripening occurred indicating that the priming process can be obtained without measurable uterine contractions. (From Forman, Ulmsten, Banyai *et al.* Ref. 24, with permission)

it is possible to induce biochemical changes in the unripe cervices similar to those seen in spontaneously ripened cervices[22,23]. In contrast, treatment with oxytocin does not affect the biochemical composition of the unripe cervix[23,24] (Figure 4.10). These observations are consistent with the biophysical findings and clinical experiences after oxytocin treatment of patients with unfavourable cervical states[25-27]. Prostaglandins, in particular PGE$_2$, appear to regulate the state of the cervix, especially in term pregnancy. The effects on the cervix may occur in parallel with those in the myometrium[28]. However, it should be emphasized that local application of prostaglandin (PGE$_2$) within the cervical canal can induce biochemical and biophysical changes resulting in a softening of the cervix without a concomitant measurable increase in myometrial activity[24] (Figure 4.11). These reported changes can also be induced in patients treated preoperatively with PGE$_2$ before termination of early pregnancy by dilatation and evacuation[29-31]. The clinical use of prostaglandins for cervical priming and/or labour induction is dealt with in detail in Chapters 10 and 11.

STRUCTURE AND FUNCTION OF SMOOTH MUSCLE COAT

A thin smooth muscle coat surrounds the outer, vaginal part of the cervix. As previously discussed, the importance of the smooth muscle for the function of the cervix is not clear. Anatomically this is a thin muscle layer, especially at the distal portio of the organ, although EM studies have shown that individual muscle cells may penetrate deeply into the stroma[20]. Whereas it was previously assumed that the smooth muscle of the cervix might act as a sphincter and serve to close the cervical canal, the observations by Danforth and co-workers have shown that the connective tissue is an important component of cervical closure mechanism in humans[1-3]. Recently, however, it has been demonstrated that the smooth muscle of the human cervix also has spontaneous activity and that this activity can be changed by drugs such as prostaglandins. Similar to the fibrous connective tissue, the smooth muscle layer seems to undergo changes during the menstrual cycle and pregnancy. In vitro studies show that prostaglandin F$_{2\alpha}$ results in either an increase or no change in cervical smooth muscle activity, whereas PGE$_2$ results in relaxation[4-6]. Most interesting are the rapid increases in intracervical pressure observed in some patients with severe primary dysmenorrhoea (Figure 4.12). Such pressure variations can hardly be caused by biochemical changes in the connective tissues, but must be due to smooth muscle contractions. Calculations comparing the myometrial and cervical contractile forces of the human uterus reveal that the cervical smooth muscle has a contractile capacity of about 10% of that of the myometrium[32].

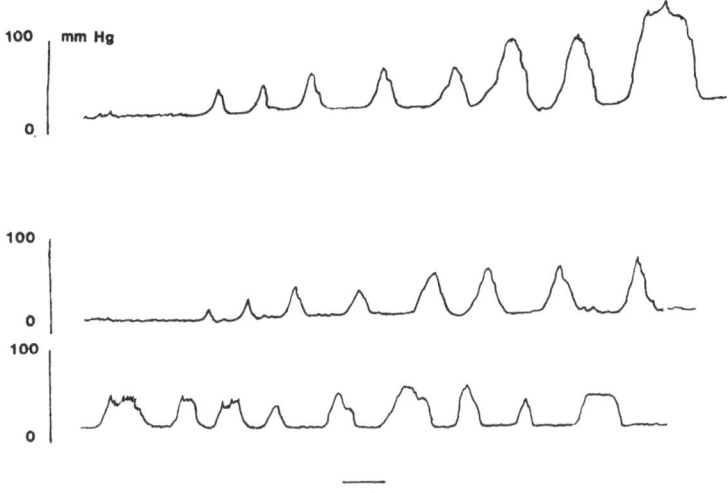

Figure 4.12 Retrograde uterine contractions elicited from the cervix. The recordings are performed with a three-channelled microtransducer catheter as shown in Figure 4.13. As seen, the contractions start within the cervix (lower tracing) and then move to the isthmus (middle tracing) ending in the fundus uteri (upper tracing). At the end of the recording a high pressure amplitude is registered within the fundus and at that time the patient reported dysmenorrhoeic pains

CERVICAL-RELATED TISSUES

In pregnancy the amniotic membranes are involved in production of substances, e.g. prostaglandins, that may affect the cervical state. Changes in the biochemical composition and physical characteristics of these membranes are less pronounced but of a similar nature to those seen in the cervical stroma[33-35]. This also appears to be the case concerning the vascular supply to the cervix and to the uterocervical supportive ligaments, although the number of investigations of these structures presently is too limited to allow definite conclusions.

METHODS TO EVALUATE CERVICAL STATE

In vivo

Various methods have been used to investigate the function of the human cervix. Several of these techniques are impractical for use in

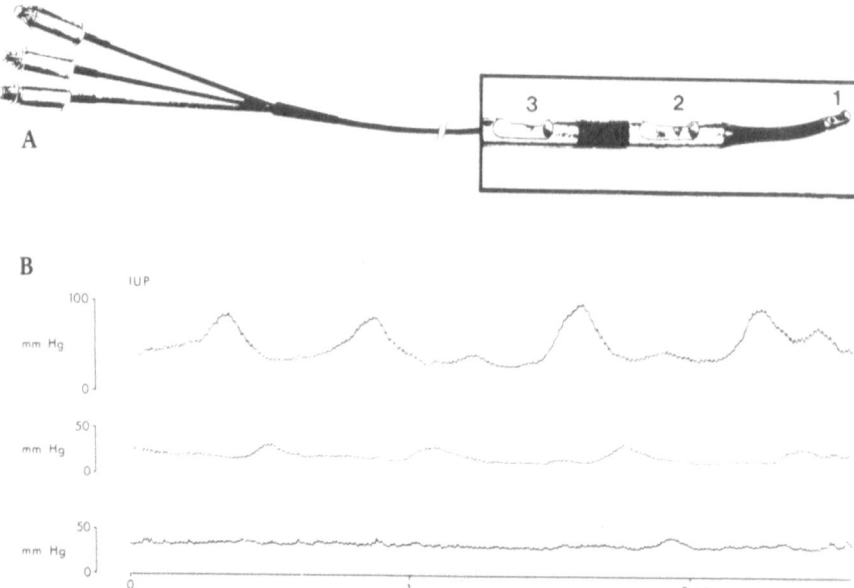

Figure 4.13 **A:** Multi-channelled microtransducer catheter used for recordings of uterine activity *in vivo*. The recordings are obtained simultaneously from three different sections of the catheter: (1) fundus; (2) isthmus; and (3) cervix uteri. **B:** Normal uterine activity as recorded by the microtransducer catheter during first day of menstruation in a healthy woman. As seen, regular antegrade contractions are present, i.e. starting in fundus (top tracing) moving towards isthmus (middle) and cervix uteri (bottom tracing). Compare with Figure 4.12. (From Ulmsten and Andersson, Ref. 36)

continuous monitoring of dynamic changes over long time periods *in vivo*.

In the non-pregnant uterus simultaneous pressure recordings from the fundus, isthmus and cervix uteri can be carried out using microtransducers[36] (Figure 4.13). Such recordings document pronounced cervical activity, most probably affecting the myometrial contractility, in some patients with dysmenorrhoea (Figure 4.12). The clinical significance of these observations requires evaluation. The cervix can also be evaluated by X-ray techniques such as hysterography. Generally this technique is used in patients with problems of infertility and in those with suspected cervical incompetence. The cervical state can also be visualized by ultrasound, a method commonly used during pregnancy[37,38]. Although relatively few investigations using this technique have been reported, in the future ultrasound may play a more important role in the investigation of cervical function *in vivo*. Figure 4.14 shows a simultaneous recording of cervical dilatation with the ultra-

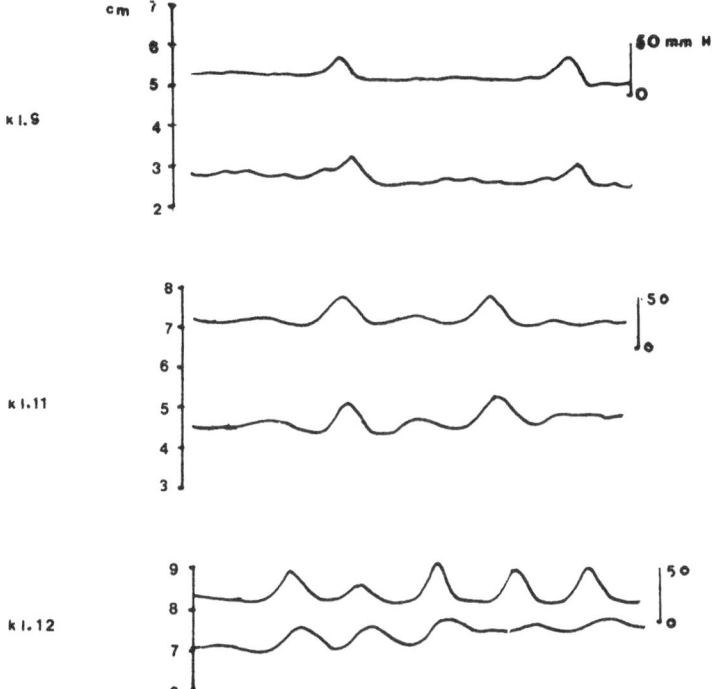

Figure 4.14 Cervical dilatation as measured by the ultrasound transit method and myometrial activity as measured by an intrauterine microtransducer catheter from a pregnant patient during progress of labour. Upper tracing = uterine contractions (mmHg); lower tracing = cervical dilatation (cm). Points of time of the recordings are indicated to the left

sound transit time method and intrauterine pressure with microtransducers in a patient whose labour was induced at term[39].

Direct measurements of cervical dilatation have also been made by different mechanical instruments[40], but these techniques are not as practical as direct clinical examination by the obstetrician. For the experienced examiner, the length of the cervix, its consistency, the degree of dilatation or opening of the cervical canal remain important features to recognize. Unfortunately, digital examination is often highly subjective and two examiners, unless they have carried out numerous consecutive examinations, often differ in their assessments of these features. To help overcome this deficiency, several cervical scoring systems have been developed. The most widely used is the Bishop score[41]. In the non-pregnant state and early pregnancy, metal sounds can be used to assess the degree of opening of the cervical canal. Figure 4.15 illustrates a modification of Hegar's sound described

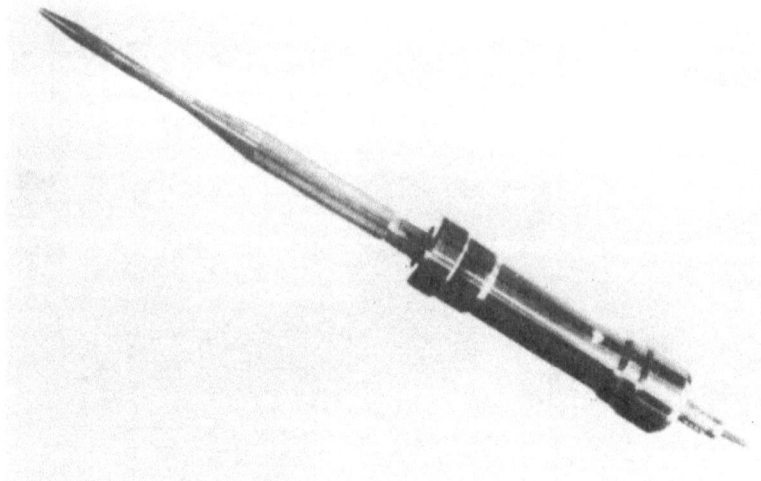

Figure 4.15   Cervical dilator probe used to measure cervical compliance *in vivo* as described by Calder. A force-sensing handle with a 6 mm cervical dilator attached. (From Calder, Ref. 26)

by Calder[26]. This instrument can be used to calculate cervical resistance in early pregnancy before abortion by dilatation and evacuation.

In vitro

In vitro techniques of assessing the morphology, biochemistry and biophysics of the cervix rely upon analyses of cervical tissue biopsies obtained with needles, clamps, scissors or scalpels. The problems encountered are those of difficulty in biopsy of a firm cervix and the risks of bleeding. To avoid these we have developed a rotating needle biopsy instrument which has facilitated tissue sampling, especially from non-pregnant cervices[17].

CONTROL OF CERVICAL STATE

Uterine effects

It was formerly believed that the cervix was only a passive inferior portion of the uterus. It is now evident that myometrial activity may affect the state of the cervix. This is seen, for example in Figure 4.13 where uterine contractions in advanced labour cause the cervix to dilate. Recent investigations, however, have indicated that the cervical

state can also be changed without significant influence of the myometrium. Figure 4.11 shows that cervical ripening can be achieved by
intracervical application of PGE_2 without measurable uterine contractions. Sometimes the opposite situation is present, i.e. the cervix affects
the state of the myometrium, as is seen in Figure 4.11 which illustrates
retrograde contraction waves emanating from the cervix of a dysmenorrhoeic patient.

In a normal situation, increased/decreased myometrial activity and
decreased/increased cervical resistance coincide. Thus, in the non-pregnant woman the cervix is relatively closed at the time of ovulation
when the myometrium is relaxed, and then cervical resistance decreases
continuously towards the end of the cycle in order to allow menstrual
blood outflow while, at the same time, myometrial activity steadily
increases towards menstruation. By the same token, the cervix is more
or less closed during the main part of pregnancy and normally does
not start to dilate until 'some weeks' before delivery. However, in
some patients this dilatation occurs at an earlier time and without
recognized uterine activity. Some of these women turn up in the delivery ward with a widely dilated cervix without having experienced
painful uterine contractions, i.e. 'travail insensible'. This is particularly
true in cases of multiple gestation. Another type of cervical dysfunction
is seen in some patients with imminent preterm deliveries due to cervical incompetence in which the uterine contractions may hardly be
recognized either by the patients or recording equipment despite considerable and progressive cervical dilatation. In these patients the biochemistry of the cervix resembles that seen in the spontaneous or
PG-induced ripened cervix at term[22,42].

Hormonal influence

Without doubt hormones such as oestrogen and progesterone as well
as prostaglandins influence the state of the non-pregnant and pregnant
uterus and its cervix. It is not within the scope of this chapter to
discuss in detail the non-pregnant cervix and its role in fertility and
infertility[43]. It is widely accepted that oestrogen in the non-pregnant
cervix increases the compliance of the non-pregnant cervix and causes
the cervical canal to dilate, whereas progesterone has opposite effects.
Progesterone also affects the composition of the cervical mucus, making it less accessible to sperm penetration and thereby acting as a
natural type of contraceptive[44,45]. Relaxin also effects sperm penetration
in the human non-pregnant cervix. In addition, prostaglandins appear
to influence the different parameters of the non-pregnant cervix. For
example, it has been suggested that they may affect the composition
and production of mucus as well as the state of the stroma and the

muscular layer[6,10,17]. PGE_2 generally seems to have relaxing effects on the stroma and smooth muscle layer, whereas $PGF_{2\alpha}$ either has no effect or stimulates the smooth muscle layer to contract[4-6]. Considering the interplay between the myometrium and cervix, it has been suggested that an 'inadequate' ratio of $PGF_{2\alpha}$ and PGE_2 might contribute to the production of dysmenorrhoeic cramps[46,47].

Recently it has been suggested that neuropeptides such as vasoactive intestinal peptide (VIP) and substance P have an effect on the human uterus[48]. Although only a few observations exist, VIP has been reported to have a relaxing effect on the cervix state, whereas the effects of substance P and vasopressin seem to vary. The clinical importance of these observations requires further study for clarification.

During pregnancy there is a common hormonal control of the state of the cervix and myometrium[49]. Though the exact mechanisms underlying the spontaneous initiation of human parturition are not yet known, it is evident that, as stated by Liggins, 'any hypothesis for the initiation of human labour is incomplete unless it includes a satisfactory explanation for the structural changes in the cervix'[50]. A direct influence of oestrogen and progesterone on the cervix in the initiation of labour has not been convincingly proven, and there are conflicting reports on cervical ripening by various oestrogens[51-53]. Similarly, progesterone treatment does not effectively normalize uterine activity and the cervical state in patients with preterm labour. Despite these observations, it has been proposed that a change in the oestrogen/progesterone ratio forms the basis for subsequent structural changes mediated by other stimuli such as prostaglandins which results in cervical dilatation and the onset of labour.

Oxytocin, widely believed to be involved in both spontaneous and induced labour, does not appear to induce significant ripening effects on the cervix before labour has been established[25-27]. This fact must be considered at the time of proposed induction of labour in patients with unfavourable cervices. Oxytocin, in contrast to prostaglandins, is ineffective in terminating early pregnancy. Despite considerable myometrial stimulation by oxytocin, the state of the cervix is generally unaffected. These clinical observations have been confirmed in sophisticated experimental investigations by Forman et al.[24].

According to recently reported observations, relaxin may also influence the cervical state. Intravaginal administration of relaxin induces significant cervical ripening before term labour induction, even in the absence of prominent uterine activity[54].

Prostaglandins

During pregnancy there is an increased production of various prostaglandins. Figure 4.16 illustrates sites of prostaglandin synthesis related
to the uterine cervix and corpus. Prostaglandin E_2 production predominates in the cervix. Although their physiological role has not been

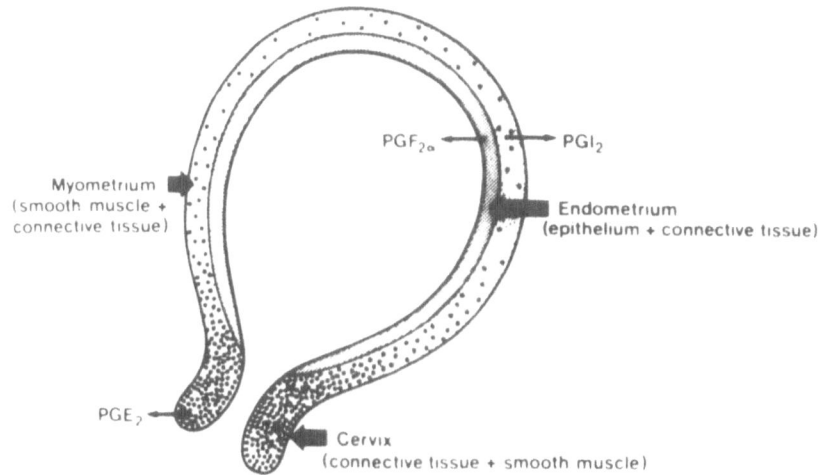

Myometrium
(smooth muscle +
connective tissue)

$PGF_{2\alpha}$ PGI_2

Endometrium
(epithelium + connective tissue)

PGE_2

Cervix
(connective tissue + smooth muscle)

Figure 4.16 Diagram illustrating the different major prostaglandins synthesized in each of the three components of the uterus. (From Liggins, Ref. 49,
with permission)

definitely established, a large number of investigations have demonstrated the efficacy of using prostaglandins to ripen the cervix and to
induce term labour[19,26,55-57]. PGE_2 is favoured before $PGF_{2\alpha}$ because
PGE_2 relaxes the musculature of the cervix while $PGF_{2\alpha}$ contracts it,
although both PGE_2 and $PGF_{2\alpha}$ can induce cervical ripening. Clinical
observations are supported by those obtained from studies *in vitro*[6,32].
Due to the presence of undesirable side-effects which occur after systemic administration of PGs, local applications of these compounds is
preferred. It is possible to achieve both cervical ripening and induction
of labour in patients with highly unfavourable cervical states using
intracervical application of small doses of PGE_2[19,57]. This technique can
also be used for preoperative dilatation of the cervix before termination of early pregnancy by D&E[31,57]. The reader is referred to Chapters
10 and 15 for a more complete review of the pharmacology of cervical
ripening. It may be that in the future newly synthesized prostaglandin
analogues might be superior to presently existing compounds for the
proper control of cervical state.

In parallel with the physiological and pharmacological effects of prostaglandins on the cervical state, where ripening and dilatation are the main aims, treatment with prostaglandin synthetase inhibitors might be valuable in patients with an excessive cervical dilatation preterm. Theoretically, PG inhibitors should be attractive in premature labour, since these compounds should normalize both an increased myometrial activity and a decreased cervical resistance. Promising clinical results have been reported using different PG inhibitors for treatment of preterm labour[58,59].

CONCLUSIONS

A major function of the cervix is to control the passage into and out of the uterine cavity which is effected mainly by the thick connective tissue of the stroma. The epithelium of the cervical canal also contributes to this function by secretion of cervical mucus which has unique rheological properties and at times acts as a selective barrier to penetration of spermatozoa and unwanted micro-organisms. Mucus production is under hormonal control, regulated by oestrogen, progesterone and probably also by prostaglandins.

The cervical stroma which constitutes the main part of the human cervix is composed of a mass of connective tissue. It undergoes dramatic biochemical and biophysical changes, especially during late pregnancy, that results in a softened and ripened cervix which is easily dilated at labour. These changes can be enhanced or induced by prostaglandins.

Normally, a close interplay exists between the cervix and the myometrium. The previously held concept that the cervix is only an inferior, passive part of the corpus uteri should be abandoned. Cervical function can be independent of uterine activity, and cervical ripening can be obtained without apparent uterine contractions. Moreover, in patients with dysmenorrhoea, increased activity in the cervix can elicit retrograde peristaltic contractions affecting the contractile activity in the myometrium.

During pregnancy and the menstrual cycle, changes in the biochemical composition of the cervical stroma are paralleled by biophysical changes of the cervical tissue. In the ripened term pregnant cervix the collagen is degraded by proteinases while simultaneously cervical distensibility or compliance is increased. Prostaglandins can mimic these changes, and this fact is of great clinical importance, especially in patients in whom cervical maturation fails to appear naturally. It has yet to be established whether treatment with prostaglandin inhibitors can normalize or prevent further deterioration of the cervical state in patients with preterm labour.

Several methods have been developed to investigate the state of the cervix *in vivo*, but all are afflicted with technical and biological disadvantages. *In vitro* analyses of tissue biopsies taken from different parts of the cervix serve as important research tools to assess the biochemical and biophysical characteristics of the cervix. The results of such analyses have improved our understanding of cervical function. In addition, in the future they will improve the possibilities of a proper pharmacological intervention in patients with cervical dysfunction.

ACKNOWLEDGEMENT

This study has in part been supported by grants from the Swedish Research Council No. B84-17X-6856-1A.

References

1 Danforth, D. N. (1954). The fibrous nature of the human cervical tissue. *Am. J. Obstet. Gynecol.*, 53, 541.

2 Danforth, D. N. (1954). The distribution and functional activity of the cervical musculature. *Am. J. Obstet. Gynecol.*, 68, 1261

3 Danforth, D. N., Buckingham, J. C. and Roddick Jr, J. W. (1960). Connective tissue changes incident to cervical effacement. *Am. J. Obstet. Gynecol.*, 80, 939

4 Najak, Z., Hillier, K. and Karim, S. M. M. (1970). The action of prostaglandins on the human isolated cervix. *Br. J. Obstet. Gynaecol.*, 77, 701

5 Cornley, M. and Hackbart, I. (1979). Die Wirkung von PGE_2 und $PGF_{2\alpha}$ auf vershiedener Uterussegmente *in vitro*. *Arch. Gynecol.*, 227, 83

6 Bryman, I., Shani, S., Norström, A. and Lindblom, B. (1984). Influence of prostaglandins on contractility of the isolated human cervical muscle. *Am. J. Obstet. Gynecol.*, 63, 280

7 Hafez, E. S. (1982). Structural and ultrastructural parameters of the uterine cervix. *Obstet. Gynecol. Survey*, 37, 507

8 Elstein, M. (1982). Cervical mucus: its physiological role and clinical significance. *Adv. Exp. Biol.*, 144, 301

9 Chantler, E. (1982). Structure and function of cervical mucus. In Chantler, E., Elder, M. and Elstein, M. (eds.) *Mucus in Health and Disease. Adv. Exp. Med. Biol.*, 144, 251-64

10 Carlstedt, I., Lindgren, H., Sheehan, J., Ulmsten, U. and Wingerup, L. (1983). Isolation and characterization of human cervical-mucus glucoproteins. *Biochem. J.*, 211, 13

11 Odeblad, E. (1968). The functional structure of human cervical mucus. *Acta Obstet. Gynecol.*, 48, 59

12 Beson, M. F., Parish, G. R., James, S. L. and Marriott, C. (1982). A scanning electron microscopic study of human cervical mucus. In Chantler, E., Elder, M. and Elstein, M. (eds.) *Mucus in Health and Disease. Adv. Exp. Med. Biol.*, 144

13 Uldbjerg, N., Ulmsten, U. and Ekman, G. (1983). The ripening of the human uterine cervix in terms of connective tissue biochemistry. In Ulmsten, U. and Ueland, K. (eds.) *The Forces of Labor: Uterine Contractions and the Resistance of the Cervix. Clin. Obstet. Gynecol.*, **26**, 14

14 Norström, A., Wilhelmsson, L. and Hamberger, L. (1981). The regulatory influence of prostaglandins on protein synthesis in the human non-pregnant cervix. *Prostaglandins*, **22**, 117

15 Scott, J. E. (1981). Dermatan sulphate-rich proteoglycan associates with rat tail-tendon collagen at the d-band in the gap region. *Biochem. J.*, **197**, 213

16 Liggins, G. C. (1983). Cervical ripening as an inflammatory reaction. In Ellwood, D. A. and Anderson, A. B. M. (eds.) *The Cervix in Pregnancy and Labor.* pp. 1–9. (Edinburgh, London, Melbourne, New York: Churchill Livingstone)

17 Uldbjerg, N., Ekman, G., Malmström, A., Olsson, K. and Ulmsten, U. (1983). Ripening of the human uterine cervix related to changes in collagen glycosaminoglycans and collagenolytic activity. *Am. J. Obstet. Gynecol.*, **167**, 662

18 Kischer, C. W., Drogemüller, W., Schetlar, M., Schvapil, M. and Vining, J. (1980). Ultrastructural changes in the architecture of collagen in the human cervix treated with urea. *Am. J. Pathol.*, **99**, 525

19 Ulmsten, U., Wingerup, L. and Ekman, G. (1983). Local application of prostaglandin for cervical ripening or induction of term labor. In Ulmsten, U. and Ueland, K. (eds.) *The Forces of Labor: Uterine Contractions and the Resistance of the Cervix. Clin. Obstet. Gynecol.*, **26**, 95

20 Uldbjerg, N., Ekman, G., Herltoft, T., Malmström, A., Ulmsten, U. and Wingerup, L. (1983). Human cervical tissue and its reaction to prostaglandin E_2. *Acta Obstet. Gynecol. Scand. Suppl.*, **113**, 163

21 Conrad, J. and Ueland, K. (1983). Physical characteristics of the cervix. In Ulsten, U. and Ueland, K. (eds.) *The Forces of Labor: Uterine Contractions and the Resistance of the Cervix. Clin. Obstet. Gynecol.*, **26**, 27

22 Ekman, G., Uldbjerg, N. and Ulmsten, U. (1984). Cervical collagen an important predictor of labor delivery time. *Am. J. Obstet. Gynecol.* (in press)

23 Ekman, G., Uldbjerg, N., Malmström, A. and Ulmsten, U. (1983). Increased postpartum collagenolytic activity in cervical connective tissue from women treated with prostaglandin E_2. *Gynecol. Obstet. Invest.*, **16**, 292

24 Forman, A., Ulmsten, U., Banyai, J., Wingerup, L. and Uldbjerg, N. (1982). Evidence for a local effect of intracervical prostaglandin E_2-gel. *Am. J. Obstet. Gynecol.*, **143**, 756

25 Valentine, B. H. (1977). Intravenous oxytocin and oral prostaglandin E_2 for ripening of the unfavourable cervix. *Br. J. Obstet. Gynaecol.*, **84**, 846

26 Calder, A. (1981). The human cervix in pregnancy – a clinical perspective.

In Elwood, D. A. and Anderson, A. B. M. (eds.) *The Cervix in Pregnancy and Labor.* p. 103. (Edinburgh, London, Melbourne, New York: Churchill Livingstone)

27 Ulmsten, U., Wingerup, L. and Andersson, K.-E. (1979). Comparison of prostaglandin E_2 and intravenous oxytocin for induction of labor. *Obstet. Gynecol.*, **54**, 581

28 Calder, A. A., Embrey, M. P. and Hillier, K. (1974). Extraamniotic prostaglandin E_2 for the induction of labour at term. *J. Obstet. Gynaecol. Br. Commonw.*, **81**, 39

29 Christensen, N. and Bygdeman, M. (1985). The effect of prostaglandins on the bioconversion of arachidonic acid in cervical tissue in early human pregnancy. *Prostaglandins* (in press)

30 Norström, A., Bryman, I., Lindblom, B. and Christensen, N. J. (1985). Effects of 9-deoxo-16,16-dimethyl-9-methylene PGE_2 on muscle contractile activity and collagen synthesis in the human cervix. *Prostaglandins* (in press)

31 Wingerup, L., Ulmsten, U., Bygdeman, M. and Hamberger, L. (1981). Prostaglandin E_2-gel for pretreatment of the cervix in nulliparous patients having a late first trimester termination of pregnancy. *Arch. Gynecol.*, **231**, 1

32 Conrad, J. T. and Ueland, K. (1976). Reduction of the stretch modulus of human cervical tissue by prostaglandin E_2. *Am. J. Obstet. Gynecol.*, **126**, 218

33 Manabe, Y. and Sagawa, N. (1983). Changes in the mechanical forces of cervical distension before and after rupture of the membranes. *Am. J. Obstet. Gynecol.*, **147**, 667

34 Okita, J. R., Johnstone, J. M. and MacDonald, P. C. (1983). Source of prostaglandin precursor in human fetal membranes. *Am. J. Obstet. Gynecol.*, **147**, 477

35 Ibrahim, M. E. A., Bou-Resli, M. N., Al-Zaid, N. S. and Bishay, L. F. (1983). Intact fetal membranes. *Acta Obstet. Gynecol. Scand.*, **62**, 481

36 Ulmsten, U. and Andersson, K.-E. (1979). Multi-channel intrauterine pressure recording by means of micro-transducers. *Acta Obstet. Gynecol. Scand.*, **58**, 115

37 Brook, I., Feingold, M., Schwartz, A. and Zakut, H. (1981). Ultrasonography in the diagnosis of cervical incompetence in pregnancy – a new diagnostic approach. *Br. J. Obstet. Gynaecol.*, **88**, 640

38 Jackson, G., Pendleton, H. J., Nichol, B. and Wittmann, B. K. (1984). Diagnostic ultrasound in the assessment of patients with incompetent cervix. *Br. J. Obstet Gynaecol.*, **91**, 232

39 Lindström, K., Persson, P.-H. and Ulmsten, U. (1978). Cervimetry based on the ultrasound transit time method. In Holmer, N. G. and Lindström, K. (eds.) *New Methods in Medical Ultrasound.* (Lund, Sweden: Studentlitteratur)

40 Bakke, T. (1974). Cervical consistency in women of fertile age measured with a new mechanical instrument. *Acta Obstet. Gynecol. Scand.*, **53**, 293

41 Bishop, E. H. (1964). Pelvic scoring for elective induction. *Obstet. Gynecol.*, **24**, 266

42 Theobald, P. W., Rath, W., Kühnle, H. and Kuhn, W. (1982). Histological

and electron microscopic examinations of collagenous connective tissue of the non-pregnant cervix, the pregnant cervix and the pregnant prostaglandin-treated cervix. *Arch. Gynecol.*, **231**, 241

43 Gaton, E., Zejdel, L., Bernstein, D., Glezerman, M., Czernobilsky, B. and Insler, V. (1982). The effect of oestrogen and gestagen on the mucus production of human endocervical cells: a histochemical study. *Fertil. Steril.*, **38**, 580

44 Van Kooij, R. J., Roelofs, H. J. M., Kathmann, G. A. N. and Kramer, M. F. (1980). Human cervical mucus and its mucous glucoprotein during the menstrual cycle. *Fertil. Steril.*, **34**, 226

45 Nirmala, V. and Thomas, J. A. (1982). Histochemical aterations in the uterine endocervical mucosa in different phases of the normal menstrual cycle and in the altered cycle. In Chantler, E., Elder, M. and Elstein, M. (eds.) *Mucus in Health and Disease. Adv. Exp. Med. Biol.*, **44**, 285

46 Bygdeman, M., Bremme, K., Gillespie, A. and Lundström, V. (1979). Effects of prostaglandins on the uterus – prostaglandins and uterine contractility. *Acta Obstet. Gynecol. Scand. Suppl.*, **87**, 33

47 Lumsden, M. A., Kelly, R. W. and Baerd, D. D. (1983). Primary dysmenorrhoea: the importance of both prostaglandins E_2 and $F_{2\alpha}$. *Br. J. Obstet. Gynaecol.*, **90**, 1135

48 Helm, G., Ottessen, B., Fahrenkrug, J., Larsen, J., Owman, C., Sjöberg, N. O., Stålberg, B., Sundler, F. and Walles, B. (1981). Vasoactive intestinal polypeptide (VIP) in the human female reproductive tract: distribution and motor effects. *Biol. Reproduct.*, **25**, 227

49 Liggins, G. C. (1983). Initiation of spontaneous labor. In Ulmsten, U. and Ueland, K. (eds.) *The Forces of Labor: Uterine Contractions and the Resistance of the Cervix. Clin. Obstet. Gynecol.*, **26**, 47

50 Liggins, G. C., Forster, C. S., Grieves, S. A. and Schwarz, A. L. (1977). Control of parturition in man. *Biol. Reprod.*, **16**, 39

51 Gordon, A. J. and Calder, A. A. (1977). Oestradiol applied locally to ripen the unfavourable cervix. *Lancet*, **2**, 1319

52 Thiery, M., De Crezelle, M., Vankets, H., Voorhoot, L., Verheugen, C., Smis, B., Gerris, J. and Marten, S. G. (1978). Extraamniotic oestrogens for the unfavourable cervix. *Lancet*, **2**, 835

53 Quinn, M. A., Murphy, A. J. and Kuhn, R. J. P. (1981). A double-blind trial of extraamniotic oestriol and prostaglandin $F_{2\alpha}$ gels in cervical ripening. *Br. J. Obstet. Gynaecol.*, **88**, 644

54 MacLennan, A. H., Green, R. C., Bryant-Greenwood, G. D., Greenwood, F. C. and Seamark, R. F. (1980). Ripening of the human cervix and induction of labour with purified porcine relaxin. *Lancet*, **1**, 220

55 MacKenzie, I. Z. (1981). Clinical studies on cervical ripening. In Ellwood, D. A. and Andersen, A. B. M. (eds.) *The Cervix in Pregnancy and Labor.* p. 163. (Edinburgh, London, Melbourne and New York: Churchill Livingstone)

56 Shepherd, J. H. (1981). The role of prostaglandins in ripening the cervix and inducing labor. *Clin. Perinatol.*, **8**, 49

57 Ulmsten, U. and Wingerup, L. (1979). Cervical ripening induced by prostaglandin E_2 in viscous gel. *Acta Obstet. Gynecol. Scand. Suppl.*, **84**, 1

58 Wiqvist, N. (1981). Preterm labor: other drug possibilities including drugs not to use. In Elder, M. B. and Hendricks, C. H. (eds.) *International Medical Review*. (Butterworth)
59 Niebyl, J. R. (1981). Prostaglandin synthetase inhibitors. *Sem. Perinat.*, 5, 274

5
The uterus

V. LUNDSTRÖM

INTRODUCTION

The uterus contains large amounts of prostaglandins which play important roles in many physiological processes. The endometrium is particularly rich in the primary prostaglandins, $PGF_{2\alpha}$ and PGE_2, which are readily absorbed by the myometrium and regulate its contractility. Other prostaglandins, such as prostacyclin and thromboxane, are also produced in the endometrium and within its vessel walls; these compounds contribute to haemostasis during menstruation. Recent research indicates that endogenous prostaglandins are of importance in the regulation of uterine activity, both normal and abnormal, human fertility, and normal menstruation as well as irregular and abnormal bleeding.

Anatomy

The architecture of the uterine muscle fibres is complex and has been studied by stepwise dissection[1]. The corpus uteri contains three separate muscle layers: (1) a thin subperitoneal layer, lying parallel to or in a spiral along the axis of the uterus; (2) an intermediate irregular layer; and (3) an internal spiral-shaped layer. In the isthmic part of the uterus the muscle fibres are organized in a circular direction (Figure 5.1). At the uterotubal junction, however, there are again three distinct muscle layers: the external is spiral or longitudinal; the intermediate is circular, and the internal is longitudinal (Figure 5.2). Each of these separate muscle layers has a specific response to the administration of different prostanoids *in vitro*[2].

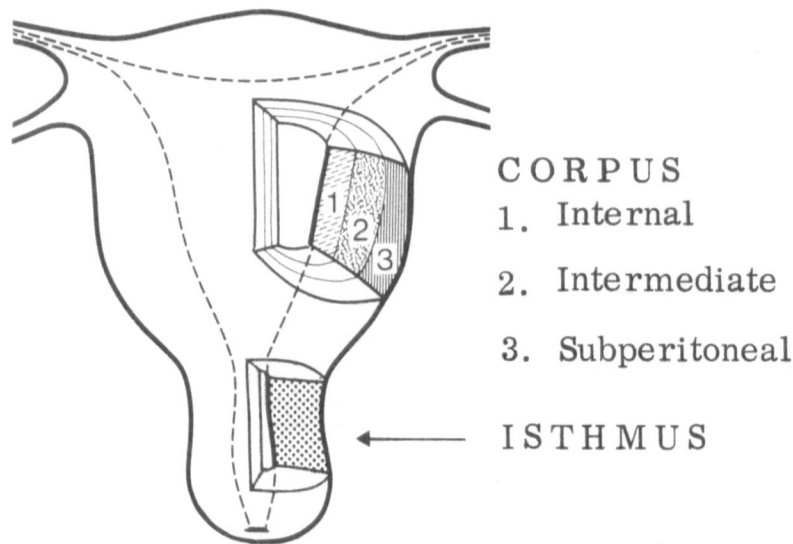

CORPUS
1. Internal

2. Intermediate

3. Subperitoneal

ISTHMUS

Figure 5.1 The corpus uteri contains an internal spiral-shaped layer, an intermediate irregular layer and a thin parallel or spiral-shaped subperitoneal layer. The isthmus contains a circular muscle fibre layer

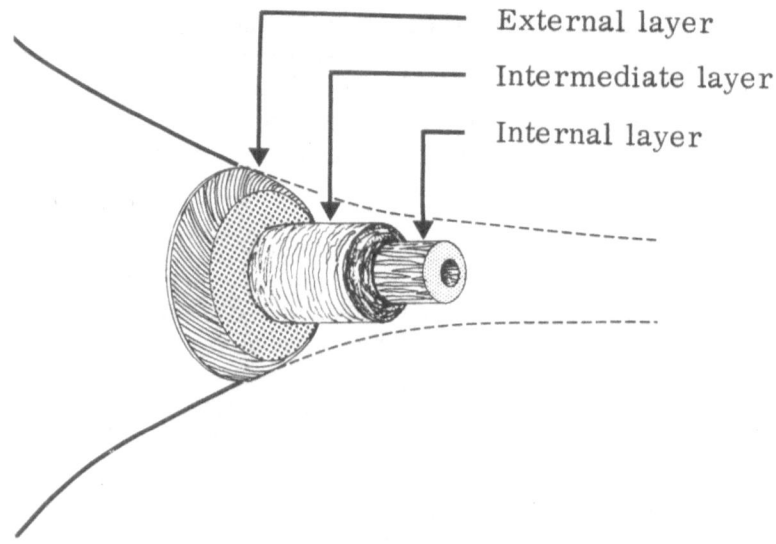

External layer

Intermediate layer

Internal layer

Figure 5.2 The uterotubal junction contains an external spiral or longitudinal muscle layer, an intermediate circular layer and an internal longitudinal layer

NON-PREGNANT UTERUS

Uterine contractility *in vitro*

The effect of the different prostaglandins in *in vitro* testing systems is as follows.

Prostaglandin E_2
PGE_2 inhibits the spontaneous contractility in the intermediate and the internal layer of the corpus uteri. However, the subperitoneal myometrial fibres generally react to low doses (1 ng/ml), although higher concentrations (100 ng/ml) induce relaxation. In contrast, the isthmus uteri appears insensitive to PGE_2. In the uterotubal junction, PGE_2 induces a stimulatory response in the external layer, while the circular intermediate layer responds with relaxation. The internal longitudinal layer shows a variable response pattern to PGE_2, depending upon the phase of the cycle: stimulation around the time of ovulation and inhibition during the other phases of the cycle.

Prostaglandin $F_{2\alpha}$
$PGF_{2\alpha}$ stimulates all layers of the corpus uteri irrespective of the phase of the menstrual cycle. All layers of the uterotubal junction are stimulated by $PGF_{2\alpha}$ in all phases of the cycle. $PGF_{2\alpha}$ does not induce changes in the contractility of the isthmus uteri, however. A mixture of $PGF_{2\alpha}$ and $PGE_{2\alpha}$ in equal concentrations stimulates all layers of the corpus uteri. When the ratio $PGE_2/PGF_{2\alpha}$ is increased to 10/1, however, a relaxation is observed in the corpus uteri.

Prostacyclin
Administration of prostacyclin (PGI_2) results in an inhibition of the contractility of all muscle layers of the myometrium irrespective of phase of the menstrual cycle. Moreover, the muscle layers of the uterotubal junction also respond to PGI_2 with inhibition. The efficacy of PGI_2 in relaxing the experimental strips is estimated to be one tenth that of $PGF_{2\alpha}$.

Endoperoxides
Moderate doses (5–100 ng/ml) of PGH_2 elicit a stimulatory response on the myometrium. On the other hand, higher doses cause a biphasic response, i.e. contraction followed by subsequent relaxation. At the uterotubal junction, the addition of PGH_2 results in a stimulatory response in the muscle strips of the external spiral-shaped as well as the internal longitudinal layers, while inhibition occurs in the intermediate circular layer.

Thromboxane
TxA_2 is the most potent prostaglandin known; it causes stimulation of all layers of the corpus uteri and of the uterotubal junction even at very low doses (0.07 ng/ml).

Contractility *in vivo*

Regulation of non-pregnant uterine contractility
The contractility of the myometrium can be recorded by open-end catheters, microballoons or microtransducers inserted through the cervical canal and placed within the uterine cavity. Characteristic contractility patterns are observed during different phases of the menstrual cycle. In the proliferative phase contractions are characterized by a small (10-30 mmHg) amplitude, a frequency of 1-3 per minute, and a resting tone of 10-25 mmHg. Around the time of ovulation, however, the tone increases to 40-60 mmHg, and the frequency increases to 3-5 per minute with a reduction of the amplitude to 5-10 mmHg. After ovulation the tone decreases to 10-30 mmHg once again with a concomitant increase in the amplitude to 80 mmHg.

Increased uterine activity during menstruation is a well recognized phenomenon. Regular contractions are seen with high (around 100-150 mmHg) amplitudes and a basal tone around 30-50 mmHg. These patterns depend on the levels of steroid hormones. The administration of exogenous steroid hormones to menopausal women elicits a pattern characteristic of the younger woman[3].

Excessive endogenous endometrial synthesis of $PGF_{2\alpha}$ is associated with dysmenorrhoea. A large amount of $PGF_{2\alpha}$ is absorbed by the myometrium and this results in a typical hypercontractility state during menstrual cramps. The administration of a potent prostaglandin biosynthesis inhibitor[4] relaxes the myometrium within 45 min. Obviously, the synthesis of endogenous PG can be affected rapidly, thus changing patterns of contractility and relieving menstrual pain. The exact relationship between steroid hormones and prostaglandin synthesis is not known at present. However, it is believed that the different prostaglandins mediate uterine contractility following steroid hormone impulses.

Administration of PG
Intravenous administration of PGE_2 and $PGF_{2\alpha}$ both result in stimulation of the myometrium; however, prostaglandin E_2 is 2-3 times more potent than $PGF_{2\alpha}$; the threshold dose of the former is 20 μg whereas it is 50 μg for the latter[5]. Continuous intravenous infusion of $PGF_{2\alpha}$ or PGE_2 over a period of several hours induces contractions of high amplitudes similar to those observed during menstruation.

Oral administration of 1.0 mg of PGE_2 in tablet form usually results in stimulation of uterine activity during both the proliferative and secretory phases of the menstrual cycle. During menstruation, however, an inhibitory response to oral administration of PGE_2 occurs[6]. Some patients with dysmenorrhoea obtain pain relief following intake of PGE_2 orally. In contrast, prostacyclin has been given intravenously in doses which cause cardiac side-effects without any effect being noted on uterine contractility[7,8].

Following systemic administration of the primary prostaglandins, the rapid metabolism of these compounds in lung and liver results in small amounts reaching the uterus. More accurate information on the sensitivity of the myometrium to prostaglandins may be obtained if they are administered directly into the uterine cavity via injection through a thin catheter. Experiments performed at different phases of the menstrual cycle have shown that the myometrium is most sensitive in the early proliferative and late secretory phases[9]. During these phases, 1 μg of PGE_2 or 2·5 μg of $PGF_{2\alpha}$ elicits a strong stimulatory response. However, around the time of ovulation the uterus becomes insensitive to PGE_2 as well as $PGF_{2\alpha}$ using doses in a range of 40–50 μg. If the doses are further increased to 100 or 200 μg, inhibition occurs[10].

At the time of menstruation, administration of PGE_2 elicits a complex response. Low doses of PGE_2 (2–5 μg) stimulate contractility, while higher doses (30–40 μg) act to relax the myometrium. $PGF_{2\alpha}$ consistently stimulates uterine contractility during menstruation.

A discrepancy in response to PGE_2 following its local administration has been observed in women with unexplained infertility compared with normally fertile women[11]. A strong stimulatory response of the myometrium is seen at midcycle in women with unexplained infertility while fertile controls respond with slight inhibition.

Prostacyclin injected locally into the uterine cavity produces gradual stimulation. Presumably this effect is not a direct effect of prostacyclin on the muscle fibres, but more likely represents a secondary effect from its strong vascular action[7,8].

Little is known about the *in vivo* effect of thromboxane and the endoperoxides.

Endometrium

The human endometrium forms many different prostaglandins; these include the endoperoxides, PGE_2 and $PGF_{2\alpha}$, thromboxane A_2 (TxA_2) and prostacyclin (PGI_2)[12]. Unfortunately, since prostaglandins are also synthesized artificially during sampling, it is very difficult to determine tissue levels. In fact, the mechanical stimulation which inevitably takes place during biopsy is one of the most powerful factors for triggering biosynthesis of prostaglandins. As a result, measured amounts of these compounds are much more likely to reflect the capacity of the tissue to synthesize the arachidonate metabolites rather than actual *in situ* prostaglandin levels. This is particularly true when the enzymatic activity in the biopsy sample is not interrupted immediately by transfer (within seconds) to an organic solvent at a very low temperature.

Because of these limitations, it seems appropriate to summarize existing data only in a general way. The results of most studies indicate a progressive increase in concentrations of $PGF_{2\alpha}$, the $F_{2\alpha}$-metabolite (15-keto-13,14-dihydro-$PGF_{2\alpha}$) and PGE_2 from the proliferative to the secretory phase with peak levels being reached during menstruation. The ratio between the F and the E prostaglandins is generally higher in the secretory and menstrual phases compared with the proliferative phase. Several conditions are associated with overproduction of PGE_2 and/or $PGF_{2\alpha}$ or alterations in the E/F ratio; these include dysmenorrhoea, dysfunctional bleeding, IUD-induced bleeding and endometrial carcinoma.

Local circulation

The functional layer of the endometrium is endowed with a unique vascular pattern, i.e. coiled spiral arteries which function as end arteries. Each spiral artery supplies a narrow but separate, longitudinal segment of the uterine mucosa and it does this without communicating with the other arteries. The venous drainage of the endometrium depends on a rich plexus of veins which show some areas of dilatation; these areas are known as venous lakes. Menstruation is preceded by prolonged vasoconstriction of the spiral arteries and under the action of an unknown vasoconstrictor substance[13]. Some investigators[14] have proposed that this substance is $PGF_{2\alpha}$.

Thromboxane A_2 is also a potent vasoconstrictor, whereas PGE_2 and prostacyclin cause vessel dilatation. The role of these prostaglandins, present in the endometrium, for vascular changes during menstruation still remains uncertain.

Haemostasis

Platelets play an important role in the process of haemostasis during menstruation. Different haemostatic mechanisms are active during the premenstrual and menstrual phases. These events have been studied in detail by Christaens and co-workers and will be summarized here[15].

In the premenstrual phase, although there is little extravasation of blood, the process of stromal disintegration begins. Some of the blood vessels lose their continuity with resultant holes in the endothelium and interruptions in the basal membrane. Even though the platelets are exposed to the subendothelial collagen, no adherence or accumulation of the platelets was observed[15].

Once menstruation has started and the tissue is being shed, the blood vessels completely lose their continuity. At this time the vessel stumps in the functional layer are filled with plugs formed by degranulated platelets and fibrin. As the shedding of tissue progresses, these plugs are shed along with the tissue, and new plugs are formed upstream. Haemostasis by plugs is not complete, and some vessels remain unoccluded. The maximum plug formation occurs 7 h after onset of menstruation, but the process of plug formation continues up to 13 h afterwards. Twenty hours after the onset of menstruation, the vessels supplying the shedding surface are constricted, and no more extravasation of red or white blood cells is seen. In summary, plug formation seems to be the main haemostatic mechanism in early menstruation, although the process is remarkably inadequate. In late menstruation, however, vasoconstriction may play the primary role in providing haemostasis.

In recent years our understanding of certain biochemical events has clarified and contributed to the understanding of haemostasis. Platelets convert arachidonic acid to prostaglandin endoperoxides via the cyclo-oxygenase pathway. These compounds are highly potent inducers of platelet aggregation. The endoperoxides are further metabolized by platelets to form thromboxane A_2. This latter compound is also an extremely potent vasoconstrictor and inducer of platelet release reaction and its importance for haemostasis is obvious. Vessel walls convert arachidonic acid almost exclusively into prostacyclin which exerts opposing actions to those caused by TxA_2; it relaxes vessel walls and inhibits and even reverses platelet aggregation[16]. It is possible that haemostasis due to plug formation during menstruation is regulated by the balance between the proaggregating and the vasoconstrictive TxA_2 generated by the platelets and the vasodilating and aggregation inhibiting prostacyclin generated by the vessel walls[17].

Continuous administration of a prostaglandin synthesis inhibitor does not interfere with the length of the menstrual cycle or the onset of menstruation[18]. Endometrial biopsies after daily administration of 1000 mg naproxen have revealed a certain stabilization of the lysosomal membranes, but no other significant differences in the size of the endometrial glands or leukocytic infiltration were observed[19].

The treatment of primary dysmenorrhoea with a prostaglandin synthetase inhibitor (NSAID) results in a relaxation of the myometrium with disappearance of ischaemic pain. In some women this change in uterine contractility is accompanied by a reduction in menstrual blood loss[20]. Obviously, a relaxation of the myometrium does not predispose a woman to menorrhagia. On the contrary, the effect of NSAID is thought to act on the prostacyclin–thromboxane axis. It is not yet known whether the human endometrium can biosynthesize leukotrienes. However, the abundant presence of various types of leukocytes in the endometrium during menstruation indicates that the presence of leukotrienes is likely. The leukotrienes may also be partly responsible for vasoconstriction.

The use of prostaglandin synthetase inhibitors has been tested in the treatment of menorrhagia. The rationale for this approach is also based on observation that PGE is increased in the endometrium of women with menorrhagia[21]. Mefenamic acid, a potent NSAID, was first used for the treatment of menorrhagia and led to a demonstrated reduction of menstrual blood loss to less than 80 ml in 17 out of 22 patients[22]. A similarly effective decrease in menstrual flow in IUD users has been reported after use of mefenamic acid[23] as well as naproxen[24].

In summary, the different events which regulate menstruation appear to depend on the balance between different prostaglandins which possess opposing effects. An inhibition of the biosynthesis of the prostaglandin endoperoxides might not interfere with the balance between these prostaglandins and therefore have no visible effect. The exact mechanism initiating the onset of menstruation is not yet known. However, the prostaglandins appear to have an important role in mediating this process.

PREGNANT UTERUS

Considerable evidence from animal studies suggests that a surge of local prostaglandin production precedes the onset of parturition[25]. This phenomenon has not been demonstrated in the human and, as yet, there has been no clear demonstration of a change in quantities or types of prostaglandins before the onset of uterine contractions in the human female.

Early on it became apparent that prostaglandins influenced the activity of human myometrium *in vitro*, albeit with different effect depending on whether the myometrium was from a pregnant or a nonpregnant woman. In general, prostaglandins of the E and F type show the most marked stimulatory effect on pregnant myometrium.

Endogenous synthesis of PG

Peripheral blood levels of PG in pregnancy
Numerous investigations have measured blood levels of prostaglandins during pregnancy, but the results have varied and often are contradictory. In general, higher levels of PFG have been found in the ante-

natal period. However, there is no agreement about changes in concentration during labour, since PGF levels have been variously reported as being highest during the first stage of labour[26] and highest at the time of delivery[27]. These reported variations may be due either to differences in assay methodology or to release of $PGE_{2\alpha}$ from platelets during collection and subsequent handling of the blood sample. These circumstances, plus the known short half-life (<1 min) of the primary prostaglandins in plasma, have impeded the establishment of a reliable body of knowledge about the endogenous production of prostaglandins in the uterus.

In contrast, more accurate information on the endogenous synthesis of prostaglandins is obtained by measuring the major metabolite of $PGF_{2\alpha}$, namely 15-keto-13,14-dihydro-$PGF_{2\alpha}$, which has a longer half-life time (8 min) and is not synthesized artificially from platelets. The endogenous levels of 15-keto-13,14-dihydro-$PGF_{2\alpha}$ range between 20 and 60 pg/ml during the last 4 weeks of normal pregnancy. These levels increase during active labour at term, reaching peak values between 300 and 600 pg/ml[28].

Amniotic fluid and PG content
The amniotic fluid can be analysed for the presence of primary prostaglandins without interference by synthesis of PG from platelets or leukocytes. Many workers have examined amniotic fluid at various gestational periods and during labour. The amniotic fluid concentrations of primary prostaglandins in early and mid-pregnancy are lower than near term[29,30], and these levels increase progressively during the course of labour.

The apparent accumulation of PGE and PGF in the amniotic fluid during labour in the human results from an increase in the intrauterine production of PGE and PGF as a result of uterine contractions. Further evidence that an increase in prostaglandin synthesis occurs during labour arises from the observation of a significant increase in the chief metabolite of $F_{2\alpha}$, namely 15-keto-13,14-dihydro-$PGF_{2\alpha}$, in amniotic fluid during labour[31].

Prostaglandin synthesis and metabolism in the uterus

Arachidonic acid in its free form is the obligatory precursor for the synthesis of PGE_2 and $PGF_{2\alpha}$ via the 'prostaglandin synthetase' group of enzymes. Fatty acids are mainly stored by incorporation into phospholipids (see Chapter 2). Studies of the bioconversion of arachidonic acid in the different tissues of the pregnant uterus document the presence of several pathways[32] (Figure 5.3). Tissue homogenates of amnion, chorion, placental arteries, placenta and myometrium obtained

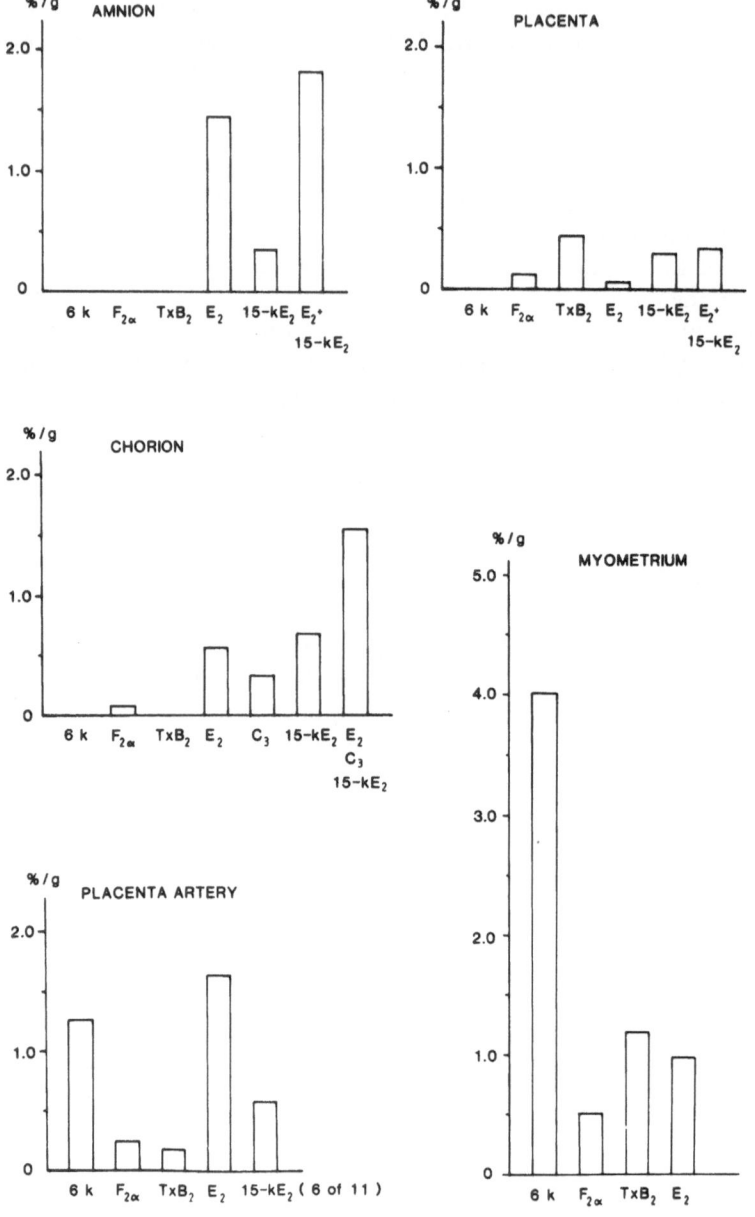

Figure 5.3 Summary of the conversion of [^{14}C]arachidonic acid in percentage of incubated radioactivity per gram wet tissue into 6-keto-PGF$_{1\alpha}$ (6 k), PGF$_{2\alpha}$, PGE$_2$, TxB$_2$ and 15-keto PGE$_2$ (15-k E$_2$) in whole cell homogenates of the different tissues obtained from Caesarean section ($n = 11$; mean values). C$_3$ = uncharacterized PGE$_2$ metabolite (Christensen, 1984)[32]

before or after labour are all capable of converting labelled arachidonic acid into one or more prostaglandins. Whereas PGE_2 is formed in all of these tissues, $PGF_{2\alpha}$ only is found in chorion, placenta, myometrium and placental arteries. Further, 6-keto-$PGF_{1\alpha}$ is found in myometrium and placental arteries, but only sporadically in chorion and placenta. Thromboxane A_2 formation takes place in placental arteries, placenta and myometrium. The highest rate of conversion of arachidonic acid to prostaglandins is reported in the myometrium, in contrast to the placenta which has the lowest capacity for achieving this conversion. The amnion shows a higher capacity for synthesizing PGE_2 than either the chorion or the myometrium.

If fetal membrane arachidonic acid is to act as a precursor for prostaglandin synthesis, then a means of liberating it from phospholipids must exist; hence phospholipase A_2 is required for liberation of the arachidonic acid. The human chorioamnion contains a high level of phospholipase A_2 activity and this enzyme is also present in the decidua. No difference in enzyme activity has been shown with the onset of labour, but evidence that the enzyme is active at that time is found in the demonstration of an 8-fold increase of phospholipase A_2 in amniotic fluid from before labour to the time of established labour[33]; the increase during labour is correlated with cervical dilatation[34]. The origin of phospholipase A_2 is believed to be the lysosomes in the decidua or the fetal membranes.

Pregnant uterine contractility and PG

In vitro
Before labour, spontaneous uterine activity *in vitro* appears more often in specimens obtained from the lower segment than from the upper segment[35]. During labour, however, there is no difference in spontaneous activity between the isthmus and the corpus.

Administration of PG
Depending upon the concentration, PGE_2 may produce either excitatory or inhibitory responses in myometrial strips obtained before labour. Low concentrations of PGE_2, in the range of 1–3 ng/ml, cause an excitatory response with an increase in amplitude and duration of contractions in isthmic and corporal specimens[35]. Higher concentrations, however, in the range of 10 ng/ml or higher, cause an initial stimulation followed by a period of quiescence lasting for more than 15 min. During this relaxation time the muscle resists further stimulation by PGE_2. Further increase in dosage to a range of 30–500 ng/ml elicits no effect or a very weak stimulation while the inhibitory response tends to increase.

Table 5.1   The effect of oxytocin and prostaglandin $F_{2\alpha}$ and E_2 on myometrial activity before and during labour (+ = stimulation, − = inhibition)

		Oxytocin	$PGF_{2\alpha}$	PGE_2 Low concentration	PGE_2 High concentration	$PFG_{2\alpha} + PGE_2$
Before labour	Fundus	+	(+)	+	−	
	Isthmus	+	+	+	−	
During labour	Fundus	+	+	+	+	+
	Isthmus	+	(+)	±	−	−

During labour, PGE_2 has a consistently stimulatory effect on strips taken from the upper uterine segment irrespective of the concentrations tested. A different response was obtained from the lower uterine segment, where inhibition occurs following administration of high concentrations (100 ng/ml).

Before labour, $PGF_{2\alpha}$ has almost no effect on myometrial strips taken from the uterine corpus. A strong excitatory action is obtained when $PGF_{2\alpha}$ is added to strips obtained from the isthmus. After labour starts, the upper segment reacts to $PGF_{2\alpha}$ by stimulation at all concentrations tested; strips from the lower uterine segment, however, show no response. A combination of PGE_2 and $PGF_{2\alpha}$ elicits a clear-cut stimulation on specimens from the upper segment and a marked inhibition from the lower uterine segment.

The different effects of PGE_2 and $PGF_{2\alpha}$ on the corpus and the isthmus before and during labour are summarized in Table 5.1. These results are compatible with the concept that the fundal myometrium changes its reaction towards consistent stimulation once labour has been initiated.

In vivo

Both the E and the F prostaglandins *in vivo* are potent stimulants of the human uterus at any stage of pregnancy. The basic response of the early and midpregnant uterus to immediate intravenous injection of the primary prostaglandins is characterized by a rapid elevation of uterine tonus that declines gradually towards the normal resting level. The increment of the increase in tone has been utilized as a satisfactory parameter to evaluate the potency and to compare the dose–response relationship between different PGs. The threshold dose of a single intravenous injection at midpregnancy is about $20\,\mu g$ of PGE and PGE_2, respectively, $100–200\,\mu g$ of $PGF_{2\alpha}$ and approximately $500\,\mu g$ of $PGF_{1\alpha}$. The potencies of PGE_1 and PGE_2 are approximately eight times higher[36] than $PGF_{2\alpha}$.

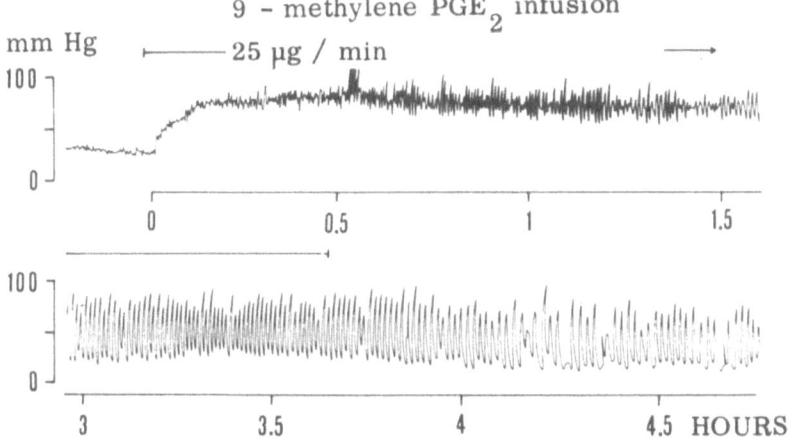

Figure 5.4 The development of uterine contractility during intravenous infusion of 25 µg/min of 9-deoxo-16,16-dimethyl-9-methylene-PGE$_2$ (Bygdeman et al., 1970 Contraception, **22**, 153–64)

The initial response of the early and midpregnant human uterus to an intravenous infusion of PGs generally resembles that following a single intravenous injection. Continuous intravenous administration maintains uterine activity with the gradual development of labour-like contractions. Continued infusion after 3 h results in a decrease in baseline reactivity towards the preinfusion level (Figure 5.4).

If either PGE$_2$ or PGF$_{2\alpha}$ is administered in stepwise increasing doses as an intravenous infusion at or near term, a contractility pattern develops identical to that of normal labour. An increase in sensitivity of the myometrium during the last half of pregnancy is shown by a reduction of threshold doses of PGs at this stage compared with midpregnancy.

If oxytocin and naturally occurring prostaglandins are compared with regard to their ability to stimulate uterine contractility, both similarities and differences in effect can be demonstrated. Labours induced at term either by oxytocin or by prostaglandin are very similar; however, the pregnant uterus is much more sensitive to prostaglandins than to oxytocin in early gestation. An increase in sensitivity takes place around the 20th week of gestation for both prostaglandins and oxytocin; this sensitivity increases until the 36th week after which it remains unchanged until the beginning of spontaneous labour. In sharp contrast, the non-pregnant uterus responds rapidly even to small doses of prostaglandins while it is quite insensitive to oxytocin.

PG-induced hormonal changes

Both PGE_2 and $PGF_{2\alpha}$ have been used for induction of labour. However, the major disadvantage with $PGF_{2\alpha}$ is that this compound must be administered intravenously and it is accompanied by gastro-intestinal side-effects such as vomiting and diarrhoea. On the other hand, the oral administration of PGE_2 has proven to be as effective as an intravenous infusion of oxytocin in inducing labour. In this regard, PGE_2 seems to be correlated with a higher incidence of atypical contractility patterns than are seen with oxytocin. Special attention has been focused on the overall changes in serum levels of circulating hormones after administration of PGE_2 and oxytocin[37]. Prostaglandin E_2 administration elicits increases in maternal serum oestriol levels, and decreases in prolactin and thyroid stimulating hormone (TSH) concentrations. Oxytocin infusion, on the other hand, results in a decrease in maternal serum progesterone levels. Among PGE_2-treated patients delivered within 4 h of treatment, serum cortisol levels are significantly elevated 2 h after the initiation of treatment. In contrast, women whose labours are induced with oxytocin do not respond with a rapid serum cortisol increase. During prolonged labour (more than 9 h), no significant increase in maternal serum cortisol levels is observed 4 h after onset of PGE_2 treatment. Based on animal studies, it has been suggested that prostaglandins stimulate steroid production in the adrenal cortex, possibly through a combined action at the adrenal[38,39], pituitary[40], and suprapituitary levels[41]. The inability of PGE_2 to override dexamethasone-induced inhibition of cortisol secretion has been interpreted to mean that PGE_2 influences cortisol at a suprapituitary level in the female. It has been shown that glucocorticoid administration results in decreased serum prolactin levels[42].

Initiation of labour

In spite of great research efforts, little is known regarding the precise order of events leading to initiation of normal parturition in the human female. However, it is generally agreed that the hormonal environment during the last months of pregnancy prepares the uterus for labour[43-45]. Cervical ripening also is an important factor which predisposes to the normal onset of labour[46].

The pituitary–adrenal axis
The activation of the fetal pituitary–adrenal axis is the trigger mechanism for the onset of parturition in sheep[47]. However, in the human female the pituitary–adrenal axis seems to be of minor importance for the initiation of labour, based on the observations that the mean ges-

tational length in pregnancies with an anencephalic fetus[48] or with adrenal hypoplasia[49] are not significantly different from that of normal fetuses. Furthermore, the administration of large doses of glucocorticoids to preterm pregnant women does not generally induce premature labour[50].

Oestrogen and progesterone

The human placenta, unlike that of many other species, possesses a low level of 17α-hydroxylase activity[51]. In the absence of this particular enzyme, it is unlikely that cortisol can exert its effect on parturition[52,53]. However, the human placenta is rich in sulphatase and aromatase which can convert the substrate dehydroepiandrosterone, which is supplied by the fetal adrenal cortex, into oestrogens[54]. The key position of the human fetal adrenal in steroid biosynthesis and the important role of oestrogens and progesterone for the onset of labour in other species tend to support the impression that the human fetus could also be of crucial importance for initiating its own parturition. To date, meticulous measurements of various oestrogens and progesterone in maternal serum have not provided evidence of any marked changes during the time preceding labour[55-57], and the administration of exogenous oestrogens does not induce labour in women[58,59].

The hypothesis that a high progesterone concentration in the pregnant human myometrium protects against myometrial activity while decreasing concentrations possibly result in accelerated contractility – the so called 'progesterone-block' theory of Csapo – initially gained support, but has subsequently been criticized because progesterone treatment is neither effective for blocking labour[60], nor for prolonging gestation[61]. In spite of this, oestrogens and progesterone still could be of importance in the local environment within the uterus while not necessarily being reflected in the serum levels. It appears that an increase in oestrogen/progesterone ratio is of importance for synchronizing uterine contractions during labour. In this respect, the gap junction formation, i.e. the low resistance connections between smooth muscle cells, seems to be regulated by oestrogens and progesterone[62].

Oxytocin

The secretion of oxytocin from the maternal posterior pituitary lobe was long thought to be the major triggering factor for the initiation of labour[63]. It was furthermore thought that an increase in oxytocinase activity during pregnancy served to prevent preterm labour[64], and that oxytocinase activity may decrease at onset of labour[65]. The activity of oxytocinase can be inhibited by prostaglandins[66]. However, the absence of definite evidence of an oxytocin surge prior to labour makes it doubtful that oxytocin is the 'key hormone' in the mechanism responsible for initiation of labour in the female[67].

Recently, it has been shown that the concentration of myometrial oxytocin receptors increases during pregnancy[68] and that this phenomenon is dependent on the presence of oestrogens. Consequently, the increase in oxytocin receptors may lower the threshold for oxytocin activation of the myometrium. Of great interest is the fact that oxytocin binds to specific receptors in the decidua which promote prostaglandin synthesis. There is also evidence for a fetal secretion of prostaglandin at term, thus indicating a possible mechanism by which the fetus may control local uterine activity. No data are yet available to substantiate a transfer of fetal oxytocin through the human placenta and membranes. An argument against the importance of oxytocin of fetal origin is that oxytocin is not detectable in the umbilical arterial and venous plasma of anencephalic fetuses delivered at term[69].

Vasopressin
Vasopressin, like oxytocin, is produced in the posterior pituitary gland and has a strong stimulating effect on the myometrium. The role of maternal vasopressin for the initiation of labour is not known. However, an arteriovenous difference in concentrations in the umbilical circulation has been demonstrated indicating fetal secretion of vasopressin[70]. It is not known whether the vasopressin peptide is transferred through the placenta into the myometrium.

Neuromuscular mechanisms
The innervation of the myometrium is generally thought to be of little importance for human parturition[71]. However, some studies indicate a decrease of sympathetic nerves in the pregnant human uterus[72] and suggest that this denervation may be responsible for keeping the uterus functionally inactive during pregnancy. It has also been shown that adrenergic denervation results in hypersensitivity of the myometrium to catecholamines[73]. The existence of denervation hypersensitivity, accompanied by increased levels of catecholamines in the amniotic fluid[74] at the end of pregnancy and the fact that the human myometrium has the capacity to convert adrenaline to noradrenaline[75], both provide a possibility for the human fetus to stimulate uterine activity.

Prostaglandins
It is generally accepted that prostaglandins play an important role in the physiology of human labour. Evidence for this concept derives from two observations: first, exogenous PGs can induce labour[76, 77] and, second, the decidua has a capacity to synthesize prostaglandins which increases during labour. Furthermore, inhibition of prostaglan-

din synthesis can interrupt premature labour[78, 79]. For example, aspirin administration during pregnancy in patients with rheumatoid arthritis is associated with a significant increase in the average length of gestation, an increase in the frequency of postmaturity and a longer duration of spontaneous labour (as well as a greater blood loss at delivery)[80].

It is presently thought likely that the last step in the complicated series of interconnected events resulting in the onset of human parturition is an increased prostaglandin production. Major events seem to occur in the following order: degeneration of the decidua, release of lysosomal enzymes and synthesis of prostaglandins, followed by uterine contractions and expulsion of the uterine content[81] (Figure 5.5) (Table 5.2).

Except for the availability of precursors, the factors which initiate or regulate the endogenous production of prostaglandin in the uterus have not been fully identified. Since oxytocin in the oestrogen-primed uterus has been shown to induce prostaglandin liberation[82, 83] and since

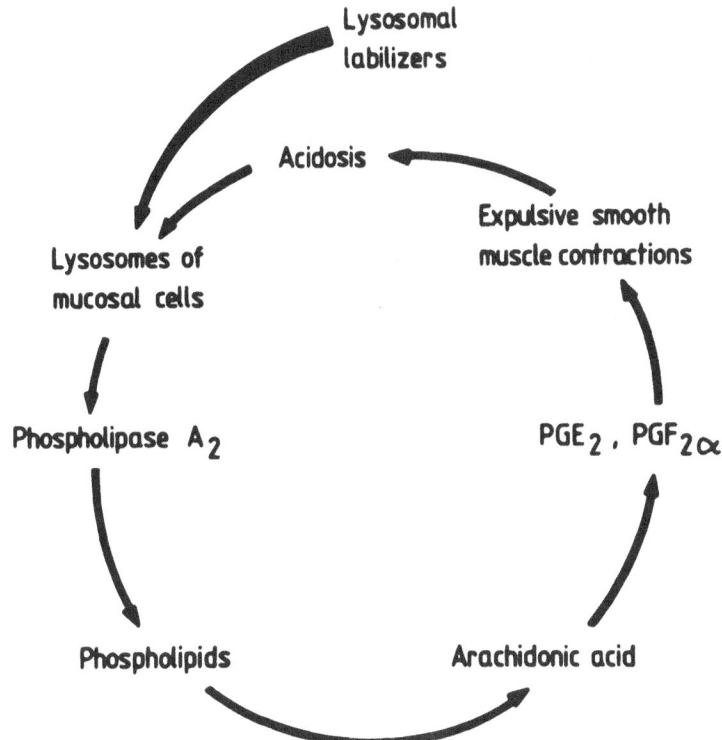

Figure 5.5 Hypothetical scheme of reactions for the mechanism triggering the onset of expulsive contractions in mucosa-lined muscles (Gustavii, 1977)[81]

Table 5.2 Factors which influence lysosomal labilizers

Inhibition	Stimulation
(1) PG biosynthesis inhibitors (2) Progesterone	(1) Abortifacients: hypertonic saline, rivanol, urea (2) Mechanical factors: intrauterine device (3) Hormones: oestrogens, oxytocin, vasopressin? (4) Fetal factors? (5) Neurogenic factors: adrenergic system

prostaglandins promote uterine contractions, a possible link has been established between a trigger substance and an effector compound. Other factors, however, such as alterations in levels of oestrogens and progesterone, mechanical factors and probably the presence of catecholamines may also influence prostaglandin release. As yet it still is not clear which of these factors is most important or whether there is still an unknown mechanism that plays an essential role for prostaglandin biosynthesis and release in connection with initiation of labour.

SUMMARY

The uterus contains large amounts of prostaglandins which regulate its contractility. Prostaglandin $F_{2\alpha}$ consistently stimulates the uterine muscle fibres both *in vitro* and *in vivo*. Prostaglandin E_2 relaxes the myometrium *in vitro* but can cause either relaxation or stimulation of the non-pregnant human uterus *in vivo* depending on the phase of the menstrual cycle. Both $PGF_{2\alpha}$ and PGE_2 stimulate the pregnant human uterus *in vivo* and may be used for induction of labour and abortion. An increase in the endogenous synthesis of $PGF_{2\alpha}$ in the endometrium has been correlated with primary dysmenorrhoea while excessive synthesis of PGE_2 is described in association with menorrhagia. The process of haemostasis during menstruation is further controlled by thromboxane and prostacyclin.

References

1 Wetzstein, R. (1965). Der Uterus Muskel. Morphologie. *Arch. Gynecol.*, **202**, 1
2 Wilhelmsson, L. (1981). Biological actions of prostaglandins on different tissues within the nonpregnant human uterus. *Dissertation*, Gothenburg University, Sweden

3 Bengtsson, L. P. and Theobald, E. W. (1966). The effects of oestrogen and gestagen on the nonpregnant human uterus. *Br. J. Obstet. Gynaecol.*, **73**, 273

4 Lundström, V., Gréen, K. and Wiqvist, N. (1976). Prostaglandins, indomethacin and dysmenorrhea. *Prostaglandins*, **11**, 893–904

5 Roth-Brandel, U., Bygdeman, M. and Wiqvist, N. (1970). Effect of intravenous administration of prostaglandin E_1 and $F_{2\alpha}$ on the contractility of the nonpregnant human uterus *in vito*. *Acta Obstet. Gynecol. Scand., Suppl.*, **5**, 19

6 Bygdeman, M., Bremme, K., Gillespie, A. and Lundström, V. (1979). Effects of the prostaglandins on the uterus. Prostaglandins and uterine contractility. *Acta Obstet. Gynecol. Scand., Suppl.*, **87**, 33–8

7 Wilhelmsson, L., Wikland, M. and Wiqvist, N. (1981). PGH_2, TXA_2 and PGI_2 have potent and differentiated actions on human uterine contractility. *Prostaglandins*, **21**, 277

8 Swahn, M. L. and Lundström, V. (1983). The effect of intravenous and intrauterine administration of prostacyclin on the non-pregnant uterine contractility *in vivo*. *Acta Obstet. Gynecol. Scand., Suppl.*, **113**, 47–50

9 Martin, J. N., Jr., Bygdeman, M. and Eneroth, P. (1975). The influence of locally administered prostaglandin E_2 and $F_{2\alpha}$ on uterine motility in the intact nonpregnant human uterus. *Acta Obstet. Gynecol. Scand.*, **57**, 141

10 Toppozada, M. (1974). *In vivo* inhibition of the human nonpregnant uterus by prostaglandin E_2. *Prostaglandins*, **8**, 401

11 Toppozada, M., Khowessah, M., Schaala, S., Essman, M. and Rahman, H. A. (1977). Aberrant uterine response to prostaglandin E_2 as a possible etiologic factor in functional infertility. *Fertil. Steril.*, **28**, 434

12 Gréen, K., Christensen, N. J. and Bygdeman, M. (1981). The chemistry and pharmacology of prostaglandins with reference to human reproduction. *J. Reprod. Fertil.*, **62**, 269

13 Markee, J. E. (1940). Menstruation in intraocular endometrial transplants in the rhesus monkey. *Contrib. Embryol.*, **177**, 221

14 Downie, J., Poyser, N. L. and Wunderlich, M. (1974). Levels of prostaglandins in human endometrium during the normal menstrual cycle. *J. Physiol.*, **236**, 465

15 Christiaens, G. C. M. L., Sixma, J. J. and Haspels, A. A. (1980). Morphology of haemostasis in menstrual endometrium. *Br. J. Obstet. Gynaecol.*, **87**, 425

16 Bunting, S., Gryglewski, R., Moncada, S. and Vane, J. R. (1976). Arterial walls generate from prostaglandin endoperoxides, a substance (prostaglandin X) which relaxes strips of mesenteric and coeliac arteries and inhibits platelet aggregation. *Prostaglandins*, **12**, 897

17 Moncada, S. and Vane, J. R. (1979). Mode of action of aspirin-like drugs. *Intern. Med.*, **24**, 1

18 Chaudhuri, G. and Elder, N. G. (1976). Lack of evidence for inhibition of ovulation by aspirin in women. *Prostaglandins*, **11**, 727

19 Lundström, V., Landgren, B. M., Eneroth, P. and Johannisson, E. (1983). The effect of a prostaglandin synthetase inhibitor on the hormonal profile and the endometrium in women. *Acta Obstet. Gynecol. Scand., Suppl.*, **113**, 77

20 Lundström, V. (1978). Treatment of primary dysmenorrhea with prostaglandin synthetase inhibitors – a promising therapeutic alternative. *Acta Obstet. Gynecol. Scand.*, **57**, 241

21 Hillier, K. and Kasonde, J. M. (1976). Prostaglandin E and F concentrations in human endometrium after insertion of intrauterine contraceptive device. *Lancet*, **1**, 15

22 Anderson, A. B. M., Haynes, P. J., Guillebaud, J. and Turnbull, A. C. (1976). Reduction of menstrual blood loss by prostaglandin synthetase inhibitors. *Lancet*, **1**, 774

23 Guillebaud, J., Anderson, A. B. M. and Turnbull, A. C. (1978). Reduction by mefenamic acid of increased menstrual blood loss associated with intrauterine contraception. *Br. J. Obstet. Gynaecol.*, **85**, 53

24 Davies, A. J., Anderson, A. B. M. and Turnbull, A. C. (1981). Reduction by naproxen of excessive menstrual bleeding in women using intrauterine devices. *Obstet. Gynecol.*, **57**, 74

25 Thorburn, G. D., Challis, J. R. C. and Currie, W. B. (1977). Control of parturition in domestic animals. *Biol. Reprod.*, **16**, 18

26 Brummer, H. C. and Craft, I. L. (1973). Prostaglandin F$_{2\alpha}$ and labour. *Acta Obstet. Gynecol. Scand.*, **52**, 273

27 Karim, S. M. M. and Devlin, J. (1967). Prostaglandin content of amniotic fluid during pregnancy and labour. *J. Obstet. Gynaecol. Br. Commonw.*, **74**, 230

28 Gréen, K., Bygdeman, M., Toppozada, M. and Wiqvist, N. (1974). The role of prostaglandin F$_{2\alpha}$ in human parturition. Endogenous plasma levels of 15-keto-13,14-dihydro-PGF$_{2\alpha}$ during labor. *Am. J. Obstet. Gynecol.*, **120**, 25

29 Dray, F. and Frydman, R. (1976). Primary prostaglandins in amniotic fluid in pregnancy and spontaneous labor. *Am. J. Obstet. Gynecol.*, **126**, 13

30 Hibbard, B. M., Sharma, S. C., Fitzpatrick, R. J. and Hamlett, J. D. (1974). Prostaglandin F$_{2\alpha}$ concentrations in amniotic fluid in late pregnancy. *J. Obstet. Gynaecol. Br. Commonw.*, **81**, 35

31 Keirse, M. J. N. C., Mitchell, M. D. and Turnbull, A. C. (1977). Changes in prostaglandin F and 13,14-dihydro-15-keto prostaglandin F concentrations in amniotic fluid at the onset of and during labour. *Br. J. Obstet. Gynaecol.*, **84**, 743

32 Christensen, N. J. (1984). Studies on the bioconversion of arachidonic acid in human pregnant reproductive tissues. *Dissertation*, Karolinska Institute, Stockholm, Sweden

33 MacDonald, P. C., Schultz, F. M., Duenhoelter, J. J., Gant, N. F., Jimenez, J. M., Pritchard, J. A., Porter, J. C. and Johnston, J. M. (1974). Inhibition of human parturition. I. Mechanism of action of arachidonic acid. *Obstet. Gynecol.*, **44**, 629

34 Keirse, M. J. N. C., Flint, A. P. F. and Turnbull, A. C. (1974). Prostaglandins in amniotic fluid during pregnancy and labour. *J. Obstet. Gynaecol. Br. Commonw.*, **81**, 131

35 Wikland, M., Lindblom, B., Wilhelmsson, L. and Wiqvist, N. (1982). Oxytocin, prostaglandins and contractility of the human uterus at term pregnancy. *Acta Obstet. Gynecol. Scand.*, **5**, 467

36 Bygdeman, M., Kwon, S. W., Mukherjee, T., Roth-Brandel, U. and Wiqvist, N. (1970). Effects of PGF compounds on contractility of human pregnant uterus. *Am. J. Obstet. Gynecol.*, **106**, 567

37 Bremme, K. and Eneroth, P. (1980). Changes in serum hormone levels during labor induced by oral PGE₂ or oxytocin infusion. *Acta Obstet. Gynecol. Scand., Suppl.*, **92**, 31

38 Flack, J. D. and Ramwell, P. W. (1972). A comparison of the effect of ACTH, cyclic AMP, dibutyryl acid AMP and PGE₂ on corticosteroidogenesis *in vitro*. *Endocrinology*, **90**, 371

39 Saruta, T. and Kaplan, N. M. (1972). Adrenocortical steroidogenesis: the effects of prostaglandins. *J. Clin. Invest.*, **51**, 2246

40 Louis, T. M., Challis, J. R. G., Robinson, J. S. and Thorburn, G. D. (1976). Rapid increase of foetal corticosteroids after prostaglandin E₂. *Nature (London)*, **264**

41 McCann, S. M., Ojeda, S. R., Harms, P. G., Wheaton, J. E., Sundberg, D. K. and Fawcett, C. P. (1976). Control of adenohypophyseal hormone secretion by prostaglandins. In Naftolin, F., Ryan, K. J. and Davies, J. (eds.) *Subcellular Mechanisms on Reproductive Neuroendocrinology*. p. 407. (Amsterdam: Elsevier Scientific Publishing Company)

42 Copinshi, G., L'Hermite, M., Lecluq, R., Goldstein, J., Vanhaelst, L., Virasoro, E. and Robyn, C. (1975). Effects of glucocorticoids on pituitary hormonal responses to hypoglycemia. Inhibition of prolactin release. *J. Clin. Endocrinol. Metab.*, **40**, 442

43 Csapo, A. (1961). Defence mechanism of pregnancy. In Wolstenholme, G. E. W. and Cameron, M. P. (eds.) *Progesterone and Defence Mechanism of Pregnancy*. CIBA Foundation Study Group No. 9. pp. 3-31. (London: Churchill Ltd)

44 Csapo, A. and Sauvage, J. (1968). The evolution of uterine activity during human pregnancy. *Acta Obstet. Gynecol. Scand.*, **47**, 181

45 Batra, S. and Bengtsson, L. P. (1978). 17-β-estradiol and progesterone concentrations in myometrium of pregnancy and their relationships to concentrations in peripheral plasma. *J. Clin. Endocrinol. Metab.*, **46**, 662

46 Calder, A. A. (1981). The human cervix in pregnancy: a clinical perspective. In Ellwood, D. A. and Anderson, A. B. M. (eds.) *The Cervix in Pregnancy and Labour*. pp. 103-9. (Edinburgh: Churchill Livingstone)

47 Liggins, G. C., Fairclough, R. J., Grieves, S. A., Forster, C. S. and Knox, B. S. (1977). Parturition in the sheep. In *The Fetus and Birth*. CIBA Foundation Symposium No. 47. pp. 5-30. (Amsterdam: Elsevier Excerpta Medica)

48 Honnebier, W. J. and Swaab, D. F. (1973). The influence of anencephaly upon uterine growth of fetus and placenta upon gestation length. *Br. J. Obstet. Gynaecol.*, **80**, 577

49 Liggins, G. C. (1974). The influence of fetal hypothalamus and pituitary on growth. In Elliot, K. and Knight, J. (eds.) *Size at Birth*. CIBA Foundation Symposium No. 27. pp. 105-83. (Amsterdam: Elsevier Excerpta Medica)

50 Liggins, G. C. and Howie, R. N. (1972). A controlled trial of antepartum glucocorticoid treatment for prevention of the respiratory distress syndrome in premature infants. *Pediatrics*, **50**, 515

51 Siiteri, P. K. and Séron-Feree, M. (1981). Some new thoughts on the feto-placental unit and parturition in primates. In Novy, M. J. and Resko, J. A. (eds.) *Fetal Endocrinology*. pp. 1–13. (New York: Academic Press)

52 Flint, A. P. F. (1979). Role of progesterone and oestrogens in the control of the onset of labour in man: a continuing controversy. In Keirse, M. J. N. C., Anderson, A. B. M. and Bennebroeck Gravenhorst, J. (eds.) *Human Parturition*. pp. 85–100. (Netherlands: Leiden University Press)

53 Liggins, G. C. (1983). Initiation of spontaneous labor. *Clin. Obstet. Gynecol.*, **26**, 47

54 Solomon, S. and Friesen, H. G. (1968). Endocrine relations between mother and fetuses. *Ann. Rev. Med.*, **19**, 399

55 Aitken, E. H., Preedy, J. R. K., Eton, B. and Short, R. V. (1958). Oestrogen and progesterone levels in foetal and maternal plasma at parturition. *Lancet*, **2**, 1096

56 Chew, P. C. T. and Ratnam, S. S. (1976). Serial levels of plasma oestradiol 17β at the approach of labour. *J. Endocrinol.*, **71**, 267

57 Batra, S., Bengtsson, L. Ph. and Inmarsson, I. (1983). The role of estradiol and progesterone in regulation of myometrial activity for the onset of labor. *Acta Obstet. Gynecol. Scand.*, **62**, 207

58 Kelly, J. V. (1961). The effect of intravenous estrogens on uterine motility. *Am. J. Obstet. Gynecol.*, **82**, 1207

59 Larsen, J. W., Hanson, T. M., Caldwell, B. V. and Speroff, L. (1973). The effect of estradiol infusion on uterine activity and peripheral levels of prostaglandin F and progesterone. *Am. J. Obstet. Gynecol.*, **117**, 276

60 Fuchs, F. and Stakeman, G. (1960). Treatment of threatened premature labor with large doses of progesterone. *Am. J. Obstet. Gynecol.*, **79**, 172

61 Csapo, A., de Sousa-Filko, M. B., de Souza, J. C. and de Souza, O. (1956). Effect of massive progestational hormone treatment on the parturient human uterus. *Fertil. Steril.*, **17**, 621

62 Garfield, R. E. and Hayashi, R. H. (1981). Appearance of gap junctions in the myometrium of women during labor. *Am. J. Obstet. Gynecol.*, **140**, 254

63 Theobald, G. W. (1968). Oxytocin reassessed. Obstet. Gynecol. Surv., **23**, 109–31

64 Klimek, R. (1968). Clinical studies on the balance between isooxytocinases in the blood of pregnant women. *Clin. Chim. Acta*, **20**, 233

65 Fuchs, F. (1971). Endocrinology of labor. In Fuchs, F. and Klopper, F. (eds.) *Endocrinology of Pregnancy*. pp. 306–27. (New York: Harper & Row)

66 Szontagh, F. E., Morvay, J. and Falkay, G. (1974). Changes in blood level of cystine aminopeptidase in abortions induced by prostaglandin $F_{2\alpha}$ *Ann. Chir. Gynaecol. Fenn.*, **63**, 198

67 Chard, T., Boyd, N. R. H., Forsling, M. L., McNeilly, A. S. and Landon, J. (1970). The development of a radioimmunoassay for oxytocin: the extraction of oxytocin from plasma and its measurement during parturition in human and goat blood. *J. Endocrinol.*, **48**, 223

68 Fuchs, A. R., Fuchs, F., Husslein, P., Soloff, M. S. and Fernstrom, M. J.

(1982). Oxytocin receptors in human parturition: a dual role for oxytocin in the initiation of labor. *Science*, **215**, 1396.

69 Otsuki, Y., Tanizawa, O., Yamaji, K., Fujita, M. and Kurachi, K. (1982). Feto-maternal plasma oxytocin levels in normal and anencephalic pregnancies. *Acta Obstet. Gynecol. Scand.*, **62**, 235

70 Chard, T., Hudson, C. N., Edwards, C. R. W. and Boyd, N. R. H. (1971). Release of oxytocin and vasopressin by the human foetus during labour. *Nature (London)*, **234**, 352.

71 Bell, C. (1972). Autonomic nervous control for reproduction: circulatory and other factors. *Pharmacol. Rev.*, **24**, 679

72 Thorbert, G., Alm, P., Björklund, A. B., Owman, C. and Sjöberg, N-O. (1979). Adrenergic innervation of the human uterus. Disappearance of the transmitter and transmitter-forming enzymes during pregnancy. *Am. J. Obstet. Gynecol.*, **135**, 223

73 Nakanishi, H., McLean, J., Wood, C. and Burnstock, G. (1969). The role of sympathetic nerves in control of the nonpregnant and pregnant human uterus. *J. Reprod. Med.*, **11**, 20

74 Philippe, M. and Ryan, K. J. (1981). Catecholamines in human amniotic fluid. *Am. J. Obstet. Gynecol.*, **139**, 204

75 Hobel, C. J., Parvez, H., Parvez, S., Lirette, M. and Papiernik, E. (1981). Enzymes for epinephrine synthesis and metabolism in the myometrium, endometrium, red blood cells and plasma of pregnant human subjects. *Am. J. Obstet. Gynecol.*, **141**, 1009

76 Karim, S. M. M., Trussell, R. R. and Hillier, K. (1969). Induction of labour with prostaglandin $F_{2\alpha}$ *J. Obstet. Gynaecol. Br. Commonw.*, **76**, 769

77 Thiery, M. and Amy, J. J. (1975). Induction of labour with prostaglandins. In Karim, S. M. M. (ed.) *Advances in Prostaglandin Research, Prostaglandins and Reproduction*. pp. 149–228. (Lancaster: MTP Press)

78 Zuckerman, H., Reiss, W. and Rubinstein, I. (1974). Inhibition of human premature labor by indomethacin. *Obstet. Gynecol.*, **44**, 787

79 Wiqvist, N., Lundström, V. and Gréen, K. (1975). Premature labor and indomethacin. *Prostaglandins*, **10**, 515

80 Lewis, R. B. and Schulman, J. D. (1973). Influence of acetylsalicyclic acid, an inhibitor of prostaglandin synthesis, on the duration of human gestation and labour. *Lancet*, **2**, 1159

81 Gustavii, B. (1977). Human decidua and uterine contractility. In *The Fetus and Birth*. CIBA Foundation Symposium No. 47. pp. 343–58. (Amsterdam: Elsevier Excerpta Medica)

82 Fuchs, A. R., Husslein, P. and Fuchs, F. (1981). Oxytocin and the initiation of human parturition. *Am. J. Obstet. Gynecol.*, **141**, 694.

83 Husslein, P., Fuchs, A-R. and Fuchs, F. (1981). Oxytocin and the initiation of human parturition. I. Prostaglandin release during induction of labor by oxytocin. *Am. J. Obstet. Gynecol.*, **141**, 688

6
The Fallopian tube

B. LINDBLOM

INTRODUCTION

Disordered function of the Fallopian tubes is epitomized in gynaecological practice by ectopic pregnancies. The fact that this phenomenon is almost entirely limited to the human female[1] probably is a
reflection of specific pathological conditions, but it may also indicate
that the human oviduct differs in certain critical anatomical and
physiological respects from those other species. This chapter will summarize current knowledge about the role of PGs in oviductal function
with emphasis on information about the human Fallopian tube.

The successful establishment of a human pregnancy, i.e. the implantation of a normal blastocyst in the uterine fundus, should be viewed
as being the result of a programmed series of complex but integrated
biological events. In this context, it is important to remember that the
tube is composed of various anatomical segments all of which differ to
considerable degrees. The infundibulum is characterized by its fimbriae, the ampulla by its wide lumen and deep mucosal folds and the
isthmus by its narrow lumen and well-developed muscle coat (Figure
6.1). The isthmus is bordered at its ideal end by the 'ampullary-isthmic
junction' (AIJ) and at its proximal limit by the 'uterotubal junction'
(UTJ).

OVUM PICK-UP

The process of ovum transport starts with the pick-up of the ovum-
cumulus mass from the surface of the ruptured follicle. This function
normally is carried out by sweeping movements of the fimbriae over

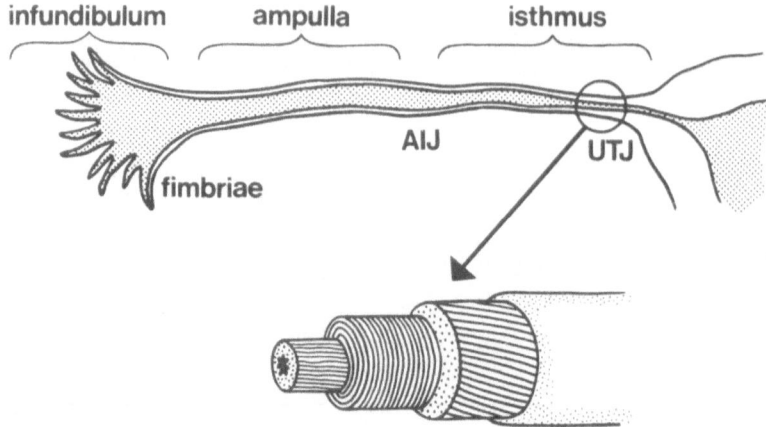

Figure 6.1   The anatomy of the human Fallopian tube (semischematic). AIJ, ampullary-isthmic junction; UTJ, uterotubal junction. In the isthmus, three different muscle layers are identified: the outer longitudinal, the intermediate circular and the innermost longitudinal layer

the ovarian surface. Such movements occur only at the time of ovulation and are accompanied by physical approximation of the ovary and the infundibulum brought about by muscular elements in the ovarian ligament and the mesosalpinx. The precise mechanisms controlling this series of movements are unknown, but it has been speculated that a substance in the follicular fluid released at the time of ovulation acts as a mediator. In fact, crude follicular fluid stimulates muscular activity in isolated human tubo-ovarian ligaments and fimbriae[2]. Since the preovulatory follicle contains considerable amounts of prostaglandins[3], PGs may be of importance in the induction of contractile effects in these structures. In fact, PGE_2, $PGF_{2\alpha}$ and PGI_2 all have been shown to stimulate contractile activity of the human tubo-ovarian ligament at the time of ovulation, whereas only $PGF_{2\alpha}$ increases the activity of the fimbriae[2]. In addition, these three PGs stimulate the contractile activity of the human mesosalpinx[4]. Thus, PGs in the follicular fluid may be involved in a chemotactic process which secures the pick-up of the ovum from the ovarian surface at the time of ovulation.

CILIARY ACTIVITY

Ovum-cumulus pick-up and subsequent transport into the midportion of the tube, i.e. the proximal part of the ampulla, take place within a few hours of follicular rupture[5]. Ciliary activity plays a central role in ampullary transport, at least in the rabbit[6-8]. In most species, the fibrial and ampullary mucosa are richly endowed with ciliated epithelial cells. In the isthmic segment, however, ciliated cells are less abundant, and secretory cells begin to dominate the mucosal surfaces. The control of ciliary activity is thought to be governed by ovarian hormones, but information concerning the role of steroids in ciliary motion is still incomplete[9]. Verdugo et al.[7] studied the influence of PGE_1, PGE_2 and $PGF_{2\alpha}$ on the beat frequency of tubal cilia in cell cultures using laser light-scattering spectroscopy. These PGs were found to induce a steady and reversible increase in frequency of ciliary beat. Maximal stimulatory activity was obtained at 0.1–1.0 μmol/l concentrations, whereas higher concentrations tended to inhibit ciliary activity. At equimolar concentrations the order of efficacy was $PGF_{2\alpha}$ $> PGE_1 > PGE_2$.

Although ciliary activity is important for ovum transport in rabbits, the physiological role of cilia in higher species remains uncertain. The observation that women with the immotile cilia syndrome may become pregnant[10,11] leads one to question the importance of ciliary activity in human ovum transport. However, it is not yet known whether the women in question possessed normal fecundability and direct motility studies in these subjects are lacking[8].

'CRITICAL' STRUCTURES

It is important to realize that no single part of the oviduct is essential to ovum transport. Pregnancy may occur in the absence of the fimbriae, the ampulla, the ampullary-isthmic junction, the uterotubal junction and even in the absence of both oviducts[12]. Pregnancies after total hysterectomy show that not even the uterus is critical in human reproduction[13]. Consequently, the question is not which anatomical portion of the oviduct is or is not essential to fertility, but rather what degree of reduction in fertility results from the absence or dysfunction of a certain structure.

Most investigators now agree that muscular activity is the primary factor controlling ovum transport in humans[14,15]. Although data concerning the influence of exogenous sex steroids on oviductal contractility are contradictory[16], endogenous oestrogens appear to stimulate tubal contractility while progesterone has the opposite effect[17]. The prostaglandins have potent actions on oviduct motility and they have been proposed to be mediators of the hormonal effects on tubal smooth muscle.

CONTRACTILITY

In vivo studies

Coutinho and Maia[18] have shown that PGE_2 decreases intraluminal pressure in the human Fallopian tube *in vivo*, whereas $PGF_{2\alpha}$ has the opposite effect. In the monkey, the oviduct response to $PGF_{2\alpha}$ is significantly increased after both spontaneous and hCG-induced ovulation. In the domestic hen, intra-aortic injection of the PG precursor arachidonic acid causes an abrupt pressure increase in all parts of the oviduct[19]. This response can be blocked by pretreatment with indomethacin and resembles that produced by PGE_2 and $PGF_{2\alpha}$. Obviously, enzyme systems in the intact hen oviduct are available for rapid transformation of arachidonic acid to stimulatory prostaglandins. In the rabbit, however, the E- and F-prostaglandins are mutually antagonistic. PGF overcomes the suppressive effect of PGE; moreover, the PGEs abolish the increase in tubal activity caused by the F-prostaglandins[20] and also inhibit spontaneous spasmodic contractions. At ovulation, there appears to be an increased response to $PGF_{2\alpha}$. Furthermore, the increase in ovarian oestradiol and progesterone secretion during the 3.5 days following coitus results in a decreased responsiveness to $PGF_{2\alpha}$ and an increased responsiveness[21] to PGE_1. It has been hypothesized that these changes might cause relaxation of isthmic tone and allow movement of eggs through the isthmus to the uterus. Further evidence of a role of PGs in this context is that the inhibitory effect of PGE_1 on the rabbit oviduct response of $PGF_{2\alpha}$ is significantly reduced 72 h after hCG administration[22]. Using intraluminal microtransducers to record changes in tubal diameter of unanaesthetized rabbits, Blair and Beck[23] found that PGE_2 blocked spontaneous activity of both the circular and longitudinal muscle layer but produced little change in the internal diameter of the tube. $PGF_{2\alpha}$, on the other hand, caused an increase in both contraction frequency and oviductal diameter. These authors suggested that $PGF_{2\alpha}$ accelerates egg transport in the isthmus by the two following mechanisms: (1) an increase in the lumenal diameter causing diminished resistance to ovum passage; and (2) an increase in the frequency of contractions providing additional propelling force for the egg.

In vitro studies

In vitro investigations of tubal motility have provided information about the interplay between different tubal segments and the properties of various muscle layers within the tube. By microscopic dissection, pieces of circular and longitudinal muscle from the human oviduct

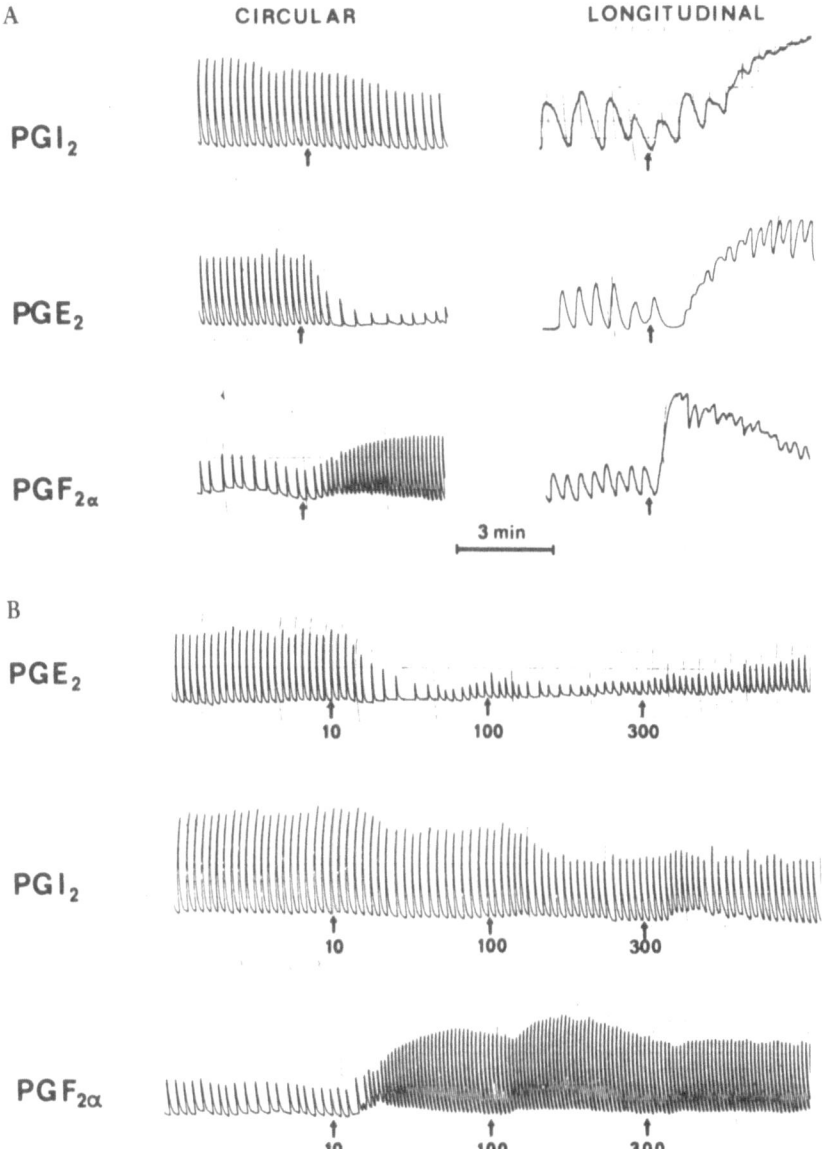

Figure 6.2   A: The effects of PGI₂, PGE₂ and PGF₂ₐ (10 ng/ml) on the contractile activity of muscle strips from the circular and longitudinal layers at the ampullary-isthmic junction of the human oviduct. **B:** The responses of three circular strips from the same patient to stepwise increased concentrations of PGE₂, PGI₂ and PGF₂ₐ (ng/ml)

have been isolated, and the contractile properties of these tissues studied in organ chambers[24]. PGE_1 and PGE_2 relax the circular muscle layer of the isthmus at concentrations lower than $1 ng/ml$, but cause contractions of the longitudinal muscle. $PGF_{2\alpha}$, on the other hand, causes a powerful stimulation of both muscle layers. Compared to these two classical PGs, PGI_2 has a comparatively weak effect (Figure 6.2)[25,26]. In the ampullary portion, PGE_2 is inhibitory to both the circular and the longitudinal muscle layers, whereas $PGF_{2\alpha}$ is excitatory to both. These findings are consistent with *in vivo* studies[18,27]. An even more complex response pattern is observed at the uterotubal junction

Table 6.1   Effects of PGE_2 on different muscle layers within the various segments of the human Fallopian tube

Segment	Inner longitudinal	Muscle layer Circular	Outer longitudinal
Ampulla	(absent)	inhibition	inhibition
Ampullary-isthmic junction	(absent)	inhibition	stimulation
Uterotubal junction	inhibition/ stimulation*	inhibition	stimulation

* Stimulation at the time of ovulation, inhibition in other cycle phases

where three different muscle layers are present. PGE_2 inhibits the activity of the innermost longitudinal layer in all menstrual cycle phases except the periovulatory phase in which a clear-cut stimulatory response occurs[28]. Irrespective of cycle phase, the circular and outer longitudinal layers show the same responses as those in the distal isthmus (Table 6.1).

INHIBITORS OF PG BIOSYNTHESIS

Although the Fallopian tube is responsive to the action of various PGs *in vitro*, data still do not prove that endogenous PGs play a role in the control of tubal motility. Such evidence, however, may be inferred by experiments using PG synthetase inhibitors. Tonpe and Lindblom[29] reported that indomethacin causes a concentration dependent and reversible inhibition of tubal contractions *in vitro*. Treatment with 5,8,11,14-eicosatetraenoic acid which acts to block the cyclo-oxygenase and the lipo-oxygenase pathways in arachidonate metabolism causes inhibition of contractile activity of human oviduct. This activity was able to be re-established in both the circular and longitudinal muscle

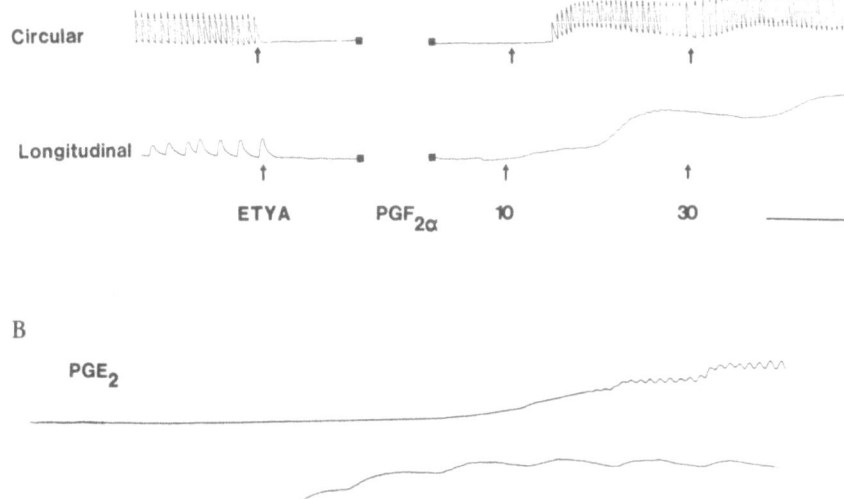

Figure 6.3 **A**: The effect of 5,8,11,14-eicosatetraynoic acid (ETYA), 30 μg/ml, on muscle layers of the human AIJ. During inhibition of activity induced by ETYA, the addition of PGF$_{2\alpha}$ leads to a gradual re-establishment of activity in both muscle layers (ng/ml). Calibrations: horizontal 3 min, vertical 4 mN). **B**: Recovery of activity after total inhibition of spontaneous activity produced by ETYA (30 μg/ml). PGF$_{2\alpha}$ is able to restore muscle tone at 3–10 ng/ml whereas PGE$_2$ is less potent (longitudinal specimens). Calibrations: horizontal 3 min, vertical 4 mN

layers of the ampullary-isthmic junction by the addition of low concentrations of PGF$_{2\alpha}$; in contrast, PGE$_2$ was able to re-establish activity only in the longitudinal layer and at considerably higher concentrations (Figure 6.3)[30]. Furthermore, arachidonic acid may induce a stimulatory response which can be blocked by 5,8,11,14-eicosatetraenoic acid (Figure 6.4). Taken all together, these data provide strong evidence that the smooth muscle of the human oviduct has the capacity to convert both endogenous and exogenous arachidonic acid to active PGs.

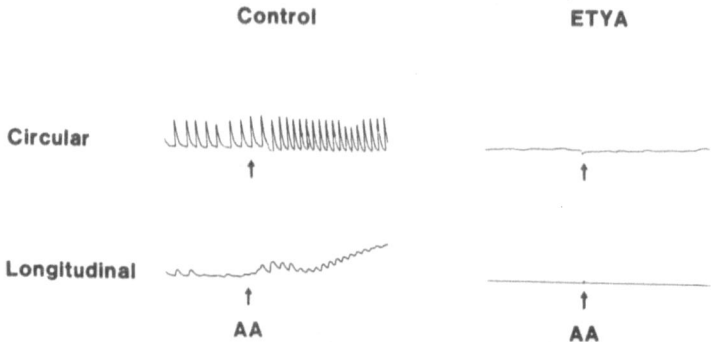

Figure 6.4   Left: The effect of arachidonic acid, 10 μg/ml on the circular and longitudinal muscle of the human AIJ. There is an increased activity in terms of a higher contraction frequency in the circular layer and increased tone and frequency in the longitudinal layer. **Right**: After treatment with ETYA (30 μg/ml) for 15 min, the response to arachidonic acid is completely abolished. Calibrations: horizontal 3 min; vertical 4 mN

PROSTAGLANDINS AND 'TUBAL LOCK'

The complex architecture of the human tubal isthmus constitutes a dynamic control system for ovum transport with an ability to close or open the 'gate' between the oviduct and the uterus depending upon the actual hormonal condition. Theoretically, a local dominance of PGF influence at the time of ovulation should be compatible with tubal closure via the contraction of all three muscle layers, whereas a dominance of PGE influence in the early luteal phase should result in opening of the isthmic 'sphincter' due to relaxation of the circular and innermost longitudinal layer. In the rabbit oviduct, an increased formation of PGF is observed under oestrogen influence[31] in addition to an increased sensitivity of tubal muscle to $PGF_{2\alpha}$. Furthermore, progesterone increases the sensitivity of tubal muscle[21] to PGE and, during the ovum transport period, the ratio of tissue concentrations of PGE to PGF is increased in a sequential manner[32].

Based on the human and animal data presented above, a hypothesis for the control of trans-isthmic ovum transport is illustrated in Figure 6.5. The central place in this hypothesis is occupied by the inner longitudinal muscle layer, since it is unique with regard to its limited extension and its hormone-dependent response to PGE_2. If the balance system between tissue concentration and sensitivity for PGE and PGF is applicable in higher species as well, then also the circular muscle layer may be involved in the control of ovum transport through the isthmus.

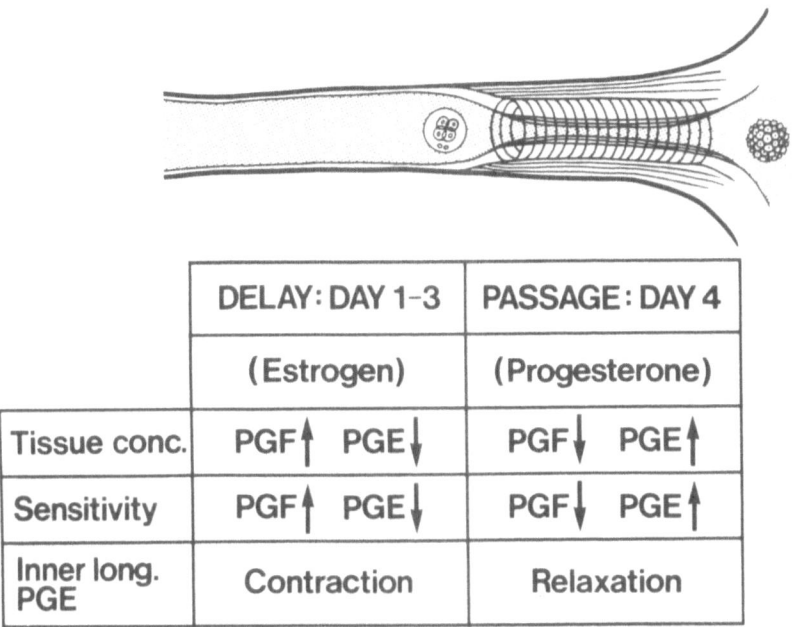

	DELAY: DAY 1–3	PASSAGE: DAY 4
	(Estrogen)	(Progesterone)
Tissue conc.	PGF↑ PGE↓	PGF↓ PGE↑
Sensitivity	PGF↑ PGE↓	PGF↓ PGE↑
Inner long. PGE	Contraction	Relaxation

Figure 6.5 An hypothesis for the regulation of trans-isthmic ovum passage. During the early period, i.e. day 1–3 following ovulation, the egg is retained in the proximal part of the ampulla or at the AIJ, where fertilization occurs. This retention is induced by high preovulatory oestrogen levels, causing an increase in PGF production and in the sensitivity of tubal muscle to PGF. This is accompanied by a contractile effect of PGE_2 on the innermost longitudinal layer. With the increased progesterone levels in the early luteal phase, there is a relative increase of PGE levels and an increased sensitivity of tubal muscle to E prostaglandins. This would induce a lowered isthmic muscle tone, including a relaxation of the innermost longitudinal layer

The compiled evidence for the involvement of PGs in the control of tubal contractility in the human female is listed in Table 6.2.

PROSTAGLANDINS AND COPPER-IUDs

Recent studies on the mechanisms responsible for the contraceptive effect of copper-bearing intrauterine devices indicate a role for PGs in the stimulatory effect of copper on contractility of the human oviduct[33]. The response of tubal muscle to copper ions is almost identical to the stimulatory effect of $PGF_{2\alpha}$ (Figure 6.6A), and this response can be completely blocked by indomethacin. Furthermore, after maximal stimulation of contractility induced by exogenous $PGF_{2\alpha}$, copper

Table 6.2   Evidence for a role of PGs in the regulation of contractile activity of various smooth muscle elements within the human oviduct

Function	Tissue	PG/PG inhibitor	Influence	Reference no.
Pick-up mechanism				
	Tubo-ovarian	E_2, $F_{2\alpha}$, I_2	stimulation	2
	ligament	indomethacin	inhibition	2
	Fimbriae	$F_{2\alpha}$	stimulation	2
		indomethacin	inhibition	2
	Mesosalpinx	E_2, $F_{2\alpha}$, I_2	stimulation	4
		indomethacin	inhibition	4
Ampullary function				
	Circular and	E_2	inhibition	27
	longitudinal	$F_{2\alpha}$, I_2	stimulation	27
	muscle layer	ETYA	inhibition	27
	In vivo intra-	E_2	inhibition	18
	luminal	$F_{2\alpha}$	stimulation	18
	pressure			
	(tubal wall)			
Isthmic function				
	Outer	E_2, $F_{2\alpha}$, I_2	stimulation	24, 26
	longitudinal	ETYA,	inhibition	29, 30
	muscle layer	indomethacin		
	Circular layer	E_2, I_2	inhibition	24, 26
		$F_{2\alpha}$	stimulation	24
		ETYA,	inhibition	29, 30
		indomethacin		
	Inner	E_2	inhibition/	28
	longitudinal		stimulation*	
	layer	$F_{2\alpha}$, I_2	stimulation	28

* Stimulation at the time of ovulation, inhibition in other cycle phases

becomes ineffective as a stimulant (Figure 6.6B) whereas other agents such as the α-adrenoceptor agonist phenylephrine seem to cause a further stimulation despite prior 'loading' of the muscle[29] with $PGF_{2\alpha}$. The conclusion derived from these experiments is that copper ions released from a copper intrauterine contraceptive device increase contractility of the oviducts by an action on the endogenous synthesis of prostaglandins, predominantly of the F-series. A moderate increase of PGF could be compatible with increased ovum transport rate, while a marked increase of PGF production could result in 'tubal lock'. However, the ultimate significance of these effects compared to other mechanisms of action that copper-IUDs may have remain to be explored.

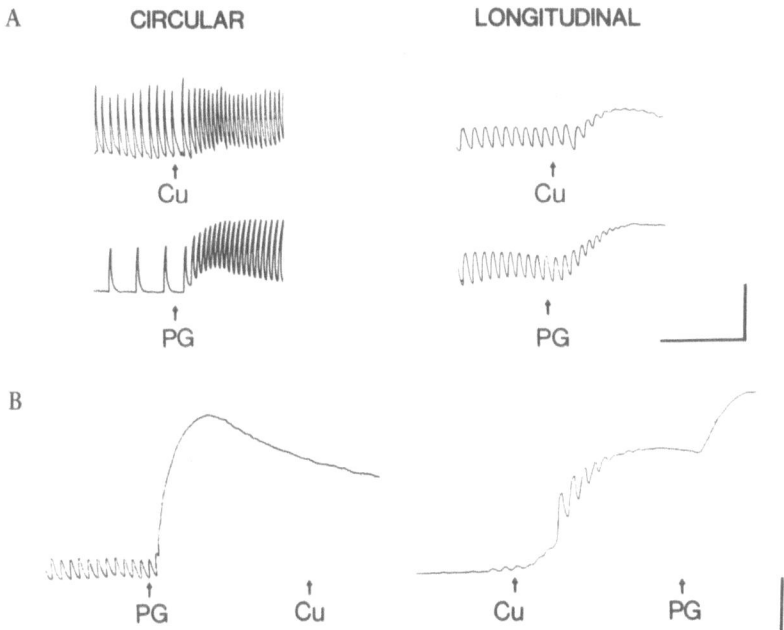

A CIRCULAR LONGITUDINAL

Figure 6.6 **A:** A visual comparison between the excitatory action of copper chloride (1 mmol/l) and $PGF_{2\alpha}$ (30 ng/ml) on strips from the human AIJ. Calibrations: horizontal 3 min, vertical 4 mN. **B:** After maximal response induced by $PGF_{2\alpha}$ (1 µg/ml), copper chloride (3 mmol/l) is unable to produce any further stimulation (left). After a maximal copper response, $PGF_{2\alpha}$ causes a further stimulation (right). Note the progressively developing response after administration of Cu and the more acute PG response. Longitudinal muscle strips, human AIJ. Calibrations: horizontal 3 min, vertical 4 mN

MODE OF ACTION

The effector systems underlying the excitatory and inhibitory effects of PGs on tubal motility is not yet clear; however, evidence is accumulating to suggest that $PGF_{2\alpha}$ exerts its action via different mechanisms from those associated with PGEs. For example, the excitatory effects of the PGs on the outer longitudinal muscle layer at the ampullary-isthmic junction occur via a divergent influence on the tissue levels of cyclic AMP[34]. In the circular muscle layer, where the effects of PGE_2 and $PGF_{2\alpha}$ are antagonistic, PGE_2 inhibits the discharge of action potentials, whereas $PGF_{2\alpha}$ increases the rate of spike discharge and hence the contraction frequency[35]. The high degree of conformity between the effects of PGE_2 and those of PGI_2 points to a possible similarity in their mechanisms of action.

An interesting difference between the circular and the longitudinal muscle coat of the oviduct with regard to both spontaneous activity and response of various drugs has recently been analysed by Forman et al.[36]. Using the calcium antagonist nifedipine, these investigators found evidence that the phasic spontaneous contractions in the circular layer depend on calcium influx via potential-sensitive membrane channels, whereas receptor-operated calcium channels seemed to be involved in the PG-induced tonic contractions observed mainly in the longitudinal muscle layer.

Blocking the α- and β-adrenoreceptors does not appear to interfere with the effects of PGs, and blocking of PG biosynthesis has no effect on the excitatory actions of catecholamines[29,37]. Furthermore, a maximal contraction evoked by $PGF_{2\alpha}$ can be further augmented by phenylephrine. These observations suggest that catecholamines and prostaglandins may stimulate smooth muscle cells within the oviduct by different mechanisms, at least in part. Both kinds of stimuli, however, seem equally sensitive to reduction of extracellular calcium and pharmacological blocking of calcium channels[29,36]. In addition, an influence of PGE_2 on noradrenaline release from adrenergic terminals has been demonstrated[38,39].

PROSTAGLANDINS AND OVUM TRANSPORT

Despite the information accumulated to date regarding the influence of endogenous and exogenous PGs on tubal motility, the precise involvement of PGs in ovum transport is far from clear. In 1972, Chang and Hunt[40] reported that a subcutaneous injection of $PGF_{2\alpha}$ at the 24th hour after insemination caused the elimination of ova from the oviducts and uteri of rabbits. Later, it was shown that both PGE_1 and $PGF_{2\alpha}$ could accelerate ovum transport and reduce the implantation rate appreciably if these agents were administered 13 h after ovulation[41], with $PGF_{2\alpha}$ having a greater effect than PGE_2. Similar treatment 4 or 9 h after ovulation was less effective. The majority of eggs in the PG-treated animals either were found in the proximal third of the oviduct and in the uterus or were not recovered.

$PGF_{2\alpha}$ administered vaginally 24 h following ovulation results in a dose-dependent reduction of implantation sites in rabbits. Salomy and Goldstein found that a dose of 25 mg $PGF_{2\alpha}$ reduced the number of implantation sites to 0.6 per test animal vs. 7.3 in the control group[42]. They suggested that tubal hypercontractility was responsible for the observed contraceptive effect, but noted also that other factors could be involved as well, both at the ovarian or at the uterine level.

Takeda et al. studied the effect of a single injection of $PGF_{2\alpha}$ on egg distribution in the reproductive tract of rabbits[43]. Maximal rates of egg recovery and lowest rates of implantation occurred concomitantly when $PGF_{2\alpha}$ was injected 27 h after mating. It appeared that the sensitivity to $PGF_{2\alpha}$ gradually decreased with the passage of the time after ovulation. This observation is particularly interesting in view of the reduced oviductal response to $PGF_{2\alpha}$ after progesterone treatment[21]. Furthermore, the uterine response to $PGF_{2\alpha}$ is decreased significantly in progesterone-treated rabbits[44]. Inhibitors of PG biosynthesis, on the other hand, have been shown to accelerate ovum transport in rabbits by antagonizing oestrogen-induced 'tube-locking'[45]. These and other findings at one time seemed promising for the development of new approaches to human contraception. However, since clinical trials using PG analogues as contraceptive agents have been unrewarding[46,47], the initial enthusiasm was subsequently lost.

CONCLUDING REMARKS

The human oviducts appear to synthesize significant amounts of PGs, especially $PGF_{2\alpha}$ and PGE_2. Both of these agents have potent actions on tubal motility. Whether or not ovum transport can be clinically influenced by PG compounds still remains uncertain. Compounds which stimulate tubal contractility do not necessarily accelerate ovum transport and if a contraceptive agent acting on ovum transport is to be efficient, then it probably would need to continue its action at the uterine level, since the premature arrival of the ovum in the uterus does not always protect against pregnancy[48].

In summary, PGs seem to be centrally involved in a number of tubal functions. In particular, the isthmic musculature appears to be influenced by PGEs and PGFs, probably operating in a balance system regulated by ovarian steroids. The clinical applications of this knowledge remain to be seen.

ACKNOWLEDGEMENTS

This work was supported by grants from the Swedish Medical Research Council (No. 2873), Magnus Bergvall's Foundation, Harald Jeansson's Foundation, and Harald and Greta Jeansson's Foundation. Thanks are also due to Ms. Bodil Banner for expert secretarial help.

References

1 Adams, C. E. (1977). Ectopic pregnancy. *Biblphy. Reprod.*, 30, 97–8
2 Sterin-Speziale, N., Gimeno, M. F., Zapata, C., Bagnati, P. E. and Gimeno, A. L. (1978). The effect of neurotransmitters, bradykinin, prostaglandins, and follicular fluid on spontaneous contractile characteristics of human fimbriae and tubo-ovarian ligaments isolated during different stages of the sexual cycle. *Int. J. Fertil.*, 23, 1–11
3 Mitchcell, M., Carson, R. and Trounson, A. (1984). An inhibitor of prostaglandin synthesis in antral fluid of human ovulatory follicles. *Society for the Study of Reproduction*, Annual meeting, (abstract)
4 Gimeno, M. F., Gimeno, A. L., Borda, El, Sterin-Speziale, N., Chaud, M. and Zapata, C. (1982). Physiological and pharmacological aspects of tubal motility. Implications of the contractile activity of the mesosalpinx (1). In Muldoon, T. G., Mahesh, V. B. and Ballester, B. P. (eds.), *Recent Advances in Fertility Research, Part B, Developments in the Management of Reproductive Disorders*. pp. 77–89. (New York: Alan R. Liss Inc.)
5 Croxatto, H. B., Ortiz, M. E., Diaz, S., Hess, R., Balmaceda, J. and Croxatto, H. D. (1978). Studies on the duration of egg transport by the human oviduct. II. Ovum location at various intervals following luteinizing hormone peak. *Am. J. Obstet. Gynecol.*, 132, 629–34
6 Halbert, S. A., Tam, P. Y. and Blandau, R. J. (1978). Egg transport in the rabbit oviduct. The roles of cilia and muscle. *Science*, 191, 1052
7 Verdugo, P., Rumery, R. E. and Tam, P. Y. (1980). Hormonal control of oviduct ciliary activity: effects of prostaglandins. *Fertil. Steril.*, 33, 193–6
8 Eddy, C. A. (1982). The role of cilia in tubal function. In Muldoon, T. G., Mahesh, V. B. and Ballester, B. P. (eds.), *Recent Advances in Fertility Research, Part B, Developments in the Management of Reproductive Disorders*. pp. 91–101. (New York: Alan R. Liss Inc.)
9 Weström, L., Mårdh, P. A., von Mecklenburg, C. and Håkansson, C. H. (1977). Studies on ciliated epithelia of the human genital tract. II. The mucociliary wave pattern of Fallopian tube epithelium. *Fertil. Steril.*, 28, 955–61
10 Afzelius, B. A., Camner, P. and Mossberg, B. (1978). On the function of cilia in the female reproductive tract. *Fertil. Steril.*, 29, 72–4
11 Jean, Y., Langlais, J., Roberts, K. D., Chapdelaine, A. and Bleau, G. (1979). Fertility of a woman with nonfunctional ciliated cells in the Fallopian tubes. *Fertil Steril.*, 31, 349–50
12 Pauerstein, C. J. and Carlton, A. E. (1979). The role of the oviduct in reproduction; our knowledge and our ignorance. *J. Reprod. Fertil.*, 55, 223–9
13 Meizner, I., Glezerman, M., Harroch, D. B. and Leventhal, H. (1984). Abdominal pregnancy following hysterectomy. *Obstet. Gynecol.*, 39, 145–6
14 Anand, S. and Guha, S. K. (1978). Mechanisms of transport of the ovum in the oviduct. *Med. Biol. Eng. Comput.*, 16, 256–61
15 Blandau, R. J. (1978). Comparative aspects of tubal anatomy and physiology as they relate to reconstructive procedures. *J. Reprod. Med.*, 21, 7–15

16 Aref, I. and Hafez, E. S. E. (1973). Utero-oviductal motility with emphasis on ova transport. *Obstet. Gynecol. Surv.*, **28**, 680-703

17 Lindblom, B., Hamberger, L. and Ljung, B. (1980). Contractile patterns of isolated oviductal smooth muscle under different hormonal conditions. *Fertil. Steril.*, **33**, 283-7

18 Coutinho, E. M. and Maia, H.S. (1971). The contractile response of the human uterus, Fallopian tubes, and ovary to prostaglandins *in vivo*. *Fertil. Steril.*, **22**, 539-43

19 Wechsung, E. and Houvenaghel, A. (1981). Effect of arachidonic acid on oviductal pressure in the domestic hen. *Biol. Reprod.*, **24**, 519-22

20 Spilman, C. H. and Harper, M. J. K. (1973). Effect of prostaglandins on oviduct motility in estrous rabbits. *Biol. Reprod.*, **9**, 36-45

21 Spilman, C. H. (1974). Oviduct response to prostaglandins: influence of estradiol and progesterone. *Prostaglandins*, **25**, 465-72

22 Maia, J., Barbosa, I., Harper, M. J. K., Hodgson, B. J. and Pauerstein, C. J. (1977). Effect of ovulation and hormonal treatment on the *in vitro* response of rabbit oviducts to prostaglandins E_1 and $F_{2\alpha}$. *Fertil. Steril.*, **28**, 91-5

23 Blair, V. D. and Beck, L. R. (1977). *In vivo* effects of prostaglandin $F_{2\alpha}$ and E_2 on contractility and diameter of the rabbit oviduct using intraluminal transducers. *Biol. Reprod.*, **16**, 122-7

24 Lindblom, B., Hamberger, L. and Wiqvist, N. (1978). Differentiated contractile effects of prostaglandins E and F on the isolated circular and longitudinal smooth muscle of the human oviduct. *Fertil. Steril.*, **30**, 553-9

25 Omini, C., Pasargiklian, R., Folco, G.C., Fano, M. and Berti, F. (1978). Pharmacological activity of PGI_2 and its metabolite 6-oxo-$PGF_{1\alpha}$ on human uterus and Fallopian tubes. *Prostaglandins*, **15**, 1045-54

26 Lindblom, B., Wilhelmsson, L. and Wiqvist, N. (1979). The action of prostacyclin (PGI_2) on the contractility of the isolated circular and longitudinal muscle layers of the human oviduct. *Prostaglandins*, **17**, 99-104

27 Caschetto, S., Lindblom, B., Wiqvist, N. and Wilhelmsson, L. (1979). Prostaglandins and the contractile function of the human oviductal ampulla. *Gynecol. Obstet. Invest.*, **10**, 212-20

28 Wilhelmsson, L., Lindblom, B. and Wiqvist, N. (1979). The human utero-tubal junction: contractile patterns of different smooth muscle layers and the influence of prostaglandin E_2, prostaglandin $F_{2\alpha}$, and prostaglandin I_2 *in vitro*. *Fertil. Steril.*, **32**, 303-7

29 Tonpe, N. and Lindblom, B. (1979). The influence of prostaglandin synthetase inhibition on the spontaneous contractile activity and induced responses of the human oviduct. *Acta Physiol. Scand.*, **107**, 181-3

30 Lindblom, B. and Andersson, A. (1985). The influence of cyclooxygenase inhibitors and arachidonic acid on contractile activity of the human Fallopian tube. *Biol. Reprod.*, **32**, 475-9

31 Saksena, S. K. and Harper, M. J. K. (1975). Relationship between concentration of prostaglandin F (PGF) in the oviduct and egg transport in rabbits. *Biol. Reprod.*, **13**, 68-76

32 Rajkumar, K., Garg, S. K. and Sharma, P. L. (1979). Possible role of pros-
    taglandins in the regulation of ovum transport in rabbits. *Indian J. Med.
    Res.*, **70**, 636-41

33 Lindblom, B. and Hamberger, L. (1981). Copper and contractility of the
    human Fallopian tube. *Am. J. Obstet. Gynecol.*, **141**, 398-402

34 Lindblom, B. and Hamberger, L. (1980). Cyclic AMP and contractility of
    the human oviduct. *Biol. Reprod.*, **22**, 173-8

35 Lindblom, B. and Wikland, M. (1982). Simultaneous recording of electrical
    and mechanical activity in isolated smooth muscle of the human oviduct.
    Biol. Reprod., **27**, 393-8

36 Forman, A., Andersson, K. E. and Ulmsten, U. (1983). Effects of calcium
    and nifedipine on noradrenaline- and $PGF_{2\alpha}$-induced activity of the
    ampullary-isthmic junction of the human oviduct *in vitro*. *J. Reprod.
    Fertil.*, **67**, 343-9

37 Lindblom, B. (1979). Hormonal and neuronal control of human oviductal
    contractility. *Thesis*, University of Göteborg

38 Paton, D. M. and Johns, A. (1975). Effects of prostaglandin E_2 and indo-
    methacin on responses of the isthmus of rabbit oviduct to norepinephrine
    and transmural stimulation. *Chem. Pathol. Pharmacol.*, **11**, 15-24

39 Moawad, A. H., Hedqvist, P. and Kim, M. H. (1976). Correlation of
    plasma estrogens and progesterone levels with the *in vitro* adrenergic
    response in the isthmus of the human oviduct. In Harper, M. J. K., Pauer-
    stein, C. J. *et al.* (eds.), *Ovum Transport and Fertility Regulation.* p. 276.
    (Copenhagen: Scriptor)

40 Chang, M. C. and Hunt, D. M. (1972). Effect of prostaglandin $F_{2\alpha}$ on the
    early pregnancy of rabbits. *Nature (London)*, **236**, 120-1

41 Ellinger, J. V. and Kirton, K. T. (1974). Ovum transport in rabbits injected
    with prostaglandin E_1 and $F_{2\alpha}$. *Biol. Reprod.*, **11**, 93-6

42 Salomy, M. and Goldstein, P. J. (1978). Prevention of pregnancy in rabbits
    using vaginal application of prostaglandin $F_{2\alpha}$. *Fertil. Steril.*, **29**, 456-8

43 Takeda, T., Tsutsumi, Y., Hara, S. and Ida, M. (1978). Effects of pros-
    taglandin $F_{2\alpha}$ on egg transport and *in vivo* egg recovery from the vaginas
    of rabbits. *Fertil. Steril.*, **30**, 79-85

44 Porter, D. G. and Behrman, H. R. (1971). Prostaglandin-induced myome-
    trial activity inhibited by progesterone. *Nature (London)*, **232**, 627-8

45 Valenzuela, G., Ross, H. D., Harper, M. J. K. and Pauerstein, C. J. (1977).
    Effect of inhibitors of prostaglandin synthesis and metabolism on ovum
    transport in the rabbit. *Fertil. Steril.*, **28**, 992-7

46 Croxatto, H. B., Ortiz, M-E., Guiloff, E., Ibarra, A., Salvatierra, A-M.,
    Croxatto, H. D. and Spilman, C. H. (1978). Effects of 15(S)-15-methyl
    prostaglandin $F_{2\alpha}$ on human oviductal motility and ovum transport. *Fertil.
    Steril.*, **30**, 408-14

47 Croxatto, H. B., Ortiz, M. E. and Hess, R. (1979). Attempts to modify
    ovum transport in women. *J. Reprod. Fertil.*, **55**, 231-7

48 Adams, C. E. (1979). Consequences of accelerated ovum transport, includ-
    ing a re-evaluation of Estes' operation. *J. Reprod. Fertil.*, **55**, 239-46

7
The ovary

L. HAMBERGER, P. O. JANSON and L. NILSSON

INTRODUCTION

When prostaglandins (PGs) were first introduced into clinical medicine, they were mainly applied to the field of reproductive biology for the purpose of inducing uterine contractility[1], pregnancy termination[2] and labour[3]. Since then, however, our knowledge concerning the role of PGs on the hypothalamic, hypophyseal and gonadal function has increased immensely, and refinements of biochemical techniques have led to the identification of other arachidonic acid compounds with biological activity within the field of reproduction. Among these are the thromboxanes and leukotrienes[4,5]. Numerous comprehensive reviews on the physiology of PGs have appeared in recent years[6-9].

This chapter discusses PGs as they relate to ovarian function, stressing the clinical implications. Data from various animal as well as human experiments will be included. The following points will be presented: follicular recruitment and growth, follicular fluid composition, follicular rupture, corpus luteum function and, finally, luteal regression.

FOLLICULAR GROWTH

The growth of antral follicles is under gonadotrophic control, in contrast to growth at earlier stages which is independent of FSH and LH but may be controlled by other factors, such as position in the ovary, blood supply, and local hormonal factors including the effects of PGs. Lamprecht *et al.*[10] have shown that PGE_2 can stimulate adenylate cyclase activity in neonatal rat ovaries at an age when these ovaries have not acquired sensitivity to LH. PGE_2 has also been shown to stimulate the adenylate cyclase cyclic AMP system in preovulatory follicles from older rats[11], as well as in small and large follicles from the human ovary.

Both the macro- and microcirculation of the ovary seem to be of importance for the recruitment and selection of specific follicles for ovulation as well as for the earlier phases of follicular development. In this context, the finding of various putative neurotransmitters such as catecholamines, acetylcholine, vasoactive intestinal peptide (VIP) and neuropeptide Y (NPY) within the ovary and ovarian pedicle[12] can be of importance in controlling the blood supply to and within the ovary. For example, the follicle destined to ovulate has a more elaborate capillary network between its thecal and granulosa layers than the other follicles[13]. Moreover, the ovary with the corpus luteum has an increased blood flow compared to the contralateral one[14], and this may also be true for the ovary bearing the dominant follicle. Bendz et al. have shown that a countercurrent transport mechanism for steroids and PGs exists between the ovarian artery and tubal veins in the human[15]. Vasoactive substances like PGs thus either act to modulate this countercurrent flow and/or transport themselves back to different parts of the ovary to exert their effects. Thus, PGs act as regulators or modulators of early follicular growth by increasing the blood supply to certain follicles and/or by inducing FSH receptors on granulosa cells of preantral follicles[16,17]. PGs produced by any ovarian compartment either act by local diffusion or are carried back to the ovary through countercurrent transport: in the latter circumstance, high local concentrations can be built up within the ovary. Such a local mechanism may be of great importance in avoiding a rapid metabolization and inactivation of biologically active but labile PGs.

FOLLICULAR FLUID COMPOSITION

LeMaire and co-workers[18] isolated rabbit follicles at 1, 5 and 9 h after treatment with hCG and found a gradual increase in the concentration of both PGE and F in the follicular fluid; the concentration was greatest at 9 h. This change was noticeable, however, only in follicles that were destined to ovulate[19]. Similar elevations in follicular fluid of PGE and PGF following LH administration in vitro were recently reported utilizing the isolated perfused rabbit ovary[20]. Preovulatory elevation of LH levels in follicular fluid has also been measured in follicles from rats[21] and pigs[22] and indicates the involvement of these compounds in the process of ovulation.

If the mechanism of action of ovulation is common to various species, including man, a rise of the PG concentrations in the follicular fluid prior to ovulation seems to be obligatory. Thus far, however, such a rise has not been satisfactorily demonstrated in the human. Darling et al.[23] reported on a small number of patients in whom the ovarian levels of $PGF_{2\alpha}$ were higher around the expected time of ovulation as compared with those found in the follicular or luteal phases of the natural cycle. More recently, two reports concerned with PGs in aspirated human follicular fluid from induced ovulation cycles showed a tendency towards higher levels of both $PGF_{2\alpha}$ and PGE_2 in follicles containing mature and fertilizable oocytes[24,25]. In the study by Chikhaoui et al.[25], the ratio between PGE_2 and $PGF_{2\alpha}$ was noted to increase with increasing maturity of the oocyte, while thromboxane B_2 remained at a low level. If these observations are correct, one may infer that PGE_2 is the most important PG for ovulation in the human. Such a view could be in accordance with the recent observation that PGE_2, but not $PGF_{2\alpha}$, interferes with collagen synthesis in the follicular wall, since PGE_2 inhibits the incorporation of radiolabelled proline into total protein of isolated specimens from the apex of preovulatory human follicles[26].

The considerable amounts of $PGF_{2\alpha}$ found in human preovulatory follicular fluid might reflect either a true formation of this prostaglandin in the follicle or a conversion from PGE_2 due to the alkaline pH of this follicular fluid at the time of ovulation. The ethical limitations of performing longitudinal studies within the same human subject present marked difficulties in studying the dynamic shifts in PGs of the follicular fluid during follicular maturation. The isolated perfused human ovary may prove to be the only way to adequately study this issue[27]. To date, most data have been obtained from ovarian perfusions in the rabbit[28] and the rat[29].

FOLLICULAR RUPTURE

Systemic administration of PG synthesis inhibitors such as aspirin and indomethacin have been shown to inhibit ovulation in various species including mice, rabbits, rhesus monkeys, marmoset monkeys and goldfish, but this blockade can be counteracted by administration of exogenous PGs[9]. The data concerning humans are inconclusive. In one investigation, aspirin administration failed to block ovulation[31], while in another study the preovulatory administration of indomethacin failed to cause entrapment of oocytes in corpora lutea[32]. Unfortunately, these studies comprise only a small number of observations, and it is possible that the doses of aspirin and indomethacin were insufficient to adequately block follicular PG synthesis.

The fact that indomethacin treatment does not affect follicular steroid production or oocyte maturation[33] implies that PG is related mainly to the rupture of the follicle and not to other processes that occur during preovulatory development. The injection of indomethacin into rabbit follicles blocks ovulation *in vivo*[34], and follicular ruptures are also inhibited by indomethacin in isolated perfused ovaries from rabbits treated with hCG[35] and LH[36]. These findings indicate that indomethacin acts locally in the follicle rather than through a blockade of gonadotrophin secretion.

The site(s) of action of $PGF_{2\alpha}$ on follicular rupture is not presently known. Various theories have been put forward proposing a local PGF-mediated metabolic action on the follicular wall. For instance, Holmes *et al.* have suggested that PGs are involved in the activation, synthesis and/or release of collagenase[37] or other proteolytic enzymes[38]. PGs have been shown to cause lability of lysosomal membranes in non-ovarian tissue[39]. Cajander and Bjersing[39] have published morphological and histochemical evidence of an increased lysosomal activity in the germinal epithelium overlying preovulatory follicles. According to their hypothesis, the release of lysosomal enzymes may initiate the break-down of the follicular wall. Another hypothesis is that PGs may diminish collagen synthesis in the follicular apex, thereby further weakening this structure[26]. It is also possible that PGs play an important role for the increased smooth muscle activity observed in the follicular wall around the time of ovulation[40]. Contractions in the follicle may thus promote the extrusion of the ovum through an enzymatically weakened follicular wall.

CORPUS LUTEUM FUNCTION

The life span of the human corpus luteum (CL) is most likely influenced both by luteotrophic and luteolytic factors. Various PGs are of importance within both these groups of compounds. During the last decade several authors have suggested a physiological role for PGs in the luteinizing process[9,41]. Channing[41,42] demonstrated that granulosa cells from the monkey cultured in the presence of PGE_1 and PGE_2 became luteinized and increased secretion of progesterone. Later, Challis *et al.*[43] showed that PGE_2 was synthesized by the human CL, and Marsh and LeMaire[44] showed that PGE_2 stimulated cyclic AMP and progesterone formation in the human CL. *In vitro* studies of the human corpus luteum have shown that PGE_2 causes an increase in both cyclic AMP and progesterone formation in the early and mid-luteal phases[9,30,45]. At least in the early luteal phase, PGE_2 may be a more potent luteotrophic factor than is either LH or hCG. In the early luteal phase $PGF_{2\alpha}$, on the other hand, apparently has no influence on these parameters. Although not conclusive, these data suggest that PGE_2 may have a luteotrophic effect in the newly formed human CL.

LUTEAL REGRESSION

As early as 1968 Pharris et al.[46] demonstrated that $PGF_{2\alpha}$ caused luteal regression in the rat. It is now clear that PGs, particularly $PGF_{2\alpha}$, cause luteal regression in many other species as well[47]. The mechanism of $PGF_{2\alpha}$ action in the CL involves an inhibition of LH-induced progesterone secretion. This action was first reported in the rat[48] and confirmed in the monkey[41]. McNatty et al.[49] demonstrated a similar mechanism in human granulosa cells luteinized in culture. A further aspect of this mechanism in the human CL has been clarified by taking the age of the CL into consideration. Dennefors et al.[30] found that $PGF_{2\alpha}$ in fresh CL slices influenced neither cyclic AMP nor progesterone formation in contrast to the 7–8th day postovulation at which time a clear-cut antigonadotrophic (LH or hCG) effect was found on both these parameters. $PGF_{2\alpha}$ does not seem to inhibit the binding of gonadotrophin to LH receptors[50]. The hypothesis that $PGF_{2\alpha}$ inhibits the functional integration of the LH receptor is consistent with the observation that $PGF_{2\alpha}$ inhibits gonadotrophin-induced cyclic AMP formation. Grinwich et al.[51] have suggested that $PGF_{2\alpha}$ may cause a down-regulation of luteal gonadotrophin receptors. Recently, Dorflinger et al.[52] proposed that an additional mediator of the $PGF_{2\alpha}$ effect may exist, since the progesterone response to cyclic AMP in intact luteal cells was reduced in the presence of $PGF_{2\alpha}$. Another theory concerning the mechanism behind the antigonadotrophic action of $PGF_{2\alpha}$ involves the interaction between prostaglandins and catecholamines in the human CL. Hamberger et al.[53] found that the addition of noradrenaline to incubated specimens of young human CL makes the CL susceptible to the antigonadotrophic effect of $PGF_{2\alpha}$ on cyclic AMP and progesterone formation. Biochemical determination of the content of catecholamines in the human CL has shown a predominance for noradrenaline in increasing concentrations during the second half of the luteal phase; in the presence of endogenous noradrenaline, $PGF_{2\alpha}$ exerts its antigonadotrophic effect[30]. If endogenous noradrenaline is blocked in the tissue by propranolol the antigonadotrophic effect of $PGF_{2\alpha}$ is completely blocked[54].

So far, only the mechanisms of action of PGs on the cellular level have been discussed, but not their production sites. Evidence exists for: (1) a local production of PGs in the CL; (2) a local production of PGs in other ovarian compartments (follicles, stroma); and (3) an extra-ovarian PG production utilizing a countercurrent vascular mechanism to reach the CL. In a large number of species, uterine $PGF_{2\alpha}$ seems to be critical for luteolysis and, at least in the sheep, it has been convincingly shown that $PGF_{2\alpha}$ is transported from the utero-ovarian vein to the ovarian artery by a countercurrent mechanism[55]. In the human, removal of the uterus does not seem to affect CL function[56]. However, $PGF_{2\alpha}$ is likely to be involved in luteolysis in the human. The source of PG production in the human may be chiefly an intra-ovarian one, although a countercurrent mechanism in the ovarian pedicle has also been described[15]. In studies on isolated CL from rhesus monkeys, exogenous administration of oestrogen induces a drop in progesterone during the luteal phase which is accompanied by a rise in PGF. Indomethacin blocks this effect[57]. Although oestrogens have been shown to reduce luteal progesterone production in the human CL *in vitro*[58], this 'luteolytic' effect does not appear to be mediated via PGs since the CL levels of PGE and PGF are unaffected by exogenous oestrogens[57]. The experimental designs used thus far for the investigation of the role of PGs for luteal regression may not properly reflect physiological conditions with respect to time dynamics and dosages of PGs. For example, Schram *et al.* recently demonstrated that pulsatile administration of ultralow doses of $PGF_{2\alpha}$ caused luteolysis in the sheep[59]. The technique of pulsatile drug administration[60] may lead to further clarification of the role of PGs in luteal function in the human.

SUMMARY

Following the discovery that prostaglandins were involved in regulating ovarian metabolism, the mechanism of this regulation has been subject to intensive investigation. In this chapter, the involvement of prostaglandins in follicular recruitment, growth and rupture, as well as corpus luteum function, and luteolysis has been briefly reviewed. Since the first reports appeared concerning direct and independent metabolic effects of a specific prostaglandin, it has gradually become apparent that this group of local hormones are of pre-eminent importance as modulators of other hormonal effects. Their short half-lives may be of special and requisite importance to meet with the needs of the dynamic shortlasting events which characterize the menstrual cycle in women.

At present the involvement of various PGs in the physiology of ovulation is unquestionable, although their clinical importance as pharmacological agents in this context is still debatable. For such purposes, specific blockers of prostaglandin synthesis seem more promising. However, the lack of specificity in blocking PG effects has also limited their clinical usefulness. More specific blockers of selective parts of the pathways of PG synthesis may lead to the development of useful pharmaceutical agents in the future.

References

1 Bygdeman, M., Kwon, S. and Wiqvist, N. (1967). The effect of prostaglandin E_1 on human pregnant myometrium *in vivo*. In Bergström, S. and Samuelsson, B. (eds.) *Nobel Symposium 2: Prostaglandins*. pp. 93-6. (Uppsala: Almqvist & Wiksell)

2 Wiqvist, N. and Bygdeman, M. (1970). Induction of therapeutic abortion with intravenous prostaglandin $F_{2\alpha}$. *Lancet*, **1**, 889

3 Karim, S. M., Trussell, R. R., Patel, R. C. and Hillier, K. (1968). Response of pregnant human uterus to prostaglandin $F_{2\alpha}$ induction of labour. *Br. Med. J.*, **4**, 621-3

4 Green, K., Christensen, N. and Bygdeman, M. (1981). The chemistry and pharmacology of prostaglandins with reference to human reproduction. *J. Reprod. Fertil.*, **62**, 269-81

5 Granström, E. (1983). Prostaglandin biochemistry, pharmacy and physiological function. *Acta Obstet. Gynecol. Scand., Suppl.*, **113**, 9-13

6 Behrman, H. R. (1979). Prostaglandins in hypothalamo-pituitary and ovarian function. *Ann. Rev. Physiol.*, **41**, 685-700

7 Armstrong, D. I. (1981). Prostaglandins and follicular function. *J. Reprod. Fertil.*, **41**, 283-91

8 Baird, D. T. (1983). Factors regulating the growth of the preovulatory follicle in the sheep and human. *J. Reprod. Fertil*, **69**, 343-52

9 Dennefors, B., Hamberger, L., Hillensjö, T., Holmes, P., Janson, P. O., Magnusson, C. and Nilsson, L. (1983). Aspects concerning the role of prostaglandins for ovarian function. *Acta Obstet. Gynecol. Scand., Suppl.*, **113**, 31-41

10 Lamprecht, S. A., Zor, U., Tsafriri, A. and Lindner, H. R. (1973). Action of prostaglandin $E_{2\alpha}$ and of luteinizing hormone on ovarian adenylate cyclase protein kinase, and ornithine decarboxylase activity during postnatal development and maturity in the rat. *J. Endocrinol.*, **57**, 217-33

11 Nilsson, L., Rosberg, S. and Ahren, K. (1974). Characteristics of the cyclic 3',5'-AMP formation in isolated ovarian follicles from PMSG-treated immature rats after stimulation *in vitro* with gonadotrophins and prostaglandins. *Acta Endocrinol. (Copenh)*, **77**, 559-74

12 Owman, C., Sjöberg, N. O. and Sjöstrand, N. O. (1974). Short adrenergic neurons, a peripheral neuro-endocrine mechanism. In Fujawara, M. and Tanaka, C. (eds.) *Amino Fluorescence Histochemistry*. pp. 47-66. (Tokyo: Tokyo Igaku Shoin, Ltd.)

13 Zeleznik, A. J., Schuler, H. M. and Reichert, L. E. (1981). Gonadotrophin binding sites in the rhesus monkey ovary: role of vasculature in the selective distribution of human chorionic gonadotrophin to the preovulatory follicle. *Endocrinology*, **109**, 356

14 Janson, P. O. (1974). Ovarian blood flow: methodological and functional studies in the rabbit. *Thesis*, Gothenburg

15 Bendz, A., Einer-Jensen, N., Lundgren, O. and Janson, P. O. (1979). Exchange of krypton-85 between the blood vessels of the human uterine adnexa. *J. Reprod. Fertil.*, **57**, 137–42

16 Peters, H. (1979). Some aspects of early follicular development. In Midgley, R. A. and Sadler, W. A. (eds.) *Ovarian Follicular Development and Function*. pp. 3–13. (New York: Raven Press)

17 Armstrong, D. T. (1981). Prostaglandins and follicular functions. *J. Reprod. Fertil.*, **62**, 283–91

18 LeMaire, W. J., Yang, N. S. T., Behrman, H. R. and Marsh, J. M. (1973). Preovulatory changes in the concentrations of prostaglandin in the rabbit Graafian follicles. *Prostaglandins*, **3**, 367–76

19 Yang, N. S. T., Marsh, J. M. and LeMaire, W. J. (1974). Postovulatory changes in the concentrations of prostaglandins in rabbit Graafian follicles. *Prostaglandins*, **6**, 37–44

20 Koos, R. D., Clark, M. R., Janson, P. O., Ahren, K. F. B. and LeMaire, W. J. (1983). Prostaglandin levels in preovulatory follicles from rabbit ovaries perfused *in vitro*. *Prostaglandins*, **25**, 715–24

21 Bauminger, S. and Lindner, H. R. (1975). Preovulatory changes in ovarian prostaglandin formation and their hormonal control in the rat. *Prostaglandins*, **9**, 737–51

22 Ainworth, L., Baker, R. D. and Armstrong, D. T. (1975). Preovulatory changes in follicular fluid prostaglandin F levels in swine. *Prostaglandins*, **9**, 915–25

23 Darling, M. R. N., Jogee, M. and Elder, M. G. (1982). Prostaglandin $F_{2\alpha}$ levels in the human ovarian follicle. *Prostaglandins*, **23**, 551–6

24 Hoppen, H-O., Calabrese, R., Voss, S. and Trounson, A. O. (1984). Prostaglandins in human follicular fluid. *Acta Endocrinol.*, **102**, suppl. 253, 135

25 Chikhaoui, Y., Nicolas, J. C., Cristol, P., Mares, P., Hedon, B., Chaintreuil, J., Flandre, O., Damon, M., Descomps, B. and Crastes de Paulet, A. (1984). Influence de l'environment hormonal de l'ovocyte humaine sur son évolution ultérieure *in vitro*. *J. Gynecol. Obstet. Biol.*, **12**, 253–8

26 Dennefors, B., Tjugum, J., Norström, A., Janson, P. O., Nilsson, L., Hamberger, L. and Wilhelmsson, L. (1982). Collagen synthesis inhibition by PGE_2 within the human follicular wall – one possible mechanism underlying ovulation. *Prostaglandins*, **24**, 295–301

27 Janson, P. O., Jansson, I., Skryten, A., Damber, J. E. and Lindstedt, G. (1981). Ovarian endocrine function in young women undergoing radiotherapy for carcinoma of the cervix. *Gynecol. Oncol.*, **11**, 218–23

28 Janson, P. O., LeMaire, W. J., Källfelt, B., Holmes, P. V., Cajander, S., Bjersing, L., Wiqvist, N. and Ahren, K. (1982). The study of ovulation in the isolated perfused rabbit ovary. I. Methodology and pattern of steroidogenesis. *Biol. Reprod.*, **26**, 456–65

29 Sogn, J., Abrahamsson, G. and Janson, P. O. (1984). The effect of luteinizing hormone on the release of cyclic AMP and progesterone from the isolated perfused luteal ovary of the PMSG treated rat. *Acta Endocrinol.*, **106**, 265–70

30 Dennefors, B., Sjögren, A. and Hamberger, L. (1982). Progesterone and adenosine 3′5′-monophosphate formation by isolated corpora lutea of different ages: influence of human chorionic gonadotrophin and prostaglandins. *J. Clin. Endocrinol. Metab.*, **55**, 102-7

31 Plunkett, E. R., Moon, Y. S., Zamecnik, J. and Armstrong, D. T. (1975). Preliminary evidence of a role for prostaglandin F in human follicular function. *Am. J. Obstet. Gynecol.*, **123**, 391-7

32 Chaduri, G. and Elder, M. G. (1976). Lack of evidence for inhibition of ovulation by aspirin in women. *Prostaglandins*, **11**, 727-35

33 Toppozada, M., El-Abd, M., El-Sokkary, H. and El-Rahman, H. A. (1979). Effect of prostaglandin inhibitor on human ovulation. *Singapore J. Obstet. Gynecol.*, **10**, 42-4

34 Armstrong, D. T. and Grinwich, D. L. (1972). Blockade of spontaneous and LH induced ovulation in rats by indomethacin, an inhibitor of prostaglandin biosynthesis. *Prostaglandins*, **1**, 21-8

35 Armstrong, D. T., Moon, Y. S. and Zamecnik, J. (1974). Evidence for a role of prostaglandins in ovulation. In Mougdal, N. R. (ed.) *Gonadotrophins and Gonadal Function*. pp. 345-369. (New York: Academic Press)

36 Wallack, E. E., Bronson, R., Hamada, Y., Wright, K. H. and Stevens, V. C. (1975). Effectiveness of prostaglandin $F_{2\alpha}$ in restoration of hMG-hCG induced ovulation in the rhesus monkey. *Prostaglandins*, **10**, 129-38

37 Holmes, P. V., Sogn, J., Källfelt, B., LeMaire, W. J., Ahren, K., Cajander, S. and Bjersing, L. (1983). Effects of $PGF_{2\alpha}$ and indomethacin on ovulation and steroid production in the isolated perfused rabbit ovary. *Acta Endocrinol.*, **104**, 223-39

38 Espey, L. L. (1974). Ovarian proteolytic enzymes and ovulation. *Biol. Reprod.*, **10**, 216-35

39 Cajander, S. and Bjersing, L. (1975). Fine structural demonstration of acid phosphatase in rabbit germinal epithelium prior to induced ovulation. *Cell. Tissue Res.*, **164**, 279-89

40 Walles, B., Edvinsson, L., Nybell, G., Owman, C. and Sjöberg, N-O. (1974). Amine-induced influence on spontaneous ovarian contractility in the guinea-pig and the cat. *Fertil. Steril.*, **25**, 602-11

41 Channing, C. P. (1972). Effects of prostaglandin inhibition 7-oxo-13-prostynoic acid and eicosa-5,8,11,14-tetra-enoic acid upon luteinization of rhesus monkey granulosa cells in culture. *Prostaglandins*, **2**, 351-67

42 Channing, C. P. (1972). Stimulatory effects of prostaglandins upon luteinization of rhesus monkey granulosa cell culture. *Prostaglandins*, **2**, 351-67

43 Challis, J. R. G., Clader, A. A., Dilley, S., Forster, C. S., Hillier, K., Hunger, D. J. S., McKenzie, I. Z. and Thorburn, G. S. (1976). Production of prostaglandins F and $F_{2\alpha}$ by corpora lutea, corpora albicantes and stroma from the human ovary. *J. Endocrinol.*, **68**, 401-8

44 Marsh, J. M. and LeMaire, W. J. (1974). Cyclic AMP accumulation and steroidogenesis in the human corpus luteum: effect of gonadotrophins and prostaglandins. *J. Clin. Endocrinol., Metab.*, **38**, 99-106

45 Hamberger, L., Nilsson, L., Dennefors, B., Khan, I. and Sjögren, A. (1979). Cyclic AMP formation of isolated human corpora lutea in response to hCG-interference by $PGF_{2\alpha}$. *Prostaglandins*, **17**, 615-21

46 Pharriss, B. B., Wyngarden, L. F. and Gutkneckt, G. D. (1968). Biological interaction between prostaglandin and luteotropins in the rat. In Rosenberg, E. (ed.) *Gonadotrophins*. pp. 121–9. (Los Altos: Geron-X)

47 Behrman, H. R. (1979). Prostaglandins in hypothalamo-pituitary and ovarian function. *Ann. Rev. Physiol.*, 41, 685–700

48 Behrman, H. R., Yoshinaga, K. and Greep, R. O. (1971). Extraluteal effects of prostaglandins. *Ann. NY Acad. Sci.*, 180, 426–35

49 McNatty, K. P., Henderson, K. M. and Sawers, R. S. (1975). Effects of prostaglandin $F_{2\alpha}$ and E_2 on the production of progesterone by human granulosa cells in tissue culture. *J. Endocrinol.*, 67, 231–40

50 Thomas, J. P., Dorflinger, L. J. and Behrman, H. R. (1978). Mechanism of the rapid antigonadotrophic action of prostaglandins in cultured luteal cells. *Proc. Natl. Acad. Sci. (USA)*, 75, 1344–8

51 Grinwich, D. L., Hichens, M. and Behrman, H. R. (1976). Control of the LH receptor by prolactin and prostaglandin $F_{2\alpha}$ in rat luteal cell. *Mol. Cell Endocrinol.*, 33, 225–41

52 Dorflinger, L. J., Luborsky, J. L., Gore, S. D. and Behrman, H. R. (1983). Inhibitory characteristics of prostaglandin E_2 in rat luteal cell. *Mol. Cell Endocrinol.*, 33, 225–41

53 Hamberger, L., Dennefors, B., Hamberger, B., Janson, P. O., Nilsson, L., Sjögren, A. and Wiqvist, N. (1980). Is vascular innervation a prerequisite for PG induced luteolysis in the human corpus luteum. In Samuelsson, B., Ramwell, P. W. and Pauletti, R. (eds.) *Advances in Prostaglandin and Thromboxane Research*. Vol. 8, pp. 1365–8. (New York: Raven Press)

54 Bennegård, B., Dennefors, B. and Hamberger, L. (1984). Interaction between catecholamines and prostaglandin $F_{2\alpha}$ in human luteolysis. *Acta Endocrinol.*, 106, 532–7

55 McCracken, J. A., Carlson, J. C., Glew, M. E., Goding, J. R., Baird, D. T., Gréen, K. and Samuelsson, B. (1972). Prostaglandin identified as a luteolytic hormone in sheep. *Nature New Biol.*, 238, 129–34

56 Janson, P. O. and Jansson, I. (1977). The acute effect of hysterectomy on ovarian blood flow. *Am. J. Obstet. Gynecol.*, 7, 349–52

57 Auletta, F. J., Agins, H. and Scommegna, A. (1978). $PGF_{2\alpha}$ mediation of the inhibitory effect of estrogen on the corpus luteum of the rhesus monkey. *Endocrinology*, 103, 1183–9

58 Thibier, M., El-Hassam, N., Clark, M. R., LeMaire, W. J. and Marsh, J. M. (1980). Inhibition by estradiol of human chorionic gonadotrophin induced progesterone accumulation in isolated human luteal cells: lack of mediation by prostaglandin. *J. Clin. Endocrinol. Metab.*, 50, 590–2

59 Schramm, W., Bovaird, L., Glew, M. E., Schramm, G. and McCracken, J. A. (1983). Corpus luteum regression induced by ultra-low pulses of $PGF_{2\alpha}$. *Prostaglandins*, 26, 347–64

60 Sandow, J. (1983). Clinical application of LHRH and its analogues. *Clin. Endocrinol.*, 18, 571–92

8
Implantation

N. L. POYSER

INTRODUCTION

The question of precisely when a human pregnancy actually commences has become the subject of increasingly intense public discussion. Is it at the time of gamete union (fertilization) or is it at the time when the blastocyst attaches itself to the endometrium (implantation)? If the former, then some argue that the intrauterine contraceptive device (IUD), the 'morning after pill' and, in some cases, the oral contraceptive pill should be reclassified as abortifacients rather than contraceptives, since they prevent the implantation of a fertilized egg but do not necessarily prevent fertilization. Some argue that such methods should therefore come under the control of abortion laws, with stringent control of their use. At present, these methods remain classified as 'contraceptive' in nature. What is not in dispute, however, are the facts that (1) a non-fertilized egg is not capable of implanting, and (2) a new human life will not be produced if a fertilized egg fails to implant. Consequently, the process of implantation is crucial in the process of the progression of a fertilized egg into a new individual.

Implantation is difficult to study in humans since it occurs before a woman has any indication that she is pregnant. Much of our present information about implantation has been obtained by studying this process in non-primate species, particularly rodents. Figure 8.1 indicates the average day of implantation in several species. In common with the mouse, hamster, rat, guinea-pig and rabbit, implantation in human occurs within a week of fertilization, whereas in ungulates (e.g. sheep, cow) and equidae (e.g. horse) implantation occurs considerably later.

This chapter will discuss the role prostaglandins may have in the

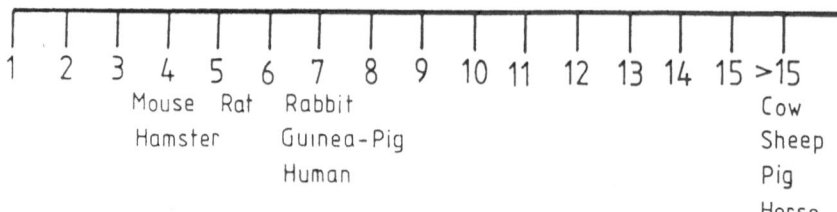

Figure 8.1 Indication of the average day of implantation in several species, where day 1 is the day of fertilization

process of implantation. Many of the studies regarding prostaglandins and implantation have been carried out on the smaller mammalian species indicated in Figure 8.1, and the relevant findings from these studies form the main experimental content of this chapter. The hormonal control of implantation will be described first, however, since the uterus has to be suitably prepared if the blastocyst is to attach itself to the endometrium.

HORMONAL CONTROL OF IMPLANTATION

In order for implantation to occur the uterus must be primed for several days with progesterone. Progesterone is secreted by the corpus luteum. Should the corpus luteum malfunction and produce insufficient quantities of progesterone, then implantation will not occur. In pregnant, ovariectomized rats maintained on progesterone, oestradiol administration is also necessary for implantation. In pregnant intact rats a surge of oestradiol from the ovary occurs in the early hours of day 4 after fertilization[1]. A similar surge occurs in the pseudopregnant rat, thus indicating that the presence of the blastocyst is not necessary for this surge of oestradiol[2]. Tamoxifen (an anti-oestrogen) prevents implantation in the rat[1]. A pre-implantation increase in plasma oestradiol concentrations also occurs in mice[3], and oestradiol is necessary for implantation in this species.

That implantation occurs in pregnant, ovariectomized rabbits maintained only on progesterone suggests that oestradiol is not required for implantation in this species. However, instillation on day 5 of the anti-oestrogenic drug, CI-628, into the uterus of pregnant, intact rabbits or of pregnant, progesterone-treated ovariectomized rabbits prevents implantation[4,5]. CI-628 also inhibits the transformation of rabbit morulae into blastocysts, as well as inhibiting several associated metabolic functions[6]. These effects of CI-628 on the development of the embryo are overcome[6] by oestradiol-17β. Since rabbit blastocysts are capable of synthesizing oestrogens[7,8], non-maternal oestrogen appears necessary for the successful development of a blastocyst and for implantation in the rabbit. Oestrogens can be synthesized by the blastocyst of the rat, mouse, hamster, pig, cow, sheep, mare and roe deer[7,9-13]. In several species the oestradiol produced by embryos appears necessary for their normal development. It is not known, however, whether oestrogens are synthesized by the blastocyst in humans. Progesterone appears necessary for implantation in all mammalian species including humans, while oestradiol of maternal and/or blastocyst origin is essential in only some species. At present it is not known whether oestradiol is essential for implantation in humans.

EFFECT OF NON-STEROIDAL ANTI-INFLAMMATORY DRUGS ON IMPLANTATION

Depending on the dose, indomethacin treatment of rats, mice, hamsters and rabbits delays, reduces or totally inhibits implantation when administered immediately prior to the expected time of this phenomenon[14-19]. Table 8.1 shows the effect of treating rats with indomethacin (3 mg/kg) at 09:00 and 17:00 on days 3 and 4 on the number of

Table 8.1 Effect of indomethacin treatment (3 mg/kg subcutaneously at 09:00 and 17:00 on days 3 and 4) on the number (mean ± SEM) of implantation sites and uterine weight (mean ± SEM) at 10:00 on day 9 in pregnant rats[19]

Treatment	No. of rats	No. of implantation sites	Uterine weight (mg)
Control	12	5.5 ± 1.2	912 ± 104
Indomethacin:			
(1)	7	0	$374 \pm 31^*$
(2)	4	$2.6 \pm 1.1^*$	$349 \pm 65^*$

* Significantly lower ($p < 0.05$) than in control group

implantation sites when examined on day 9 of pregnancy[19]. In seven (64%) of 11 treated rats, implantation was completely prevented, while in the remaining four (39%) the number of implantation sites was reduced to one half that of untreated controls. Since indomethacin inhibits prostaglandin synthesis, these studies suggest that prostaglandins may be involved in the process of implantation. However, the administration of prostaglandin (PG) E_2 or $PGF_{2\alpha}$ to mice and rabbits only partially overcomes the block of implantation produced by indomethacin[17,20]. Thus, indomethacin may have other actions besides its antiprostaglandin effect which inhibit implantation.

PROSTAGLANDIN PRODUCTION BY THE EARLY PREGNANT UTERUS

The rat

Figure 8.2 shows the amounts of 6-oxo-PGF$_{1\alpha}$(the chemically hydrated product of PGI$_2$ which reflects PGI$_2$ production), PGE$_2$ and PGF$_{2\alpha}$ synthesized by homogenates of the rat uterus during early pregnancy[19]. The production of 6-oxo-PGE$_{1\alpha}$ and PGE$_2$ peaks on day 5 (the day of implantation), suggesting that PGI$_2$ and/or PGE$_2$ have a role in implantation. A similar pattern is seen in pseudopregnant rats indicating that this increase in uterine prostaglandin synthetase activity is independent of the presence of the blastocyst, although uterine PGF$_{2\alpha}$ production on day 5 is lower in pregnant rats than in pseudopregnant rats[19,21]. Tamoxifen prevents the increase in uterine prostaglandin synthesizing capacity in pseudopregnant rats (pregnant rats have not been

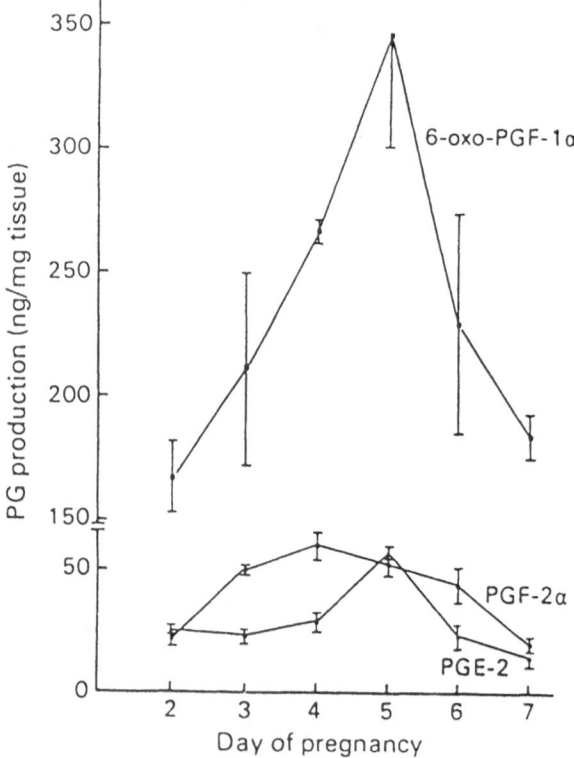

Figure 8.2   Prostaglandin production (mean±SEM, $n=5$) by rat uterine homogenates during early pregnancy[19]

studied in this respect), suggesting that it is oestradiol secreted by the ovary which is responsible for the increase in uterine PG synthetase activity, probably the cyclo-oxygenase component according to Fenwick et al.[21]. In ovariectomized rats treated with progesterone and oestradiol to suitably prime the uterus for an implantation response, the synthesizing capacity of the uterus for PGE_2, $PGF_{2\alpha}$ and thromboxane (Tx) B_2, but not for 6-oxo-$PGF_{1\alpha}$, is increased[22]. It thus appears that these steroidal hormones, although having some stimulatory effect, are not solely responsible for the changes in prostaglandin synthetase activity in the rat uterus.

In these studies using rat uterine homogenates, arachidonic acid availability is not rate limiting, since sufficient free endogenous arachidonic acid is released during the homogenization and incubation processes. However, in the intact rat uterus in vivo, arachidonic acid availability is rate limiting. It is generally believed that arachidonic acid for prostaglandin synthesis is released from phospholipids by phospholipase A_2 (PLA_2). Oestradiol treatment of ovariectomized rats greatly increases uterine PLA_2 activity, whereas progesterone, although having no effect on its own, reduces the stimulatory action of oestradiol[23]. During early pregnancy in rats, uterine PLA_2 activity is highest on the morning of day 4, some 24 to 36 h before implantation. On the morning of day 5, uterine PLA_2 is 64% lower than on the morning of day 4, although uterine PLA_2 activity increases two-fold from the morning to the afternoon of day 5. This increased activity on the afternoon of day 5 is similar to the activities of PLA_2 in the uterus on the mornings of days 2 and 3[24]. Consequently, the absolute activity of PLA_2 does not appear to be the factor controlling prostaglandin production by the rat uterus during early pregnancy. Indeed, there is no difference in the overall amounts of PGE_2, $PGF_{2\alpha}$ or 6-oxo-$PGF_{1\alpha}$ contained in the rat uterus on days 3, 4 and 5 of pregnancy[19]. Although oestradiol and progesterone treatment of ovariectomized rats increases uterine PLA_2 and prostaglandin synthetase activities, such steroid treatment does not increase outputs of PGE_2 or $PGF_{2\alpha}$ (6-oxo-$PGF_{1\alpha}$ was not studied) from the uterus superfused in vitro[22]. There is, however, a two-fold increase in the uterine outputs of PGE_2, $PGF_{2\alpha}$ and 6-oxo-$PGF_{1\alpha}$ between days 2 and 5 of pregnancy in rats[25], suggesting that the changes in activities of the enzymes connected with prostaglandin synthesis stimulate a general increase in prostaglandin output from the intact rat uterus at the time of implantation.

The rabbit

In uterine flushings from pregnant rabbits, $PGF_{2\alpha}$ levels decrease and PGE_2 levels increase between days 3 and 7 and the ratio of $PGF_{2\alpha}$ to

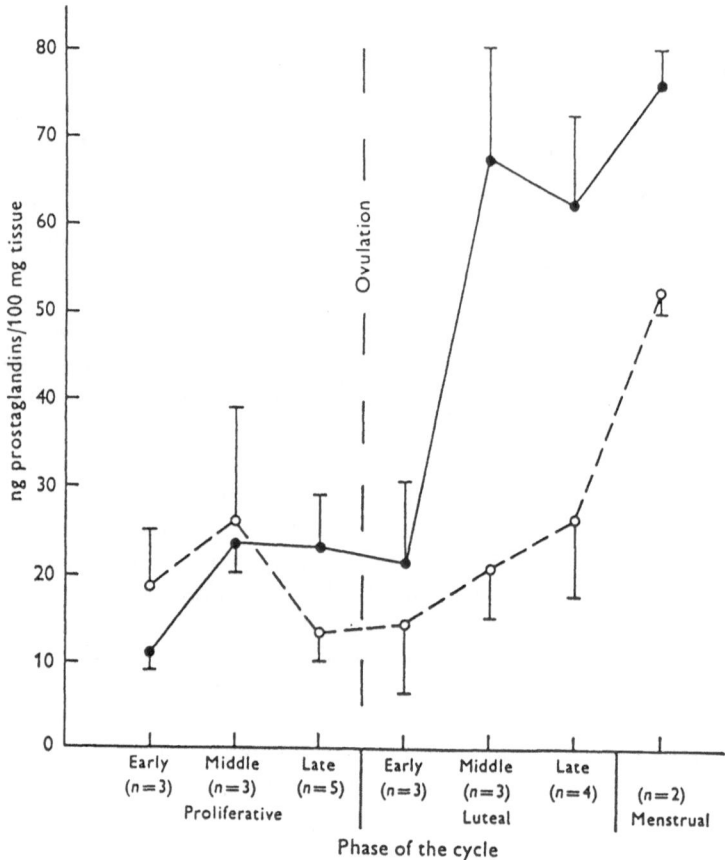

Figure 8.3 Concentrations of prostaglandin $F_{2\alpha}$ (●——●) and prostaglandin E_2 (○---○) in the human endometrium during the menstrual cycle[29]

PGE$_2$ is reversed. In the endometrium and uterine venous plasma of pregnant rabbits, $PGF_{2\alpha}$ levels remain constant while PGE_2 levels increase several-fold between days 3 and 7[26]. PGE_2 and $PGF_{2\alpha}$ output from the pregnant rabbit uterus cultured *in vitro* tends to be higher on days 6 and 7 than on days 4 and 5[27]. These studies suggest that there is increased uterine synthesis release by prostaglandins, particularly PGE_2, at the time of implantation in the rabbit.

Other species

In the guinea-pig, the outputs of PGE_2 and $PGF_{2\alpha}$, but not of 6-oxo-$PGF_{1\alpha}$, are approximately 50% lower on day 7 of pregnancy than on day 7 of the oestrous cycle. Endometrial $PGF_{2\alpha}$ synthesizing capacity is similarly 50% lower, whereas endometrial PGE_2 synthetizing capacity is increased two-fold in the pregnant compared to the non-pregnant guinea-pig on day 7[28]. However, it is clear that no general increase in uterine PG synthesis and release occurs at the time of implantation in the guinea-pig.

In non-pregnant women, using no contraceptive hormones, the endometrial levels of $PGF_{2\alpha}$ but not of PGE_2, are high at the mid-luteal phase, the time at which implantation is expected to occur (Figure 8.3)[29]. It is not known whether similar changes occur in endometrial prostaglandin levels in women when the ovum is fertilized.

PROSTAGLANDIN PRODUCTION BY THE BLASTOCYST

Rabbit blastocysts contain appreciable quantities of PGE_2 and $PGF_{2\alpha}$ just before implantation, and rabbit blastocysts are capable of synthesizing PGE_2 and $PGF_{2\alpha}$ in greater quantities on days 6 and 7 than on earlier days of pregnancy[26,27,30-33]. Prostaglandins are also synthesized by the blastocyst of the cow, sheep and pig[10,34-39]. These findings suggest that prostaglandins produced by the blastocyst may be involved in implantation in addition to or in place of prostaglandins of uterine origin. In contrast, prostaglandin production by rat and mouse blastocysts is low[33,40]. No information is known concerning prostaglandin production by human blastocysts at the time of implantation.

PROSTAGLANDIN E_2 BINDING SITES IN THE ENDOMETRIUM

In endometrial membrane preparations from pseudopregnant rats, specific binding of PGE_2 is not detectable on day 2, is low on days 3 and 4, increases markedly on day 5 and reaches a maximum on day 6[41]. High affinity binding sites for PGE_2 in rat endometrium are not detectable during the oestrous cycle[42]. By treating ovariectomized rats with ovarian steroids, it was determined that the production of endometrial PGE_2 sites in the rat is progesterone dependent. These binding sites are located in the stromal cells and not the epithelial cells of progesterone-treated, ovariectomized rats[42]. The increase in PGE_2 binding sites in the endometrium of pseudopregnant rats follows the increase in plasma progesterone concentrations. Presumably, a similar relationship occurs in the pregnant rat, with PGE_2 binding to the endometrial stromal cells being high at the time of implantation.

INVOLVEMENT OF PROSTAGLANDINS IN THE VARIOUS STAGES OF IMPLANTATION

Shedding of the zona pellucida

Indomethacin, meclofenamate and various proposed antagonists of prostaglandin action (e.g. 7-oxa-prostynoic acid) prevent shedding of the zona pellucida from mouse blastocysts *in vitro*[43-45]. Indomethacin also delays shedding of the zona pellucida in hamsters *in vivo* when administered shortly before the expected time of implantation[46]. It appears that prostaglandins are involved in the shedding of the zona pellucida, and it has been suggested that, since PGE_2 promotes water movement across epithelia, PGE_2 produced by the blastocyst may promote the passage of water across the trophectoderm, causing the blastocyst to swell and the zona pellucida to burst[43]. However, two findings cast doubt on this hypothesis. First, in contrast to rabbit blastocysts, mouse blastocysts do not synthesize prostaglandins from endogenous arachidonic acid released by the application of a calcium ionophore; this observation raises doubts as to whether mouse blastocysts are capable of synthesizing prostaglandins[33]. Second, PGE_2 administered to hamsters fails to overcome the delay in the shedding of the zona pellucida produced by indomethacin. Consequently, other actions of indomethacin, such as inhibition of calcium uptake, may account for the inhibitory effect of indomethacin on the shedding of the zona pellucida. However, it is more difficult to envisage how supposedly specific inhibitors of prostaglandin action could prevent shedding of the zona pellucida[44] if prostaglandins are not involved. More studies are required to clarify the possible role of prostaglandins in the shedding of the zona pellucida.

Increase in uterine vasculature permeability

Prior to the blastocyst attaching itself to the uterus, the uterine capillaries in the immediate vicinity of the blastocyst become increasingly permeable to water. These sites may be visualized by injecting blue dye into the bloodstream, since leakage of the dye from the uterine vasculature into the surrounding tissue only occurs at the implantation sites. Pregnant rats injected with 1 mg indomethacin at 08:00 and 13:00 on day 5 failed to exhibit the uterine dye site reaction when examined on the evening of the same day, compared to a 100% response in control animals[15]. In a similar study, pregnant rats injected with indomethacin (3 mg/kg body wt.) at 09:00 and 17:00 on days 3 and 4 failed to exhibit the uterine dye site reaction when examined at 02:00 on day 6, whereas the response was observed in 100% of control rats

Table 8.2 Effect of indomethacin treatment (3 mg/kg subcutaneously at 09:00 and 17:00 on days 3 and 4) on the uterine 'dye site reaction' on day 6 in pregnant rats ($n = 5$ per group)[19]

Treatment	Time of examination (h)	No. of uterine dye sites (mean ± SEM)
Control	02:00	5.6 ± 0.6
Indomethacin	02:00	0
Control	10:00	5.5 ± 0.4
Indomethacin	10:00	5.8 ± 0.4

at this time. However, pregnant rats treated similarly with indomethacin and examined at 10:00 on day 6 had the same number of uterine dye site reactions as control rats (Table 8.2)[19]. Similarly, rats injected with 1 mg indomethacin at 08:00 and 13:00 on day 5 exhibited the uterine dye site reaction when examined between 06:00 and 07:00 on day 6[15]. These results indicate that indomethacin delays rather than abolishes the increase in uterine vascular permeability of the implantation sites in rats. Indomethacin treatment causes a similar delay in pregnant rabbits and hamsters[16,18].

The concentrations of PGE_2, $PGF_{2\alpha}$ and 6-oxo-$PGF_{1\alpha}$ in the rat uterus are higher at these sites of increased capillary permeability than in the surrounding uterus[15,47]. PGE_2 concentrations are also higher at these sites of increased capillary permeability in the pregnant hamster uterus[16]. PGE_2, but not PGI_2 or $PGF_{2\alpha}$, stimulates an increase in vascular permeability in the rat uterus, suggesting that PGE_2 produced locally within the uterus is a mediator of this process at the implantation site[48,49]. However, PGE_2 is not generally considered to have potent activity with regard to increasing capillary permeability on its own. PGE_2 probably potentiates the action of other more potent mediators, such as histamine, by its vasodilator action. It is difficult to understand why PGE_2, but not PGI_2, exhibits this action in the uterus, since PGI_2 is also a vasodilator.

In rabbits, PGE_2 and $PGF_{2\alpha}$ concentrations are also higher at the implantation sites than in areas between implantation sites[26,32]. Removal of the blastocysts from the implantation site results in a considerable decline in the concentration of PGE_2 and $PGF_{2\alpha}$ at the implantation site, suggesting that the blastocyst is the main source of these prostaglandins[32].

Decidual cell reaction

The endometrium adjacent to the implanting blastocyst exhibits a decidual cell reaction. In species such as the rat and mouse, decidualization spreads progressively from the implantation site until a considerable mass of endometrium is involved in each implantation chamber. In other species, the decidual cell reaction is minimal or absent. In humans, the endometrium during the secretory phase of the cycle consists of decidual-like cells regardless of whether the egg has been fertilized or not.

Decidualization can be produced in a uterus which has been suitably primed with progesterone and oestradiol by the application of an artificial stimulus, such as trauma or vegetable oil within the uterine cavity. Indomethacin prevents decidualization of the suitably primed rat and mouse uterus following the application of such an artificial stimulus[50-55]. PGE_2, $PGF_{2\alpha}$, PGI_2 or various prostaglandin and thromboxane analogues induce decidualization when instilled into a suitably primed rat uterus[52,56,57]. In addition, arachidonic acid (the precursor of the series of prostaglandins with two double bonds in the side-chains) but not ω-dihomo-γ-linolenic acid (the precursor of the series of prostaglandins with one double bond in the side-chain), oleic acid and palmitic acid applied to a suitably primed rat uterus induce decidualization. This action is prevented by indomethacin[52]. In pseudopregnant rabbits, the insertion into the uterus on day 6 of implants containing PGE_2, $PGF_{2\alpha}$ or arachidonic acid produces a decidual cell reaction. Control implants have no effect[58].

Oil-induced or trauma-induced decidualization of the rat or mouse uterus is followed by a rapid rise (<5 min) in the concentrations of PGE_2 and $PGF_{2\alpha}$ in the uterus[51,57,59-61]. Prostaglandin synthetase activity in trauma-induced decidualized tissue of rat uterus is seven- to ten-fold higher than in non-decidualized tissue[59]. All these studies indicate that prostaglandins produced by the uterus of the rat, mouse and rabbit are involved in the production of the decidual cell reaction. It is unknown whether the increase in endometrial $PGF_{2\alpha}$ levels during the mid-secretory phase is the reason for the human secretory endometrium consisting of decidual-like cells.

Peleg and Lindner[55] have shown that indomethacin treatment of the day 4 pseudopregnant rat uterus impedes the progesterone-induced translocation of the cytoplasmic progesterone receptor into the nucleus. PGE_2 or $PGF_{2\alpha}$ administration overcomes this blockade. Peleg and Lindner[55] suggest that 'prostaglandins may modulate the rate of translocation of progesterone receptors into the nucleus and this action of prostaglandins may be important for the decidual cell reaction'. In addition, PGE_2 inhibits and indomethacin stimulates DNA synthesis in rat decidual cells, suggesting that 'the rate of DNA synthesis in the deciduoma at the proliferative phase may be influenced by steroid-induced changes in PGE content of the uterus'[62].

$PGF_{2\alpha}$, PGE_2, arachidonic acid or vegetable oil cannot induce decidualization of a suitably primed rat uterus if the uterine luminal epithelium has been removed. Detachment of the epithelium *per se* without its removal from the uterus also prevents the decidual cell reaction[63]. An intact luminal epithelium is necessary if implantation is to occur. The application of vegetable oil into the mouse uterus which has not been primed for implantation still induces an increase in uterine $PGF_{2\alpha}$ concentration but a decidual cell reaction is not produced[61]. In mice suitably primed for implantation, tranylcypromine blocks the oil-induced decidual cell reaction but does not prevent the increase in uterine PGE_2 and $PGF_{2\alpha}$ concentrations. These studies indicate that increased uterine PGE_2 and $PGF_{2\alpha}$ levels are insufficient by themselves to induce a decidual cell reaction, even in a uterus suitably primed for this response. Since tranylcypromine is an inhibitor of PGI_2 synthesis by pig aorta[64], the inhibition of decidualization by tranylcypromine suggests that PGI_2 is the most important endogenous prostaglandin involved in this process. However, tranylcypromine actually stimulates PGI_2 production by the intact pregnant rat uterus at term[65]. It thus appears that tranylcypromine has other actions which prevent decidualization of the uterus. The effect of tranylcypromine on decidualization induced by exogenous PGE_2, $PFG_{2\alpha}$ and PGI_2 merits further investigation.

Early fetal and placental development

In the study of Kennedy[15], treating rats on day 5 with indomethacin delayed the increase in uterine vasculature permeability, but there was no significant reduction in the number of implantation sites in another group of rats similarly treated with indomethacin and examined on day 8. However, the implantation sites weighed less, indicating that development of the fetus and placenta had been slowed, and the duration of pregnancy was slightly prolonged. In the study of Phillips and Poyser[19], treating rats with indomethacin on days 3 and 4 also delayed the increase in uterine vasculature permeability, and implantation sites were absent or reduced in number in another group of rats similarly treated with indomethacin and examined on day 9 (Table 8.1). The blastocysts which implanted exhibited retarded development. In the indomethacin-treated rats in which there was a lack of implanted blastocysts, it is not known whether the blastocysts implanted late and then were re-absorbed, or whether they failed to implant due to the delayed increase in uterine vasculature permeability. It is clear, however, that when implantation does take place in the rat following indomethacin treatment immediately prior to the expected time of implantation, the subsequent development of the fetus and placenta is adversely affected.

In rabbits immunized against PGE_2 or $PGF_{2\alpha}$, implantation is not affected. However, pregnant rabbits immunized against $PGF_{2\alpha}$ abort about day 22, whereas those immunized against PGE_2 suddenly die around the same time. Postmortem examination of these rabbits demonstrated fetal death having occurred between days 12 and 21 of the pregnancy and that the placentae were small with haemorrhage and haemolysis present. The primary effect of the antibodies was directed towards the placenta[66]. Indomethacin treatment of rabbits on days 6-8 also has deleterious effects in fetal and maternal tissues and on fetal viability when examined on day 14[67]. The uterine venous plasma of pregnant rabbits contains very high concentrations of PGE_2 but not of $PGF_{2\alpha}$ shortly after the establishment of the placenta[68]; the probable source of the PGE_2 is the fetal placenta[69]. It should be understood that the pregnant rabbit is unusual in having such high concentrations of PGE_2 in its uterine venous blood. Nevertheless, these findings in rats and rabbits suggest that prostaglandins are involved in fetal and placental development immediately following implantation.

ROLE OF HISTAMINE

Histamine was implicated in the implantation process before prostaglandins were postulated to have any effect[70]. Sodium cromoglycate (Intal) inhibits histamine release from mast cells by preventing calcium uptake. Sodium cromoglycate instilled into the rabbit uterus on days 5 or 6 of pregnancy prevents implantation, an action that is overcome by the simultaneous action of histamine[71]. The intraluminal administration of DL-α-methylhistidine dihydrochloride, a specific inhibitor of the enzyme histidine decarboxylase which converts histidine to histamine, to rabbits on days 4 or 5 of pregnancy also reduces the number of implantation sites, an effect partly overcome by the simultaneous administration of histidine[72,73]. This histamine synthesis inhibitor also reduced the viability of embryos which did implant[73]. Mepyramine (H_1 histamine receptor antagonist) and cimetidine (H_2 histamine receptor antagonist), given alone or in combination to pregnant rabbits on day 7, inhibit the increase in uterine vasculature permeability[67]. It may appear somewhat surprising that both classes of histamine antagonist should prevent this increase in permeability, since rabbit endometrial cells contain only H_1 receptors and it is the blastocysts which contain H_2 receptors[74]. What is important, however, is the class of histamine receptor present in the uterine capillaries. Studies in inflammation have shown that the capillaries contain H_1 and H_2 receptors and that stimulation of these receptors leads to increased capillary permeability. It is reasonable to postulate that the uterine capillaries contain H_1 and H_2 receptors as well and this postulation would explain why both types of histamine antagonist are effective. These studies also suggest that histamine is involved in the increase in uterine vasculature permeability at the time of implantation. However, in contrast to prostaglandins, histamine produces little decidualization of the rabbit uterus[58]. Histidine decarboxylase activity is not detectable in rabbit endometrium but is detectable in rabbit blastocysts; maximum activity occurs on day 6[74]. Therefore, the blastocyst is the source of histamine for implantation in the rabbit.

Mouse morulae, recovered from the uterus on day 3, develop into blastocysts within 48 h after being placed in tissue culture. This transformation is inhibited by DL-α-methylhistidine and the inhibition is overcome by histidine[75]. Mouse blastocysts recovered from the uterus on day 4 are able to produce histamine[75], indicating that histamine synthesized by the mouse embryo is necessary for its normal development. In pigs, histamine appears to be necessary for the normal migration of the embryo[76].

The administration of an H_1 receptor antagonist and an H_2 receptor antagonist in combination, but not alone, to rats over the implantation period prevents or reduces the increase in uterine vasculature permeability[77]. Histamine augments the increase in uterine vasculature permeability produced by suboptimal doses of oestradiol administered to ovariectomized rats maintained on a progesterone analogue. The effect of the histamine is prevented by a combination of H_1 and H_2 receptor antagonist[78]. Dexamethasone reduces the oestradiol-induced increase in uterine vasculature permeability in progesterone primed rats, an inhibition[79] which is overcome by histamine or PGE_2. From this study, Johnston and Dey[79] suggest that part of the effect of histamine in augmenting implantation induced by oestrogen involves an increase in prostaglandin synthesis. However, glucocorticoids such as dexamethasone inhibit prostaglandin synthesis in intact tissues by causing the release and synthesis of macrocortin which blocks PLA_2 activity[80]. Consequently, it is questionable whether or not histamine stimulates prostaglandin synthesis by the uterus of rats treated with dexamethasone.

In indomethacin-treated pregnant mice, the administration of PGE_2 and/or $PGF_{2\alpha}$ results in implantation occurring in only about 60% of the mice[20]. In contrast, if the indomethacin-treated pregnant mice are also treated with $PGF_{2\alpha}$ and histamine, then the success rate of implantation is 100%[14]. Histamine is unlikely to stimulate prostaglandin synthesis by the uterus following indomethacin treatment. Consequently, the main action of histamine would appear to be independent of any potential to stimulate uterine prostaglandin synthesis. In conclusion, it appears that a combination of the actions of histamine and prostaglandins is normally involved in the increase in uterine vasculature permeability that occurs at the implantation site.

ROLE OF CYCLIC AMP

In pregnant, ovariectomized mice primed with progesterone and oestradiol, cyclic AMP (or its dibutyryl derivative) induces implantation[81,82]. However, by day 14 the development of the fetuses is 1 day behind normal[81] and the fetuses die prior to their expected time of birth[82]. In pregnant rats, cyclic AMP levels are slightly but significantly increased at the sites of increased uterine vasculature permeability[83]. The treatment of ovariectomized rats with progesterone and oestradiol also causes a small increase in uterine cyclic AMP levels. However, the application of any stimulus which induces the decidual cell reaction causes a much greater rise in uterine cyclic AMP levels[84]. Consequently, decidualization of the rat uterus causes an increase in both uterine cyclic AMP and prostaglandin levels. The increase in cyclic AMP is blocked by indomethacin, suggesting that this increase is dependent upon prostaglandin production[25]. Prostaglandins, particularly PGE_2, increase cyclic AMP levels in many tissues. However, cyclic AMP or its dibutyryl derivative does not increase the uterine vasculature permeability in suitably primed rats[85], nor does it cause decidualization of the uterus in suitably primed mice and rabbits[58,86]. Although cyclic AMP is capable of causing implantation to take place, it apparently does not induce the normal changes associated with implantation, and this may be the reason why the fetuses are not carried successfully to term. It is not clear whether an increase in uterine cyclic AMP is normally necessary for implantation.

EFFECTS OF EXOGENOUS PROSTAGLANDINS ON IMPLANTATION

As described earlier, implantation is dependent upon ovarian progesterone. $PGF_{2\alpha}$ administered in suitable doses to the early pregnant rat, mouse, rabbit and hamster causes plasma progesterone levels to fall due to its luteolytic action, and implantation fails. This antifertility effect of $PGF_{2\alpha}$ is completely reversed by suitable progesterone treatment[87]. PGE_2 has a similar effect to $PGF_{2\alpha}$ in rats and hamsters although it is less potent[88]. PGE_2 is not classified as being luteolytic, but the enzyme 9-keto-prostaglandin reductase in the rat ovary converts PGE_2 to $PGF_{2\alpha}$ in vivo[89]. This may explain the apparent luteolytic action of PGE_2 in the rat.

PGE$_2$ or PGF$_{2\alpha}$ instilled into the early pregnant rat uterus not only inhibits the number of implantation sites but also adversely affects the development of the embryos which do implant[90]. Treatment of rats with PGF$_{2\alpha}$ on day 4 also affects the development of the embryo and reduces the decidual cell reaction around the implantation chamber[91]. Whether these are direct detrimental effects of PGE$_2$ and PGF$_{2\alpha}$ or are secondary due to a fall in plasma progesterone levels is not presently known. However, it is clear that while endogenous prostaglandins may be involved in implantation and in small doses exogenous prostaglandins may aid implantation in certain circumstances, larger doses of exogenous prostaglandins have a detrimental effect on the implantation process in those species studied to date.

CLINICAL IMPLICATIONS

Adverse effects of prostaglandin synthesis inhibition

Prostaglandins are probably involved in several processes concerned with the establishment of pregnancy, including luteinizing hormone (LH) release (see Ref. 92), ovulation (see Chapter 7), and implantation (Table 8.3). In several non-primate species all three processes can be inhibited by compounds which block the synthesis of prostaglandin.

Table 8.3 Hormones and local mediators involved in implantation. A question mark indicates that their involvement in humans is not yet certain

Classification	Substance
Hormone	Progesterone
	Oestradiol (?)
Local uterine mediator	Prostaglandins (?)
	Histamine (?)
	Cyclic AMP (?)

In monkeys, oestradiol-induced LH release and gonadotrophin-induced ovulation are both blocked by indomethacin[93,94], indicating that prostaglandins are involved in LH release and ovulation in primates also. It is not known whether the inhibitors of prostaglandin synthesis exert such effects in humans, except for two limited studies. Aspirin taken in large doses during the menstrual cycle failed to prevent the ovulatory surge of LH and ovulation except in one woman who was taking the highest dose[95,96]. It should be noted, however, that aspirin is not one of the more potent inhibitors of the cyclo-oxygenase enzyme. It has not been reported whether women taking high doses of other and more potent cyclo-oxygenase inhibitors for chronic inflammatory conditions are fertile or not. Since continuous NSAID administration failed to inhibit ovulation, it is unlikely that intermittent aspirin ingestion has any untoward effect on ovulation.

Contraception and pregnancy termination

If prostaglandins are involved in LH release, ovulation and/or implantation in humans, it is possible that inhibitors of prostaglandin synthesis could be used to block one or more of these processes and thus prevent pregnancy. It is doubtful whether any of the presently available non-steroidal anti-inflammatory drugs could be used in this manner. Not only would large doses be required to ensure that they were highly effective, but side-effects (particularly on the gastrointestinal tract) would probably be a problem. The future development of more potent inhibitors of prostaglandin synthesis with fewer side-effects, however, may open up a new approach to contraception.

Since a decline in progesterone levels before the mid-secretory stage of the cycle would prevent implantation, it is possible that a $PGF_{2\alpha}$ analogue may be developed which is luteolytic in humans and which could be administered as a 'morning after pill' or even a 'several mornings after pill'. Such treatment would also induce the menstrual flow due to the decline in progesterone levels. This 'premature menstruation' could be looked upon as a disadvantage, although some women may be prepared to tolerate it if the possibility of a pregnancy were avoided. Alternatively, it might be looked upon as an advantage, since women would not have to wait as long as with the current 'morning after pill' to know that a possible pregnancy has been avoided.

One of the disappointing aspects in the clinical application of prostaglandin therapy is the lack of sound, specific inhibitors of prostaglandin activity at the receptor level. The experimental use of potent specific PGE_2 and specific $PGF_{2\alpha}$ antagonists could document whether or not prostaglandins are involved in LH release, ovulation and implantation in humans. If prostaglandins are indeed involved, then prostaglandin antagonists could be clinically useful to prevent pregnancy. If these agents were used solely to prevent LH release and/or ovulation, they would obviously be contraceptive in their action. However, if they are used to prevent implantation, it would still be appropriate to ask if they act as contraceptives or abortifacients. This question would also apply to inhibitors of prostaglandin synthesis and to a luteolytic $PGF_{2\alpha}$ analogue if such were used to prevent implantation. At present, this is a hypothetical question which does not require an answer. However, if such drugs are developed and are shown to be effective and clinically useful in preventing implantation in humans, the question will probably require an answer in the future, at least on the part of individuals who concern themselves with such issues.

References

1 Watson, J., Anderson, F. B., Alam, M., O'Grady, J. E. and Heald, P. J. (1975). Plasma hormones and pituitary luteinizing hormone in the rat during the early stages of pregnancy and after post-coital treatment tamoxifen (ICI 46,474). *J. Endocrinol.*, 65, 7–17

2 Shaikh, A. A. and Abraham, G. E. (1969). Measurement of estrogen surge during pseudopregnancy in rats by radioimmunoassay. *Biol. Reprod.*, 1, 378–80

3 Gerber, G. B., Jacquet, P., Leonard, A. and Maes, J. (1979). Evaluation des taux d'oestradiol, de progesterone et de prostaglandines E et $F_{2\alpha}$ durant le gestation chez la Souris. *C.R. Soc. Biol.*, 173, 644–59

4 Bhatt, B. M. and Bullock, D. W. (1974). Binding of oestradiol to rabbit blastocysts and its possible role in implantation. *J. Reprod. Fertil.*, 39, 65–70

5 Dey, S. K., Dickman, Z. and Sen Gupta, J. (1976). Evidence that the maintenance of early pregnancy in the rabbit requires 'blastocyst estrogen'. *Steroids*, 28, 481–5

6 Paria, B. C., Sen Gupta, J. and Manchanda, S. K. (1984). Role of embryonic oestrogen in rabbit blastocyst development and metabolism. *J. Reprod. Fertil.*, 70, 429–36

7 Dickmann, Z., Dey, S. K. and Sen Gupta, J. (1975). Steroidogenesis in rabbit preimplantation embryos. *Proc. Natl. Acad. Sci. USA*, 72, 298–300

8 Wu, J. T. and Lin, G.-M. (1982) Effect of aromatase inhibitor on oestrogen production in rabbit blastocysts. *J. Reprod. Fertil.*, 66, 655–62

9 Dey, S. K. and Dickmann, Z. (1974). Δ^{5}-$^{3}\beta$-Hydroxysteroid dehydrogenase activity in rat embryos on days 1 through 7 of pregnancy. *Endocrinology*, 95, 321–2

10 Shemesh, M., Milaguir, F., Ayalon, N. and Hansel, W. (1979). Steroidogenesis and prostaglandin synthesis by cultured bovine blastocysts. *J. Reprod. Fertil.*, 56, 181–5

11 Flood, P. F., Betteridge, K. J. and Irvine, D. S. (1979). Oestrogens and androgens in blastocaelic fluid and cultures of cells from equine conceptuses of 10–22 days gestation. *J. Reprod. Fertil. Suppl.*, 27, 413–20

12 Gadsby, J. E., Heap, R. B. and Burton, R. D. (1980). Oestrogen production by blastocyst and early embryonic tissue of various species. *J. Reprod. Fertil.*, 60, 409–17

13 Flood, P. F. and Ghazi, R. (1981). Hydroxysteroid dehydrogenases in the ovine endometrium and trophoblast. *J. Reprod. Fertil.*, 61, 47–52

14 Saksena, S. K., Lau, I. F. and Chang, M. C. (1976). Relationship between oestrogen, prostaglandin $F_{2\alpha}$ and histamine in delayed implantation in the mouse. *Acta Endocrinol.*, 81, 801–7

15 Kennedy, T. G. (1977). Evidence for a role of prostaglandins in the initiation of blastocyst implantation in the rat. *Biol. Reprod.*, **16**, 286–91

16 Evans, C. A. and Kennedy, T. G. (1978). The importance of prostaglandin synthesis for the initiation of blastocyst implantation in the hamster. *J. Reprod. Fertil.*, **54**, 255–61

17 Hoffman, L. H. (1978). Antifertility effects of indomethacin during early pregnancy in the rabbit. *Biol. Reprod.*, **18**, 148–53

18 Hoffman, L. H., Dipietro, D. L. and McKenna, T. J. (1978). Effects of indomethacin on uterine capillary permeability and blastocyst development in rabbits. *Prostaglandins*, **15**, 823–8

19 Phillips, C. A. and Poyser, N. L. (1981). Studies on the involvement of prostaglandins in implantation in the rat. *J. Reprod. Fertil.*, **62**, 73–81

20 Lau, I. F., Saksena, S. K. and Chang, M. C. (1973). Pregnancy blockade by indomethacin, an inhibitor of prostaglandin synthesis: its reversal by prostaglandin and progesterone in mice. *Prostaglandins*, **4**, 795–803

21 Fenwick, L., Jones, R. L., Naylor, B., Poyser, N. L. and Wilson, N. H. (1977). Production of prostaglandins by the pseudopregnant rat uterus, *in vitro*, and the effect of tamoxifen with the identification of 6-keto-prostaglandin $F_{1\alpha}$ as a major product formed. *Br. J. Pharmacol.*, **59**, 191–9

22 Brown, C., Gosden, R. G. and Poyser, N. L. (1984). Effects of age and steroid treatment in the rat uterus in relation to implantation. *J. Reprod. Fertil.*, **70**, 649–56

23 Pakrasi, P. L., Cheng, H. C. and Dey, S. K. (1983). Prostaglandins in the uterus: modulation by steroid hormones. *Prostaglandins*, **26**, 991–1009

24 Cox, C., Cheng, H. C. and Dey, S. K. (1982). Phospholipase A_2 activity in the rat uterus during early pregnancy. *Prostaglandins, Leukotrienes & Medicine*, **8**, 375–81

25 Poyser, N. L. (1985). Prostaglandin output from the early pregnant rat uterus superfused *in vitro*, and the effect of A23187 and trifluoperazine. *J. Reprod. Fertil.*, **74**, 271–7

26 Sharma, S. C. (1980). Temporal changes in PGE, $PGF_{2\alpha}$ oestradiol 17β and progesterone in uterine venous plasma and endometrium during early pregnancy. In Crastes de Paulet, A., Thaler-Dao, H. and Dray, F. (eds.) *Prostaglandins and Reproductive Physiology*. pp. 243–63. (Paris: INSERM)

27 Harper, M. J. K., Norris, C. J. and Rajkumar, K. (1983). Prostaglandin release by zygotes and endometria of pregnant rabbits. *Biol. Reprod.*, **28**, 350–62

28 Poyser, N. L. (1984). Prostaglandin production by the early pregnant guinea-pig uterus in relation to implantation and luteal maintenance, and the effect of oestradiol. *J. Reprod. Fertil.*, **72**, 117–27

29 Downie, J., Poyser, N. L. and Wunderlich, M. (1974). Levels of prostaglandins in human endometrium during the normal menstrual cycle. *J. Physiol. (London)*, **236**, 465–72

30 Dickmann, Z. and Spilman, C. H. (1975). Prostaglandins in rabbit blastocysts. *Science*, **190**, 997

31 Dey, S. K., Chien, S. M., Cox, C. L. and Cristi, R. D. (1980). Prostaglandin synthesis in the rabbit blastocyst. *Prostaglandins*, **19**, 449–53

32 Pakrasi, P. L. and Dey, S. K. (1982). Blastocyst is the source of prostaglandins in the implantation site in the rabbit levels. *Prostaglandins*, 24, 73-7

33 Racowsky, C. and Biggers, J. D. (1983). Are blastocyst prostaglandins produced endogenously? *Biol. Reprod.*, 29, 379-88

34 Marcus, G. J. (1981). Prostaglandin formation by the sheep embryo and endometrium as an indication of maternal recognition of pregnancy. *Biol. Reprod.*, 25, 56-64

35 Hyland, J. H., Manns, J. G. and Humphrey, W. D. (1982). Prostaglandin production by ovine embryos and endometrium *in vivo*. *J. Reprod. Fertil.*, 65, 299-304

36 Lewis, G. S., Thatcher, W. W., Bazer, F. W. and Curl, J. S. (1982). Metabolism of arachidonic acid *in vitro* by bovine blastocysts and endometrium. *Biol. Reprod.*, 27, 431-9

37 Lewis, G. S. and Waterman, R. A. (1983). Metabolism of arachidonic acid *in vitro* by porcine blastocysts and endometrium. *Prostaglandins*, 25, 871-80

38 Lewis, G. S. and Waterman, R. A. (1983). Effects of endometrium on metabolism of arachidonic acid by bovine blastocysts *in vitro*. *Prostaglandins*, 25, 881-9

39 Davis, D. L., Pakrasi, P. L. and Dey, S. K. (1983). Prostaglandins in swine blastocysts. *Biol. Reprod.*, 128, 1114-18

40 Kennedy, T. G. and Armstrong, D. T. (1981). The role of prostaglandins in endometrial vascular changes at implantation. In Glasser, S. R. and Bullick, D. W. (eds.) *Cellular and Modular Aspects of Implantation*. pp. 349-58. (New York: Plenum Press)

41 Kennedy, T. G., Martel, D. and Psychoyos, A. (1983). Endometrial prostaglandin E_2 binding: characterization in rats sensitized for the decidual cell reaction and changes during pseudopregnancy. *Biol. Reprod.*, 29, 556-64

42 Kennedy, T. G., Martel, D. and Psychoyos, A. (1983). Endometrial prostaglandin E_2 binding during the estrous cycle and its hormonal control in ovariectomized rats. *Biol. Reprod.*, 29, 565-71

43 Biggers, J. D., Leonov, B. V., Baskar, J. F. and Fried, J. (1978). Inhibition of hatching of mouse blastocysts *in vitro* by prostaglandin antagonists. *Biol. Reprod.*, 19, 519-33

44 Baskar, J. F., Torchiana, D. F., Biggers, J. D., Corey, E. J., Andersen, N. H. and Subramanian, N. (1981). Inhibition of hatching of mouse blastocysts *in vitro* by various prostaglandin antagonists. *J. Reprod. Fertil.*, 63, 359-63

45 Hurst, P. R. and MacFarlane, D. W. (1981). Further effects of non-steroidal anti-inflammatory compounds on blastocyst hatching *in vitro* and implantation rates in the mouse. *Biol. Reprod.*, 25, 777-84

46 Terranova, P. F. and Dey, S. K. (1982). Indomethacin delays zona-shedding and implantation in the ovariectomized progesterone-treated hamster. *Prostaglandins*, 24, 165-72

47 Kennedy, T. G. and Zamecnik, J. (1978). Concentration of 6-keto-prostaglandin $F_{1\alpha}$ is markedly elevated at site of blastocyst implantation in rat. *Prostaglandins*, 16, 599-605

48 Kennedy, T. G. (1979). Prostaglandins and increased endometrial vascular permeability resulting from the application of an artificial stimulus to the uterus of the rat sensitized for the decidual cell reaction. *Biol. Reprod.*, **120**, 560-6

49 Kennedy, T. G. (1979). Does prostaglandin I_2 (PGI_2) mediate the increased endometrial vascular permeability which results from the application of a deciduogenic stimulus to the sensitised rat uterus? *Biol. Reprod.*, **20**, Suppl. 1, 99A

50 Tobert, J. A. (1976). A study of the possible role of prostaglandins in decidualization using a nonsurgical method for the instillation of fluids into the rat uterine lumen. *J. Reprod. Fertil.*, **47**, 391-3

51 Rankin, J. C., Ledford, B. E., Jonsson, H. T. Jr. and Baggett, B. (1979). Prostaglandins, indomethacin and decidual cell reaction in the mouse uterus. *Biol. Reprod.*, **20**, 399-404

52 Sananes, N., Baulieu, E.-E. and Le Goascogne, C. (1981). A role for prostaglandins in decidualization of the rat uterus. *J. Endocrinol.*, **89**, 25-33

53 Miller, M. M. and Morchoe, C. C. C. (1982). Inhibition of artificially induced decidual cell reaction by indomethacin in the mature oophorectomized rat. *Anat. Rec.*, **204**, 223-30

54 Buxton, L. E. and Murdoch, R. N. (1982). Lectins, calcium ionophore A23187 and peanut oil as deciduogenic agents in the uterus of pseudopregnant mice: effects of tranylcypromine, indomethacin, iproniazid and propranolol. *Aust. J. Biol. Sci.*, **35**, 63-72

55 Peleg, S. and Lindner, H. R. (1982). The effect of prostaglandins on progestin receptor translocation and on decidual cell reaction *in vivo* and *in vitro*. *Endocrinology*, **110**, 1647-52

56 Miller, M. M. and Morchoe, C. C. C. (1982). Decidual cell reaction induced by prostaglandin $F_{2\alpha}$ in the mature oophorectomized rat. *Cell Tissue Res.*, **225**, 189-99

57 Kennedy, T. G. and Lukash, L. A. (1982). Induction of decidualization in rats by the intrauterine infusion of prostaglandins. *Biol. Reprod.*, **27**, 253-60

58 Hoffman, L. H., Strong, G. B., Davenport, G. R. and Frolich, J. C. (1977). Deciduogenic effects of prostaglandins in the pseudopregnant rabbit. *J. Reprod. Fertil.*, **50**, 231-7

59 Anteby, S. O., Bauminger, S., Zor, U. and Lindner, H. R. (1975). Prostaglandin synthesis in decidual tissue of the rat uterus. *Prostaglandins*, **10**, 991-9

60 Jonsson, H. T., Rankin, J. C., Ledford, B. E. and Baggett, B. (1979). Uterine prostaglandin levels following stimulation of the decidual cell reaction: effects of indomethacin and tranylcypromine. *Prostaglandins*, **18**, 847-57

61 Milligan, S. R. and Lytton, F. D. C. (1983). Changes in prostaglandin levels in the sensitized and non-sensitized uterus of the mouse after the intrauterine instillation of oil or saline. *J. Reprod. Fertil.*, **67**, 373-7

62 Peleg, S. (1983). The modulation of decidual cells, proliferation and differentiation by progesterone and prostaglandins. *J. Steroid. Biochem.*, **19**, 283-9

63 Jejeune, B., Van Hoeck, J. and Leroy, F. (1981). Transmitter role of the luminal uterine epithelium in the induction of decidualization in rats. *J. Reprod. Fertil.*, **61**, 235–40

64 Gryglewski, R. J., Bunting, S., Moncada, S., Flower, R. J. and Vane, J. R. (1976). Arterial walls are protected against deposition of platelet thrombi by a substance (prostaglandin-X) which they make from prostaglandin endoperoxides. *Prostaglandins*, **12**, 685–713

65 Phillips, C. A. and Poyser, N. L. (1981). Prostaglandins, thromboxanes and the pregnant rat uterus at term. *Br. J. Pharmacol.*, **73**, 75–80

66 Elzayat, S. and Stylos, W. A. (1974). The effect of circulating prostaglandin antibodies on reproduction in rabbits with special reference to the placenta. *Endocrinology*, **95**, 1642–8

67 Hoos, P. C. and Hoffman, L. H. (1983). Effect of histamine receptor antagonists and indomethacin on implantation in the rabbit. *Biol. Reprod.*, **29**, 833–40

68 Lytton, F. D. C. and Poyser, N. L. (1982). Concentrations of $PGF_{2\alpha}$ and PGE_2 in the uterine venous blood of rabbits during pseudopregnancy and pregnancy. *J. Reprod. Fertil.*, **64**, 421–9

69 Lytton, F. D. C. and Poyser, N. L. (1982). Prostaglandin production by the rabbit uterus and placenta *in vitro*. *J. Reprod. Fertil.*, **66**, 591–9

70 Ferrando, G. and Nalbandov, A. V. (1968). Relative importance of histamine and estrogen on implantation in the rat. *Endocrinology*, **83**, 933–7

71 Dey, S. K., Villaneueva, C., Chien, S. M. and Crist, R. D. (1978). The role of histamine in implantation in the rabbit. *J. Reprod. Fertil.*, **53**, 23–6

72 Dey, S. K., Johnson, D. C. and Santos, J. G. (1979). Is histamine production by the blastocysts required for implantation in the rabbit? *Biol. Reprod.*, **21**, 1169–73

73 Dey, S. K. (1981). Role of histamine in implantation: inhibition of histidine decarboxylase induces delayed implantation in the rabbit. *Biol. Reprod.*, **24**, 867–9

74 Dey, S. K., Villanueva, C. and Abdou, N. I. (1979). Histamine receptors on rabbit blastocyst and endometrial cell membranes. *Nature (London)*, **278**, 648–9

75 Dey, S. K. and Johnson, D. C. (1980). Histamine formation by mouse preimplantation embryos. *J. Reprod. Fertil.*, **60**, 457–60

76 Pope, W. F., Maurer, R. R. and Stormshak, F. (1982). Intrauterine migration of the porcine embryo: influence of estradiol 17β and histamine. *Biol. Reprod.*, **27**, 575–9

77 Brandon, J. M. and Wallis, R. M. (1977). Effect of mepyramine, a histamine H_1-, and burimamide, a histamine H_2-receptor antagonist, on ovum implantation in the rat. *J. Reprod. Fertil.*, **50**, 251–4

78 Brandon, J. M. and Raval, P. J. (1979). Interaction of estrogen and histamine during ovum implantation in the rat. *Eur. J. Pharmacol.*, **57**, 171–7

79 Johnson, D. C. and Dey, S. K. (1980). Role of histamine in implantation: dexamethasone inhibits estradiol induced implantation in the rat. *Biol. Reprod.*, **22**, 1136–41

80 Blackwell, G. J., Carnuccio, R., Di Rosa, M., Flower, R. J., Parente, L.

and Persico, P. (1980). Macrocortin: a polypeptide causing the anti-phospholipase effect of glucocorticoids. *Nature (London)*, **287**, 147-9

81 Webb, F. T. G. (1975). Implantation of ovariectomized mice treated with dibutyryl adenosine 3'5'-monophosphate (dibutyryl cyclic AMP). *J. Reprod. Fertil.*, **42**, 511-17

82 Holmes, P. V. and Bergstrom, S. (1975). Induction of blastocyst implantation in mice by cyclic AMP. *J. Reprod. Fertil.*, **43**, 329-32

83 Vilar-Rojas, C., Castro-Osuna, G. and Hicks, J. J. (1982). Cyclic AMP and cyclic GMP on the implantation site of the rat. *Int. J. Fertil.*, **27**, 56-9

84 Rankin, J. C., Ledford, B. E. and Baggett, B. (1977). Early involvement of cyclic nucleotides in the artificially stimulated decidual cell reaction in the mouse uterus. *Biol. Reprod.*, **17**, 549-54

85 Kennedy, T. G. (1983). Prostaglandin E_2, adenosine 3':5'-cyclic monophosphate and changes in endometrial vascular permeability in rat uteri sensitized for the decidual cell reaction. *Biol. Reprod.*, **29**, 1069-76

86 Leroy, F., Vansande, J., Shetgen, G. and Brasseur, D. (1974). Cyclic AMP and the triggering of the decidual reaction. *J. Reprod. Fertil.*, **39**, 207-11

87 Horton, E. W. and Poyser, N. L. (1976). Uterine luteolytic hormone: A physiological role for prostaglandin $F_{2\alpha}$. *Physiol. Rev.*, **56**, 595-651

88 Labhsetwar, A. P. (1973). A comparative study of some effects of prostaglandin $F_{2\alpha}$, $F_{1\alpha}$, E_2 and E_1 on reproductive processes of rats and hamsters. In Bergstrom, S. (ed.) *Advances in the Biosciences*. pp. 641-4. (Brausehweig: Vieweg & Son, GmbH)

89 Watson, J., Shepherd, T. S. and Dodson, K. S. (1979). Prostaglandin E-2-9-keto-reductase in ovarian tissues. *J. Reprod.*, **57**, 489-96

90 Batta, S. K. and Martini, L. (1975). Anti-implantation effects of prostaglandins in the rat. *Prostaglandins*, **10**, 1075-86

91 Scott, J. E. and Persaud, T. V. N. (1981). Prostaglandin $F_{2\alpha}$ and conceptus-endometrial interaction during early gestation in the mouse. *Prostaglandins & Medicine*, **7**, 133-47

92 Poyser, N. L. (1981). *Prostaglandins in Reproduction*. (Chichester: Research Studies Press, A Division of John Wiley & Sons Ltd)

93 Carlson, J. C., Wong, A. P. and Perrin, D. G. (1977). Luteinizing hormone secretion in the rhesus monkey and a possible role for prostaglandins. *Biol. Reprod.*, **16**, 622-6

94 Wallach, E. E., De La Cruz, A., Hunt, J., Wright, K. H. and Stevens, V. C. (1975). The effect of indomethacin on HMG-HCG induced ovulation in the rhesus monkey. *Prostaglandins*, **9**, 645-58

95 Chaudhuri, G. and Elder, M. G. (1976). Lack of evidence for inhibition of ovulation by aspirin in women. *Prostaglandins*, **11**, 727-35

96 Greenway, F. L. and Swerdloff, R. S. (1978). Effect of aspirin (prostaglandin synthetase inhibitor) on ovulation. *Fertil. Steril.*, **30**, 364-5

9
Male fertility

M. BYGDEMAN

INTRODUCTION

The existence of a substance or a group of substances with smooth muscle stimulating and vasodepressive properties in human semen, prostate and seminal vesicles was first described by von Euler[1,2] and Goldblatt[3] more than 50 years ago. The active principle was named prostaglandin (PG) by von Euler[2]. It was soon realized that semen was the richest natural source of prostaglandins, and seminal extracts were used for many of the early experiments that attempted to ascertain the properties of prostaglandins. Although a long time has elapsed since the discovery of these compounds, the physiological function(s) of PGs in human semen is still far from clear. The aim of this chapter is to describe their possible roles in relation to fertility.

PGs PRESENT IN SEMINAL FLUID

Human seminal fluid contains large amounts of a number of prostaglandins. The compounds identified first[4] were PGE_1, PGE_2, PGE_3, $PGF_{1\alpha}$ and $PGF_{2\alpha}$. More recently 19-hydroxy-PGE_1, 19-hydroxy-PGE_2, 19-hydroxy-$PGF_{1\alpha}$ and 19-hydroxy-$PGF_{2\alpha}$ have also been identified[5-7]. The 8β-isomers of all these compounds are also present.

The presence of PGA and PGB compounds as well as their 19-hydroxylated derivatives has also been described[8,9]. Today, most evidence indicates that these compounds are largely artifactual and formed non-enzymatically, since they are found only after incubation or storage and not in fresh semen if analysed immediately after ejaculation[8,9].

If prostacyclin or thromboxane A_2 indeed are present in human seminal fluid, their concentration is far below that of the previously mentioned compounds[10].

Amounts of prostaglandins comparable to those found in the human male have also been identified in subhuman primates, i.e. the sheep and the goat[11,12]. Semen from other farm animals and laboratory animals also contains prostaglandins, but in considerably smaller amounts. However, the relation between the quantity of the different compounds varies considerably between different species.

Different quantitative methods have been used to measure the prostaglandins present in human seminal fluid. Among these are bioassay which measures the total amount of smooth muscle contractile activity of the semen, radioimmunoassay, measurement of ultraviolet absorption after alkali treatment, gas chromatography and mass spectrometry with selected ion monitoring. All except for the bioassay may be used to determine the different groups of prostaglandins in single semen samples. Nonetheless, prostaglandin measurement is not as yet considered a routine procedure for the evaluation of the infertile couple. What is important in any measurement is to prevent the breakdown of PGEs and 19-hydroxy-PGEs into PGAs and PGBs. Since this degradation may be a rapid, variable and uncontrolled process, it may account, at least in part, for discrepancies observed in previously reported levels of seminal PGs. This process of degradation is not inhibited even by storage of the semen sample at $-20°C$.

A fairly simple procedure for quantification of the PGs which avoids the potential for degradation is to convert the PGE and 19-hydroxy-PGE to the corresponding PGB compounds by alkali treatment and measure their ultraviolet absorption after purification by chromatography. The PGFs and the 19-hydroxy-PGFs can be assayed by gas chromatography which also allows measurement of the 8β-isomers. This method has the advantage of measuring sperm characteristics and semen prostaglandins in a single semen sample after liquefaction[13].

In comparison with other body fluids the total prostaglandin concentration in semen obtained from fertile men may be as high as 1·0 mg per ejaculate. The concentration of the different prostaglandins is approximately 60–75 μg/ml for the PGEs, 2–3 μg/ml for the PGFs, 270–350 μg/ml for the 19-hydroxy-PGEs and 12–20 μg/ml for the 19-hydroxy-PGFs[13,16] (Table 9.1). According to Bygdeman and Samuelsson[17] the mean distribution of the PGEs is 25 μg/ml PGE_1, 23 μg/ml PGE_2 and 5·5 μg/ml PGE_3[17]. As noted in Table 9.1, however, the PG concentration varies considerably between different individuals. Over time the variation in the same individual is much less marked, provided that sperm characteristics remain unchanged[18].

Table 9.1 The concentration of prostaglandins in human seminal fluid from fertile men. Mean values (μg/ml) and range (if not otherwise indicated)

Reference	PGE	PGF	19-hydroxy-PGE	19-hydroxy-PGF
Svanborg et al.[13]	62 (15–44)	2·8 (1·0–5·3)	326 (155–638)	14·9 (7·0–20·6)
Templeton et al.[14]	73 (2–272)	2·1 (0·1–7·0)	267 (53–1094)	18·3 (3·0–62·0)
Tussel and Gelpi[15]	64 (9–164)	2·6 (1·0–6·6)	593 (142–1047)	12·7 (4·0–19·0)
Gerozissis and Dray[15]	21·2	3·6	350 (± 139)*	22·0 (± 9)*

* Standard error of mean

ORIGIN AND REGULATION OF PROSTAGLANDINS IN SEMEN

Although minor amounts (ng/g tissue) of PGE and PGF have been found in the testis and prostate[19,20], the seminal vesicles clearly are the main source of PGs in human semen. The seminal vesicles convert radioactive [^{14}C]eicosatetraenoic acid to PGE_1 in significant amounts[21]. Using a split ejaculate technique, the highest prostaglandin concentration is found in the same fraction as is fructose[22,23]. It also appears that the concentration of the different prostaglandins is essentially unchanged before and after vasectomy[23,24]. In one patient, a silicon prosthesis was introduced into the straight part of the ductus deferens, and secretory products from the testis, epididymis, and the ejaculate, representing mainly the secretion from the seminal vesicles and the prostate gland, could be collected separately. The secretions from the testis and epididymis contained no detectable amounts of prostaglandins whereas the concentration remained within normal limits in the ejaculate from the same patient[24].

Factors regulating the concentration of prostaglandins in human seminal plasma are poorly understood. Testosterone seems to stimulate PG production. Sturde[25] observed a significant increase in the prostaglandin concentration measured by bioassay after androgen treatment for 6–12 weeks. In hypogonadal men with a low seminal PG concentration, testosterone replacement therapy gives a similar result. The effect on 19-hydroxy-PGE concentration is more marked than that on PGE concentration[26].

Previous studies have shown that aspirin treatment results in a reduction of PGE and PGF concentration[27]. If a strong PG biosynthesis inhibitor such as naproxen is used, the content of all prostaglandins is reduced to between 12 and 32% of the pretreatment concentration within a week. Sperm characteristics remain unchanged, however[28]. As will be described later, an optimal PG concentration may be a prerequisite for normal sperm function. It is not known whether treatment with anti-inflammatory compounds affects fertility. However, therapy with indomethacin for 3 months has been reported to increase sperm count and sperm motility in men with oligozoospermia[29].

PROSTAGLANDINS AND MALE FERTILITY

The high concentration of prostaglandins in the seminal fluid of men, as compared with most tissues and body fluids, has led to the logical assumption that these compounds are intimately involved in reproductive processes. In some reports[30,31], a relationship between seminal prostaglandin levels and the number of children has been found. Data indicate that men whose wives had more than one pregnancy every other year had significantly higher levels of seminal prostaglandins than men in infertile marriages.

These early results based on bioassay[30,31] have later been supported by data obtained from quantitative chemical determinations. Human males who are infertile for no apparent reason possessed significantly lower concentrations of seminal prostaglandins, especially PGEs, than men of normal fertility[32-34]. Bygdeman and co-workers[33] showed that seminal fluid from fertile men contained about 55 µg/ml of PGE, an amount that was significantly higher than the corresponding value, 18 µg/ml, found in men of infertile marriages in whom no other abnormalities could be detected. Men in infertile marriages also have, in general, a lower concentration of 19-hydroxy-PGE and a higher concentration of 19-hydroxy-PGF in their semen than do fertile men[13] (Table 9.2). Evidence for a correlation between fertility and seminal prostaglandin concentration also exists in sheep. Following artificial insemination, the fertility of the rams increased by more than 15% if

Table 9.2 Prostaglandin concentration in human seminal fluid from fertile and infertile men[13]

Group of men	Mean value (µg/ml) and range			
	PGE	PGF	19-hydroxy-PGE	19-hydroxy-PGF
Infertile	43	2·6	260	17·0
	(7–117)	(1·4–4·8)	(113–427)	(9·4–34·1)
Fertile	62	2·8	326	14·9
	(15–144)	(1·0–5·3)	(155–638)	(7·0–20·6)

PGE_2 and $PGF_{2\alpha}$ were added to the diluted ram semen in amounts which restored the total amounts of PGs to normal[35].

The precise mechanism by which seminal prostaglandins influence fertility is not clear. In fact, the correlation need not be one of cause and effect, since male infertility may be due to other factors which also produce low seminal prostaglandin levels. If the high prostaglandin concentration is of importance for normal fertility, at least three modes of action are possible. Prostaglandins may act on the sperm itself, on the male reproductive tract at or shortly before ejaculation, or on the female reproductive tract after ejaculation.

Relation between PG concentration, seminal fluid and sperm characteristics

A significant correlation exists between the density of sperm and the PG concentration[18,36]. In men with polyzoospermia ($> 250 \times 10^6$/ml) the mean content of PGE is low, $17\,\mu$g/ml; in contrast, if the sperm density is normal ($20–250 \times 10^6$/ml), the mean PGE concentration is $69\,\mu$g/ml. The corresponding values for 19-hydroxy-PGE are 153 and $262\,\mu$g/ml, respectively[18]. In some patients oligozoospermia is associated with a high PGE and 19-hydroxy-PGE concentration[18]. It is possible that the low sperm density and reduced fertility is compensated by this high PG concentration.

The PG concentration is also related to sperm motility. The concentration of 19-hydroxy-PGE is higher and that of 19-hydroxy-PGF lower in ejaculates with normal sperm motility compared to those with an abnormal motility[18] (Table 9.3). The PGE concentration in this study was unrelated to sperm motility in contrast to the finding of Consentino et al.[37] Addition of physiological amounts of 19-hydroxy-PGF (but not PGE or $PGF_{2\alpha}$) to semen samples with normal sperm motility results in a decreased sperm activity and a reduction of the capacity of the sperm to penetrate cervical mucus in vitro (Gottlieb et al., personal communication). If large amounts of $PGF_{2\alpha}$ ($250\,\mu$g) are added, this compound also reduces sperm motility[38]. These data suggest that one important function of seminal prostaglandins is to stimulate the kinetic activity and motility of the sperm at the time of ejaculation. The PGs may also act by altering Ca^{2+} flux and thereby control intracellular cyclic AMP levels and sperm metabolism. Optimal seminal PG concentration may therefore be of importance for the functional requisites of the sperm.

The PGE and PGF concentration in human seminal fluid also varied with other different seminal compounds which are regarded to be of importance for a normal fertility. A high PGF concentration coincides with a high calcium concentration and a high percentage of tapered

Table 9.3   Relation between prostaglandin concentration and sperm motility[18]

No. of patients	Sperm motility	PGE (μg/ml)	19-hydroxy-PGE (μg/ml)	19-hydroxy-PGF (μg/ml)	19-hydroxy-PGE / 19-hydroxy-PGF
15	Abnormal*	71±37	228±81	19·7±7	12±3
13	Normal*	70±35	289±76	10·9±3	27±6
		NS	$p<0.05$	$p<0.01$	$p<0.001$

* According to WHO definition: progressive motility in at least 60% of the sperms

sperm and inversely with a low zinc concentration. A high PGE concentration is found in semen samples with a high calcium concentration and a high sperm motility score. Plasma testosterone concentration is also inversely related to seminal PGE levels[37].

PGs and ejaculation

Von Euler[2] first suggested that prostaglandins might be involved in the contractile events of ejaculation, since the epididymis, vas deferens and seminal vesicles contain smooth muscle, and PGs have a potent action on such tissue. This suggestion has been reinforced more recently by the finding that prostaglandins can influence neurotransmission in the sex accessory tissues of the male. Hedquist and von Euler[39] measured the contractile response of the vas deferens of several laboratory animals to noradrenaline and to nerve stimulation in the absence or presence of varying concentrations of PGE_1, PGE_2 and $PGF_{2\alpha}$. PGE_1 and PGE_2 potentiated noradrenaline-induced smooth muscle contractile responses of the isolated guinea-pig vas deferens and seminal vesicles. However, the contractile response of the vas deferens to postganglionic nerve stimulation was inhibited by low doses of PGE_1 and potentiated by high doses. In animal experiments PGE_2 may also cause relaxation of the epididymis[39]. Some of our preliminary data support the suggestion of von Euler: for example, single dose oral administration of PGE_2 to male volunteers results in a shortlived but significant decrease in sperm density. Such an effect may be explained by a relaxation of epididymis and vas deferens such as occurs in laboratory animals.

PGs and passive sperm transport

It is possible that changes in uterine contractility during or shortly after intercourse may facilitate sperm transport. During sexual stimulation there is a marked increase in uterine activity; following female orgasm inhibition of contraction occurs. A pressure gradient between the vagina and the uterus may favour passive sperm transport. Although it is tempting to postulate that changes in uterine and vaginal contractility might be influenced by seminal prostaglandins, experimental evidence to support this assumption is still lacking in the human. Evidence that PGE_1, PGE_2 and $PGF_{2\alpha}$ enhance sperm migration and fertilization has been provided in experiments in rabbit, however[40]. As described elsewhere in more detail, both PGEs and PGFs generally have a stimulatory effect on human uterine contractility following intravenous, intrauterine and vaginal administration (see Chapter 5). For both compounds the threshold dose following intrauterine admin-

Table 9.4 Prostaglandins in human seminal fluid. Origin, regulation and physiological function

	Origin	Regulation		Related to				
		Testosterone	PG biosynthesis inhibitors	Fertility	Sperm density	Sperm motility	Ejaculation process	Passive sperm transport
PGE	Seminal vesicles	↑	↓	Yes	Yes	?	(Yes)	(Yes)
PGF	,,	↑	↓	Yes	No	?	?	(Yes)
19-hydroxy-PGE	,,	↑	↓	Yes	No	Yes	?	(Yes)
19-hydroxy-PGF	,,	↑	↓	Yes	No	Yes	?	(Yes)

↑ Stimulation   ↓ Inhibition   ? Unclear at present   (Yes) Possible but not proved

istration is at the microgram level[41] and this may be the explanation for the increased uterine contractility and pain sometimes observed following intrauterine insemination.

The amount of PGE and PGF compounds present in single ejaculates (approximately 150–250 μg) is less than the vaginal dose required to induce an apparent uterine response. However, an effect on uterine contractility is still possible, since during intercourse uterine reactivity to prostaglandins may be increased because of different hormonal and nervous stimuli associated with sexual stimulation. Moreover, the vaginal absorption may be enhanced by local changes in the vaginal epithelium. In addition, the importance of the 19-hydroxy compounds that are present in almost milligram amounts in the ejaculate is unknown.

SUMMARY AND CONCLUSIONS

Human seminal fluid contains a number of prostaglandins: PGE_1, PGE_2, PGE_3, $PGF_{1\alpha}$, $PGF_{2\alpha}$, 19-hydroxy-PGE_1, 19-hydroxy-PGE_2, 19-hydroxy-$PGF_{1\alpha}$ and 19-hydroxy-$PGF_{2\alpha}$ and their 8β-isomers in large amounts (see Table 9.4). The total PG content in a single ejaculate is approximately 1·0 mg. The concentration of specific compounds varies considerably between individuals, whereas the variation in concentration in the same individual with time is less marked, provided that sperm characteristics remain constant. All the PGs most likely originate from the seminal vesicles. The contributions from the testis or the epididymis are minimal, if any. Most likely, testosterone stimulates PG production, while treatment with anti-inflammatory drugs significantly reduces semen PG concentration. The concentrations of at least the PGEs, 19-hydroxy-PGEs and 19-hydroxy-PGFs appear to be related to fertility. A correlation between PG concentration and sperm density, sperm motility, certain other compounds in seminal fluid and sperm penetration capacity in cervical mucus *in vitro* can be demonstrated.

It is likely that the functional requisites of sperm depend upon optimal seminal prostaglandin concentration. It is also possible that the PGEs are of importance in the ejaculatory process and for the transport of sperm within the female genital tract. At present, measurements of seminal prostaglandins are not part of the routine evaluation of the infertile couple. The situation may change, however, in the future, when the precise role of PG in male fertility becomes more clear and methods to influence seminal PG concentration may be available.

ACKNOWLEDGEMENTS

The studies performed by our group and referred to in this chapter were supported by the Swedish Medical Research Council (Project No. B84-17X-05696-05).

References

1 von Euler, U.S. (1935). The specific blood pressure lowering substance in human prostate and seminal vesicles secretions. *Klin. Wochenschr.*, **14**, 1182–3

2 von Euler, U.S. (1936). On the specific vasodilating and plain muscle stimulating substances from accessory genital glands in man and certain animals (prostaglandin and vesiglandin). *J. Physiol.*, **881**, 213–34

3 Goldblatt, M.W. (1933). A depressor substance in seminal fluid. *J. Soc. Chem. Ind.*, **52**, 1056–7

4 Samuelsson, B. (1963). Isolation and identification of prostaglandins from human seminal fluid. *J. Biol. Chem.*, **238**, 3229–34

5 Taylor, P.L. (1979). The 8-iso prostaglandins: evidence for eight compounds in human semen. *Prostaglandins*, **17**, 259–67

6 Taylor, P.L. and Kelly, R.W. (1974). 19-hydroxylated E prostaglandins as the major prostaglandins in human semen. *Nature (London)*, **250**, 665–7

7 Taylor, P.L. and Kelly, R.W. (1975). The occurrence of 19-hydroxy-F prostaglandins in human semen. *FEBS Lett.*, **57**, 22–5

8 Cooper, J. and Kelly, R.W. (1975). The measurement of E and 19-hydroxy E prostaglandins in human seminal fluid. *Prostaglandins*, **10**, 507–14

9 Middleditch, B.S. (1975). PGA: fact or artifact? *Prostaglandins*, **9**, 409–11

10 Gerozissis, K. and Dray, F. (1981). Radioimmunoassay of prostaglandins in the semen of fertile men. *J. Reprod. Fertil.*, **61**, 487–90

11 Bygdeman, M. and Holmberg, O. (1966). Isolation and identification of prostaglandins from ram seminal plasma. *Acta Chem. Scand.*, **20**, 2308–10

12 Kelly, R.W., Taylor, P.L., Hearn, J.P., Short, R.V., Martin, D.E. and Marston, J.H. (1976). 19-hydroxy prostaglandin E as a major component of the semen of primates. *Nature (London)*, **260**, 544–5

13 Svanborg, K., Bygdeman, M., Eneroth, P. and Bendvold, E. (1982). Quantification of prostaglandins in human seminal fluid. *Prostaglandins*, **24**, 363–75

14 Templeton, A.A., Cooper, J. and Kelly, R.W. (1978). Prostaglandin concentrations in the semen of fertile men. *J. Reprod. Fertil.*, **52**, 147–50

15 Tussel, J.M. and Gelpi, E. (1980). Prostaglandins E and F, and 19-hydroxylated E and F (series I and II) in semen of fertile men. Gas and liquid chromatographic separation with selected ion detection. *J. Chromatogr.*, **181**, 295–310

16 Gerozissis, K. and Dray, F. (1981). Radioimmunoassay of prostaglandins in the semen of fertile men. *J. Reprod. Fertil.*, **61**, 487–90

17 Bygdeman, M. and Samuelsson, B. (1966). Analyses of prostaglandins in

human semen. Prostaglandins and related factors. *Clin. Chem. Acta*, **13**, 465–74

18 Bendvold, E., Svanborg, K., Eneroth, P., Gottlieb, C. and Bygdeman, M. (1984). The natural variations in prostaglandin concentration in human seminal fluid and its relation to sperm quality. *Fertil. Steril.*, **41**, 743–7

19 Carpenter, M. P., Robinson, R. D. and Thuy, L. P. (1978). Prostaglandin metabolism in human semen. *FEBS Lett.*, **57**, 22–5

20 Conte, D., Laguzzi, G., Boniforti, L., Cantafora, A., DiSilverio, F., Latino, C., Lalloni, G., Mesolella, V. and Isidori, A. (1980). Prostaglandin content and metabolic activity of the human prostate. In Crastes de Paulet, A., Thaler-Dao, H. and Dray, F. (eds.) *Prostaglandin and Reproductive Physiology.* pp. 89–96. (Paris: INSERM)

21 Hamberg, M. (1976). Biosynthesis of prostaglandin E by human seminal vesicles. *Lipids*, **11**, 249–50

22 Eliasson, R. (1959). Studies on prostaglandins. *Acta Physiol. Scand.*, **Suppl. 158**, 1–73

23 Gerozissis, K., Jouannet, P., Soufir, J. C. and Dray, F. (1982). Origin of prostaglandins in human semen. *J. Reprod. Fertil.*, **65**, 401–6

24 Bendvold, E., Svanborg, K., Bygdeman, M. and Norén, S. (1985). On the origin of prostaglandins in human seminal fluid. *Int. J. Androl.*, **8**, 37–43

25 Sturde, H. C. (1971). Behaviour of sperm prostaglandins under therapy with androgen. *Arzneim. Forsch.*, **21**, 1302–7

26 Skakkebaeck, N. E., Kelly, R. W. and Cocker, C. S. (1976) Prostaglandin concentrations in the semen of hypogonadal men during treatment with testosterone. *J. Reprod. Fertil.*, **47**, 119–21

27 Horton, E. W., Jones, R. L. and Marr, C. G. (1973). Effects of aspirin on prostaglandin and fructose levels in human semen. *J. Reprod. Fertil.*, **33**, 385–93

28 Bendvold, E., Gottlieb, C., Svanborg, K., Bygdeman, M., Eneroth, P. and Cai, Q. H. (1985). The effect of naproxen on the concentration of prostaglandin in human seminal fluid. *Fertil. Steril.*, **43**, 922–6

29 Barkay, J., Harpaz-Lerpel, S., Ben-Ezra, S., Gordon, S. and Zuckerman, H. (1984). The prostaglandin inhibitor effect of antiinflammatory drugs in the therapy of male infertility. *Fertil. Steril.*, **42**, 406–11

30 Asplund, J. A. (1947). A quantitative determination of the content of contractive substances in human sperm and their significance for the motility and vitality of the spermatozoa. *Acta Physiol. Scand.*, **13**, 103–8

31 Hawkins, O. F. and Labrum, A. H. (1981). Semen prostaglandin levels in 50 patients attending a fertility clinic. *Acta Physiol. Scand.*, **13**, 1–10

32 Brummer, H. C. and Gillespie, A. (1972). Seminal prostaglandins and fertility. *Clin. Endocrinol.*, **1**, 363–8

33 Bygdeman, M., Fredericsson, B., Svanborg, K. and Samuelsson, B. (1970). The relation between fertility and prostaglandin content of seminal fluid in man. *Fertil. Steril.*, **26**, 622–9

34 Collier, J. C., Flower, R. L. and Stanton, S. L. (1975). Seminal prostaglandins in infertile men. *Fertil. Steril.*, **26**, 868–71

35 Dimov, V. and Georgiev, G. (1977). Ram semen prostaglandin concentration and its effect on fertility. *J. Anim. Sci.*, **44**, 1050–4

36 Kelly, R. W., Cooper, I. and Templeton, A. A. (1979). Reduced prostaglandin levels in the semen of men with very high sperm concentrations. *J. Reprod. Fertil.*, **56**, 195-9

37 Cosentino, M. J., Emilson, L. B. V. and Cockett, A. T. K. (1984). Prostaglandins in semen and their relation to male fertility: a study of 145 men. *Fertil. Steril.*, **41**, 88-94

38 Cohen, M. S., Colin, M. J., Golimbu, M. and Hotchkiss, R. S. (1977). The effects of prostaglandins on sperm motility. *Fertil. Steril.*, **28**, 78-85

39 Hedquist, P. and von Euler, U. A. (1972). Prostaglandin induced neurotransmission failure in the field stimulated vas deferens. *Neuropharmacology*, **11**, 177-87

40 Spilman, H. C., Finn, A. E. and Norland, J. F. (1973). Effect of prostaglandins on sperm transport and fertilization in the rabbit. *Prostaglandins*, **4**, 57-64

41 Martin, J. N., Jr., Bygdeman, M. and Eneroth, P. (1978). The influence of locally administered prostaglandin E_2 and $F_{2\alpha}$ on uterine motility in the intact nonpregnant human uterus. *Acta Obstet. Gynecol. Scand.*, **57**, 141-7

SECTION III
CLINICAL APPLICATIONS

10
Cervical ripening

A. A. CALDER

INTRODUCTION

Looked at in its simplest form, childbirth consists of the process whereby the uterus drives the fetus through the birth canal to the outside world. Hitherto, the uterus has had to remain quiescent, so that the fetus could grow to the requisite point for extrauterine survival. With adaptations for this new role, nature has arranged that the fetus should now travel from the comparative safety of the uterus to the hostile world outside. This journey through the birth canal has been described as the most hazardous journey any of us ever makes and, in truth, it may result in serious injury and even death. However, the journey is not the same for all passengers; some face the prospect of a long, arduous and dangerous journey, while for others it seems to be accomplished with ease. The prime object of obstetric care is to make the journey as safe as possible for all concerned.

Numerous factors contribute to the ultimate success or failure of human labour. Problems may arise from abnormalities in the size and attitude of the fetal head or from the capacity and shape of the mother's bony pelvis. The fetus may begin its journey in a poor state which will inevitably worsen along the way. A sudden calamity, such as prolapse of the umbilical cord or placental abruption may bring immediate peril to the passenger and, in some instances, its mother. Fetal malpresentation or malposition may lead to serious intrapartum complications.

In the absence of complicating factors, however, labour consists simply of the opening of the birth canal as the fetus is passively driven through it by the myometrial contraction and retraction.

Whilst the uterus generates its powerful contractions, the birth canal must of necessity yield and allow the fetus to pass; but just as the uterus has changed its role from quiescence to activity, the birth canal also must experience a *volte-face*. The success of the pregnancy thus far has depended to a large extent on the ability of the fetus to remain in the uterus. The natural ability of the myometrium to contract strongly has been suppressed. Almost as important as this myometrial quiescence, however, is the ability of the uterine cervix to remain closed, barring the fetus from gaining exit from the uterus. Failure of the cervix to carry out its appointed role may lead to protrusion and premature rupture of the fetal membranes, spontaneous abortion or preterm delivery. In some instances, the underlying pathophysiology of these events is described as cervical incompetence.

The capacity of the cervix to remain closed depends to a great degree on its rigidity, a property which derives from the collagen fibres which make up the bulk of the cervical stroma. Just as the myometrium must undergo a reversal of function with the onset of labour, so must the cervix. If, at the start of labour, the cervix was as rigid as it were in the non-pregnant or early pregnant state, then successful labour would be impossible. A radical modification of the physical state of the cervix is an essential prerequisite to successful labour, and this modification is the process called cervical ripening.

The factors which control cervical ripening have been reviewed by Ulmsten in Chapter 4 of this volume and will not be restated here. Suffice it to say, however, that the evidence that prostaglandins play a central role in this physiological process is now beyond dispute.

CERVICAL RIPENING AND PARTURITION

Applied physiology

If we ignore for the moment that in a minority of instances mechanical complications such as feto-pelvic disproportion have an important bearing on the course of human labour and delivery, we then can consider the physiology of parturition in terms of the process whereby the expulsive forces of the myometrium overcome the resistance of soft tissues of the birth canal, particularly those of the cervix. Crucial to an understanding of this process is the recognition that, unlike the rest of the uterus, the cervix is not primarily a muscular structure but rather one composed predominantly of connective tissue[1]. The overwhelming component of this connective tissue is collagen, although there is some elastic tissue. Muscle fibres represent no more than a sixth of the formed element component[2,3].

The process of cervical ripening consists of a fundamental change

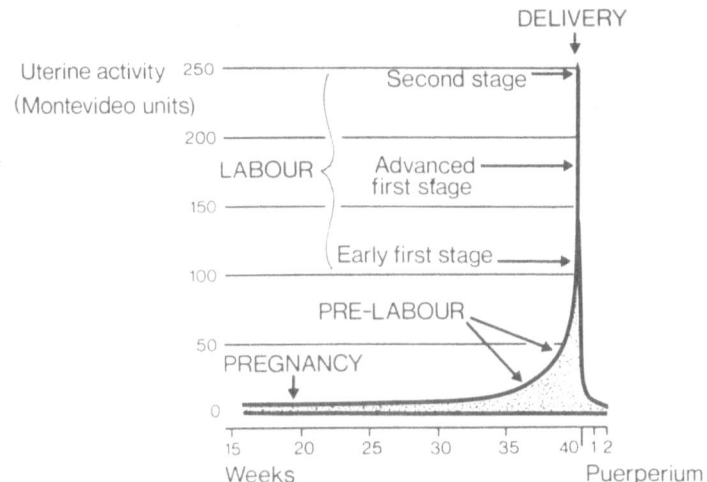

Figure 10.1 Pattern of myometrial contractility through the course of pregnancy, prelabour, labour and delivery (after Caldeyro-Barcia[5])

in the arrangement of the collagen fibres as a result of which the tensile strength of the tissue is dramatically reduced. This change is imperative for easy effacement and dilatation. The factors which bring about these changes and the accompanying modification of the ground substance have already been reviewed by Ulmsten (Chapter 4).

Parturition requires that the process of cervical ripening be highly developed as a prelude to efficient and rapid dilatation during labour; ripening is thus seen as an event of late pregnancy taking place before the onset of labour[4]. In normal cases 'maturation' of the cervix may be said to accompany maturation of the fetus.

It may seem articificial to separate cervical ripening from cervical dilatation when they are really parts of the same continuum. However, this is no different from the mental separation we make between the uterine contractions of labour and those which immediately precede its onset. As Caldeyro-Barcia has demonstrated[5], uterine contractility builds up gradually in late pregnancy (during 'prelabour') until the point at which labour becomes established (Figure 10.1). This is not unlike an aircraft becoming airborne following its preflight phase of developing increasing velocity along the runway. At the point of take-off, the aircraft must have enough speed and enough wind resistance to leave the ground. In the case of labour, the cervix must have ripened sufficiently so that the generation of powerful uterine contractions can bring about the physical changes necessary to allow the fetus to leave the uterus.

A successful outcome of pregnancy requires that these important

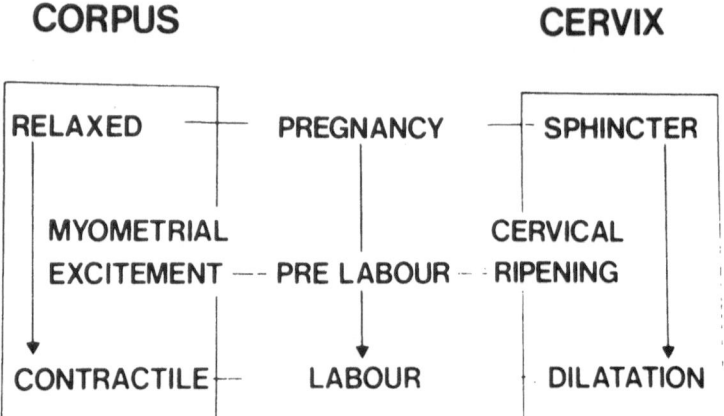

Figure 10.2 Diagrammatic representation of the synchronous relationship between the corpus uteri and the cervix uteri which is necessary for normal transition from pregnancy to labour

physiological events – cervical ripening, cervical dilatation and myometrial contractility – occur in concert with one another and with appropriate timing. A proper synchrony between myometrial and cervical function during pregnancy, prelabour and labour is essential if the normal course is to be followed (Figure 10.2). Complications arise when departures occur from the normal pattern of timing and synchrony.

Clinical problems

Reference has already been made to the contribution of cervical incompetence to spontaneous midtrimester abortion or preterm delivery. At the other side of the spectrum is the patient in whom failure of cervical ripening is associated with prolongation of pregnancy and prolonged and/or incoordinate labour. The most obvious manifestation of lack of appropriate synchrony between the uterus and the cervix is seen when uterine contractility is stimulated in an effort to induce labour at a time when the cervix is still unripe. The clinical circumstances in which this arises include patients at term in whom cervical ripening has failed to take place and patients in whom delivery is necessary prior to term, i.e. at a stage of pregnancy when cervical ripening would not yet have been expected. The maternal and fetal morbidity associated with induction of labour by amniotomy and intravenous oxytocin infusion, especially in primigravid patients[6], is inversely proportional to the degree of cervical ripening present at the time of induction (Table 10.1). Induction by this method, when the cervix is unripe,

Table 10.1 Influence of cervical ripeness on morbidity in induced labour

Cervical score	Number	Length of induced labour (hours; mean ± SD)	Caesarean section	Pyrexia in labour	Apgar score < 5 at 1 minute
0–2 (unripe)	16	16·1 ± 5·1	8 (50%)	8 (50%)	6 (37·5%)
3–5 (intermediate)	48	11·7 ± 4·2	5 (10·4%)	4 (8·3%)	5 (10·4%)
6–11 (ripe)	61	7·4 ± 2·9	0 (0%)	0 (0%)	0 (10%)
0–11 (all)	125	10·2 ± 5·0	13 (10·4%)	12 (9·6%)	11 (8·8%)

leads to prolonged, unproductive labour and an unacceptably high rate of maternal distress, pyrexia, fetal acidosis and Caesarean section.

Measurement of cervical ripeness

Clearly an objective measure of cervical ripeness is an essential pre-requisite to any investigation of this subject, and the introduction by Bishop in 1964 of his scoring system has been of enormous value in this respect. The Bishop system[7] takes account of the five factors which may reflect cervical ripeness, i.e. dilatation, effacement, consistency and position, as well as the level of the presenting part of the fetus relative to the bony pelvis of the mother. The slight modification[8] of Bishop's original system which we have used since 1974 is illustrated in Figure 10.3.

The use of such a scoring system allows comparable groups of patients to be studied in the kind of randomized controlled trials which are essential to investigations of cervical ripening. The objectivity which this brings to comparison of the groups of patients treated in different ways has greatly improved the understanding of the influence of the cervix in the total obstetric performance. Nevertheless, individual clinical interpretations of such scoring systems involve a considerable degree of observer variability which complicates our attempt to directly compare different studies carried out under different conditions and by different investigators.

The single most important additional variable in this regard is parity. O'Driscoll et al.[9] have suggested that when it comes to performance in labour nulliparous and multiparous patients are so different as to almost belong to different species. Nowhere is this difference more obviously seen than in matters relating to the cervix. The cervix which

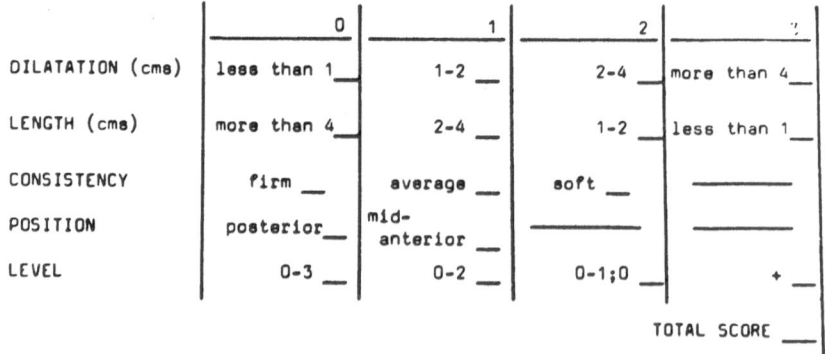

Figure 10.3 Cervical score (Calder et al.[8]) modified from the 'pelvic score' of Bishop[7]

has not previously undergone the softening, effacement and dilatation necessary for parturition behaves quite differently from the cervix which has undergone these activities. Unfortunately, in many of the studies which have purported to investigate matters relating to cervical ripening, the authors have failed to recognize this and have studied groups of patients of mixed parity. As a result, the value of their studies has been greatly diminished.

An objective assessment of cervical ripeness and recording of a score is an essential part of the management of any patient in whom intervention to achieve delivery is deemed necessary.

Clinical importance of cervical ripening

The induction of labour by inappropriate methods when the cervix is unripe generally produces poor results. There is no merit in such interventions, if the resultant morbidity is comparable to or greater than the morbidity which the intervention was designed to prevent.

The decision to intervene in a given pregnancy must take account of the mental application of the so-called 'obstetric balance'[10] (Figure 10.4). In the great majority of instances, this balance will remain tipped against intervention throughout the pregnancy, and the spontaneous onset of labour at term will preclude the need for intervention of this sort. In a significant minority of patients, however, delivery becomes necessary when assessment of the wellbeing of the fetus, the efficiency of the placenta, the health of the mother or the presence of other obstetric risk factors appear to result in the balance changing toward delivery, i.e. when the evidence suggests that the fetus *in utero* should become the neonate in the nursery (or the neonatal intensive care unit).

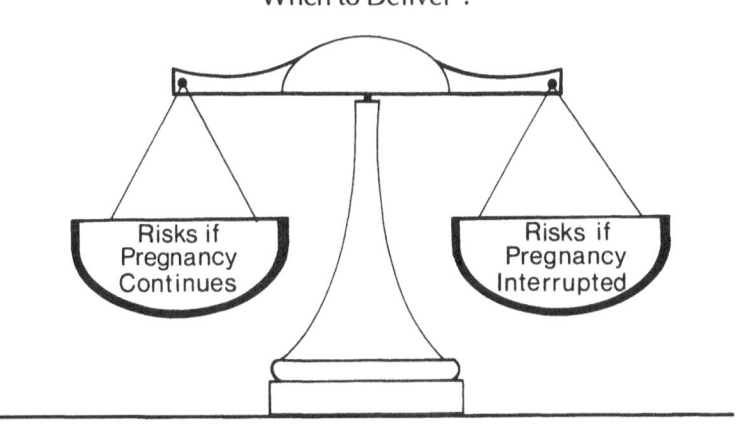

Figure 10.4   The obstetric balance (Calder[10])

The weight of the factors on either side of the balance may vary considerably and, in the most difficult cases, both sides carry grave risks. In other words, the neonate may face precarious survival and the need for highly skilled intensive care, but this may appear preferable to the hostile environment of the uterus.

In assessing the obstetric balance, however, it often is necessary to consider the additional influence of induced labour, and this assessment depends primarily on the degree of cervical ripeness. Pelvic examination to determine the cervical score gives a good prediction of the type of labour which can be expected with induction. If the cervix is very ripe, a short easy labour which brings little additional stress to the fetus may be anticipated. However, if the cervix is very unripe the converse applies. In the latter instance the clinician has three options. Firstly, he may defer the decision to deliver the patient in the hope that spontaneous improvement in cervical ripening may simplify management. This is at best a temporary expedient. The kind of obstetric factors giving rise to the concern for expedient delivery rarely recede and usually worsen.

The second option is to choose delivery by Caesarean section in an attempt to minimize stressful influences on the fetus. Often this is the appropriate choice, but it introduces additional risk factors for the mother, both in the short term and in subsequent pregnancies. The third choice is to try to achieve cervical ripening as a prelude to induction of labour. Historically, a variety of different methods has been used to achieve this objective, but in recent years it has become clear that the prostaglandins possess pre-eminent advantages when used for this purpose.

Traditional methods of cervical ripening

Before prostaglandins were available, a variety of techniques, mostly mechanical in nature, were used to try to ripen the cervix. The introduction of bougies or laminaria[11] and the use of balloons, such as those on the Foley catheter[12], were popular, as were digital stretching of the cervix and stripping of the membranes[13]. It is interesting to postulate that any influence that these methods had on the process of cervical ripening probably resulted from the release of endogenous prostaglandins as a result of the mechanical stimulus[14].

Pharmacological procedures

A variety of pharmacological agents have been used to try to achieve cervical ripening. Prolonged intravenous infusion of oxytocin has long been popular for this purpose, but there is little objective evidence that this tedious practice is of any real benefit[15]. A variety of other substances, including oestrogens[16-18], relaxin[19,20], corticosteroids[21] and various enzymes[22,23] have also been administered but the results of these therapies have been inconsistent, often unconvincing and, in no instance, superior to those achieved using prostaglandins.

PROSTAGLANDINS FOR CERVICAL RIPENING

The prostaglandins have had a substantial impact in their role as agents for pre-induction cervical ripening. Numerous studies have been conducted and reported using various prostaglandins in different vehicles administered by diverse routes, over different time schedules and achieving very different results.

Which prostaglandin?

The only two prostaglandins studied extensively for cervical ripening have been prostaglandin $F_{2\alpha}$ and prostaglandin E_2. To date, there have been no reports of the use of synthetic prostaglandin analogues for this purpose and, for reasons of toxicology and because of problems inherent to product licensing, it may be some time before any such studies are conducted. The question of efficacy of the different compounds is therefore simple – which of the two classic prostaglandins (PGE_2 or $PGF_{2\alpha}$) is the more effective for cervical ripening? This question can be considered from two different standpoints: on the one hand, the involvement of both these substances in the physiological control of labour and cervical ripening and, on the other, the evidence concerning clinical efficacy.

It is clear that both prostaglandin E_2 and prostaglandin $F_{2\alpha}$ are intimately involved in the control of human parturition[24]. Both substances appear in the amniotic fluid in increasing concentrations as labour progresses[25], and the metabolites of both appear in the peripheral circulation in a similar manner[26]. Confusion has arisen, however, as to the role of prostaglandin E_2 in controlling myometrial activity, and it has often been claimed that the main influence of this compound is to relax the myometrium[27]. This claim is hardly borne out by clinical experience[28,29], as this agent is widely and successfully used for the induction of labour. It may be that the biological conversion of prostaglandin E_2 to prostaglandin $F_{2\alpha}$ is necessary for myometrial stimulation and that such conversion occurs under the influence of the enzyme, 9-ketoreductase[30]. It is now established[31] that cervical tissue synthesizes prostaglandin E_2 and that such production is increased at the time of parturition. The cervical production of prosta-glandin $F_{2\alpha}$, on the other hand, seems less abundant.

Two studies have suggested that prostaglandin E_2 is the dominant influence in early labour, whereas prostaglandin $F_{2\alpha}$ dominates as labour progresses[32,33]. If this is so, it then seems quite probable that prostaglandin E_2 is more concerned with cervical change and prosta-glandin $F_{2\alpha}$ with uterine contractility. While this approach may be a gross simplification, it is a useful working hypothesis, and it seems to correlate well with the clinical influence of these agents.

Although numerous studies of prostaglandins for cervical ripening have been reported, very few of them have made a direct comparison[34] of prostaglandin E_2 and prostaglandin $F_{2\alpha}$. Direct comparisons are complicated by the difficulty of comparing equivalent doses of these two agents. Prostaglandin E_2 generally is considered to be more potent by a factor of 5–10 than is prostaglandin $F_{2\alpha}$ when used for induction of abortion or induction of labour. However, this difference in potency may vary at different stages of pregnancy and depend on the relative importance of softening the cervix and stimulating the myometrium in a given clinical situation. Given these complexities, a fair comparison of the two agents is virtually impossible, and clinicians must rely on rather crude clinical evaluations. The consensus is, taking account of factors such as side-effects in addition to clinical efficacy, that prosta-glandin E_2 generally is preferred to prostaglandin $F_{2\alpha}$.

Routes of administration

Since the prostaglandins began to have clinical applications, at least ten different routes of administration have been tried. Major systemic routes include oral, intramuscular and intravenous. The fundamental problem surrounding the clinical application of prostaglandins derives from the fact that these substances behave very differently from other biological substances, notably the hormones. Hormones such as in-sulin, thyroxine and the corticosteroids normally are given systemi-cally to obtain their diverse systemic effects. Others, notably the pitui-tary hormones including oxytocin and the gonadotrophins, are given systemically for their effect on one particular target organ.

In contrast, the prostaglandins are synthesized and released at or close to their sites of action. A single agent may have a myriad of biological effects, depending on its target. Moreover, the prostaglan-dins are rapidly inactivated upon reaching the systemic circulation. As a result, if prostaglandins are administered systemically with the object of producing a response in a target organ such as the uterus or the cervix, two major obstacles must be considered. First, a large dose is required to overcome the problem of metabolic breakdown. Second, such large doses given systemically will inevitably produce unwanted effects on other tissues and organ systems, and these side-effects may be sufficient to nullify the benefit of therapy.

For these reasons local routes of administration have been favoured and, in this regard, the female genital tract is unusually accessible to such therapy. Just as the intra-amniotic and extra-amniotic routes were found to be well suited for various techniques of therapeutic abortion, so the routes of administration surrounding the cervix have proven best where the object has been cervical ripening. For this purpose, prostaglandins have been administered above the cervix (extra-amniotically), below the cervix (vaginally), within the cervical tissues (intracervically) or within the cervical canal (endocervically). Techniques have also been developed using devices such as cervical caps to hold pharmacologically active agents against the ectocervix.

From a theoretical point of view, it is clear that the more directly prostaglandins can be delivered to the target organ, in this instance the cervix, the purer the response is likely to be. Clinical experience supports this premise and has shown that where there is a 'spill-over', side-effects may be encountered. This is particularly true in one regard. Because the agents used to date have a potent stimulating effect on the myometrium, it is not surprising that almost invariably cervical ripening with prostaglandins is associated with a greater or lesser degree of uterine contractility. Since the ultimate object of treatment is to achieve a successful induction of labour, this may not be a serious objection. However, clearly the aim should be to achieve a correct balance between the modification of the cervical resistance, on the one hand, and stimulating uterine contractility on the other. Too much myometrial stimulation, together with too little increase in cervical compliance, represents a failure of the technique. Apart from the inferior results that this produces for the patients, clinicians are unlikely to consider a method popular if their patients have their labour induced at inconvenient and unpredictable times. Moreover, unwanted uterine contractility may constitute a significant hazard if the fetus is already compromised and likely to sustain hypoxia during the cervical ripening phase.

As local routes of administration were developed, it was necessary to experiment with a variety of different vehicles in which to administer the prostaglandins. Numerous pessaries and gels have been devised in order to allow delivery of the active agent to the appropriate site and allow a suitable pattern of release and absorption. Individual routes of administration will now be considered together along with the different vehicle formulations and dosage schedules which have been found to be effective.

Oral route

Although a number of early studies[35-38] suggested that prostaglandin E$_2$ tablets given orally on a variety of schedules of dosage and duration of therapy were of clinical use, the benefit appeared to be at the best marginal and this route has now been supplanted by the more effective local routes.

Intravenous route

Prolonged intravenous infusion of PGE$_2$ may produce useful cervical ripening, but side-effects are likely to be troublesome[39]. Thiery[40] has, however, reported good results with premature rupture of the membranes combined with an unfavourable cervix.

Intramuscular route

This route has not been extensively investigated for cervical ripening. There is nothing to suggest that it would have any value unless prostaglandin analogues with a highly selective cervical ripening effect are developed. For the moment, there seems no likelihood of this happening.

Extra-amniotic route

The first local route to be studied with specific reference to the problem of the unripe cervix was the extra-amniotic route[41]. The initial technique described by Calder and Embrey involved passage of a transcervical catheter in a similar fashion to that which had been previously described for continuous extra-amniotic infusion of prostaglandin E$_2$ in order to induce abortion in the second trimester of pregnancy[42]. In the initial studies a dilute solution of PGE$_2$ in normal saline was infused. While this yielded good results, the technique was cumbersome, required infusion equipment and restricted the patient's freedom of movement. Consequently, again recalling techniques of therapeutic abortion[43], the viscous gel Tylose was used as a vehicle for the delivery of a bolus extra-amniotic dose of prostaglandin E$_2$ for cervical ripening[44]. A Foley catheter was passed into the extra-amniotic space where it was retained above the cervix by inflating the balloon with 20 ml of water. 5 ml of 5% Tylose (methylhydroxy-ethylcellulose) containing 400–500 mg PGE$_2$ was then instilled. The technique, undoubtedly enhanced by the presence of the Foley catheter, was highly effective. A variable degree of uterine contractility was observed in addition to the expected cervical ripening, and over the course of 4–6 h the cervical status improved in almost every case. Unfortunately, labour became established in a minority of patients. This generally occurred among those patients with higher cervical scores, although this was not invariably the case. The Foley catheter was generally extruded within 6 h,

Table 10.2 Results of cervical ripening with PGE_2 in viscous gel prior to induction of labour by amniotomy and intravenous oxytocin infusion

	Pretreatment with PGE_2 in viscous gel	Control patients (no pretreatment)
No. of patients	106	31
Induction–delivery interval (h; mean $\pm$ SD)	$11 \cdot 1 \pm 4 \cdot 8$	$14 \cdot 9 \pm 5 \cdot 5$
Caesarean section rate	9%	32%
Maternal pyrexia rate ($> 38\,°C$)	4%	32%
Birth asphyxia rate (at 1 min, Apgar < 4)	6%	23%

after which the majority of patients failed to progress into labour and thus required formal labour induction the following day after a night's sleep. The results obtained with this method[45] are outlined in Table 10.2. The length of labour and the incidence of maternal and fetal complications were dramatically reduced among primigravid subjects with cervical scores of 3 or less compared to those encountered among untreated controls. This technique has been subject to much more extensive clinical trials, including randomized controlled studies notably by Thiery[46-48] and others[49,50], and our initial favourable results have generally been confirmed.

While the extra-amniotic route is undoubtedly effective (one might even suggest it is the most effective route), there are a number of practical and theoretical objections to its use. Many clinicians find it unattractive because of the theoretical risk of introducing infection into the intra-uterine cavity. In practice this has not been a problem. A second objection lies in the fear of inducing uterine hyperstimulation. This has been seen on rare occasions; in our experience its occurrence invariably coincided with bleeding in the chorio-decidual space following insertion of the catheter. In reality, this phenomenon should preclude the administration of prostaglandins. Perhaps the most important objection, however, lies in the fact that patients may find the insertion of the catheter disagreeable and thereby find other local routes of administration more acceptable. Nevertheless, we hold to the view that this technique is to be preferred for the most unripe and difficult induction prospects[51].

Vaginal route
Following the success of the extra-amniotic route, McKenzie and Embrey[52] considered the vaginal administration of PGE_2 in viscous gel. Although requiring a substantially larger dose of PGE_2 (i.e. 2–5 mg),

results were seemingly comparable to those previously obtained via the extra-amniotic route. Because much of the absorption from vaginal administration is systemic, some uterine stimulation is inevitable with this route as is a greater potential for side-effects, although in practice these have not been unduly troublesome. Undoubtedly this route is well suited to the intermediately unripe cervix, although we have found[51] that cases with very low cervical scores may be better served by extra-amniotic therapy.

A variety of different vehicles have been employed for vaginal therapy including wax based and lactic acid pessaries[53,54] and sophisticated hydrogels for sustained release[55]. The main advantage of these vehicles lies in their ease of administration and increased scope for repeated administration.

Endocervical and pericervical routes

In 1978 Ulmsten and associates[56] found that the use of a cross-linked starch polymer produced a gel which had the capacity to remain within the endocervical canal. The results reported by this group[56,57] and other workers[58-60] are apparently comparable with those achieved with extra-amniotic or vaginal therapy, and the dose required (500 μg PGE_2 or less) is small. The technique therefore has the attraction of combining the low dose capacity of the extra-amniotic route with a slightly less invasive technique.

Comparison of techniques

In the past 7 or 8 years a bewildering variety of reports have appeared describing series of cases where cervical ripening has been sought using the techniques described above. Unfortunately, almost every study used a different protocol regarding indications for induction, gestational age, cervical score criteria, parity and method of induction following ripening. Assessment of the relative efficacy of each technique has therefore been extremely difficult. Clearly, if a particular study included more patients with relatively higher cervical scores than did another study, the results of the former are likely to be very much better than those of the latter. In addition, many authors failed to recognize the important influence of parity on the outcome. The problem of a high morbidity in association with labour induction when the cervix is unripe is very largely a problem seen among primigravidae. An unripe cervix in a multigravid subject is of much less serious import[61]. Over the years different academic departments have tended to pursue those techniques which they found successful in their own hands.

Although a small number of randomized comparisons of different techniques have been published, these also have been fraught with difficulties. For instance, it is almost impossible to design a protocol in which two techniques can be fairly compared, as the biggest obstacle is the matter of dosage. Obviously, the vaginal route requires a higher dosage than does the extra-amniotic route or the endocervical route, but the question now becomes how much larger a dose? Studies which have tried to compare the relative efficacy of PGE_2 and $PGF_{2\alpha}$ have been plagued by similar constraints.

Our own experience has led us to conclude that vaginal prostaglandin E_2 therapy using a dosage of 3 mg in either a viscous gel or a vaginal tablet is simple, acceptable and effective for the majority of cases requiring cervical ripening. In a minority of cases, however, it

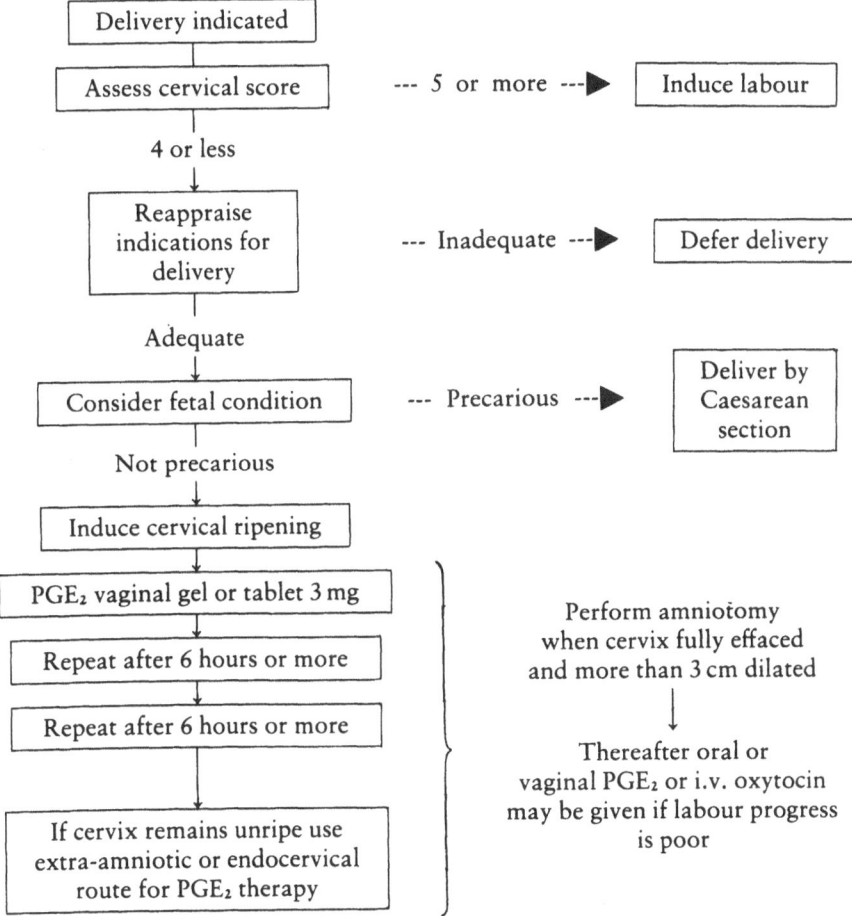

Figure 10.5   Flow diagram indicating a clinical approach to cervical ripening

is ineffective, and we believe this generally occurs among patients with the lowest cervical scores and reflects the most difficult ripening challenge. Such subjects may be identified on the basis of their failure to respond to two or three doses of 3 mg PGE_2 administered vaginally at 6 hourly intervals. In those circumstances, nothing has been lost other than time, and recourse may then be made to extra-amniotic or endocervical therapy. A flow diagram of a recommended step by step procedure is shown in Figure 10.5.

Precautions

The biggest single danger in the use of prostaglandins for cervical ripening lies in the potential for uterine hyperstimulation. With this in mind, the lowest effective dose should always be chosen. Opinions vary concerning the need for continuous fetal heart rate monitoring during cervical ripening therapy, although clearly when the indication for induction derives from fetal or placental compromise, this activity must be regarded as mandatory. At least two fetal deaths have been reported following cervical ripening with $PGF_{2\alpha}$ therapy[62].

It need hardly be stated that if cervical ripening is required as part of induction of labour one can hardly expect perfect results in every case. The strength of the indications for intervention must therefore be weighed against the likely morbidity resulting from intervention. Cervical ripening can certainly reduce such morbidity, but it cannot be recommended in circumstances where the indications for delivery are trivial.

A PHILOSOPHY FOR CERVICAL RIPENING AND INDUCTION OF LABOUR

The prospects for induction of labour are difficult in expectant mothers, especially primigravidae, in whom the cervix is found to be unripe in late pregnancy. In contrast, induction of labour is easy and almost invariably successful if the cervix is very ripe. If the cervix is very ripe, spontaneous labour should be regarded as imminent. Conversely, when the cervix is very unripe, spontaneous labour remains a distant prospect. In short, the closer the mother is to the spontaneous onset of labour, the easier and more successful will be an intervention to induce labour, almost regardless of the technique chosen.

The prospective mother with an unripe cervix requires the most careful consideration, as the unripe cervix probably reflects a general lack of 'labour readiness'. If it is felt necessary to deliver such a patient, then the time taken using prostaglandins to ripen the cervix will pay dividends. In most instances, it is possible to bring about an improvement in the cervical status so that when labour becomes established its course may more closely approximate the normal pattern. Because local applications of prostaglandins have proved effective, safe and acceptable for induction of labour when the cervix is ripe[63,64] we feel that the correct approach to induction of labour when the cervix is unripe is to administer local prostaglandins and to observe the cervical response. Therapy should be repeated if necessary until the cervix is soft, effaced and 3-4 cm dilated. When that point has been reached, amniotomy may be the only further requirement for successful induction.

References

1 Danforth, D. N. (1947). The fibrous nature of the human cervix, and its relation to the isthmic segment in gravid and nongravid uteri. *Am. J. Obstet. Gynecol.*, 53, 541-60

2 Danforth, D. N. (1954). The distribution and functional significance of the cervical musculature. *Am. J. Obstet. Gynecol.*, 65, 1261-70

3 Danforth, D. N. (1980). Early studies of the anatomy and physiology of the human cervix – and implications for the future. In Naftolin, F. and Stubblefield, P. G. (eds.) *Dilatation of the Uterine Cervix*. pp. 3-15. (New York: Raven Press)

4 Anderson A. B. M. and Turnbull, A. C. (1969). Relationship between length of gestation and cervical dilatation, uterine contractility and other factors during pregnancy. *Am. J. Obstet. Gynecol.*, 105, 1207

5 Caldeyro-Barcia, R. (1958). Uterine contractility in obstetrics. In *Proceedings of the Second International Congress of Gynecology and Obstetrics*, Montreal. Vol. 1, p. 65.

6 Calder, A. A. (1981). The human cervix in pregnancy and labour: a clinical perspective. In Ellwood, D. A. and Anderson, A. B. M. (eds.) *The Cervix in Pregnancy and Labour*. pp. 103-22. (Edinburgh: Churchill Livingstone)

7 Bishop. E. H. (1964). Pelvic scoring for elective induction. *Obstet. Gynecol.*, 24, 266.

8 Calder, A. A., Embrey, M. P. and Hillier, K. (1974). Extra-amniotic prostaglandin E₂ for the induction of labour at term. *J. Obstet. Gynaecol. Br. Commonw.*, 81, 39

9 O'Driscoll, K., Jackson, R. J. A. and Gallagher, J. T. (1970). Active management of labour and cephalo-pelvic disproportion. *J. Obstet. Gynaecol. Br. Commonw.*, 77, 385-9

10 Calder, A. A. (1983). Methods of induction of labour. In Studd, J. (ed.) *Progress in Obstetrics and Gynaecology*. pp. 86-100. (Edinburgh: Churchill Livingstone)

11 Stubblefield, P. G. (1980). Present techniques for cervical dilatation. In Naftolin, F. and Stubblefield, P. G. (eds.) *Dilatation of the Uterine Cervix*. pp. 335-42. (New York: Raven Press)

12 Embrey, M. P. and Mollison, B. G. (1967). The unfavourable cervix and induction of labour using a cervical balloon. *J. Obstet. Gynaecol. Br. Commonw.*, 74, 44

13 Myerscough, P. R. (1982). *Munro Kerr's Operative Obstetrics*. p. 271. (London: Baillière Tindall)

14 Mitchell, M. D., Flint, A. P. F., Bibby, J., Brunt, J., Arnold, J. M., Anderson, A. B. M. and Turnbull, A. C. (1977). Rapid increases in plasma prostaglandin concentrations after vaginal examination and amniotomy. *Br. Med. J.*, 2, 1183

15 Lilienthal, C. and Ward, J. (1971). Medical induction of labour. *J. Obstet. Gynaecol. Br. Commonw.*, 78, 317

16 Gordon, A. and Calder, A. A. (1977). Estradiol applied locally to ripen the unfavourable cervix. *Lancet*, 1319

17 Tromans, P., Beazley, J. and Shenouda, P. (1981). Comparative study of oestradiol and PGE_2 vaginal gel for ripening the unfavourable cervix before induction of labour. *Br. Med. J.*, 282, 678

18 Thiery, M., De Gezelle, H., Van Kets, H., Voorhoof, L., Verheugen, C. *et al.* (1979). The effect of locally administered estrogens on the human cervix. *Geburtsh. Perinat.*, 183, 448

19 Maclennan, A., Green, R., Bryant-Greenwood, G., Greenwood, F. and Seamark, R. (1980). Ripening of the human cervix and induction of labour with purified porcine relaxin. *Lancet*, 1, 220

20 Maclennam, A., Green, R., Bryant-Greenwood, G., Greenwood, F. and Seamark, R. (1981). Cervical ripening with combinations of vaginal $PGF_{2\alpha}$, estradiol and relaxin. *Obstet. Gynecol.*, 58, 601

21 Katz, Z., Lancet, M. and Levavi, E. (1979). The efficacy of intraamniotic steroids for induction of labour. *Obstet. Gynecol.*, 54, 31

22 Droegemueller, W., Chvapil, M., Vining, J., Whitaker, L. and Christian, C. (1978). Urea and dilation of the cervix. *Am. J. Obstet. Gynecol.*, 132, 775

23 Green, P. (1968). Intracervical injection of hyaluronidase. *Am. J. Obstet. Gynecol.*, 99, 337

24 Keirse, M. J. N. C. (1979). Endogenous prostaglandins in human parturition. In Keirse, M. J. N. C., Anderson, A. B. M. and Bennebroek Gravenhorst, J. (eds.) *Human Parturition*. pp. 101–42. (Leiden University Press)

25 Hillier, K., Calder, A. A. and Embrey, M. P. (1974). Concentrations of prostaglandin $F_{2\alpha}$ in amniotic fluid and plasma in spontaneous and induced labours. *J. Obstet. Gynaecol. Br. Commonw.*, 81, 257–63

26 Green, K., Bygdeman, M., Toppozada, M. and Wiqvist, N. (1974). The role of prostaglandin $F_{2\alpha}$ in human parturition. Endogenous plasma levels of 15-keto-13,14-dihydro-prostaglandin $F_{2\alpha}$ during labour. *Am. J. Obstet. Gynecol.*, 120, 25–31

27 Huszar, G. (1981) Biology and biochemistry of myometrial contractility and cervical maturation. *Sem. Perinatol.*, 5, 216–35

28 Elder, M. (1980). Induction of labour with prostaglandins. In Sakamoto, K. (ed.) *Gynecology and Obstetrics*. pp. 296–301 (Amsterdam: Excepta Medica)

29 Embrey, M. (1979). Induction of labour. *Lancet*, 2, 793

30 Mitchell, M. D. (1984). Role of prostaglandins in parturition. *Prostagl. Perspect.*, **1**, 1–4

31 Elwood, D., Mitchell, M., Anderson, A. and Turnbull, A. (1979). Prostaglandin production by the cervix. Observations *in vitro* and *in vivo*. *Br. J. Obstet. Gynaecol.*, **86**, 826

32 Keirse, M. J. N. C. and Turnbull, A. C. (1973). E prostaglandins in amniotic fluid during late pregnancy and labour. *J. Obstet. Gynaecol. Br. Commonw.*, **80**, 970–3

33 Keirse, M. J. N. C., Flint, A. P. F. and Turnbull, A. C. (1974). F prostaglandins in amniotic fluid during pregnancy and labour. *J. Obstet. Gynaecol. Br. Commonw.*, **81**, 131–5

34 Mackenzie, I. and Embrey, M. (1978). A comparison of PGE_2 and $PGF_{2\alpha}$ vaginal gel upon subsequent labour. *Br. J. Obstet. Gynaecol.*, **85**, 657

35 Friedman, E. and Sachtleben, M. (1975). Preinduction priming with oral PGE_2. *Am. J. Obstet. Gynecol.*, **121**, 521

36 Golbus, M. and Creasy, R. (1977). Uterine priming with oral PGE_2 prior to elective induction with oxytocin. *Prostaglandins*, **14**, 577

37 Lauersen, N., Mackenzie, I. and Embrey, M. (1980). Preinduction cervical priming with oral PGE_2. *Am. J. Obstet. Gynecol.*, **135**, 1057

38 Pearce, D. (1977). Preinduction priming of the uterine cervix with oral PGE_2 and a placebo. *Prostaglandins*, **14**, 571

39 Thiery, M. and Amy, J (1977). Spontaneous and induced labor: two roles for the PGs. In Wynn, R. M. (ed.) *Obstetrics and Gynaecology Annual*. Vol. 6, p. 127. (Connecticut: Appleton-Century-Crofts)

40 Thiery, M. (1983). Preinduction cervical ripening. In Wynnn, R. M. (ed.) *Obstetrics and Gynaecology Annual*. pp. 103–46. (Connecticut: Appleton-Century-Crofts)

41 Calder, A. A. and Embrey, M. (1973). Prostaglandins and the unfavourable cervix. *Lancet*, **2**, 1322

42 Miller, A., Calder, A. A. and Macnaughton, M. C. (1972). Termination of pregnancy by continuous intrauterine infusion of prostaglandins. *Lancet*, **2**, 5

43 Lippert, T. and Modly, T. (1973). Induction of abortion by the extraamniotic administration of prostaglandin gels. *J. Obstet. Gynaecol. Br. Commonw.*, **80**, 1025

44 Calder, A. A., Hillier, K. and Embrey, M. (1976). Prostaglandin therapy for cervical ripening prior to induction of labor. In Samuelsson, B. and Paoletti, R. (eds.) *Advances in Prostaglandin and Thromboxane Research*. p. 993. (New York: Raven Press)

45 Calder, A. A. (1980). Pharmacological management of the unripe cervix in the human. In Naftolin, F. and Stubblefield, P. G. (eds.) *Dilatation of the Uterine Cervix*. pp. 317–33. (New York: Raven Press)

46 Thiery, M., Defoort, P., Benijis, G., Derom, R., Mariens, G. *et al.* (1978). Fetal effects of cervical ripening with extraamniotic prostaglandin E_2 in gel. *Prostaglandins*, **15**, 175

47 Thiery, M., Defoort, P., Benijis, G., van Eyck, J., Hennay, T. *et al.* (1977). Effectiveness of extraovular injection of prostaglandin E_2 in tylose gel to ripen the cervix prior to elective induction of labour at term. *Prostaglandins*, **14**, 381

48 Thiery, M., Parewijck, W., De Gezelle, H., Van Kets, H., Smis, B. *et al.* (1978). Preinduction ripening of the cervix with extraamniotic prostaglandin E₂ in gel. *Z. Geburtsh. Perinat.*, **182**, 352

49 Wilson, P. (1978). A comparison of four methods of ripening the unfavourable cervix. *Br. J. Obstet. Gynaecol.*, **85**, 941

50 Hutchon, D., Geirsson, R. and Patel N. (1980). A double-blind controlled trial of prostaglandin E₂ gel in cervical ripening. *Int. J. Gynaecol. Obstet.*, **17**, 604

51 Stewart, P., Kennedy, J.H., Hillan, E. and Calder, A.A. (1983). The unripe cervix: management with vaginal or extra-amniotic prostaglandin E₂. *J. Obstet. Gynecol.*, **4**, 90–3

52 Mackenzie, I. and Embrey, M. (1977). Cervical ripening with intravaginal prostaglandin E₂ gel. *Br. Med. J.*, **2**, 1381

53 Pearce, J., Shepherd, J. and Sims, C. (1979). Prostaglandin E₂ pessaries for induction of labour. *Lancet*, **1**, 572

54 Kennedy, J.H., Gordon-Wright, A.P., Stewart, P., Calder, A.A. and Elder, M.G. (1982). Induction of labour with a stable based prostaglandin E₂ vaginal tablet. *Eur. J. Obstet. Gynecol. Reprod. Biol.*, **14**, 203

55 Embrey, M., Graham, N. and Mcneill, M. (1980). Induction of labour with a sustained-release prostaglandin E₂ vaginal pessary. *Br. Med. J.*, **281**, 901

56 Ulmsten, U. (1979). A new gel for intracervical application of PGE₂. *Lancet*, **1**, 377

57 Ulmsten, U., Kirstein-Pedersen, A., Stenberg, P. and Wingerup, L. (1979). A new gel for intracervical application of prostaglandin E₂. *Acta Obstet. Gynecol. Scand. Suppl.*, **84**, 19

58 Heinzl, S., Ramzin, P., Schneider, M. and Luescher, K. (1980). Priming der zervix mit PG-gel bei unreifer geburtssituation am termin. *A. Geburtsh. Perinat.*, **184**, 395

59 Steiner, H., Zahradnik, H., Breckwoldt, H., Robrecht, D. and Hillemans, H. (1979). Cervical ripening prior to induction of labour. Intracervical application of a prostaglandin E₂ viscous gel. *Prostaglandins*, **17**, 125

60 Goeschen, K. (1981). Induktion der zervixreife mit PGE₂-gel bei risikoschwangerschaften. *M.D. Thesis*, Berlin

61 Gordon, A. and Calder, A.A. (1983). Cervical ripening. *Br. J. Hosp. Med.*, **30**, 52–60

62 Quinn, M. and Murphy, A. (1981). Fetal death following extraamniotic prostaglandin F₂α gel. Report of two cases. *Brit. J. Obstet. Gynaecol.*, **88**, 650

63 Mellows, H.J., Sims, C.D. and Craft, I. (1977). Prostaglandin E₂ in tylose for induction of labour in patients with a favourable cervix. *Proc. R. Soc. Med.*, **70**, 537–8

64 Kennedy, J.H., Stewart, P., Barlow, D.H., Hillan, E. and Calder, A.A. (1982). Induction of labour: a comparison of a single prostaglandin E₂ vaginal tablet with amniotomy and intravenous oxytocin. *Br. J. Obstet. Gynaecol.*, **89**, 704–7

11
Induction of labour

A. P. LANGE

INTRODUCTION

Together with Caesarean section, induction of labour is the obstetricians' method of terminating pregnancy at or near term. These two procedures represent the most effective tools for the treatment or prevention of complications towards the end of pregnancy.

This chapter will evaluate the place of prostaglandins as agents for the induction of labour. Whilst prostaglandins are now well-established as effective drugs for inducing abortion during the second trimester, they are not equally well regarded for induction of labour. Much depends on the obstetrician's attitude towards induction of labour. It is not the author's intent to suggest that a specific agent should be used in a given clinical situation, but rather to present the reader with a comprehensive survey that should assist in arriving at an appropriate clinical course of action.

FACTORS INFLUENCING THE COURSE OF INDUCED LABOUR

Uterine sensitivity

As pregnancy proceeds, the uterine muscles become more and more irritable, and the pregnant woman feels more and more uterine contractions during the last months before birth. At the same time, the sensitivity of the uterine muscles to oxytocin is increased[1]. An oxytocin sensitivity test for the quantitative recording of the uterine muscle's irritability prior to induction of labour has been suggested[2], but this test has not been employed to any great extent, probably due to the fact that the sensitivity of the uterine musculature to oxytocin can also be related to the degree of ripeness of the cervix[1,3].

Cervical ripeness

During the last months and weeks prior to labour, the cervix undergoes biochemical as well as structural changes (see Chapter 4). These changes can be detected clinically[4,5], and the degree of cervical ripeness at the time when active labour starts is of decisive importance with regard to its course and duration[6,7]. Various scores[7-12] have been proposed for the quantitative assessment of cervical ripeness. The best known of these is Bishop's pelvic score[9]; this system uses a scale from 0 to 13 (see Table 11.1a) to assess the station of the presenting fetal part and the cervical length, dilatation, consistency and position in the birth

Table 11.1a   Bishop's pelvic score

Features	Score			
	0	1	2	3
Station in relation to spines (cm)	−3	−2	−1–0	+1–+2
Cervix dilatation (cm)	0	1–2	3–4	4
length (cm)	3	2	1	0
consistency	firm	medium	soft	
position	posterior	mid	anterior	

Table 11.1b   The author's modification of Bishop's pelvic score (reprinted with permission from the American College of Obstetricians and Gynecologists (*Obstet. Gynecol.*, 1982, **60**, 145))

	−3 cm	−2 cm	−1–0 cm	1–2 cm
STATION in relation to the spines	0	1	2	3

	0 cm	1–2 cm	3–4 cm	>4 cm
DILATATION of the cervix	0	2	4	6

	3 cm	2 cm	1 cm	0 cm
LENGTH of the cervix	0	1	2	3

canal. These five factors are then weighted as shown in Table 11.1a. The author and his co-workers have used Bishop's score for a number of years. As a result of our experience[7], we recently have suggested a modification of the original Bishop system (Table 11.1b). In addition to containing a more correct weighting of the influence of the individual factors, it is easier to remember and to use. Moreover, in all probability it will be more reproducible, as only three factors are included, and the factor having the greatest weight, i.e. cervical dilatation, is that which can be determined with the greatest exactitude[13].

Parity

Primiparae have a longer duration of labour than multiparae in both the first and the second stage. Measured in Bishop score units, the difference in duration of labour between primi- and multiparae is 3–4 points[7]. It does not appear, however, to be more difficult to induce labour in primiparae, provided, of course, that the cervix is ripe[7]. Unfortunately, the cervix in primiparae is more often unripe than in multiparae.

Other factors

It is unclear whether the age of the mother has any influence on the course of labour[8,14]. According to some authorities, this is doubtful[7,8]; however, it is recognized that the cervical condition in elderly primiparae will be more often unripe than in younger primiparae. Neither the size nor the gestational age of the fetus seems to have direct influence on either the duration or the inducibility of labour[7,15].

EVALUATION OF CASES PRIOR TO INDUCTION

Before deciding whether to carry out induction of labour, the following points must be taken into consideration (Figure 11.1):

(1) *Fetal maturity* Is the gestational age established? Is the calculated term correct? Is there agreement between gestational age and the cervical state? If doubt exists as to the gestational age and induction of labour indeed is indicated, then it may be advisable to determine the lecithin/sphingomyelin ratio in the amniotic fluid.

(2) *Cervical ripeness* The condition of the cervix is decisive for the success of an induction attempt. If the cervical score is 6 or more (Table 11.1b), then the case is inducible and primary amniotomy should nearly always be possible. With a score of less than 6, on

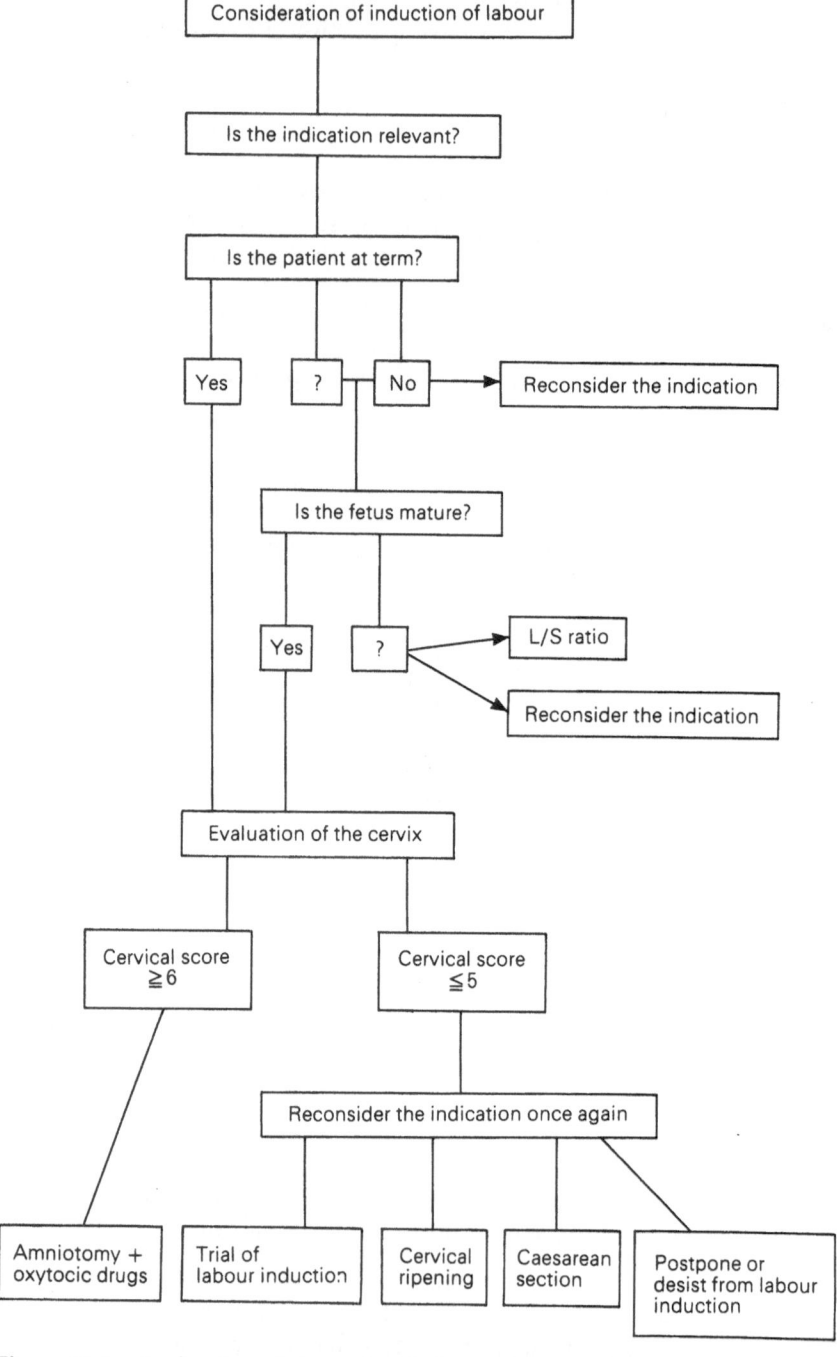

Figure 11.1 Evaluation of the patient before induction of labour

the other hand, it is more doubtful that labour can be induced. The less ripe the cervix, the more consideration should be given to either cervical ripening or Caesarean section.

(3) *Indication* After having considered the first two points, it may be necessary to re-evaluate the case to determine whether the indication is reasonable. In some instances, the indication for terminating the pregnancy may be so pressing that Caesarean section is reasonable. In others, as with elective inductions, for example, the procedure can be postponed if the score is less than 6 or there is some doubt as to the gestational age.

METHODS OF INDUCTION OF LABOUR

Amniotomy

This is the most widely practised of the older methods, and it is still widely used either alone or in combination with oxytocics. However, the efficacy of amniotomy alone depends on the state of the cervix. Turnbull and Anderson[16] found that amniotomy without the use of oxytocin could induce labour in 75% of patients within 24 h. In other studies[17,18], amniotomy alone was capable of inducing uterine contractions in 53% of patients within 2-4 h. Amniotomy has achieved its greatest success in combination with oxytocic agents. Correctly used, the method is safe and acceptable. The risks attached to amniotomy are intrauterine infection or prolapse of the umbilical cord; these risks are minimal, however, provided that the fetal presenting part is engaged and that the cervix is sufficiently ripe. It has been suggested that amniotomy gives rise to an increased occurrence of early fetal heart rate-decelerations[19], but this does not appear to be of any clinical importance[20].

Oxytocin

The basis for using oxytocin as an effective and safe drug was reported by Theobald *et al.*[21] in 1948 using an intravenous drop infusion of posterior-pituitary extract. Synthetic oxytocin became available in 1955 and, with the work of Turnbull and Anderson[16,22,23], the aim of finding the ideal dose regime for oxytocin was achieved. This method, a combination of amniotomy and intravenous drop infusion of oxytocin, in which the dose is gradually increased depending on the uterine response, has demonstrated its efficacy in numerous studies.

In contrast, intramuscular administration of oxytocin for labour induction has long been abandoned due to the risk of hyperstimulation[24]. Attempts have also been made to administer oxytocin and its derivatives intranasally, sublingually or buccally. These latter forms of administration have certain advantages with regard to acceptability; clearly, they are easier to administer and the patient can better be kept mobile, but they offer no advantages with regard to efficacy[25]. With respect to safety, the single dose represents a risk factor which in itself necessitates the use of relatively small doses that may yield insufficient stimulation. With the correct application of modern infusion equipment and a well-trained and experienced staff, intravenous infusion of oxytocin, in combination with amniotomy, is an effective, safe and acceptable method for the induction of labour.

There are no contraindications to this method apart from those applying to vaginal delivery. The infusion equipment results in certain limitations in patient ambulation, however. The method has no inherent side-effects, although the antidiuretic effect of oxytocin should be borne in mind with prolonged infusion, as should the possibility of uterine rupture as a result of inappropriate dosage and/or infusion rates.

Amniotomy and intravenous drop infusion of oxytocin in increasing doses related to uterine response is the method of induction of labour with which obstetricians have the most experience. Until the appearance of the prostaglandins, this was the method which complied most with the ideal requirements of efficacy, safety and acceptability. New methods for the induction of labour should therefore be compared to this method in terms of efficacy, safety and acceptability.

Prostaglandins

In 1968, the same year in which the optimal method for the use of oxytocin for labour induction was first reported[23], Karim et al.[26] reported the use of prostaglandins for the induction of labour. These workers induced labour in ten cases at or near term by the intravenous infusion of $PGF_{2\alpha}$. This preliminary report was augmented by 35 new cases in 1969[27]. In 1970 Beazley et al.[28], Karim et al.[29] and Embrey[30] published almost simultaneously the first clinical reports regarding labour induction with intravenous infusions of PGE_2. Karim and Sharma[31] successfully used orally administered prostaglandins for the same purpose in 1971. Since 1973[32] attempts have been made to use various forms of local application of PGE_2, partly for the induction of labour and partly for 'cervical ripening'. The later indication subsequently has become an independent field for the use of PGE_2. After these pioneering studies, innumerable publications have appeared on the use of prostaglandins for labour induction. Now, some 15 years later, it should be possible to develop a reliable concept of the value of prostaglandins for the induction of labour. This chapter will discuss the role of prostaglandins in labour induction near or at term when the fetus is viable.

The reader should be aware, however, that for a number of reasons it is difficult to directly compare the results of various labour induction studies. First, the patient population varies, owing to the use of different selection and exclusion criteria. Second, numerous studies are without controls and, in some of those where controls are presented, there are differences in the quality of the patients selected. Third, the principles regarding the use of primary amniotomy differ. Fourth, in some reports treatment of the control groups would be considered insufficient by many workers. Last, the results of the studies are reported in various ways; for example, the success rate is stated in percentage indicating the number of deliveries within 12, 24 or more hours, or the number of labours established within 6, 8, 12 or more hours, or the number of patients with cervix dilatation of 6 cm or more after a certain number of hours, etc. Unlike specific aspects of gynaecology, such as IUD performance, about which internationally accepted outcome criteria exist, the absence of such criteria impedes the value of conclusions drawn from a comparison of published studies.

Obviously, it would be advantageous if there were standardized rules for the conduct of labour induction investigations or to describe outcome events (success or failure). Such a set of working definitions and criteria have been suggested almost a decade ago[33], but there seems to be little clinical adherence to these or other standardized criteria as yet.

INTRAVENOUS ADMINISTRATION OF PROSTAGLANDINS

In the early investigations of this method, intravenous $PGF_{2\alpha}$ was given at a constant infusion rate of about 0.05 μg kg^{-1} min^{-1}, whereas PGE_2 was administered in an initial dose of about $5–11$ ng kg^{-1} min^{-1}. It was known from earlier studies[34] that the sensitivity of uterine muscles to prostaglandins was far less dependent on gestational age than was the case with oxytocin, and the results of these preliminary investigations were promising. Although considerable expectations were present with regard to the use of prostaglandins as agents for induction of labour, especially in those cases where labour might be difficult to induce by means of high dose oxytocin, comparative studies[35-39] were soon reported and were unable to fulfil these initial expectations.

Table 11.2 Studies comparing intravenous prostaglandins and oxytocin

References	Type of study	No. of cases PG/ox	Dose schedule PGF$_{2\alpha}$ or PGE$_2$ (μg/min)	Dose schedule oxytocin (mU/min)	Success ratio (%) PG/ox	Uterine hypertonia PG/ox No. of cases	Side-effects Vomiting PG/ox	No. of cases Phlebitis PG/ox
Beazley and Gillespie, 1971[35]	Double-blind	150/150	0·21-6·70 E$_2$ Doubled/60 min	2·1-67 Doubled/60 min	73/73	0/0	—	'occasionally'
Craft et. al., 1971[36]	Randomized	15/15	0·5-4·0 E$_2$ Increased/15 min	1-128 Doubled/20 min	100/60	2/1	7/0	'many'
Rangarajan et al., 1971[37]	Matched	20/20	2·5-20 F$_{2\alpha}$ Doubled/15-30 min	1·25-10 Doubled/15-30 min	80/95	8/1	—	—
Anderson et al., 1972[38]	Double-blind	52/33	2·5-40 F$_{2\alpha}$ Doubled/30-240 min	1-16 Doubled/30-240 min	71/76	5/0	—	—
Vakhariya and Sherman, 1972[39]	Double-blind	50/50	2-40 F$_{2\alpha}$ Increased/30 min	1-16 Increased/30 min	94/96	8/1	1/0	1/0
Spellacy et al., 1973[40]	Double-blind	115/107	2·5-40 F$_{2\alpha}$ Doubled/60-240 min	0·5-8 Doubled/60-240 min	74·5/66·4	24/9	12/6	7/0
Brown et al., 1973[41]	Matched	32/32	2·5-7·5 F$_{2\alpha}$ Increased/30-240 min	2·8-49 Increased/15 min	69/87·5	0	1/0	4/0
Brown et al., 1973[41]	Matched	53/53	0·25-0·75 E$_1$ Increased/30-240 min	2·8-49 Increased/15 min	83/89	0	0	6/0
Calder and Embrey, 1975[42]	Double-blind	50/50	0·1-4·0 E$_2$ Doubled/30 min	1-64 Doubled/30 min	98/90	2/2	—	23/1

PGE_2 and $PGF_{2\alpha}$ in comparison to oxytocin

Table 11.2 shows some of the comparative investigations reported between 1971 and 1975[35-42]. The majority of studies were randomized, double-blind investigations, although two had matched control groups. The Table illustrates the dose regimes, the efficacy measured in terms of the 'success ratio' and, finally, the number of cases of uterine hypertonia and side-effects.

Efficacy

The overall impression obtained from these eight comparative studies is that these two prostaglandins, given intravenously, are perhaps as effective as intravenous oxytocin, but not more effective. To a certain extent, the existing differences can be attributed to the considerable differences in the dose regimes employed. Moreover, if one looks closely at these differences, particularly with regard to the doses of oxytocin, it is somewhat surprising that the difference in efficacy is not greater. It thus appears that both prostaglandins and oxytocin administered intravenously in suitable doses related to the clinical responses can initiate labours quite indistinguishable from each other, irrespective of whether the patients are nulliparae or multiparae, have a ripe or unripe cervix or whether amniotomy has or has not been carried out. At present there is no clear evidence to suggest that this is not the case if one considers pregnancies with a living fetus at or near term. The question remains, however, what exactly is the ideal dose regime of i.v. prostaglandins.

Safety

In the majority of the studies referred to above, uterine hypertonia was reported in association with i.v. administration of $PGF_{2\alpha}$[37-40] or PGE_2[36,38,42]. At present no uniform or clear definition of this term exists; however, in most studies it is used to indicate an increase in the resting uterine tone lasting for several minutes. Tables 11.2, 11.3 and 11.4 depict among other data the wide variations in the frequency of occurrence of hypertonia in a number of investigations where intravenous $PGF_{2\alpha}$ and PGE_2 have been used.

Apart from some cases in the study of Spellacy et al.[40] (Table 11.2), nearly all episodes of uterine hypertonia disappeared rapidly after cessation of the prostaglandin infusion, a change in the patient position or in one case following the administration of a β-sympathomimetic agent. The variation in incidence of uterine hypertonia results in part from the various definitions employed to describe it and in part from

Table 11.3 Labour induction studies of intravenous PGF$_{2\alpha}$

References	No. of cases	Dose schedule i.v. PGF$_{2\alpha}$ (µg/min)		Uterine hypertonia No. of cases	Side-effects: no. of cases	
					Vomiting	Phlebitis
Karim et al., 1969[27]	35	3	Constant dose	0	—	—
Witting et al., 1973[48]	60	0·5–16	Constant dose	8	—	—
Johnson et al., 1974[50]	115	4–24	Doubled/30 min	3	15	0
Sharma et al., 1975[49]	90	2·5–20	Doubled/1–4h	2	16	1
Lindmark et al., 1975[47]	49	3–24	Increased 3 µg min^{-1} h^{-1}	5	3	4
Beck and Lilling, 1976[51]	100	2·5–40	Doubled/30–60 min	4	29	1
Thiery et al., 1977[44]	64	2–30	Increased/30 min	21	10	0
Kerekes and Domokos, 1977[52]	104	2·5–5·0–7·5	Increased/30 min	3	8	—
Lindberg, 1977[45]	106	3–18	Increased 3 µg min^{-1} h^{-1}	1	5	7
O'Sullivan et al., 1978[53]	50	0·5–25	Doubled/hour	6	12	4
Allen et al., 1983[46]	100	1·5–6	Doubled/30 min	1	1	—

Table 11.4 Labour induction studies of intravenous PGE$_2$

References	No. of cases	Dose schedule i.v. PGE$_2$ (µg/min)		Uterine hypertonia No. of cases	Side-effects: no. of cases	
					Vomiting	Phlebitis
Bygdeman et al., 1968[34]	7	1·0–8·0	Constant dose (20 min)	4	0	0
Beazley et al., 1970[28]	40	0·4–4·0	Doubled when necessary	1	—	0
Karim et al., 1970[29]	50	0·5	Constant dose	—	1	—
Embrey, 1970[30]	25	0·5–6·0	Increased when necessary	0	—	—
Roberts and Turnbull, 1971[43]	18	0·75–3·0	Doubled when necessary	4	—	—
Anderson et al., 1972[38]	15	0·3–4·8	Doubled/30–60 min	2	0	—
Thiery et al., 1977[44]	53	0·2–2·0	Increased/30 min	26	7	0

the fact that the intrauterine pressure was measured in some studies but not in others.

In some of the investigations comparing prostaglandins with oxytocin[40,42], instances of uterine hypertonia were also observed after administration of oxytocin. Thus, Spellacy et al.[40] recorded nine cases of uterine hypertonia in their oxytocin group, none of which required obstetric intervention, in contrast to the necessity to perform Caesarean section in six of the 24 cases registered in the $PGF_{2\alpha}$ group. Calder and Embrey[42] observed only two incidents of uterine hypertonia in both the oxytocin and PGE_2 groups, even though the dose of oxytocics was increased rapidly and, in respect to PGE_2, to a magnitude corresponding to that employed by Spellacy et al.[40] for $PGF_{2\alpha}$. Calder and Embrey[42] defined hypertonia as a resting tone of more than 30 mmHg, whereas Spellacy et al.[40] used a limit of 15 mmHg. Apart from the two cases of uterine hypertonia noted by Calder and Embrey[42], a considerable number of cases in the PGE_2 group were noted to have strong contractions followed by a number of weaker contractions before the resting tone was regained compared to the oxytocin group. There seems little doubt that the incidence of abnormal contractile patterns is increased in patients treated with i.v. prostaglandins compared to oxytocin.

The magnitude of the prostaglandin dose per minute, as well as the dose regime employed, represent the two factors which have the greatest influence on hypertonia. Bygdeman et al.[34] were the first to use i.v. PGE_1 and PGE_2 for short infusions in near term pregnant women. These workers employed rather high doses (Table 11.4) and, in two of their seven cases, therapy had to be discontinued due to uterine hypertonia.

Comparatively low doses were used in those studies in which no cases of uterine hypertonia were observed (Table 11.2[41], Table 11.3[27], Table 11.4[29]). By way of contrast, in those investigations in which the dose was increased at short intervals and/or in large increments, numerous cases of uterine hypertonia were seen (Table 11.2[39,40], Table 11.4[43,44]). Only a small number of cases of uterine hypertonia were recorded in those studies in which a small initial dose was increased in small increments (Table 11.3[45,46]). Two of the studies mentioned in Tables 11.2 and 11.3[36,47] also report uncoordinated uterine activity and other abnormal contractile patterns of the type previously described by Roth-Brandel and Adams[54]. It seems likely that the increased occurrence of cases of uterine hypertonia is mainly related to the doses of prostaglandins selected. The occurrence of other forms of abnormal contraction patterns can also, to a certain extent, be satisfactorily explained in the same manner, but this raises the question as to whether the contractions brought about by prostaglandins are identical to those arising from spontaneous or oxytocin-induced labour.

Anderson and Schooley[55] compared labour contractions by assessing their intensity, duration and frequency during spontaneous, i.v. oxytocin-induced or i.v. $PGF_{2\alpha}$-induced labour. They found no differences with regard to the intensity or frequency of contractions, but patients treated with oxytocin differed in that single contractions were of a significantly shorter duration. Roux et al.[56] studied 'the initial contraction duration' and 'terminal relaxation phase', and could not demonstrate any difference in contractions induced by i.v. oxytocin, i.v. $PGF_{2\alpha}$ or oral PGE_2. Seitchik et al.[57] studied intrauterine pressure wave form characteristics, and found that contractions during spontaneous and $PGF_{2\alpha}$-induced labour did not differ, whereas both differed from those induced by oxytocin inasmuch as the latter were of shorter duration and the ratio of the maximum tension developed to maximum rate of tension development in the oxytocin group was significantly lower. This observation indicates that oxytocin-induced contractions are accompanied by an altered inotropic state different from that seen with contractions in spontaneous as well as $PGF_{2\alpha}$-induced labour.

As previously stated, there appears to be no clinical difference between the oxytocin- or prostaglandin-induced labour – provided the drugs are given in suitable doses. A certain degree of risk of uterine hypertonia and other types of abnormal contractions will always exist when oxytocics are used. This risk leads to the question, 'When does hypertonia become of clinical significance?', i.e. when does it constitute a real danger to the life of the fetus. Among 1617 births induced by i.v. prostaglandins, a total of 140 cases of hypertonia were recorded, corresponding to an incidence of 8·7% (Tables 11.2, 11.3, 11.4). Serious problems were observed in only one study[40] in which Caesarean section was considered necessary in six instances.

The increased occurrence of hypertonia and other abnormalities of contractions in the studies in which prostaglandins were compared to oxytocin[36-40,42,44] indicate that i.v. prostaglandins are less safe than i.v. oxytocin unless the incidence of these abnormalities can be reduced to a level corresponding to those known to occur with i.v. oxytocin administration.

How should intravenous prostaglandins be administered?

The questions of the efficacy and safety of intravenous prostaglandins as compared to intravenous oxytocin are closely related to the dose employed and the dose increments administered per time interval. The therapeutic range for i.v. oxytocin is well known; it varies from 1 to far above 100 mU/min. This dose, depending on the uterine response, can normally be doubled up to several times an hour without any greater risk of uterine hypertonia provided this is done under qualified supervision. In contrast, several studies have demonstrated that this dose regimen cannot be directly transferred to i.v. prostaglandins[37,39,40,43]. Uterine sensitivity to prostaglandins varies from case to case, and this sensitivity is related to a number of factors, as is the case with oxytocin. However, the therapeutic range of prostaglandins is narrower, and the difference between the therapeutic dose and the dose leading to hypertonia is considerably less. The dose range for i.v. $PGF_{2\alpha}$ lies in the area 1–40 μg/min and for PGE_2 0.1–4.0 μg/min.

In their first studies, Karim et al. employed constant doses of approximately 3 μg $PGF_{2\alpha}$/min[27] and roughly 0.5 μg PGE_2/min[29] with a high success ratio and no cases of hypertonia. These studies were mainly carried out on Bantustan Negroes. Other investigators have been less successful in achieving corresponding results on Caucasians with such a low dose. Witting et al.[48] found that a constant but slightly higher dose of 4 μg/min of $PGF_{2\alpha}$ provided a safe and effective treatment, while lower doses were ineffective.

Due to the individual variation in the sensitivity of uterine muscle to oxytocics, it appears reasonable to commence with small doses of prostaglandins, for example 1–3 μg $PGF_{2\alpha}$/min or 0.2 μg PGE_2/min, and thereafter to increase the dose roughly every 30 min until an adequate response is seen. Afterwards, the intervals between the dose increments should be hourly or longer. It is questionable as to how important the length of these intervals is with regard to efficacy. Sharma et al.[49] compared a low dose with a high dose therapeutic regimen. The dose of $PGF_{2\alpha}$ was increased at different rates, so that the low dose group reached a maximum of 20 μg $PGF_{2\alpha}$/min after only 9 h of drug infusion, while the high dose group reached the same maximum value after only 3 h of infusion. There was no difference in the success ratio achieved by the two doses.

When dealing with prostaglandins, the procedure of doubling the dose, as is often done with the administration of oxytocin, should be avoided. It is far more prudent to increase the dose by the addition of 3–5 μg $PGF_{2\alpha}$/min or 0.2–0.5 μg PGE_2/min. In studies where such cautions have been utilized, it has been possible to maintain maximum doses not exceeding 6 μg $PGF_{2\alpha}$/min[46], 7.5 μg $PGF_{2\alpha}$/min[41] or 0.75 μg PGE_2/min[41] and achieve satisfactory results. The more rapidly and the larger the increments by which the dose of prostaglandins is increased, the greater the risk of uterine hypertonia. It should be mentioned, however, that several investigators have obtained good results with rapid dose increments of either $PGF_{2\alpha}$[58] or PGE_2[42] in conjunction with the use of an automatic Cardiff infusion pump. The concomitant use of primary amniotomy and prostaglandins means that smaller doses can be used[48]. In this respect there seems to be no difference between

oxytocin and prostaglandins. If the membranes are intact, the prostaglandin dose should be reduced when rupture occurs, as uterine sensitivity to prostaglandins may change considerably in connection with
membrane rupture[59].

It is thus advisable that i.v. prostaglandins for labour induction
should be given in slow incremental dose regimes in which the increments are carried out *by addition* and *not by multiplication*. With this
form of administration, it will probably be difficult to distinguish
clinically between i.v. prostaglandins and i.v. oxytocin in respect to
efficacy and safety. On the other hand, there seems to be little doubt
that i.v. prostaglandins are more difficult to administer and at present
more unpredictable than i.v. oxytocin infusion. It is therefore doubtful
whether i.v. prostaglandin therapy will reach the same level of acceptance by clinicians as is the case with i.v. oxytocin.

Acceptability

The discomfort attached to the intravenous administration of prostaglandins does not differ from that of intravenous oxytocin. The work
of the staff is also similar. The problem of side-effects, however, is a
different matter altogether. Side-effects seen with the intravenous administration of prostaglandins include nausea, vomiting, diarrhoea and
local phlebitis.

Tables 11.2, 11.3 and 11.4 show the number of cases of vomiting
recorded in individual studies. The gastrointestinal side-effects were
rarely of serious character. Among the somewhat more than 1600 cases
in these Tables, cessation of treatment was required in less than ten
cases. Signs of local phlebitis were noted at the infusion site in the
majority of studies. However, it appears that this local inflammatory
reaction to prostaglandins is dose-related, and it can be counteracted
by the use of intravenous catheters which reach the larger veins.

Based on the data in Tables 11.2–11.4, it is realistic to consider
side-effects of minor importance, particularly in instances where a low
incremental dose regime is used. However, intravenous oxytocin has
no such side-effects whatsoever, and therefore may be considered more
acceptable than intravenous prostaglandins in some clinical settings.

Conclusion

Intravenous $PGF_{2\alpha}$ and PGE_2 administered according to a low incremental dose augmentation regime and under qualified supervision is
an effective method of induction of labour at or near term. The main
factor responsible for choosing a certain prostaglandin has been the
availability of these drugs. Whereas PGE_2 is not commercially available

for intravenous use, $PGF_{2\alpha}$ is now available for intravenous use all over the world. Thus, pharmaceutical availability has been significant in the present utilization of $PGF_{2\alpha}$. With or without amniotomy, intravenous prostaglandins are equally as effective as intravenous oxytocin, but do not appear to be any better, neither when considering patients in general nor when patients are compared according to parity or cervical ripeness. The administration of intravenous prostaglandins should, with the previously mentioned reservations, be associated with a high degree of safety that is difficult to distinguish clinically from that known to be associated with i.v. oxytocin. With regard to acceptability, however, prostaglandins are less acceptable than oxytocin as the latter has no known side-effects. Thus, there is little evidence to support the proposition that i.v. prostaglandins are preferable to i.v. oxytocin for labour induction. Nonetheless, even though i.v. prostaglandins have not achieved the degree of superiority once hoped for, they are still used at certain centres for routine induction of labour[46,60]. It would be desirable to have a larger controlled study of i.v. $PGF_{2\alpha}$ in a low dose incremental regime compared to i.v. oxytocin before i.v. $PGF_{2\alpha}$ is finally abandoned.

ORAL ADMINISTRATION OF PROSTAGLANDINS

Karim and Sharma[31] used an aqueous solution of PGE_2 and $PGF_{2\alpha}$ given orally to 80 and 20 patients, respectively, in their first study; 79 of the PGE_2 group and 16 of the $PGF_{2\alpha}$ group were successfully induced. Either $0.5-1.5$ mg PGE_2 or $5-15$ mg $PGF_{2\alpha}$ were administered at 2 hourly intervals. The method seemed safe; no case of uterine hypertonia was observed. One year later, Karim and Sharma[61] reported their experience in Kampala with the routine use of oral PGE_2 given to 1000 patients, 764 of whom were labour induction cases. Amniotomy was not routinely performed, and 85% of the patients had a Bishop score of less than 5. No case of uterine hypertonia was recorded but there was one case of uterine rupture with maternal death. It was subsequently determined that the patient had previously had a myomectomy and thus was originally a poor candidate for stimulation. Of the 764 labour induction cases, 561 were delivered within 24 h, 17 had an induction–delivery time that surpassed 48 h, and it was necessary to discontinue prostaglandin therapy in two cases due to vomiting. These results were remarkably good, as were those obtained with i.v. prostaglandins from the same clinic[27,29]. Part of the explanation may be a racial difference in response to prostaglandins.

As oral $PGF_{2\alpha}$ was found to be less effective than PGE_2[31] and was accompanied by a higher incidence of gastrointestinal side-effects[62], the

Table 11.5 Studies comparing oral PGE₂ with intravenous oxytocin

References	Type of study	No. of cases PGE₂/ox	Dose schedule PGE₂ (mg)	Dose schedule oxytocin (mU/min)		Success ratio (%) PGE₂/ox	Uterine hyper-tonia PGE₂/ox	Side-effects PGE₂/ox	
								Vomiting	Diarrhoea
Kelly et al., 1973[63]	Matched control	49/49	0·5–2·0/2 h	—	Doubled every 30 min	95·9/93·9	0/0	15/6	3/0
Read and Martin, 1974[44]	Randomized	99/88	0·5–2·0/2 h	5–80	Doubled every 30 min	100/100	—	2/0	—
Ratnam et al., 1974[65]	Randomized	107/100	0·5–2·0/h	—	Increased every 30 min	916/95	0/0	2/0	8/0
Beard et al., 1975[66]	Randomized	22/20	0·5–2·0/1–2 h	1–128	Doubled every 10 min	91/100	5/6	1/0	—
Miller et al., 1975[67]	'Randomized' (treatment changed weekly)	(1) 57/51 (2) 43/51	(1) 0·5/h Constant dose (2) 0·5–2·0/h Increased every 2 h	2–	Doubled every 15 min	(1) 80·7/96 (2) 86/96	—	(1) 1/0 (2) 4/0	0/0
Elder, 1975[68]	Partly matched	65/41	0·5/h Constant dose	2–64	Doubled every 15 min	77/100	0/0	10/11	—
Cunningham et al., 1976[69]	Matched	84/84	0·5–1·5/h Increased every hour	—	Increased every 15–30 min	86/70	1	30/—	3/—
Nelson and Bryans, 1976[70]	Partly matched	49/49	0·5–1·5/h Increased every 2 h	2–16	Doubled hourly	82/65	1/1	5/3	8/0
Lykkesfeldt and Osler, 1979[71]	Partly matched	69/92	0·5/30 min – amniotomy	—	+ amniotomy Cardiff inf.	74/100	—	6/0	—
Bremme and Bygdeman, 1980[72]	Randomized	104/96	0·5–1·0/h	5–20	Increased hourly	100/100	1/0	11/7	2/0
Lichtenegger et al., 1981[73]	Not stated	153/172	0·5–1·5/h	2–20	—	see text	0/0	8/0	0/0
Lange et al., 1981[74]	Randomized	99/102	0·5–1·5/h Increased every 2 h	5–45	Increased every 30 min	96/100	0/0	8/1	3/0
Secher et al., 1981[17]	Randomized	182/165	0·5–1·5/h Increased every 2 h	5–45	Increased every 30 min	72/71·5	0/0	13/3	1/0

use of oral PGE_2 soon became standard. A large number of reports on labour induction with oral PGE_2 have appeared. Tablets containing 0·5 mg of PGE_2 are now commercially available over most of the world. Tables 11.5 and 11.6 contain a survey of some of these studies in which oral PGE_2 has been employed for labour induction; the dose schedules are given, as are the reported number of cases of uterine hypertonia and gastrointestinal side-effects in each study.

PGE_2 compared to i.v. oxytocin

Table 11.5 contains data from a number of studies where oral PGE_2 has been compared to i.v. oxytocin[17,63-74]. The design of these studies differs in several respects; six are randomized and the others more or less matched control studies. In one study[73], there is no indication of the design. The latter differs from all the others in that oral PGE_2 is claimed to be significantly better than i.v. oxytocin. The difference seems to be due to different policies in the two patient groups with regard to the time at which amniotomy was carried out.

Efficacy

The differences between the efficacy of oral PGE_2 and i.v. oxytocin in the 13 studies cited in Table 11.5 were not very great, especially when two of them[71,73] are disregarded due to methodological differences making the study groups noncomparable with respect to efficacy. Only in a very few instances is it possible to discern a statistically significant difference, for example, between the induction–delivery times[72,74]. Evaluated as a whole, oral administration of PGE_2 is an effective oxytocic agent capable of standing comparison to i.v. oxytocin, but which is no more effective than oxytocin, neither as a whole nor when the patients are divided into primiparae/multiparae, into those with mature/immature cervices or those with intact or ruptured membranes. If only the induction–delivery time is considered, the majority of investigators[17,66,72,74] have determined that oxytocin use is of shorter duration than PGE_2, with no corresponding differences in the success ratio. This shorter induction–delivery duration must be attributed to the mode of administration. It is possible to increase the dose of oxytocin relatively quickly, while this is impracticable with oral PGE_2.

Safety

The studies listed in Tables 11.5 and 11.6 include 4403 births induced with oral PGE_2. There were 34 cases of uterine hypertonia, corresponding to an incidence of 0·77%. Some of the studies in Table 11.6

Table 11.6 Labour induction studies of oral PGE$_2$

References	No. of cases	Dose schedule, oral PGE$_2$ (mg)	Uterine hypertonia No. of cases	Side-effects Vomiting	Side-effects Diarrhoea
Karim and Sharma, 1971[31]	80	0·5-1·5/2 h	0	2	—
Craft, 1972[75]	50	0·5-3·0/1 h	0	15	8
Karim and Sharma, 1972[61]	764	0·5-2·0/2 h	—	65	34
Laursen and Wilson, 1974[76]	52	0·5-3·0/30-60 min	1	29	10
Murnaghan et al., 1974[77]	50	0·5-2·0/30-120 min	0	18	3
Friedman and Sachtleben, 1974[78]	45	(1) 0·5/h (2) 1·0/1 h (3) 0·5-3·0/30 min	3	12	8
Elder and Stone, 1974[79]	70	0·5/1 h	—	17	—
Tsakok et al., 1975[80]	370	0·5-1·5/h Dose increased every 4 h	0	18	5
Basu and Rajan, 1975[81]	169	0·5-2·0/2 h	3	27	18
Friedman et al., 1975[82]	100	(1) 0·5/1 h (2) 1·0/1 h	6	19	16
Alaily and Morewood, 1975[83]	100	0·5-1·5/1 h Dose increased every 4 h	0	18	1
Gabert et al., 1976[84]	100	0·5-1·5/h	0	26	
Visscher et al., 1977[85]	50	0·5-1·0/h	2	4	0
Ang et al., 1978[86]	153	0·5-1·0/h Dose increased after 4 h	0	7	3
Gordon-Wright et al., 1979[87]	144	0·5-1·5/h Dose increased every 2 h	3	33%	
Nelson and Bryans, 1978[88]	144	0·5-1·5/h	3	28	
Squires and Masson, 1980[89]	680	0·5-1·5/h Dose increased every hour	6	191	29

evaluated different doses and dose regimes; this may explain the slightly higher occurrence of uterine hypertonia in this Table compared to Table 11.5. Fourteen cases of uterine hypertonia were reported by two groups (Table 11.5[66], Table 11.6[78,82]). Both groups of investigators used intrauterine pressure registration. Beard et al.[66] defined uterine hypertonia as a resting tone > 15 mmHg for 5 min, whereas Friedman and Sachtleben[78] did not define uterine hypertonia but reported three cases, two of which had tetanic contractions while the third only developed an elevation of the resting tone. One of the instances of tetanic contractions occurred after a maximum dose of 3 mg had been given; in this case there was no doubt that hyperstimulation was present. In the other series, Friedman et al. (Table 11.6[82]) described six cases of uterine hypertonia of which only one occurred in the group given a 0·5 mg dose, whereas the other five all occurred in the group given a 1·0 mg dose. The cases were described partly as coupled contractions with elevated resting tone and partly as tetanic contractions; all were uncomplicated, as were the majority of cases of hypertonia observed in the study of Beard et al.[66]. The comparatively higher number of cases in their control group were also uncomplicated.

Just as many cases of uterine hypertonia were recorded in the control materials as in the PGE$_2$ materials reported in Table 11.5. Uterine hypertonia, tetanic contractions and other abnormalities of labour contractions occur in normal unstimulated labours. Friedman and Sachtleben[90] reviewed 2834 monitoring records retrospectively; of these 948 were from births induced or stimulated by i.v. oxytocin and 99 induced by oral PGE$_2$. Tetanic-like contractions occurred in 10·4% of all 2834 births. The frequency among unstimulated births was 8·3%, among oxytocin-induced births 13·2%, and oral PGE$_2$-induced births 17·2%. At the same time, it was noted that there were significantly more variable and late decelerations in the fetal heart rate (FHR) among the stimulated labours and that these occurred mainly among those parturients treated with oxytocin; in contrast, the incidence of FHR-decelerations among those parturients given oral PGE$_2$ corresponded to the incidence among unstimulated patients. The authors interpreted these findings as confirmation that there is more similarity between spontaneous and PGE$_2$-induced labour contractions than with those induced by oxytocin, as described by Seitchik et al.[57].

Even though it does not appear that oral PGE$_2$ gives rise to a higher incidence of uterine hypertonia than does i.v. oxytocin, it is still not certain that oral PGE$_2$ can be considered as safe as i.v. oxytocin. The question still remains, 'To what extent does the individual case of uterine hypertonia represent a risk to the fetus?' Case reports[88,91,92] have shown that uterine hypertonia can occur even with low dose PGE$_2$ therapy and at times long after the last dose has been admin-

istered. This situation is far more difficult than when hypertonia occurs
in connection with i.v. oxytocin. In the latter circumstance, it is merely
a question of discontinuing the oxytocin infusion, whereas with oral
PGE_2 therapy it is impossible to stop the action of the drug.

Although Table 11.5 shows no difference in the safety of the two
drugs and as far as is known no fetal deaths have taken place which
could be attributed to hyperstimulation during oral PGE_2 treatment, it
must still be concluded that i.v. oxytocin has a greater safety margin
than oral PGE_2 due to its mode of administration. Thus oral PGE_2
must be administered with caution and care, and this treatment necessitates careful surveillance of the patient, as is the case with i.v. oxytocin.

How should oral PGE_2 be administered?

Oral PGE_2 was initially administered in a dose range of 0·5–2·0 mg in
single doses at intervals of 2 h. It was soon found that this interval
could be reduced to 1 h. A few obstetricians have used up to 3·0 mg of
PGE_2 every 30–60 min (Table 11.6[75,76,78]), but such high doses were
accompanied by severe gastrointestinal side-effects[75,76,78] and a higher
risk of uterine hypertonia[78]. In their pilot study, Friedman and Sachtleben[78] rejected the high dose regime of 0·5–3·0 mg/h for this reason.
On the other hand, Lauersen and Wilson[76] stated that they had occasionally given 3·0 mg every 30 min for several hours without complications.

Friedman et al.[82] continued their investigations by comparing the
two dose regimes with a constant dose of 0·5 and 1·0 mg/h, respectively. The success rate was comparable, but the induction–delivery
time was slightly longer in the group given 0·5 mg/h. As mentioned
earlier, there was a considerable excess of cases of uterine hypertonia
in the high dose group, and gastrointestinal side-effects occurred in
approximately one half of the patients compared with a single case in
the low dose group. In a similar investigation, Gabert and Herbertson[93]
found comparable results to Friedman et al.[82] with respect to efficacy,
but there were no cases of uterine hypertonia and the incidence of
side-effects was low in both patient groups. Basu and Rajan[94] compared two matched groups of patients treated with constant doses of
1·0 mg every other hour and 0·5 mg every hour, respectively, and found
no differences between the groups.

The majority of investigators have used an incremental dose regime
with the augmentation varying from 0·5 mg every 30–60 min to 0·5 mg
every fourth hour. Secher et al.[17] and Lange et al. (Table 11.5[74]) increased the individual doses by 0·5 mg every other hour, so that the
maximum dose of 1·5 mg was only reached after 4 h of treatment. In

contrast, Westergaard et al.[18,95], reporting from the same group, increased the dose every hour so that the maximum dose was achieved by 2 h. Even though no direct comparison was made between the two dose regimes from the same group, the change in dose regime does not appear to result in any improvement with regard to efficacy[96] but does increase the occurrence of side-effects[95,96].

Oral PGE_2 is effective irrespective of the magnitude of the single dose or the dose regime, and there appears to be little difference in the results obtained from various investigations, particularly after amniotomy. Thus, Tsakok et al.[80], Alaily and Morewood[83], and Ang et al.[86], for example, all achieved good results with a low dose incremental regime, in which the dose was increased by 0·5 mg every fourth hour to a maximum of 1·5 mg. No case of uterine hypertonia was recorded in any of these studies (Table 11.6). Bremme and Bygdeman (Table 11.5[72]) utilized 1·0 mg PGE_2/h as the maximum dose and obtained similar results to those of our group where the maximum dose was 1·5 mg/h[17,74]. In their pilot study, Bremme and Bygdeman[97] found that 1·0 mg PGE_2/h resulted in a stable plasma concentration with little risk of accumulation of the drug.

Based on their experience with 50 multiparae who had a Bishop score of 7 or more, Visscher et al. (Table 11.6[85]) have suggested some guidelines for labour induction with oral PGE_2. Because many of their labours ran rapid courses and there were several cases of abnormal uterine contractions (including two cases of uterine hypertonia), they suggested, among other things, that amniotomy be performed only after contractions had commenced. In my opinion, this recommendation is definitely inadvisable. The best test for the irritability of the uterus is to carry out primary amniotomy, where possible, and every labour induction case with a Bishop score of 7 or more should have this performed before commencing oxytocic therapy. If one waits up to 2 h after primary amniotomy, many cases no longer require additional stimulation[25]. Moreover, the range between the therapeutic and toxic dose of oxytocics becomes smaller with increasing uterine muscular sensitivity which is also correlated to cervical ripeness. If oral PGE_2 therapy is started before primary amniotomy in patients with a very ripe cervix, there is a considerable risk of hyperstimulation[82,85]. On the other hand, primary amniotomy in conjunction with oral PGE_2 gives, similar to i.v. oxytocin, better results in the form of a more rapid labour[81] and higher success rate[25,71].

Summarizing, oral PGE_2 in the form of tablets of 0·5 mg should be preceded by primary amniotomy if the cervical condition permits this. If primary amniotomy is not followed by labour contractions within 1 h, then there is good evidence that the uterine muscles are not hypersensitive to oxytocics, and oral PGE_2 may be started at a dose of

0·5 mg/h. Single doses should not be given more frequently than once every hour, and the dose should be increased slowly, at one to several hours interval, depending on uterine response. It is inadvisable to exceed a single dose of 1·5 mg/h. Using this dose regime one can expect a success rate comparable with that obtained from a combination of amniotomy and i.v. oxytocin and a degree of safety, measured in the number of cases of uterine hypertonia, which does not differ from that following the use of i.v. oxytocin – provided the treatment is carried out under qualified surveillance and that the increase in dose is adjusted to the uterine response.

Acceptability

The great advantage of oral PGE_2 is its form of administration. It is attractive in that the patient merely has to take tablets rather than being fixed to an i.v. infusion. Although the treatment is simple to administer, it requires, as does i.v. oxytocin, constant surveillance of the patient. The problem of surveillance can be overcome in either circumstance by monitoring via a telemetric system[98]. The incidence of gastrointestinal side-effects varies between 10 and 20% (Tables 11.5 and 11.6), and the occurrence of nausea, vomiting and diarrhoea are clearly increased compared to i.v. prostaglandins. The incidence of side-effects is related to the dose[78,82,87], but even 0·5 mg PGE_2/h has produced such pronounced side-effects that it has been necessary to stop treatment. Even though a number of episodes of vomiting and diarrhoea in the individual cases are modest as a rule, they make oral PGE_2 less acceptable than oxytocin which is completely free from side-effects.

Conclusion

Oral PGE_2 appears just as safe as i.v. oxytocin provided it is carefully administered. It compares to i.v. oxytocin in success ratio with or without amniotomy. However, it is not more effective and a slightly longer induction–delivery time must be expected. The greater risk of uterine hypertonia which, at least in theory, should occur with oral PGE_2 therapy has not been observed in clinical practice. Thus, it is a question of weighing the advantages of ease of administration against the disadvantages of gastrointestinal side-effects when deciding whether to use oral PGE_2 or i.v. oxytocin. An additional consideration in making this choice is that at present oral PGE_2 treatment is considerably more expensive than i.v. oxytocin.

LOCAL ADMINISTRATION OF PROSTAGLANDINS

Karim and Sharma[99] successfully used intravaginal applications of 2 mg PGE_2 or 5 mg $PGF_{2\alpha}$ every other hour for labour induction as early as 1971. They found, however, that oral administration was preferable. A large number of reports have appeared since 1973[32] in which PGE_2 has been employed extra-amniotically, intracervically or intravaginally, partly for labour induction and partly for 'cervical ripening' (see Chapter 10). In this section, the local administration of prostaglandins will only be mentioned briefly and evaluated in relation to amniotomy + i.v. oxytocin.

Extra-amniotic PGE_2

Calder et al.[100] infused 20–150 μg PGE_2 in saline per hour via a Nélaton catheter placed in the extra-amniotic space 10–12 cm from the internal orifice. Labour was induced in all the patients without complications and with only a few instances of vomiting. Miller and Mack[101] achieved similar results, but concomitantly recorded one case of uterine hypertonia and two of intrapartum fetal death. These deaths were the result of intrauterine asphyxia and their cause could hardly be attributed to the PGE_2. Sims et al.[102] used 350 μg PGE_2 given as a single dose in 7 ml of viscous Tylose gel. In more than one half of the 285 patients with a ripe cervix and in 48% of multiparae and 24% of nulliparae with an unripe cervix, this treatment alone was sufficient to induce labour. No cases of gastrointestinal side-effects were observed, and only one case of uncomplicated uterine hypertonia was noted.

Thiery et al.[103] compared extra-amniotic PGE_2 (500 μg) followed by i.v. oxytocin with primary amniotomy + i.v. oxytocin in a randomized investigation; they found that PGE_2 + oxytocin was as effective but not superior to amniotomy + oxytocin when the Bishop score was 5 or more. As extra-amniotic application of PGE_2 is more difficult and does not appear to have any important advantages compared to the intra-vaginal method[104], it has been replaced by the latter in most centres.

Intracervical PGE_2

This method was developed by a group in Malmö who produced a special viscous gel capable of retaining its consistency in the cervix and in which PGE_2 remains in stable form[105]. In a randomized investigation, 0·5 mg of PGE_2 in 4 ml of gel was compared to i.v. oxytocin for labour induction[106]. No difference was found between the methods if the cervix was ripe. With a cervical score of between 1 and 5, however, labour could be induced in significantly more patients with intracervical PGE_2. Oxytocin was given in a low dose and increased very slowly; amniotomy was not carried out before the patients were in labour and the cervix dilated to at least 4 cm. The effect was confirmed in a double-blind placebo controlled investigation with 0·5 mg PGE_2 in 2 ml of gel among nulliparae with an unripe cervix[107]. Recently, the Malmö group compared 0·5 mg PGE_2 intracervically with 4 mg PGE_2 intravaginally[108]. In patients with a cervical score of 4-5, no difference was found; on the other hand, with a cervical score of 1-3 intracervical PGE_2 was superior. Furthermore, the intravaginal, in contrast to the intracervical method, was accompanied by some degrees of side-effects and significantly increased myometric activity.

Intravaginal PGE_2

This form of administration is far more simple and appeals to both patient and staff. At the Queen Charlotte's Hospital, where approximately 20% of all births are induced, 3 mg of PGE_2 intravaginally is used as the routine method in all cases. Only 40% of patients with an unripe cervix and 22% of patients with a ripe cervix required supplementary stimulation with i.v. oxytocin[109,110]. Following the introduction of this method the frequency of Caesarean section due to failed induction declined drastically. Apart from this intriguing statement, no control material has been presented by the authors.

Vaginal tablets containing 3 mg PGE_2 are now commercially available in the UK and appear to be widely used[111-113], even in cases of previous Caesarean section[113].

Several controlled investigations of intravaginal prostaglandin therapy have been reported[111,114-116]. Kennedy et al.[111] found intravaginal PGE_2 more acceptable to patients with a Bishop score >4 than amniotomy + i.v. oxytocin, even though the induction–delivery time was longer in the PGE_2 group than in the oxytocin group. Larsen et al.[114] found vaginal PGE_2 superior to oxytocin, but the latter drug was administered intranasally, a method known to be rather ineffective[25]. Lange et al. compared patients with a Bishop score <6 and found a failure rate of 4% after intravaginal PGE_2 followed by amniotomy and oxytocin compared to a failure rate of 13% after amniotomy + oxytocin alone. Macer et al.[116] studied patients with a Bishop score >4 and found no difference in efficacy between vaginal PGE_2 followed by oxytocin compared to i.v. oxytocin without primary amniotomy.

Efficacy of local PGE_2

Based on the relatively few controlled investigations presently available, it can be concluded that 0·5 mg PGE_2 intracervically or 3 mg PGE_2

intravaginally are effective methods for the induction of labour. In particular, they seem to be advantageous as the initial treatment in cases where the cervix is so unripe that primary amniotomy is inadvisable. In contrast, these methods do not appear as effective as primary amniotomy + i.v. oxytocin in those cases where primary amniotomy is possible, i.e. with a Bishop score of 5-6 or more.

Safety of local PGE_2

Intracervical administration is safe, based on the present limited information. The Malmö group used it in over 400 cases and observed only two cases of uterine hypertonia, one of which occurred after only 1·0 mg of PGE_2 had been given[117]. The other case occurred 6-7 h after the application of 0·5 mg PGE_2; in this instance the hypertonia was accompanied by fetal bradycardia which disappeared following the administration of a β-sympathomimetic agent[107]. The final evaluation of the safety of this method requires further work in other centres.

In contrast, there now seems to be sufficient data to permit a preliminary evaluation of the safety of the intravaginal method. At the Queen Charlotte's Hospital, only one case of uterine hypertonia was observed among 1000 inductions[110], but several reports from other institutions contain 16 instances of uterine hypertonia among 500 cases[112,116,118-121]. In nine of these it was necessary to carry out emergency Caesarean section; five others required treatment with β-sympathomimetic agents. 5 mg of PGE_2 had been given in three cases[118,120], 25 mg $PGF_{2\alpha}$ in one case[118], and the remainder received 3 mg PGE_2. In only two cases did the patients have a ripe cervix, while the others had a low Bishop score. In recent years, several cases have been reported of intrapartum fetal death in connection with labour induction by the local application of prostaglandins[122-125]. Even though it is not possible to implicate the method of labour induction as the direct cause of death in all five instances, case reports such as these have not been seen since intramuscular oxytocin was abandoned 25 years ago[24]. It should also be mentioned that three cases of uterine rupture have been described after intravaginal treatment with 20 mg PGE_2 repeated several times[126-128]. All inductions following intrauterine fetal demise were accompanied by few symptoms. In one case, the fetus was found lying freely in the peritoneal cavity[126]; in the other two the rupture was discovered because of severe bleeding at the time of delivery[127,128]. Delivery via a posterior cervico-vaginal fistula in a para III has also been reported after 3 mg PGE_2[116].

It is difficult to compare the incidence of uterine hypertonia after local compared to i.v. or oral administration of prostaglandins. Patients treated with local PGE_2 have rarely been studied with intrauterine pressure measurements as has been the case with i.v. or oral administration. In contrast to the cases described above, the majority of cases of uterine hypertonia recorded in connection with the systemic administration of prostaglandins have not presented any major problems. For this reason, it is difficult to accept vaginal application of protaglandins as a safe and reliable method of labour induction. The use of effective single doses of oxytocics will always have an inherent risk, as the sensitivity of the uterus is unknown before the drug is administered. Even though some of the vaginally applied PGE_2 can be removed in cases of uterine hypertonia, it is not possible to completely avoid complications.

Intravaginal PGE_2, therefore, should be used with caution. To use this method rather than amniotomy + i.v. oxytocin in cases where the latter therapy is possible would expose the fetus to unnecessary risk. In those cases where the indication for labour induction implicates a poorly functioning placenta (placental insufficiency, IUGR, hypertensive disorders) this method should be avoided or employed only with great caution.

Acceptability of local PGE_2

The great advantage of local methods, especially vaginal, is the simplicity of their administration. Only a very few cases of gastrointestinal side-effects have been described. The extra-amniotic form of administration seems to have given way to the vaginal method. The intracervical method appears troublesome and requires the availability of a special gel; even when this is used it would appear that it is difficult to avoid the gel either reaching the extra-amniotic space or running backward to the vagina.

Conclusion

Local administration of PGE_2 is effective; it is especially recommended in cases with a low Bishop score for which induction of labour is indicated, but it should be avoided in cases with reduced placental function. In those cases where primary amniotomy can be carried out, local application has no advantage over i.v. oxytocin with regard to efficacy. These methods are by no means as safe as i.v. oxytocin, and the advantages with regard to acceptability cannot outweigh this fact.

In all probability a greater degree of safety can be achieved by reducing the single dose and eventually repeating it, but this change reduces the efficacy[129]. The problem is similar to the administration of oxytocin in single doses. If effective doses are used, as with intramuscular injection, then safety is at stake. On the other hand, if relatively low doses are used for reasons of safety, as with buccal or intranasal administration, then the effect is unsatisfactory[25].

The intracervical method is attractive, as only 0·5 mg PGE_2 is necessary, and the systemic effect, including stimulation of the myometrium, appears to be slight. Apart from Malmö, however, the method has been used in very few centres, and one can only hope that this method will prove more safe than the intravaginal method.

OTHER ASPECTS OF PROSTAGLANDINS USED FOR LABOUR INDUCTION

In the majority of studies of labour induction using prostaglandins, vital signs were observed during treatment, and numerous blood analyses were carried out before and after treatment in many of these investigations. As of this time, no undesirable effects of prostaglandins have been demonstrated. Neither have clinical nor biochemical changes been found in the newborns[130,131]. Even neonatal psychomotor development has been followed for several years and no abnormality has been found[132,133].

The increased occurrence of neonatal jaundice after labour induction is related only to fetal maturity and there is no difference between oxytocin and prostaglandins in this respect[134].

Since the eventual thermogenic effect of prostaglandins[135] appears to be of no importance with the doses employed for labour induction[136], an eventual rise in temperature during prostaglandin-induced labour cannot without a further elucidation be attributed to the treatment.

The fact that prostaglandins do not have an antidiuretic effect has been used as an argument for their use rather than oxytocin in cases of pre-eclampsia and kidney disease[137]. There are, as far as is known to the author, no reports to the effect that i.v. oxytocin has given rise to complications such as disturbances in electrolyte balance or cardiovascular problems. Cases of cramp resulting from water intoxication have only been reported following prolonged infusion of large doses for abortion[138]. Although there is hardly any reason for considering oxytocin to be contraindicated in cases of pre-eclampsia or renal disease, prolonged infusion of oxytocin should be avoided in such cases or withheld at intervals and accompanied by restriction of fluid intake.

One report suggests that induced labours have a higher incidence of postpartum bleeding compared to spontaneous labours[139]. Other investigators have found that this only applies to oxytocin-induced labour and not prostaglandins[111,140]. This question should be the object of controlled prospective investigations in the future.

Because of its bronchoconstrictive effect, lung diseases and asthma have long been considered contraindications to the use of prostaglandins. However, this does not seem valid with the doses employed for labour induction[131].

PRESENT AND FUTURE STATE OF PROSTAGLANDINS IN LABOUR INDUCTION

History repeats itself. All the older methods, i.e. oxytocin, Spartein[15,25] and now the various forms of prostaglandins, have been through the same cycle[141]. First, there are the preliminary observations which are promising and give rise to great expectations. Thereafter, the first large studies on the efficacy of the method and its advantages are put forward by authors who are enthusiastic about the new method. Then follow case reports and larger controlled investigations from different centres pointing out the less fortunate aspects of the method which often does not live up to the expectations created by the initial publications. Finally, this procedure may be followed by several years of routine use in various centres. Thus, it can take many years before a specific method for labour induction has passed through the whole cycle and the time arrives for a final evaluation of its true place among other methods. The only method which has passed through this entire cycle and still can be termed second to none is the combination of amniotomy and i.v. oxytocin.

The various forms of administration of prostaglandins for labour induction are presently in diverse stages of the 'historical cycle'; some of them therefore are hardly ready for final evaluation. However, it appears reasonable to conclude that i.v. and oral administration of prostaglandins have no advantages over i.v. oxytocin, and it is doubtful if we will see these forms of administration employed to any greater extent in the future.

It is a matter of concern that intravaginal PGE_2 in single doses of 3 mg is becoming widely used, even in cases where primary amniotomy is possible. One might fear that more case reports of serious complications will appear before the method is completely evaluated. There is a need for more controlled investigations with smaller, eventually repeated, single doses of PGE_2, perhaps administered via cervical caps, portio-adaptors[142] or the like. Until such investigations are performed the intravaginal method using 3 mg of PGE_2 should be used with some degree of reticence; furthermore, it should only be used where the cervical conditions are unfavourable.

The intracervical method has only reached the halfway mark in the 'historical cycle' and is therefore not ready enough for final evaluation.

It is one thing to have effective methods for labour induction, but another to know when they should be used. The need for any method of labour induction and particularly for new methods suitable in cases of an unripe cervix depends on the frequency of induction in the individual centre. The frequency with which induction of labour is carried out varies considerably, and, in some instances this variation depends more on obstetric traditions than on scientific knowledge[143]. If it is thought that the frequency of induction should be 20–30% then a considerable need will arise for methods that can be used with an unripe cervix. On the other hand, if the opinion is held that induction of labour with an unripe cervix should only be carried out because of important indications, i.e. placental insufficiency, intrauterine growth retardation or severe hypertensive disorders, then the need for local application methods would be very small. In many such cases the local application of PGE_2 would be contraindicated anyway, and clinicians would prefer elective Caesarean section.

There are approximately 1800 labours per year in the author's department. Labour induction is carried out only on relevant obstetrical or medical indications. We use oxytocin preceded, as far as possible, by primary amniotomy. In those cases where the prelabour evaluation shows that the cervical score is less than 6 (see Table 11.1b), the indication is re-evaluated. In the majority of cases, induction is either not carried out or the procedure is postponed until the cervical conditions have become more favourable. If delivery is necessary, Caesarean section is performed. Using this protocol, the induction rate in the past year was 3·5%. The elective Caesarean section rate was 2·1% excluding breech presentations. During the past year, in only nine cases did delivery not occur on the same day as induction was commenced. Our need for methods that are particularly well suited to an unripe cervix thus appears to be less than ten cases per year. In the past year the perinatal mortality rate was 6·0 per thousand. This has nothing to do with the induction policy of the department. It is merely mentioned, because a high induction frequency is considered by some to be a prerequisite for a low perinatal mortality rate[144].

There is a need for continued research into new methods of labour induction, particularly in cases with an unripe cervix. However, there appears to be an even greater need for research into the indications for induction and when this should be carried out. It is unacceptable to continue to permit these vital questions to be based on untimely traditions.

ACKNOWLEDGEMENT

This work was supported by the Danish Medical Research Council.

References

1 Caldeyro-Barcia, R. and Sereno, J. A. (1961). The response of the human uterus to oxytocin throughout pregnancy. In Caldeyro-Barcia, R. and Heller, H. (eds.) *Oxytocin*. pp. 177–200. (Oxford: Pergamon Press)

2 Smyth, C. N. (1958). Uterine irritability. *Lancet*, 1, 237–9

3 Embrey, M. P. and Anselmo, J. F. (1962). The effects of intravenous oxytocin on uterine contractility. *J. Obstet. Gynaecol. Br. Commonw.*, 69, 918–23

4 Anderson, A. B. M. and Turnbull, A. C. (1969). Relationship between length of gestation and cervical dilatation, uterine contractility, and other factors during pregnancy. *Am. J. Obstet. Gynecol.*, 105, 1207–14

5 Hendricks, C. H., Brenner, W. E. and Kraus, G. (1970). Normal cervical dilatation pattern in late pregnancy and labor. *Am. J. Obstet. Gynecol.*, 106, 1065–80

6 Bishop, E. H. (1955). Elective induction of labor. *Obstet. Gynecol.*, 5, 519–27

7 Lange, A. P., Secher, N. J., Westergaard, J. G. and Skovgård, I. (1982). Prelabor evaluation of inducibility. *Obstet. Gynecol.*, 60, 137–47

8 Burnhill, M. S., Danezis, J. and Cohen, J. (1962). Uterine contractility during labor studied by intra-amniotic fluid pressure recordings. *Am. J. Obstet. Gynecol.*, 83, 561–71

9 Bishop, E. H. (1964). Pelvic scoring for elective induction. *Obstet. Gynecol.*, 24, 266–8

10 Burnett, J. E. (1966). Preinduction scoring: an objective approach to induction of labor. *Obstet. Gynecol.*, 28, 479–83

11 Friedman, E. A., Niswander, K. R., Bayonet-Rivera, N. P. and Sachtleben, M. R. (1966). Relation of prelabor evaluation to inducibility and the course of labor. *Obstet. Gynecol.*, 28, 495–501

12 Friedman, E. A., Niswander, K. R., Bayonet-Rivera, N. P. and Sachtleben, M. R. (1967). Prelabor status evaluation. *Obstet. Gynecol.*, 29, 539–44

13 Beazley, J. M. and Alderman, B. (1976). The 'inductograph' – a graph describing the limits of the latent phase of induced labour in low risk situations. *Br. J. Obstet. Gynaecol.*, 83, 513–17

14 Mukherjee, S. and Biswas, S. (1959). Duration of labour and its relationship to maternal age and parity – a statistical approach. *J. Indian Med. Assoc.*, 33, 173–8

15 Cibils, L. A. (1972). Enhancement and induction of labor. In Aladjem, S. (ed.) *Risks in the Practice of Modern Obstetrics*. pp. 126–53. (Saint Louis: C. V. Mosby)

16 Turnbull, A. C. and Anderson, A. B. M. (1967). Induction of labour. *J. Obstet. Gynaecol. Br. Commonw.*, 74, 849–54

17 Secher, N. J., Lange, A. P., Hassing Nielsen, F., Thomsen-Pedersen, G. and Westergaard, J. G. (1981). Induction of labor with and without primary amniotomy. *Acta Obstet. Gynecol. Scand.*, **60**, 237–41

18 Westergaard, J. G., Lange, A. P., Thomsen-Pedersen, G. and Secher, N. J. (1983). Oral oxytocics for induction of labor. *Acta Obstet. Gynecol. Scand.*, **62**, 103–10

19 Caldeyro-Barcia, R., Schwarcz, R., Belizan, J. M., Martell, M., Nieto, F., Sabatino, H. and Tenzer, S. M. (1974). Adverse perinatal effects of early amniotomy during labour. In Gluck, L. (ed.) *Modern Perinatal Medicine*. pp. 431–49. (Chicago: Year Book Medical Publishers)

20 Stewart, P., Kennedy, J. H. and Calder, A. A. (1982). Spontaneous labour: when should the membranes be ruptured? *Br. J. Obstet. Gynaecol.*, **89**, 39–43

21 Theobald, G. W., Graham, A., Campbell, J., Gange, P. D. and Driscoll, W. J. (1948). The use of post-pituitary extract in physiological amounts in obstetrics. *Br. Med. J.*, **2**, 123–7

22 Turnbull, A. C. and Anderson, A. B. M. (1968). Induction of labour. *J. Obstet. Gynaecol. Br. Commonw.*, **75**, 24–31

23 Turnbull, A. C. and Anderson, A. B. M. (1968). Induction of labour *J. Obstet. Gynaecol. Br. Commonw.*, **75**, 32–41

24 Wrigley, A. J. (1959). The place of oxytocic drugs before the birth of the child. *J. Obstet. Gynaecol. Br. Emp.*, **66**, 857–9

25 Lange, A. P. (1984). Induction of labour. *Dan. Med. Bull.*, **31**, 89–108

26 Karim, S. M. M., Trussell, R. R., Patel, R. C. and Hillier, K. (1968). Response of pregnant human uterus to prostaglandin-$F_{2\alpha}$-induction of labour. *Br. Med. J.*, **4**, 621–3

27 Karim, S. M. M., Trussell, R. R., Hillier, K. and Patel, R. C. (1969). Induction of labour with prostaglandin $F_{2\alpha}$. *J. Obstet. Gynaecol. Br. Commonw.*, **76**, 769–82

28 Beazley, J. M., Dewhurst, C. J. and Gillespie, A. (1970). The induction of labour with prostaglandin E_2. *J. Obstet. Gynaecol. Br. Commonw.*, **77**, 193–9

29 Karim, S. M. M., Hillier, K., Trussell, R. R., Patel, R. C. and Tamusange, S. (1970). Induction of labour with prostaglandin E_2. *J. Obstet. Gynaecol. Br. Commonw.*, **77**, 200–10

30 Embrey, M. P. (1970). Induction of labour with prostaglandins E_1 and E_2. *Br. Med. J.*, **2**, 256–8

31 Karim, S. M. M. and Sharma, S. D. (1971). Oral administration of prostaglandins for the induction of labour. *Br. Med. J.*, **1**, 260–2

32 Calder, A. and Embrey, M. P. (1973) Prostaglandins and the unfavourable cervix. *Lancet*, **2**, 1322–3

33 Amy, J. J., Thiery, M., Crawford, J. S., Kerenyi, T. D., Bygdeman, M. and Karim, S. M. M. (1976). A suggested set of working definitions and criteria applicable to induction of labour. *Int. J. Gynaecol. Obstet.*, **14**, 379–83

34 Bygdeman, M., Kwon, S. U., Mukherjee, T. and Wiqvist, N. (1968). Effect of intravenous infusion of prostaglandin E_1 and E_2 on motility of the pregnant human uterus. *Am. J. Obstet. Gynecol.*, **102**, 317–26

35 Beazley, J. M. and Gillespie, A. (1971). Double-blind trial of prostaglandin E_2 and oxytocin in induction of labour. *Lancet*, 1, 152-5

36 Craft, I. L., Cullum, A. R., May, D. T. L., Noble, A. D. and Thomas D. J. (1971). Prostaglandin E_2 compared with oxytocin for the induction of labour. *Br. Med. J.*, 3, 276-9

37 Rangarajan, N. S., LaCroix, G. E. and Moghissi, K. S. (1971). Induction of labor with prostaglandin. *Obstet. Gynecol.*, 38, 546-50

38 Anderson, G. G., Hobbins, J. C. and Speroff, L. (1972). Intravenous prostaglandins E_2 and $F_{2\alpha}$ for the induction of term labor. *Am. J. Obstet. Gynecol.*, 112, 382-6

39 Vakhariya, V. R. and Sherman, A. I. (1972). Prostaglandin $F_{2\alpha}$ for induction of labor. *Am. J. Obstet. Gynecol.*, 113, 212-20

40 Spellacy, W. N., Gall, S. A., Shevach, A. B. and Holsinger, K. K. (1973). The induction of labor at term. Comparisons between prostaglandin $F_{2\alpha}$ and oxytocin infusions. *Obstet. Gynecol.*, 41, 14-21

41 Brown, A. A., Hamlett, J. D., Hibbard, B. M. and Howe, P. D. (1973). Induction of labour by amniotomy and intravenous infusions of oxytocic drugs - a comparison between prostaglandins and oxytocin. *J. Obstet. Gynaecol. Br. Commonw.*, 80, 111-15

42 Calder, A. A. and Embrey, M. P. (1975). Comparison of intravenous oxytocin and prostaglandin E_2 for induction of labour using automatic and non-automatic infusion techniques. *Br. J. Obstet. Gynaecol.*, 82, 728-33

43 Roberts, G. and Turnbull, A. C. (1971). Uterine hypertonus during labour induced by prostaglandins. *Br. Med. J.*, 1, 702-5

44 Thiery, M., Vroman, S., de Hemptinne, D., Yo Le Sian, A., Vanderheyden, K., Van Kets, H., Martens, G., Derom, R. and Rolly, G. (1977). Elective induction of labor conducted under lumbar epidural block. *Eur. J. Obstet. Gynecol. Reprod. Biol.*, 7, 181-200

45 Lindberg, B. (1977). The induction of labour by the intravenous infusion of prostaglandin $F_{2\alpha}$. *Prostaglandins*, 14, 993-1004

46 Allen, J., Forman, A., Maigaard, S. and Ulmsten, U. (1983). Low dose i.v. infusion of prostaglandin $F_{2\alpha}$ for induction of labor at term. *Acta Obstet. Gynecol. Scand. Suppl.*, 113, 187-8

47 Lindmark, G., Zador, G. and Nilsson, B. A. (1975). The induction of labour with prostaglandin $F_{2\alpha}$ by intravenous infusion. *Acta Obstet. Gynecol. Scand. Suppl.*, 37, 17-26

48 Witting, W. C., Laros, R. K. and Work, B. A. (1973). Uterine response to prostaglandin $F_{2\alpha}$ infusion in term human pregnancy. *Obstet. Gynecol.*, 42, 581-8

49 Sharma, S. D., Hale, R. W. and Muller, J. P. (1975). Induction of term labor with intravenous prostaglandin $F_{2\alpha}$. *Prostaglandins*, 10, 1019-27

50 Johnson, A., Hyatt, D. and Newton, J. (1974). Experience with prostaglandin $F_{2\alpha}$ (free acid) for the induction of labour. *Prostaglandins*, 7, 487-500

51 Beck, P. and Lilling, M. I. (1976). Induction of labor with intravenous prostaglandin. *Am. J. Obstet. Gynecol.*, 125, 648-54

52 Kerekes, L. and Domokos, N. (1977). Geburtseinleitung mit Prostaglan-

din F$_{2\alpha}$ (eine vergleichende Untersuchung unter Anwendung von Oxytozin und Prostaglandin). *Zbl. Gynäkol.*, **99**, 971–8

53 O'Sullivan, M. J., Stone, M. L., Gugliucci, C. L., Shah, G. and Stephens, M. E. (1978). Induction of labor in the high-risk pregnancy with PGF$_{2\alpha}$. *Obstet. Gynecol.*, **51**, 77–80

54 Roth-Brandel, U. and Adams, M. (1970). An evaluation of the possible use of prostaglandin E$_1$, E$_2$ and F$_{2\alpha}$ for induction of labour. *Acta Obstet. Gynecol. Scand. Suppl.*, **5**, 9–17

55 Anderson, G. G. and Schooley, G. L. (1975). Comparison of uterine contractions in spontaneous and oxytocin- or PGF$_{2\alpha}$-induced labors. *Obstet. Gynecol.*, **45**, 284–6

56 Roux, J. F., Mofid, M., Moss, P. L. and Dmytrus, K. C. (1977). Effect of elective induction of labor with prostaglandins F$_{2\alpha}$ and E$_2$ and oxytocin on uterine contraction and relaxation. *Am. J. Obstet. Gynecol.*, **127**, 718–22

57 Seitchik, J., Chatkoff, M. L. and Hayashi, R. H. (1977). Intrauterine pressure waveform characteristics of spontaneous and oxytocin- or prostaglandin F$_{2\alpha}$-induced active labor. *Am. J. Obstet. Gynecol.*, **127**, 223–7

58 Johnson, A. and Newton, J. R. (1974). Labour induced by prostaglandin F$_2$ using the Cardiff infusion apparatus. *Lancet*, **1**, 1253–4

59 Dewhurst, C. J. (1972). Discussion to Vakhariya, V. R. and Sherman, A. I. Prostaglandin F$_{2\alpha}$ for induction of labor. *Am. J. Obstet. Gynecol.*, **113**, 221–2

60 Thiery, M., Parewijek, W., and Martens G. (1982). Intravenous infusion of prostaglandin E$_2$ for management of premature rupture of membranes. *Z. Geburtsh. Perinat.*, **186**, 87–8

61 Karim, S. M. M. and Sharma, S. D. (1972). Oral administration of prostaglandin E$_2$ for the induction and acceleration of labor. *J. Reprod. Med.*, **9**, 346–52

62 Barr, W. and Naismith, W. C. M. K. (1972). Oral prostaglandins in the induction of labour. *Br. Med. J.*, **2**, 188–91

63 Kelly, J., Flynn, A. M. and Bertrand, P. V. (1973). A comparison of oral prostaglandin E$_2$ and intravenous syntocinon in the induction of labour. *J. Obstet. Gynaecol. Br. Commonw.*, **80**, 923–6

64 Read, M. D. and Martin, R. H. (1974). A comparison between intravenous oxytocin and oral prostaglandin E$_2$ for the induction of labour in parous patients. *Curr. Med. Res. Opin.*, **2**, 236–9

65 Ratnam, S. S., Khew, K. S., Chen, C. and Lim, T. C. (1974). Oral prostaglandin E$_2$ in induction of labour. *Aust. NZ. J. Obstet. Gynaecol.*, **14**, 26–30

66 Beard, R. J., Harrison, R., Kiriakidis, J., Underhill, R. and Craft, I. (1975). A clinical and biochemical assessment of the use of oral prostaglandin E$_2$ compared with intravenous oxytocin for labor induction in multiparous patients. *Eur. J. Obstet. Gynecol. Reprod. Biol.*, **5**, 203–7

67 Miller, J. F., Welply, G. A. and Elstein, M. (1975). Prostaglandin E$_2$ tablets compared with intravenous oxytocin in induction of labour. *Br. Med. J.*, **1**, 14–16

68 Elder, M. G. (1975) Uterine action after induction of labour with oral prostaglandin E$_2$ tablets compared with intravenous oxytocin. *Br. J. Obstet. Gynaecol.*, **82**, 674–81

69 Cunningham, F. G., Cox, K., Hauth, J. C., Strong, J. D. and Whalley, P. J. (1976). Oral prostaglandin E$_2$ for labor induction in high-risk pregnancy. *Am. J. Obstet. Gynecol.*, **125**, 881–8

70 Nelson, G. H. and Bryans, C. I. (1976). A comparison of oral prostaglandin E$_2$ and intravenous oxytocin for induction of labor in normal and high-risk pregnancies. *Am. J. Obstet. Gynecol.*, **126**, 549–53

71 Lykkesfeldt, G. and Ostler, M. (1979). A comparison of three methods for inducing labor. *Acta Obstet. Gynecol. Scand.*, **58**, 321–5

72 Bremme, K. and Bygdeman, M. (1980). Induction of labor by oxytocin or prostaglandin E$_2$. *Acta Obstet. Gynecol. Scand. Suppl.*, **92**, 11–21

73 Lichtenegger, W., Lahousen, M. and Kraemer, H. (1981). Comparison of oral prostaglandin E$_2$ and intravenous oxytocin for induction of labor. *Gynecol. Obstet. Invest.*, **12**, 197–202

74 Lange, A. P., Secher, N. J., Hassing Nielsen, F. and Thomsen-Pedersen, G. (1981). Stimulation of labor in cases of premature rupture of the membranes at or near term. *Acta Obstet. Gynecol. Scand.*, **60**, 207–10

75 Craft, I. (1972). Amniotomy and oral prostaglandin E$_2$ titration for induction of labour. *Br. Med. J.*, **2**, 191–4

76 Lauersen, N. H. and Wilson, K. H. (1974). Induction of labor with oral prostaglandin E$_2$. *Obstet. Gynecol.*, **44**, 793–801

77 Murnaghan, G. A., Lamki, H., Rashid, S. and Pinkerton, J. H. M. (1974). Induction of labour with oral prostaglandin E$_2$. *J. Obstet. Gynaecol. Br. Commonw.* **81**, 141–5

78 Friedman, E. A. and Sachtleben, M. R. (1974). Oral prostaglandin E$_2$ for induction of labor at term. *Obstet. Gynecol.*, **43**, 178–85

79 Elder, M. G. and Stone, M. (1974). Induction of labour by low amniotomy and oral administration of a solution compared to a tablet of prostaglandin E$_2$. *Prostaglandins*, **6**, 427–32

80 Tsakok, F. H. M., Grudzinskas, J. G., Karim, S. M. M. and Ratnam, S. S. (1975). The routine use of oral prostaglandin E$_2$ in induction of labour. *Br. J. Obstet. Gynaecol.*, **82**, 894–8

81 Basu, H. K. and Rajan, K. T. J. (1975). The role of amniotomy in induction of labour with oral administration of prostaglandin E$_2$. *Curr. Med. Res. Opin*, **3**, 397–406

82 Friedman, E. A., Sachtleben, M. R. and Green, W. (1975). Oral prostaglandin E$_2$ for induction of labor at term. II. Comparison of two low-dosage regimens. *Am. J. Obstet. Gynecol.*, **123**, 671–4

83 Alaily, A. and Morewood, G. A. (1975). Titrated prostaglandin E$_2$ tablets for induction of labour. *Br. Med. J.*, **2**, 731

84 Gabert, H. A., Brinton, J. and Brown, B. (1976). Induction of labor with oral prostaglandin E$_2$. *Am. J. Obstet. Gynecol.*, **125**, 333–8

85 Visscher, R. D., Struyk, C. D. and Visscher, H. C. (1977). Guidelines for the elective induction of labor with oral prostaglandin E$_2$. *Obstet. Gynecol.*, **49**, 15–19

86 Ang. L. T., Ng, K. H., Sivanesaratnam, V., Sinnathuray, T. A. and Yusof, K. (1978). Inducing labor with oral prostaglandin E₂ tablets. *Int. J. Gynaecol. Obstet.*, 15, 415–18

87 Gordon-Wright, A. P., Dutt, T. P. and Elder, M.G. (1979). The routine use of oral prostaglandin E₂ tablets for induction or augmentation of labour. *Acta Obstet. Gynecol., Scand.*, 58, 23–6

88 Nelson, G. H. and Bryans, C. I. (1978). Induction of labor with oral prostaglandin E₂ in normal and high-risk pregnancies. *Am J. Obstet. Gynecol.*, 132, 642–8

89 Squires, D. J. P. and Masson, E. L. (1980). Induction of labour with oral prostaglandin E₂: a Canadian multicentre study. *J. Int. Med. Res.*, 8, 175–9

90 Friedman, E. A. and Sachtleben, M. R. (1978). Effect of oxytocin and oral prostaglandin E₂ on uterine contractility and fetal heart rate patterns. *Am. J. Obstet. Gynecol.*, 130, 403–7

91 Fraser, I. S. (1974). Uterine hypertonus after oral prostaglandin E₂. *Lancet*, 2, 162

92 Thiery, M. and Amy, J. J. (1976). Uterine hypertonus after induction of labour with prostaglandin E₂ tablets. *Br. Med. J.*, 1, 958

93 Gabert, H. A. and Herbertson, R. M. (1976). The use of oral prostaglandin E₂ to induce labor at term. *J. Reprod. Med.*, 16, 276–80

94 Basu, H. K. and Rajan, K. T. J. (1975). Induction of labour with prostaglandin E₂ tablets. *J. Int. Med. Res.*, 3, 73–6

95 Westergaard, J. G., Lange, A. P., Thomsen-Pedersen, G. and Secher, N. J. (1983). Use of oral oxytocics for stimulation of labor in cases of premature rupture of the membranes at term. *Acta Obstet. Gynecol. Scand.*, 62, 111–16

96 Lange, A. P., Westergaard, J. G., Secher, N. J. and Thomsen-Pedersen, G. (1983). Labor induction with prostaglandins. *Acta Obstet. Gynecol. Scand. Suppl.*, 113, 177–85

97 Bremme, K., Kindahl, H. and Svanborg, K. (1980). Induction of labor by oral PGE₂ administration – evaluation of different dose schedules. *Acta Obstet. Gynecol. Scand. Suppl.*, 92, 5–10

98 Steiner, H., Robrecht, D., Deichsel, W., Breckwoldt, M. and Hillemanns, H. G. (1980). The use of two-channel telemetric systems in obstetrics: an ideal monitoring method for oral PGE₂ induction of labor. In Samuelsson, B., Ramwell, P. W. and Paoletti, R. (eds.) *Advances in Prostaglandin and Thromboxane Research*. Vol. 8, pp. 1483–6. (New York: Raven Press)

99 Karim, S. M. M. and Sharma, S. D. (1971). Therapeutic abortion and induction of labour by the intravaginal administration of prostaglandins E₂ and F₂ₐ. *J. Obstet. Gynaecol. Br. Commonw.*, 78, 294–300

100 Calder, A. A., Embrey, M. P. and Hillier, K. (1974). Extra-amniotic prostaglandin E₂ for the induction of labour at term. *J. Obstet. Gynaecol. Br. Commonw.*, 81, 39–46

101 Miller, A. W. F. and Mack, D. S. (1974). Induction of labour by extra-amniotic prostaglandins. *J. Obstet. Gynaecol. Br. Commonw.*, 81, 706–8

102 Sims, C. D., Mellows, H. J., Spencer, P. J. and Craft, I. L. (1979). Routine

induction of labour with extra-amniotic prostaglandin E_2 in a viscous gel. *Br. J. Obstet. Gynaecol.*, **86**, 529-32

103 Thiery, M., Parewijck, W., Martens, G., Derom, R. and van Kets, H. (1981). Extra-amniotic prostaglandin E_2 gel vs. amniotomy for elective induction of labour. *Z. Geburtsh. Perinat.*, **185**, 323-6

104 Clarke, G. A., Letchworth, A. T. and Noble, A. D. (1980). Comparative trial of extra-amniotic and vaginal prostaglandin E_2 in tylose gel for induction of labor. *J. Perinat. Med.*, **8**, 236-40

105 Ulmsten, U., Kirstein-Pedersen, A., Stenberg, P. and Wingerup, L. (1979). A new gel for intracervical application of prostaglandin E_2. *Acta Obstet. Gynecol. Scand. Suppl.*, **84**, 19-21

106 Ulmsten, U., Wingerup, L. and Andersson, K.-E. (1979). Comparison of prostaglandin E_2 and intravenous oxytocin for induction of labor. *Obstet. Gynecol.*. **54**, 581-4

107 Ulmsten, U., Wingerup, L., Belfrage, P., Ekman, G. and Wiqvist, N. (1982). Intracervical application of prostaglandin gel for induction of term labor. *Obstet. Gynecol.*, **59**, 336-9

108 Ekman, G., Forman, A., Marsal, K. and Ulmsten, U. (1983). Intravaginal versus intracervical application of prostaglandin E_2 in viscous gel for cervical priming and induction of labor at term in patients with an unfavorable cervical state. *Am. J. Obstet. Gynecol.*, **147**, 657-61

109 Shepherd, J., Pearce, J. M. and Sims, C. D. (1979). Induction of labour using prostaglandin E_2 pessaries. *Br. Med. J.*, **2**, 108-10

110 Shepherd, J. H., Bennett, M. J., Laurence, D., Moore, F. and Sims, C. D. (1981) Prostaglandin vaginal suppositories: a simple and safe approach to the induction of labor. *Obstet. Gynecol.*, **58**, 596-600

111 Kennedy, J. H., Stewart, P., Barlow, D. H., Hillan, E. and Calder, A. A. (1982). Induction of labour: a comparison of a single prostaglandin E_2 vaginal tablet with amniotomy and intravenous oxytocin. *Br. J. Obstet. Gynaecol.*, **89**, 704-7

112 Hunter, I. W. E., Cato, E. and Knox Ritchie, J. W. (1984). Induction of labor using high-dose or low-dose prostaglandin vaginal pessaries. *Obstet. Gynecol.*, **63**, 418-20

113 MacKenzie, I. Z., Bradley, S. and Embrey, M. P. (1984). Vaginal prostaglandins and labour induction for patients previously delivered by caesarean section. *Br. J. Obstet. Gynaecol.*, **91**, 7-10

114 Larsen, J., Andreasson, B. and Bock, J. E. (1983). Igangsættelse af fødsel. *Ugeskr. Læg.*, **145**, 2588-90

115 Lange, I. R., Collister, C., Johnson, J., Cote, D., Torchia, M., Freund, G. and Manning, F. A. (1984). The effect of vaginal prostaglandin E_2 pessaries on induction of labor. *Am. J. Obstet. Gynecol.*, **148**, 621-5

116 Macer, J., Buchanan, D. and Lynn Yonekura, M. (1984). Induction of labor with prostaglandin E_2 vaginal suppositories. *Obstet. Gynecol.*, **63**, 664-8

117 Wingerup, L., Andersson, K.-E. and Ulmsten, U. (1979). Ripening of the cervix and induction of labor in patients at term by single intracervical application of prostaglandin E_2 in viscous gel. *Acta Obstet. Gynecol. Scand. Suppl.*, **84**, 11-14

118 MacKenzie, I. Z. and Embrey, M. P. (1979). A comparison of PGE₂ and PGF₂α vaginal gel for ripening the cervix before induction of labour. *Br. J. Obstet. Gynaecol.*, **86**, 167-70

119 Hefni, M. A. and Lewis, G. A. (1980). Induction of labour with vaginal prostaglandin E₂ pessaries. *Br. J. Obstet. Gynaecol.*, **87**, 199-202

120 Varma, T. R. and Norman, J. (1984). A comparison of three dosages of prostaglandin E₂ pessaries for ripening the unfavourable cervix prior to induction of labor. *Acta Obstet. Gynecol. Scand.*, **63**, 17-21

121 Buchanan, D., Macer, J. and Yonekura, M. L. (1984). Cervical ripening with prostaglandin E₂ vaginal suppositories. *Obstet. Gynecol.*, **63**, 659-63

122 Quinn, M. A. and Murphy, A. J. (1981). Fetal death following extra-amniotic prostaglandin gel. *Br. J. Obstet. Gynaecol.*, **88**, 650-1

123 Lindholm, P. (1981). Partus provocatus. En sammenlignende undersøgelse mellem cervikalt placeret prostaglandin-gel og parenteralt oksytocin. *Ugesker. Læg.*, **143**, 878-81

124 Beck, I. and Clayton, J. K. (1982). Hazards of prostaglandin pessaries in postmaturity. *Lancet*, **2**, 161

125 Simmons, K. and Savage, W. (1984). Neonatal death associated with induction of labour with intravaginal prostaglandin E₂. Case report. *Br. J. Obstet. Gynaecol.*, **91**, 598-9

126 Sandler, R. Z., Knutzen, V. K., Milano, C. M. and Gleicher, N. (1979). Uterine rupture with the use of vaginal prostaglandin E₂ suppositories. *Am. J. Obstet. Gynecol.*, **134**, 348-9

127 Valenzuela, G., Hayashi, R. H., Lackritz, R. M. and Soriero, O. M. (1980). Uterine rupture at term with vaginal prostaglandin E₂. *Am. J. Obstet. Gynecol.*, **138**, 1223-4

128 Sawyer, M. M., Lipshitz, J., Anderson, G. D. and Dilts, P. V. (1981). Third-trimester uterine rupture associated with vaginal prostaglandin E₂. *Am. J. Obstet. Gynecol.*, **140**, 710-11

129 Liggins, G. C. (1979). Controlled trial of induction of labor by vaginal suppositories containing prostaglandin E₂. *Prostaglandins*, **18**, 167-72

130 Blackburn, M. G., Mancusi-Ungaro, H. R., Orzalesi, M. M., Hobbins, J. C. and Anderson, G. G. (1973). Effects on the neonate of the induction of labor with prostaglandin F₂α and oxytocin. *Am. J. Obstet. Gynecol.*, **116**, 847-53

131 Thiery, M. and Amy, J. J. (1978). Perinatal effects of natural prostaglandins used for labor induction. In Coceani, F. and Olley, P. M. (eds.) *Advances in Prostaglandin and Thromboxane Research*. Vol. 4, pp. 307-24 (New York: Raven Press)

132 De Coster, W., Goethals, A., Vandierendonck, A., Thiery, M. and Derom, R. (1976). Labor induction with prostaglandin F₂α. Influence on psychomotor evolution of the child in the first 30 months. *Prostaglandins*, **12**, 559-64

133 Friedman, E. A., Sachtleben, M. R. and Wallace, A. K. (1979). Infant outcome following labor induction. *Am. J. Obstet. Gynecol.*, **133**, 718-22

134 Lange, A. P., Secher, N. J., Westergaard, J. G. and Skovgård, I. (1982). Neonatal jaundice after labour induced or stimulated by prostaglandin E₂ or oxytocin. *Lancet*, **1**, 991-4

135 Phelan, J.P., Meguiar, R.V., Matey, D. and Newman, C. (1978). Dramatic pyrexic and cardiovascular response to intravaginal prostaglandin E₂. *Am. J. Obstet. Gynecol.*, **132**, 28-32

136 Nelson, G.H. and Bryans, C.I. (1979). Body temperature recordings during oral prostaglandin E₂ induction of labor. *Obstet. Gynecol.*, **54**, 585-7

137 Roberts, G., Anderson, A., McGarry, J. and Turnbull, A.C. (1970). Absence of antidiuresis during administration of prostaglandin $F_{2\alpha}$. *Br. Med. J.*, **2**, 152-4

138 Lange, A.P. (1983) Prostaglandins as abortifacients in Denmark. *Acta Obstet. Gynecol. Scand. Suppl.*, **113**, 117-24

139 Brinsden, P.R.S. and Clark, A.D. (1978). Postpartum haemorrhage after induced and spontaneous labour. *Br. Med. J.*, **2**, 855-6

140 MacKenzie, I.Z. (1979). Induction of labour and postpartum haemorrhage. *Br. Med. J.*, **1**, 750

141 Chassar Moir, J. (1964). The obstetrician bids, and the uterus contracts. *Br. Med. J.*, **2**, 1025-9

142 Grünberger, W., Huber, J. and Husslein, P. (1984). Local application of PGE₂ by means of a portio-adapter. *Acta Obstet. Gynecol. Scand.*, **63**, 293-7

143 Bergsjø, P., Schmidt, E. and Pusch, D. (1983). Differences in the reported frequencies of some obstetrical interventions in Europe. *Br. J. Obstet. Gynaecol.*, **90**, 628-32

144 Howie, P.W. (1977). Induction of labour. In Chard, T. and Richards, M. (eds.) *Benefits and Hazards of the New Obstetrics.* pp. 83-99. (London: W. Heinemann)

12
Inhibition of labour

M. THIERY and J. J. AMY

INTRODUCTION

To prolong gestation, even for only the few days required to achieve pulmonary maturity, tocolysis must be initiated at the earliest possible stage after the onset of labour. Unfortunately, it is difficult to make the diagnosis of preterm labour. In theory, this diagnosis should be based on the detection of changes in cervical length, dilatation and/or consistency brought about by an increase in myometrial activity rather than by the patient's subjective perception of the strength of her contractions. It is particularly important to distinguish cervical dilatation resulting from isthmic (cervical) incompetence from genuine preterm labour, as these two entities call for very different treatments. Moreover, whenever chronic tocolysis is considered, membranes should be intact and subclinical intrauterine infection should not be present, as both adversely affect the efficacy of tocolysis[1]. When present, intrauterine infection is a threat to both the fetus and the mother, which requires delivery without delay. Potent tocolytic (uterorelaxant) drugs are now available for chronic treatment of preterm labour, as well as for acute short-term suppression of uterine activity.

LONG-TERM (CHRONIC) TOCOLYSIS

General measures

According to Mathews et al.[2] bedrest (preferably in the lateral or semi-Fowler position for prevention of vena cava compression) is *the* most efficacious of all measures that can be taken to arrest untimely uterine activity. Postponement of delivery by 1 week or more is achieved in 25-50% of patients with intact membranes who were confined to bed[3,4]. However, the wide range of success reported in the literature probably reflects differences in the criteria applied for diagnosing preterm labour[5]. Until recently, loading with intravenous fluids was almost routinely initiated in women with signs of preterm labour in the USA. This procedure aimed at inhibiting endogenous oxytocin release. Because of the risk of pulmonary oedema, particularly when applied in conjunction with the administration of a β-mimetic agent and a corticosteroid, fluid expansion therapy has largely been abandoned. Factors such as frequent vaginal examination[6] which may promote the synthesis or release of endogenous uterotonic compounds (e.g. prostaglandins) should be avoided.

Pharmacological agents

Tocolytics belong to a wide variety of pharmacological agents, having a variety of mechanisms of action. The combined use of different types of these drugs may have a greater efficacy.

Ethyl alcohol

Treatment with intravenous alcohol[7] aims at inhibiting the release of oxytocin from the maternal and fetal pituitary glands. At high concentrations, alcohol may also exert a direct inhibition on myometrial activity[8]. For the purpose of tocolysis, a 9·5% solution of ethanol in 5% dextrose in water solution is administered at 7·5 ml/kg body weight for 2 h as a loading dose, then at 1·5 ml/kg for a maintenance dosage of up to 10–12 h. With this regimen, blood levels of alcohol of 0·09–0·16% are obtained[7]. Contractions often cease within the first few hours. In about 30% of the cases, however, the treatment must be repeated. Since an average of 10 h is required for the patient to eliminate this amount of ethanol, the loading dose must be reduced if treatment has to be resumed within 10 h of discontinuation. If contractions recur for the second time, further alcohol administration should be withheld[9,10].

Several authors[9-11] have found i.v. alcohol effective in postponing delivery; others have not[12,13]. When membranes are ruptured, this therapy is of little value. In comparative trials, ethanol was shown to be more effective than placebo but less effective than i.v. ritodrine[14] or parenteral[15] $MgSO_4$ in arresting preterm labour. At the dose needed for delaying delivery, ethanol produces palpitations, nausea and vomiting (due to gastritis), restlessness and inebriation. Alcohol inhibits the release of vasopressin (ADH), and care should be taken not to cause dehydration due to excessive diuresis. Nursing personnel must be available to supervise the patient, since she is likely to become intoxicated. Concomitant with ethanol therapy, gastric acid secretion is stimulated, thereby increasing the risk of regurgitation of gastric contents should general anaesthesia be used at delivery. Sedatives should be administered sparingly, as they potentiate the CNS-depressing effect of ethanol. At high doses, alcohol may adversely affect platelet function and consequently the process of haemostasis[16,17]. Alcohol administration also causes an increase in lactate–pyruvate ratio in the mother[15] which may lead to fetal acidosis as has been observed in sheep[18].

Ethanol freely passes the placenta, and the fetus is exposed to the same concentration of alcohol as is the mother. Alcohol is metabolized primarily in the liver, but the capacity of the fetal liver to oxidize this compound is much less than that of the adult. Neonatal depression is seen when treatment has been pursued until delivery.

Although animal experiments have shown a toxic effect of ethanol on the fetal liver which may persist into adulthood[19], no deleterious effect has been detected in children after maternal administration of ethanol for inhibition of preterm labour[20,21] unless birth had taken place within 15 h of termination of the infusion. In the latter event, alterations in development and personality have been noted at the age of 4–7 years[22].

Magnesium sulphate
Despite the fact that parenteral magnesium sulphate has been used for treatment of eclampsia for over 50 years, its uterodepressant effect was not recognized until relatively recently[23]. It is now commonly used as a tocolytic[24]. Magnesium ions inhibit the secretion of acetylcholine at the neuromuscular junction and compete with calcium ions at the level of the uterine muscle cell. A decrease in myometrial activity is noted as soon as the magnesaemia reaches a level of 6–8 mmol/l[24,25]. This effect can be obtained by slow i.v. injection (over 10 min) of 4 g $MgSO_4$, followed by continuous infusion of 2 g/h. Intramuscular injections are painful and should be avoided if possible.

Intravenous $MgSO_4$ is as effective as i.v. terbutaline[26] and more effective than i.v. ethanol[15] in delaying delivery. The combination of i.v. $MgSO_4$ with a β-mimetic administered orally (isoxsuprine[27] or ritodrine[28]) gives good results, but some authors[29] fear that it may expose the patient to a greater risk of cardiovascular complication.

At dosages currently applied for chronic tocolysis, $MgSO_4$ causes minimal side-effects (mainly nausea and flushing[30]) and hardly affects the process of maternal haemostasis[31]. Nonetheless, the drug should be given cautiously. Monitoring of the maternal respiratory rate and patellar reflexes is mandatory, as both are simpler than repeat assessment of magnesaemia. Cardiovascular and pulmonary intolerance have occasionally necessitated discontinuation of MgSO$_4$ administration[30], and pulmonary oedema has been reported[30] in women receiving i.v. MgSO$_4$, a number of whom had been given betamethasone simultaneously[31]. At dosages higher than 3 g/h, MgSO$_4$ may affect myocardial contractility[32] and cardiac arrest has been reported[33]. MgSO$_4$ causes chest pain and tightness less frequently than does ritodrine[30,34]. When used in combination, however, both drugs cause precordial pain (often associated with ECG changes indicative of myocardial ischaemia) more often than either one administered alone[29]. One maternal death due to respiratory arrest in an eclamptic patient administered MgSO$_4$ is on record[35]. MgSO$_4$ treatment produces fewer fetal metabolic changes than i.v. ethanol[31]. In sheep, an increase in uteroplacental blood flow has been reported[36]. Likewise, i.v. infusion of magnesium aspartate hydrochloride in man causes a rise in the blood levels of hPL. This increase probably reflects an improvement in placental function[37]. Orally administered magnesium salts have been shown to decrease the dose of β-adrenergic drug required to effect tocolysis[37,38].

Calcium entry blockers

Influx of calcium into the cytosol of the cell is essential for activation of smooth contractile proteins. Calcium ions enter the cell through specific channels in the membrane. This process is inhibited by a heterogeneous group of drugs called *calcium antagonists* or *calcium entry (channel) blockers*. Examples include verapamil (Isoptin®), compound D 600, nifedipine (Adalat®) and nicardipine, all of which are currently used in cardiology[39].

In obstetrics, calcium antagonists were first used with the aim of counteracting the cardiotoxicity of β-mimetics[40]. However, the cardioprotective effect of calcium entry blockers has recently been questioned[41-43]; moreover, verapamil is thought to be hepatotoxic[44]. The tocolytic action of calcium entry blockers has not been studied in depth. Verapamil has little efficacy[43,45], whereas nifedipine has a definite tocolytic effect when given orally in a dose of 10–20 mg every 8 h to women in preterm labour or to those undergoing termination of pregnancy[46] with intra-amniotic prostaglandin F$_{2\alpha}$. Nifedipine should be used with caution, however. A severe hypotonic haemorrhage may occur following its administration up to the day of delivery. The uterus in those cases is unresponsive to oxytocin, PGF$_{2\alpha}$ and ergot derivatives, but contracts after rapid intra-uterine lavage with 1000–2000 ml of physiological saline at 50 °C.

β-Mimetic agents

β-Adrenergic agonists are the most widely used tocolytic substances.

Table 12.1   Effects on stimulation of β-adrenoreceptors

β_1-effects	β_2-effects
Positive chronotropic and inotropic effects on the myocardium	Coronary vasodilatation
	Arteriolar dilatation
	Bronchodilatation
Decreased intestinal motility	Uterine relaxation
	Tremor
Lipolysis	Glycogenolysis

β-Adrenergic receptors – mainly of the β_2-type – are found in human myometrium[47]. The rather non-selective β-mimetic agents, isoxsuprine and orciprenaline, were used initially and caused important maternal side-effects. At present, agonists with greater β_2-specificity are being used: these include ritodrine, fenoterol and salbutamol. These latter drugs are of comparable tocolytic efficacy and show no notable differences in terms of side-effects. Indeed, none of the β-mimetic compounds currently used for tocolysis is entirely free of β_1-activity[48], a fact which explains some of their unwanted effects, particularly those of a cardiovascular nature. Stimulation of β_2-receptors other than in the uterus is responsible for hepatic gluconeogenesis, insulin release, tremor, vasodilatation and bronchodilatation observed during β-mimetic treatment (Table 12.1).

The β-adrenergic agonist binds to a β_2-receptor on the cell membrane to form a complex which activates adenylate cyclase. The rise in intracellular cAMP concentration that follows this process induces phosphorylation of cAMP-dependent protein kinase and subsequently causes sequestration of calcium in the mitochondria and the sarcoplasmic reticulum and a decrease in the activity of myosin light-chain kinase. This results in relaxation of the myometrial cell[49,50].

β-Mimetic agonists are broken down in the intestinal tract and in the liver to physiologically inactive metabolites which are eliminated with the parent compound via the urine. Mono-amino-oxidase activity in the placenta and the fetal intestines and liver explains further metabolism of the β-adrenergic drug after transplacental passage. Differences in metabolic enzyme activity are responsible for the individual variations in blood concentration, duration of action and efficacy of the β-mimetic agent during chronic tocolysis.

Because it is rapidly metabolized, oral doses of up to 20 mg ritodrine need to be repeated at 2–4 hourly intervals for chronic tocolysis[51]. It has been suggested that oral administration of β-adrenergic drugs for chronic tocolysis lacks efficacy[52], and that the i.v. route should be used in preference. Effective i.v. infusion rates vary between 50 and 400 μg/min, but the higher dose levels are often poorly tolerated.

β-Mimetic agents cross the placenta freely[52]. The significant difference observed between concentrations in maternal peripheral and uterine vein blood reflects placental inactivation of these compounds[52]. β-Mimetics decrease the intensity and the frequency of contractions of the non-gravid, gravid and puerperal myometrium[47], and they counteract the stimulant effect of oxytocin and prostaglandins[53,54].

Many clinical trials for assessment of chronic tocolysis with β-adrenergic drugs were poorly designed and are difficult to interpret. Of the better ones, all concerned ritodrine except one trial conducted with terbutaline[5]. Wesselius-de Casparis et al.[3] reported postponement of delivery by at least 1 week in 80% of the women receiving ritodrine i.v. as compared to 48% of those women given a placebo. In patients with intact membranes, delivery was delayed by a mean of 28 days in the study group, as compared to 17 days in the control group; this difference was significant. Ritodrine has also been shown to be more efficacious in postponing delivery than chlordiazepoxide (Librium®)[55] or ethanol[14,56].

Excessive myometrial inhibition leading to post-term gestation, dysfunctional or hypotonic postpartum haemorrhage has not been reported[57]. Nevertheless, active management of the third stage is advocated when i.v. β-mimetic administration has been pursued into the late second stage of labour[57]. At present, little is known about the effect of β-agonists on the lower uterine segment and the cervix[49].

The most conspicuous reaction to i.v. β-mimetic therapy is an increase in maternal heart rate. Dose-related tachycardia results from the combined activation of myocardial β_1-receptors and the reflex adjustment of the heart to peripheral arteriolar dilatation[58]. Although diastolic arterial pressure decreases only slightly, cardiac output is markedly increased during i.v. infusion of ritodrine[59]. In the lungs, β-mimetics cause bronchodilatation, vasodilatation and an increase in microvascular permeability.

β-Adrenergic drugs have been widely used in Europe for more than a decade, but serious cardiopulmonary complications have only recently been reported. The latter include arrhythmia, myocardial ischaemia, congestive heart failure, cardiomyopathy and, most typically, pulmonary oedema[60-66]. However, many of these patients had had pre-existing heart disease[67] and the majority had received other drugs (e.g. corticosteroids) as well[63-65].

The aetiology of β-mimetic induced pulmonary oedema is unclear[68]. Most clinicians consider as the most likely explanation left ventricular failure precipitated by hypervolaemia as a result of excess hydration and the antidiuretic effect of the drug superimposed on the physiological hypervolaemia of pregnancy. Some authors are of the opinion that increased pulmonary capillary permeability also plays a role[68-70].

β-Mimetic induced pulmonary oedema is acute and reversible, pro-vided of course that treatment is instituted early. Administration of the drug must be stopped at once; fluid intake must be restricted and oxygen should be administered[68]. Katz *et al.*[64] have cautioned that diuretics should be given by the i.v. route only if hypoxaemia persists despite the aforementioned measures as a precaution against causing (or aggravating) hypokaliaemia. Due to the likelihood of ventricular extrasystole and hypokaliaemia, the use of digitalis is hazardous.

A characteristic myocardial lesion (*diffuse micronecrosis*) is found in maternal and infant β-mimetic related deaths[40,71]. This is different from genuine myocardial infarction, but is comparable to both the 'infarct-like lesion' experimentally induced in rats with isoproterenol[72] and the 'adrenergic myocarditis' of phaeochromocytoma[73]. Reliable blood tests for diagnosis of myocardial damage induced by β-mimetic administra-tion are not yet available[74].

β-Adrenergic drugs stimulate glycogenolysis in the liver and in muscles and insulin release from the pancreas resulting in maternal hyperglycaemia and hypokaliaemia[75,76]. Muscle cells lack glucose-6-phosphatase activity, and the end-product of muscle glycogenolysis is lactate. Therefore, hyperlacticaemia and slight metabolic acidosis occur during i.v. β-mimetic therapy. Maternal acidosis may be aggra-vated by release of free fatty acids (lipolysis) from adipocytes due to β_1-activation[76].

Intravenous β-mimetic infusion inhibits diuresis in a manner closely similar to the action of ADH. The antidiuretic effect of these com-pounds may facilitate the occurrence of pulmonary oedema[69]. Meta-bolic alterations associated with i.v. β-mimetic therapy are generally mild and transient[75]. Nonetheless, it is wise to regularly assess women undergoing chronic tocolysis for hypokaliaemia. Diabetics require careful monitoring of plasma glucose; the dose of insulin must usually be increased during therapy.

Indirect evidence suggests that β-adrenergic drugs augment uteropla-cental perfusion, especially when the latter is compromised[77]. A signi-ficant improvement in the perinatal outcome is seen in the offspring of ritodrine-treated mothers compared to those women treated with either ethanol or placebo[56]. Part of the administered β-mimetic com-pound crosses the placenta and may therefore produce direct effects in the fetus. Indirect fetal effects result from metabolic changes in the maternal compartment[78]. During β-mimetic therapy, the fetal heart rate remains unaltered or accelerates slightly. Beat-to-beat variation is either not influenced[79] or somewhat enhanced[58].

Congestive heart failure has been reported in neonates exposed *in utero* to β-adrenergic drugs. In fatal cases, myocardial micronecrosis was found by some authors[80-82] but not by others[83]. In the rabbit fetus, high doses of fenoterol produce myocardial micronecrosis, the incidence and degree of which are inversely related to fetal heart weight[84]. Growth-retarded and very preterm babies should be considered at risk for myocardial dysfunction following exposure *in utero* to a β-mimetic, and they should be carefully assessed in this regard[82].

Acceleration of fetal pulmonary maturation following chronic tocolysis with a β-adrenergic compound was reported by Esteban-Altarriba[85] but was not confirmed by other workers[86,87]. Prolonged maternal administration of ritodrine does not alter function or morphology of the fetal endocrine pancreas[88], but transient hypoglycaemia, especially in preterm neonates and when the treatment–delivery interval was shorter than 2 days, has been observed by some authors[89,90] but not by others[91].

Clinical experience suggests that β-mimetic therapy of long duration is seldom hazardous to the infant[92]. Due to placental transfer of glucose and lactate, fetal and maternal blood levels of these substances rise in parallel, and this may be beneficial to the distressed fetus[93]. However, in the presence of fetal hypoxia, accumulation of lactic acid may take place in the brain[94].

During i.v. β-mimetic tocolysis cardiac function must be assessed periodically, and hypokaliaemia, hypomagnesaemia and fluid overload must be avoided. Corticosteroids are best not administered for acceleration of fetal lung maturation. The cardioprotective effect of calcium entry blockers is no longer accepted[95,96].

The highly specific β1-adrenergic blocker metoprolol has been given in association with i.v. infused β2-mimetics in an attempt to suppress the positive inotropic and chronotropic cardiac effects of these latter medications, but clinical experience is still limited[97–99]. Metoprolol does not diminish the tocolytic effect of fenoterol, has no influence on maternal metabolism and does not affect the neonate[97,98].

Contraindications for β-mimetic treatment are few; they include overt cardiac disease, shock and severe toxaemia[96,100,101].

Phosphodiesterase inhibitors

Phosphodiesterase physiologically hydrolyses cAMP to 5-AMP that is biologically inactive. Drugs that inhibit phosphodiesterase activity, such as papaverine, theophylline and aminophylline, all raise the intracellular levels of cAMP and thereby cause the myometrium to relax[102,103].

Recurrences of myometrial activity seen in patients chronically treated with a β-mimetic drug are frequently thought to be due to desensitization of the β-adrenergic receptor[104]. Gravidae with premature uterine activity may therefore benefit from combined or alternated treatment with a β-agonist and a phosphodiesterase inhibitor. However, phosphodiesterase inhibitors are not devoid of toxicity[105], the latter possibly being more marked when they are given in combination with a β-mimetic drug[106]. Furthermore, aminophylline readily crosses the placenta[107], and its potential fetal toxicity needs to be studied.

At the University Hospital in Brussels, gravidae with premature uterine activity are first treated on an ambulatory basis with a combination of papaverine (100 mg × 4 daily) and magnesium salts (Magnecaps®, 4 daily) given orally. This regimen is well tolerated, except for rare instances of gastric upset. It has decreased the need for other tocolytics (in particular, β-mimetics) and for hospitalization in these patients (Amy, unpublished observation).

Diazoxide

Diazoxide (Hyperstat®) is currently used for treatment of acute hypertension and insulinoma[108]. Although structurally related to thiazides, it promotes retention of sodium and water[109]. Diazoxide inhibits smooth muscle contractions, possibly by activation of adenylate cyclase[110]. It is a potent tocolytic, but is not routinely used as such. The i.v. injection of a bolus of 300 mg of diazoxide, practised until a few years ago for treatment of acute hypertensive episodes, occasionally caused sudden and severe hypotension, that could be complicated by coronary insufficiency or blindness. In gravid patients, a sudden lowering of maternal blood pressure poses a serious threat to the fetus[111]. Moreover, diazoxide readily crosses the placenta[112] and neonatal hypoglycaemia[113] and alopecia[114] have been described in infants after intrauterine exposure to this medication.

MacLean et al.[111] have shown that smaller bolus doses (75–150 mg) of diazoxide given i.v. have the same tocolytic potency without producing profound hypotension. Other authors[115,116] have eliminated the occurrence of severe hypotension by infusing 500 ml of normal saline before infusion of 300 mg diazoxide over 5–30 min. With the latter protocol, results in terms of prolongation of gestation were similar to those reported by other investigators using β-mimetic agents or $MgSO_4$ intravenously. In patients with intact membranes and less than 4 cm cervical dilatation, delivery was delayed by 72 h or more in 83% of the women. The authors stressed the ease of administration, the rapidity of action and the paucity of adverse responses.

Progestins

Based on Csapo's *progesterone block theory*[117], several investigators have attempted – unsuccessfully – to prevent or to arrest preterm labour by administration of progesterone[118-120], medroxyprogesterone acetate (MPA)[121] or α-methyl-17α-acetoxyprogesterone[122] given by various routes. Only injections of MPA directly into the myometrium[123] had any tocolytic effect, but this procedure is clinically inapplicable. Orally administered micronized natural progesterone increases the hormone concentration in the myometrium[124] but, to the best of our knowledge, clinical trials of tocolysis with this compound have not been undertaken.

Johnson et al.[125] have claimed a delay in delivery and a reduction of perinatal mortality after administration of a progestin from an early stage of pregnancy onwards, to women presenting a high risk for preterm labour. Possibly progestins potentiate the tocolytic effect of other substances[126].

Prostaglandin synthesis inhibitors

PGSIs, also known as non-steroidal anti-inflammatory drugs (NSAIDs) or aspirin-like drugs constitute a heterogeneous group of chemicals that bind in a reversible or irreversible way to cyclo-oxygenase. Inactivation of this key-enzyme in the arachidonic acid cascade prevents conversion of this lipidic substance into the endoperoxides PGG_2 and PGH_2, and into the prostanoates derived therefrom, namely prostacyclin, thromboxane A_2, PGD_2, PGE_2 and $PGF_{2\alpha}$.

Reversible inhibitors such as indomethacin interfere with PG biosynthesis only as long as their tissue concentration is sufficiently high. On the contrary, irreversible inhibitors such as aspirin cause prolonged reduction of synthetase activity, even when the concentration of the PGSI has dropped below an effective competitive level. In the latter instance, synthesis of new cyclo-oxygenase is needed to reactivate the production of prostanoates.

Beside their inhibition of prostaglandin synthesis, PGSIs may also act at the uterine level by decreasing phosphodiesterase activity, thereby raising the intracellular content in cAMP and thus contributing to myometrial relaxation[127]. The efficacy of PGSIs in terms of delaying delivery is probably also partly due to their arresting the structural changes that occur in the cervical stroma during late pregnancy that lead to effacement, softening and dilatation of the cervix[128].

Indomethacin is the PGSI most frequently used for tocolysis and the pharmacokinetics[129,130] of only this PGSI will be summarized.

Given orally, the drug is rapidly and completely absorbed; peak plasma concentrations are achieved within 1–2 h. Simultaneous ingestion of food reduces and delays emergence of peak plasma levels without affecting the amount of drug being absorbed. Rectal administration of indomethacin is associated with earlier but lower peak plasma concentrations and incomplete absorption.

After administration of therapeutic doses, at least 90% of the indomethacin in plasma is bound to proteins. 60% of an oral dose is eventually excreted in the urine, predominantly in the glucuronized form; the remainder is eliminated in the faeces via the biliary tract. A large amount of the dose undergoes an enterohepatic cycle. In adults, indomethacin has a biological half-life of 5-10 h.

Indomethacin crosses the placenta and is also excreted in the maternal milk. Fetal plasma levels are at least as great as maternal levels. Moreover, the volume of distribution of the drug in the fetus and the neonate is comparatively larger than in the older child and the adult. In view of this and its deficient glucuronidation and renal excretion processes, the neonate is only able to eliminate the PGSI very slowly, in particular if it is preterm. In preterm infants, the half-life of indomethacin is inversely correlated with gestational age.

In rat myometrial strips, PGSIs simultaneously suppress PG production and spontaneous contractile activity[131,132]. Some of these agents also antagonize the oxytocic effect of PGs on human and rabbit myometrium[133]. When given during the latter part of pregnancy to rats[134], rabbits[135], hamsters[136] or Rhesus monkeys[127], PGSIs cause a significant delay in the onset of parturition and prolongation of labour.

In man, Waltman et al.[137] noted a marked lengthening of the instillation-abortion interval in women aborted by intra-amniotic instillation of hypertonic saline when they had received aspirin or indomethacin. Reiss et al.[138] reported having stopped or considerably prolonged established labour at term by rectal administration of a single suppository of 100 mg indomethacin.

Epidemiological data and clinical trials have confirmed the tocolytic action of PGSIs. In a retrospective study, Lewis and Schulman[139] found that gestation was lengthened by an average of 7 days ($p < 0.025$) in women having taken daily 3·25 g or more of aspirin during the last months of pregnancy. In 42% of these patients, gestation lasted 42 weeks or more as compared to only 3% in women having taken no PGSI ($p < 0.001$). In treated patients, labour lasted 12 h vs. 7 h in controls ($p < 0.005$). The prospective study conducted by Collins and Turner[140] corroborated these findings.

The first reports on the deliberate use of a PGSI for arresting preterm labour are those of Zuckerman et al.[141] and Wiqvist et al.[142], published approximately a decade ago. The latter group[142] administered 25 mg of indomethacin orally every 6 h during 5 days to nine women with persistent uterine activity despite bedrest and isoxsuprine administration. Treatment was repeated, at 5-10 day intervals, until 35 weeks' gestation. Labour was arrested in eight of the nine gravidae treated. Zuckerman et al.[143] expanded their initial uncontrolled study[141] to examine a total of 297 women at 24-34 weeks' gestation presenting with regular, painful contractions occurring with a frequency of 2/10 min or more. Rupture of the membranes was diagnosed in 45 of the cases. Patients with other complications were excluded from the study. These women were given one dose and, if necessary, a second dose 1 h later of 100 mg indomethacin rectally, followed by 25 mg of the same medication orally at 6 hourly intervals until contractions ceased for 24 h. The total dose administered varied between 200 and 300 mg per patient. No other tocolytic agent was given. As a rule, 30 min after initiation of therapy, a decrease in the frequency and the intensity of uterine contractions was noted; complete myometrial quiescence occurred after 2–3 h. Repeat treatment was required in 48 patients in whom contractions recurred after a 2 week interval. In 246 of the cases (83%) delivery was delayed by 1–12 weeks; in an additional 30 cases (10%) delivery was postponed by 2–7 days. The treatment was less effective in patients with a cervical dilatation of 4 cm or more or if the membranes were previously ruptured. 51 infants were born before 37 completed weeks of gestation; 15 of these died of hyaline membrane disease.

The most convincing studies were conducted in a double-blind or randomized fashion and compared the efficacy of a placebo, a β-adrenergic agent or ethanol and indomethacin, either given alone or in conjunction with one of the aforementioned tocolytics[144-147]. These controlled clinical trials established beyond doubt that indomethacin possesses a powerful tocolytic effect when administered during pregnancy. 36 women at 25–35 weeks gestation with painful contractions occurring with a frequency of 2/10 min or more or with contractions having caused cervical effacement and a dilatation of 1–4 cm were allocated in a double-blind fashion to a treatment group and a control group. Patients with ruptured membranes or with other complicating factors were excluded from the study. The 18 women receiving indomethacin were given 1 or 2 doses (interval: 1 h) of 100 mg of the drug rectally, followed by 25 mg orally at 6 hourly intervals for a total of 4 doses. The total dose given to each patient amounted to 200–300 mg. The 18 women belonging to the control group were administered a placebo according to the same protocol. If in any of the 36 women studied labour was thought to continue after insertion of the second rectal suppository, ritodrine was administered. The route of administration and dosage of the β-mimetic are not specified in the paper. 15 of 18 women (83%) given indomethacin, as compared to 3 of the 18 control subjects (16·5%), delivered more than 1 week after initiation of therapy. Delivery was postponed by 3–7 days in 2 (11%) of the remaining treated women and in 1 (5·5%) of the control subjects. Only 1 patient given indomethacin (5·5%), as compared to 14 (78%) receiving the placebo, delivered less than 3 days after onset of treatment. Duration of pregnancy averaged $36·4 \pm 0·7$ weeks in women treated with indomethacin vs. $31·2 \pm 0·7$ weeks in controls ($p < 0·001$). Likewise, the average birthweight was $2·833 \pm 117$ g in the former group compared to only $2·028 \pm 123$ g in the latter.

Other pilot studies have shown that naproxen sodium (275 mg orally, 4–6 times daily)[148] and flufenamic acid (200–400 mg orally, every 1–2 h, followed by 800–1200 mg/day during 4 days)[149] may also be effective tocolytics.

Due to the brief duration of the treatment courses, maternal tolerance to PGSIs administered for tocolysis is generally excellent. Adverse effects occasionally encountered during therapy with indomethacin include nausea, vomiting and dyspepsia, vertigo, tinnitus and maculopapular rash. All subside within 1 day following discontinuation of drug administration. Since aspirin notably impairs haemostasis, it may cause increased intrapartum and postpartum bleeding even when the last dose has been taken several days before delivery[150]. However, aspirin is rarely if ever used for the purpose of tocolysis during the second half of pregnancy.

Reimann and Frölich[151] have cautioned against combining a PGSI with a β-adrenergic drug for treatment of preterm labour. According to these authors, several cases of pulmonary oedema have been reported in patients so treated. They attributed the increased risk of pulmonary oedema to the fact that both medications had an antidiuretic effect *per se* and that their combination led to a much greater degree of water retention.

The most serious complication encountered following intrauterine exposure to a PGSI during late pregnancy is related to the induction of premature closure of the ductus arteriosus (DA). The patency of the fetal DA is maintained by the low partial pressure of oxygen (PaO_2) in the blood and the high level of local, intramural production of PGE_2. The elevated circulating concentrations of PGE_2 that are characteristic of the fetus may play an accessory role in maintaining the patency of the ductus[152]. Administration of indomethacin to the pregnant rat[153], the ewe[154], the rabbit[155], or directly to the fetus near term causes a narrowing of the DA. The lumen of the vessel may be reduced to 20% of its original diameter. The constriction of the DA in turn causes (1) an increase in pulmonary arterial blood flow and, if the latter is maintained, (2) hypertrophy of the media of the small pulmonary arteries, and (3) an increase in pulmonary vascular resistance. Subsequently, there develops: (a) an increase in right ventricular and diastolic pressure, (b) in some cases, subendocardial ischaemia of the right side of the heart, and (c) tricuspid insufficiency.

Constriction of the DA is seen with great consistency after intrauterine exposure to a PGSI during late pregnancy in the animal species mentioned above. Momma and Takeuchi[156] studied the effects of 24 different PGSIs administered to full-term pregnant rats. All acidic compounds constricted the fetal DA in a dose-dependent manner; however, considerable differences in the intensity of the effects were noticed with doses comparable to therapeutic doses in man. Interestingly, six of the eight basic PGSIs (benzydamine HCl, perisoxal citrate, mepirizole, tinoridine HCl, MK-447 and ONO-3144) did not constrict the fetal ductus even when given at 50–100 times the equivalent of the usual clinical dose[156].

In man, partial closure of the DA accompanied by pulmonary hypertension is rarely seen in newborns exposed *in utero* to a PGSI. As of 1980, less than 20 such cases had been reported[157]. Clinically, these infants present with cyanosis, resistant hypoxaemia and, in some cases, congestive heart failure. A murmur of tricuspid insufficiency is heard in 50% of the cases. The chest X-ray and the electrocardiogram are of little value for the diagnosis. The most valuable tests are the echocardiogram and the cardiac catheterization; the latter shows a right-to-left shunt through the foramen ovale and/or the DA. In a small number of cases this complication is fatal. Most of the other affected infants recover due to a gradual decrease of pulmonary vascular resistance and resolution of the tricuspid insufficiency, but intensive therapy may be required to support these infants during their first days of life[158].

Wiqvist[157] assembled from the literature a series of 730 women given a PGSI for tocolysis during the late second or the third trimester of pregnancy. The incidence of pulmonary hypertension in the neonates was less than 3%, but it varied considerably among the various centres. From our own analysis of the literature, it appears that there is no clear-cut correlation between the expression of the syndrome and the dose of PGSI administered, the gestational age at the time of treatment or the interval between the termination of treatment and delivery[128]. Zuckerman *et al.*[143,147] did not encounter a single case of persistent pulmonary hypertension among the infants born to 315 women treated with indomethacin for tocolysis at 24–35 weeks gestation.

Total neonatal mortality in the group surveyed by Wiqvist[157] amounted to 7%. Among the 17 infants presenting with pulmonary hypertension, 14 recovered with therapy. The risk of neonatal death occurring as a consequence of persistent pulmonary hypertension following intrauterine exposure to a PGSI given for tocolysis would therefore amount to less than 0·5%. However, on the basis of available reports, it is not possible to determine whether or not there is an increased incidence of death *in utero* due to ductal closure.

Other effects of PGSIs on the fetal circulation include a decrease in blood flow to the brain, liver, intestines and kidneys, and an increase in perfusion of the heart, adrenals and placenta[158,159]. Short-term administration of PGSI has no effect on fetal heart rate, fetal systemic arterial blood pressure and fetal arterial blood pH[127,160]. Maternal uteroplacental blood flow is unaffected as well[127,160].

PGSIs also affect intrauterine breathing. Fetal lambs breathe about 40% of the time and only when the electroencephalogram shows low voltages–fast activity consistent with rapid eye movement (REM) sleep. Exposure of the fetus to a PGSI strongly stimulates its respiratory movements which increase in frequency and in amplitude, and occur continuously, regardless of the pattern of cerebral activity[160].

Stuart et al.[150] reported bleeding tendencies in nine out of ten neonates exposed in utero to aspirin during the 5 day period preceding delivery. These haemostatic abnormalities due to the irreversible binding of aspirin to platelet cyclo-oxygenase included numerous petechiae over the presenting part, haematuria, one cephalhaematoma, one subconjunctival haemorrhage and one case of bleeding from a circumcision wound. Reversible inhibitors, such as indomethacin, on the other hand, have seldom caused disturbances of coagulation in the neonate even when administered to the mother during the last days of pregnancy.

Chronic administration of very high doses of indomethacin to pregnant Rhesus monkeys resulted in a considerable lengthening of gestation (180 ± 2 vs. 167 ± 0.4 days). Fetuses exposed to indomethacin in utero and delivered by elective Caesarean section on day 180 or later were significantly heavier than the 160-day untreated fetuses, indicating continued growth beyond term despite exposure to the PGSI. However, some fetal organs (e.g. the kidneys) were smaller, and others (e.g. the liver) were larger than expected. Chronic administration of indomethacin was also associated with oligohydramnios, meconium staining and a 50% fetal mortality. It was suggested that the PGSI, by means of a decreased renal PGE_2 synthesis, caused renal vasoconstriction and antidiuresis in the fetus[127]. Indomethacin was also shown to increase the number of areas of neuronal micronecrosis in the fetal rat brain[161]. Further experimental investigation of this problem is required.

In man, neonates are rarely adversely affected by intrauterine exposure to a PGSI (other than aspirin) given to the mother for the purpose of tocolysis. Although theoretically a predisposing factor for neonatal hyperbilirubinaemia (the PGSI binds to serum albumin), only very sick infants with a syndrome of persistent pulmonary hypertension have been reported to have high levels of bilirubin. Babies fare significantly better after this type of maternal treatment (greater birthweight, better Apgar scores, lesser frequency of hyaline membrane disease, lesser mortality) than infants born to untreated mothers[147]. Furthermore, long-term (5 years) follow-up of these children has demonstrated normal development[143].

The contraindications to the use of PGSIs for tocolysis are listed in

Table 12.2   Contraindications to the use of PGSIs for tocolysis

Allergy to the PGSI
History of drug-induced asthmatic attacks
Hepatic or renal insufficiency
Gastric or duodenal ulcer
Haemorrhagic diathesis
?Rupture of the membranes?

Table 12.2. Although PGSIs undoubtedly induce profound readjustments in the biology of the fetus, deleterious perinatal effects have been seen in a minority of cases only. The very careful use of a PGSI appears justified in those cases of threatened preterm delivery that do not respond to bedrest and to more commonly used tocolytics (e.g. β-mimetic drugs, MgSO$_4$). The lowest effective dose, given in an intermittent fashion when possible, should be used[128,162].

SHORT-TERM (ACUTE) TOCOLYSIS

Indications for short-term relaxation of the uterus[78,163-165] are listed in Table 12.3.

β-Adrenergic agents are either given by slow i.v. bolus injection (e.g. up to 400 μg/min) or infused i.v. (e.g. ritodrine 100–400 μg/min) for the duration required. All currently used β-mimetic drugs are of comparable efficacy, and they act equally rapidly. For acute intrapartum tocolysis, preference should be given to a compound with short uterine action (e.g. fenoterol) to avoid prolonged hypotony[74]. The clinician should be aware that a bolus of β-agonist given i.v. will exert a strong positive inotropic effect, which may unmask undiagnosed heart disease and further complicate an emergency situation.

Good results have also been reported with the use of MgSO$_4$ i.v.[170,171] (slow bolus injection of 2–4 g, followed – if necessary – by an infusion at the rate of 2 g/h). Finally, as shown by Reiss et al.[138], even spontaneous progressive term labour can, in a majority of cases, be either arrested or markedly slowed down with a single rectal suppository of indomethacin. Tolerance to this drug is generally excellent, but obviously its tocolytic effect is more protracted than that of MgSO$_4$ or a β-mimetic drug given i.v. over a short period of time.

The uterorelaxant effect of all three types of tocolytics (β-agonists, MgSO$_4$, PGSIs) can be rapidly neutralized[53,54] by administration of oxytocin (i.v.), prostaglandins (i.v.; postpartum: also intramyometrial, e.g. PGF$_{2\alpha}$ 1–2 mg) or an ergot derivative (only postpartum). Adverse effects due to MgSO$_4$ can be controlled by slow i.v. injection of an ampoule of calcium gluconate.

Table 12.3 Short-term (acute) tocolysis

Indication	Agent of choice, comment
Prophylactic: at cervical (e.g. cerclage), uterine (e.g. fetoscopy) or abdominal surgery	PGSI, single (or multiple) dose(s)
Postponement of delivery × 24 h for use of glucocorticoids (acceleration fetal lung maturation)	Avoid β-mimetics i.v.: greater risk of pulmonary oedema
– premature rupture of membranes	MgSO₄ i.v. (PGSI + corticoid: greater risk of infection?)
– preterm labour	PGSI (+ MgSO₄ i.v.)
Facilitation external cephalic version (breech presentation)[165–168]	β-mimetic i.v.: 36% transient FHR decelerations[164], 2% partial separation of placenta or cord presentation[166]
Transfer of labouring woman to operating room or to (other) hospital[164]	PGSI (+ MgSO₄ i.v.) or β-mimetic i.v.
Acute fetal distress during labour: intrauterine resuscitation[164,169,170]	β-mimetic i.v.[164] (but: ↑FHR and ↑fetal acidosis) or MgSO₄ i.v.[169,170]
Threatening uterine rupture[162]	β-mimetic i.v.
Facilitation internal podalic version; mobilization locked twins[171]	β-mimetic i.v.
Postpartum: manual removal retained placenta[162]: correction inverted uterus[172]	β-mimetic i.v.

Calcium entry blockers, in particular nifedipine, should not be used shortly (24 h or less) before delivery because of the very real risk of inducing postpartum atony which is resistant to the action of oxytocics (Amy, unpublished observation). The use of anaesthetics such as halothane or other ether derivates for acute tocolysis is considered obsolete.

SUMMARY

Although the intricate mechanisms involved in the initiation and maintenance of myometrial activity during human parturition are not fully understood[49,50], acquaintance with available data is essential for safe and successful tocolysis. Indications and contraindications for, as well as limitations to, the administration of any tocolytic alone or in combination must be kept in mind. The present trend is to resort more rapidly to a combination of two (or more) tocolytics having different mechanisms of action to increase the efficacy of treatment. However, experience is still lacking and caution must be exerted if adverse effects are to be kept to a minimum.

Endogenous prostaglandins play an extremely important role in parturition[50]. Not only do they cause intracellular release of activator calcium in the myometrium (as does oxytocin), but they also induce formation of gap junctions between myometrial cells. These structures are essential for the spread of action potentials from cell to cell and the generalization of the contraction to the entire myometrium. Finally, prostaglandins mediate cervical ripening that takes place before labour and during early labour. PGSIs, by suppressing each of these events, provide a rational basis for their use as tocolytics. Their efficacy is unquestioned. When a serious threat of preterm delivery exists, their proper use encompasses a risk that is quite acceptable to our mind. They also have a number of valid indications in the area of acute or short-term tocolysis.

ACKNOWLEDGEMENTS

The authors are indebted to Ms Bea Pion, Ms Marleen Van der Helst, Ms Jacqueline Pyfferoen and to Ms Jet Sagerman for expert secretarial assistance.

References

1 Handwerker, S. M., Tejani, N. A., Verma, U. L. and Archbald, F. (1984). Correlation of maternal C-reactive protein with outcome of tocolysis. *Obstet. Gynecol.*, **63**, 220-4

2 Mathews, D. D., Agarwal, V. and Shuttleworth, T. P. (1982). A randomized controlled trial of complete bed rest vs. ambulation in the management of proteinuric hypertension during pregnancy. *Br. J. Obstet. Gynaecol.*, **89**, 128-31

3 Wesselius-de Casparis, A., Thiery, M., Yo Le Sian, A., Baumgarten, K., Brosens, I., Gamissans, O., Stolk, J. and Vivier, W. (1971). Results of double blind multicentre study with ritodrine in premature labour. *Br. Med. J.*, **3**, 144-7

4 Zlatnik, F. J. and Fuchs, F. (1972). A controlled study of ethanol in threatened premature labor. *Am. J. Obstet. Gynecol.*, **112**, 610-12

5 Ingemarsson, I. (1976). Effect of terbutaline on premature labor. *Am. J. Obstet. Gynecol.*, **125**, 520–4

6 Mitchell, M. D., Keirse, M. J., Anderson, A. B. and Turnbull, A. C. (1977). Evidence for a local control of PGs within the pregnant human uterus. *Br. J. Obstet. Gynaecol.*, **84**, 35–8

7 Fuchs, F., Fuchs, A. R., Poblete, V. F. and Risk, A. (1967). Effect of alcohol on threatened premature labor. *Am. J. Obstet. Gynecol.*, **99**, 627–37

8 Fuchs, A. R. and Fuchs, F. (1973). Possible mechanisms of the inhibition of labor by ethanol. In Josimovich, J. D. (ed.) *Uterine Contraction.* pp. 287–300 (New York: Wiley)

9 Fuchs, F. (1976). Prevention of prematurity. *Am. J. Obstet. Gynecol.*, **126**, 809–20

10 Fuchs, F., Fuchs, A. R., Lauersen, N. H. and Zesvoudakis, I. A. (1980). Inhibition of premature labor with ethanol. In Sakamoto, S., Tojo, S. and Nakayama, T. (eds.) *Proceedings of the IXth World Congress of Gynecology and Obstetrics*, Tokyo 1979. pp. 919–22. (Amsterdam: Excerpta Medica)

11 Mehra, P., Raghvan, K. S., Devi, P. K. and Chaudhury, R. (1970). Effect of intravenous alcohol on premature labour. *Int. J. Gynaecol Obstet.*, **8**, 160

12 Graff, G. (1971). Failure to prevent premature labor with ethanol. *Am. J. Obstet. Gynecol.*, **110**, 878–80

13 Watring, W. G., Benson, W. L. and Wiebe, R. A. (1976). Intravenous alcohol. A single blind study in the prevention of premature delivery. *J. Reprod. Med.*, **16**, 35–9

14 Lauersen, N., Merkatz, I., Tejani, N., Wilson, K., Robertson, A., Mann, L. and Fuchs, F. (1977). Inhibition of premature labor. A multicenter comparison of ritodrine and ethanol. *Am. J. Obstet. Gynecol.*, **127**, 837–45

15 Steer, C. and Petrie, R. (1977). A comparison of magnesium sulfate and alcohol for the prevention of premature labor. *Am. J. Obstet. Gynecol.*, **129**, 1–4

16 Elmer, O., Göransson, G. and Zoucas, E. (1981). Alcohol and blood loss. *Lancet*, **2**, 639

17 Mikhailidis, D. P., Jeremy, J. Y., Barradas, M., Green, N. and Dandona, P. (1983). Effect of ethanol on vascular prostacyclin (prostaglandin I_2) synthesis, platelet aggregation, and platelet thromboxane release. *Br. Med. J.*, **287**, 1495–8

18 Mann, L. I., Bhakthavathsalan, A., Liu, M. and Makowski, P. (1975). Placental transport of alcohol and its effect on maternal and fetal acid-base balance. *Am. J. Obstet. Gynecol.*, **122**, 837–44

19 Amankwah, K., Kaufman, R. and Weberg, A. (1982). The effect of maternal alcohol ingestion upon fetal liver. Presented at the *29th Annual Meeting of the Society of Gynecologic Investigation*, Dallas, abstract 144

20 Wagner, L., Wagner, G. and Guerrero, J. (1970). Effect of alcohol on premature newborn infants. *Am. J. Obstet. Gynecol.*, **108**, 308–15

21 Waltman, R. and Iniquez, E. (1972). Placental transfer of ethanol and its elimination at term. *Obstet. Gynecol.*, **40**, 180–5

22 Sisenwein, F. E., Tejani, N. A., Boxer, H. S. and DiGiuseppe, R. (1983). Effects of maternal ethanol infusion during pregnancy on the growth and development of children at four to seven years of age. *Am. J. Obstet. Gynecol.*, **147**, 52–6

23 Hall, D. G., McGaughey, H. S., Corey, E. L. and Thornton, W. N. (1959). The effects of magnesium therapy on the duration of labor. *Am. J. Obstet. Gynecol.*, **78**, 27–32

24 Harbert, G. M., Cornell, G. W. and Thornton, W. M. (1969). Effect of toxemia therapy on uterine dynamics. *Am. J. Obstet. Gynecol.*, **105**, 94–104

25 Petrie, R. H., Wu, R., Miller, F. C., Sacks, D., Sugarman, R., Paul, R. and Hon, E. (1976). The effects of drugs on uterine activity. *Obstet. Gynecol.*, **48**, 431–5

26 Miller, J., Keane, M. and Horger, E. (1982). A comparison of magnesium sulfate and terbutaline for the arrest of premature labor. *J. Reprod. Med.*, **27**, 348–51

27 Spisso, K., Harbert, G. and Thiagarajah, S. (1982). The use of magnesium sulfate as the primary tocolytic agent to prevent premature delivery. *Am. J. Obstet. Gynecol.*, **142**, 840–5

28 Tchilinguirian, N. G., Najem, R., Sullivan, G. B. and Craparo, F. J. (1984). The use of ritodrine and magnesium sulfate in the arrest of premature labor. *Int. J. Gynaecol. Obstet.*, **22**, 117–23

29 Ferguson, J. E., Hensleigh, P. A. and Kredenster, D. (1984). Adjunctive use of MgSO₄ with ritodrine for preterm labor tocolysis. *Am. J. Obstet. Gynecol.*, **148**, 166–71

30 Elliott, J. P. (1983). Magnesium sulfate as a tocolytic agent. *Am. J. Obstet. Gynecol.*, **147**, 277–84

31 Elliott, J. P., O'Keeffe, D. F., Greenberg, P. and Freeman, R. K. (1979). Pulmonary edema associated with magnesium sulfate and betamethasone administration. *Am. J. Obstet. Gynecol.*, **134**, 717–19

32 Critelli, G., Ferro, G., Peschl, C., Perticone, F. R., Ringo, F. R. and Condorelli, M. (1977). Myocardial contractility after injection or prolonged infusions of magnesium sulfate. *Acta Cardiol.*, **32**, 65–73

33 Engbaek, L. (1952). The pharmacological actions of magnesium ions with particular reference to the neuromuscular and the cardiovascular system. *Pharmacol. Rev.*, **4**, 396–424

34 Barden, T. P., Peter, J. B. and Merkatz, I. R. (1980). Ritodrine hydrochloride: a betamimetic agent for use in preterm labor. *Obstet. Gynecol.*, **56**, 1–6

35 Pritchard, J. A., Cunningham, F. G. and Pritchard, S. A. (1984). The Parkland Memorial Hospital protocol for treatment of eclampsia: evaluation of 245 cases. *Am. J. Obstet. Gynecol.*, **148**, 951–63

36 Dandavino, A., Woods, J. R., Murayama, K., Brinkman, C. R. and Assali, N. S. (1977). Circulatory effects of magnesium sulfate in normotensive and renal hypertensive pregnant sheep. *Am. J. Obstet. Gynecol.*, **127**, 769–74

37 Spätling, L. (1984). Magnesiumzusatztherapie zur Tokolyse: klinisch-chemische Ueberwachungsparameter. *Geburtsh. Frauenheilk.*, **44**, 19-24

38 Spätling, L., Eulenburg, R. and Mohr, K. (1981). Verringerung von vorzeitiger Wehentätigkeit durch orale Magnesiumgabe. *Arch. Gynäkol.*, **232**, 514-5

39 Godfraind, T., Herman, A. G. and Wellens, D. (eds.) (1984). *Calcium Entry Blockers in Cardiovascular and Cerebral Dysfunctions.* (Boston: Martinus Nijhoff)

40 Lechner, W., Dienstl, F., Poewe, W. and Daxenbichler, G. (1981). Myoglobinanstieg unter Tokolyse. *Arch. Gynäkol.*, **232**, 500-1

41 Trolp, R., Irmer, M., Bernius, U., Pohl, C. and Hallemans, H. (1981). Vergleichende Untersuchung zur Tokolysewirksamkeit bei Fenoterolmonotherapie und Kombinationstherapie Fenoterol/Verapamil. *Arch. Gynäkol.*, **232**, 521-3

42 Zandvoort, J. A. and Schellekens, L. A. (1975). Bestrijding van de door beta-adrenerge sympathicomimetische veroorzaakte cardiovasculaire bijwerkingen met behulp van Ca^{++} antagonisten. *Nederl. T. Geneesk.*, **118**, 736

43 Gummerus, M. (1977). Die Behandlung der vorzeitigen Wehentätigkeit und Antagonisierung der Nebenwirkungen der tokolytischen Therapie mit Verapamil. *Z. Geburtsh. Perinatol.*, **181**, 334-40

44 Guarascio, P., D'Amato, C., Sette, P., Conte, A. and Visco, G. (1984). Liver damage from verapamil. *Br. Med. J.*, **288**, 362

45 Mosler, K. H. and Rosenboom, H. G. (1972). Neuere Möglichkeiten einer tokolytischen Behandlung in der Geburtshilfe. *Z. Geburtsh. Perinatol.*, **176**, 85-96

46 Forman, A., Anderson, K. E. and Ulmsten, U. (1981). Inhibition of myometrial activity by calcium antagonists. *Semin. Perinatol.*, **5**, 288-94

47 Calixto, J. B., De Medeiros, Y. S. and Rae, G. A. (1984). Adrenergic control of uterine function. In Hafez, E. S. E. (ed.) *Spontaneous Abortion.* pp. 45-57. (Lancaster: MTP Press)

48 Lands, A., Arnold, A., McAuliff, J., Luduena, F. and Brown, T. (1967). Differentiation of receptor systems activated by sympathicomimetic amines. *Nature (London)*, **214**, 597-8

49 Huszar, G. and Naftolin, F. (1984). The myometrium and uterine cervix in normal and preterm labor. *N. Engl. J. Med.*, **311**, 571-81

50 Amy, J. J., De Brucker, O. and Merckx, M. (1984). Control of uterine activity in pregnancy. In Hafez, E. S. E. (ed.) *Spontaneous Abortion.* pp. 27-34. (Lancaster: MTP Press)

51 Van Lierde, M. and Thomas, K. (1982). Ritodrine concentrations in maternal and fetal serum and amniotic fluid. *J. Perinat. Med.*, **10**, 119-24

52 Smit, D. A. (1983). Efficacy of orally administered ritodrine after initial intravenous therapy. *MD Thesis*, University of Maastricht

53 Lipschitz, J. (1978). The tocolytic and cardiovascular effects of fenoterol and hexoprenaline in PGF$_{2\alpha}$-induced labor. Presented at the *25th Annual Meeting of the Society of Gynecologic Investigation*, Atlanta, abstract 192

54 Brabec, M. (1981). Vergleich der tokolytische Wirksamheit von Hexo-prenalin und Ritodrin bei Prostaglandinwehen. *Arch. Gynäkol.*, **232**, 506–8

55 Sivasamboo, R. (1972). Premature labor. In Baumgarten, R. and Wesselius-de Casparis, A. (eds.) *Proceedings of the International Symposium on the Treatment of Fetal Risks*, Vienna. pp. 16–20.

56 Merkatz, I. (1980). Evidence of efficacy. In Barden, T. P. (ed.) *Tocolytic Agents*. pp. 29–40. (New York: Biomedical Information Corporation)

57 Essed, G., Martin, C., Crevels, A. and Eskes, T. (1982). The influence of long term β-mimetic drug administration during pregnancy on blood loss post-partum. *Eur. J. Obstet. Gynecol. Reprod. Biol.*, **13**, 159–68

58 Lipschitz, J. and Baillie, P. (1976). The uterine and cardiovascular effects of β_2-selective sympathicomimetic drugs administered in an I.V. infusion. *S. Afr. Med. J.*, **50**, 1973–6

59 Bieniarz, J., Ivankovich, A. and Scommenga, A. (1974). Cardial output during ritodrine treatment in premature labor. *Am. J. Obstet. Gynecol.*, **118**, 910–20

60 Brodey, P., Fisch, A. and Huffaker, J. (1981). Acute pulmonary edema resulting from treatment for premature labor. *Radiology*, **140**, 631–5

61 Davies, A. and Robertson, M. (1980). Pulmonary oedema after administration of intravenous salbutamol and ergometrine. *Br. J. Obstet. Gynaecol.*, **87**, 539–41

62 Anderson, A. B. M. (1981). Second thoughts on stopping labour. In Studd, J. (ed.) *Progress in Obstetrics and Gynaecology*. Vol. 1, pp. 125–38. (Edinburgh: Churchill Livingstone)

63 Ingemarsson, I. (1982). Cardiovascular complications of terbutaline for preterm labor. *Am. J. Obstet. Gynecol.*, **142**, 117–18

64 Katz, M., Robertson, P. and Creasy, R. (1981). Cardiovascular complications associated with terbutaline treatment for preterm labor. *Am. J. Obstet. Gynecol.*, **139**, 605–8

65 Nagey, D. and Crenshaw, C. (1982). Pulmonary complications of isoxsuprine therapy in the gravida. *Obstet. Gynecol.*, **59**, 38–42

66 Stubblefield, P. (1978). Pulmonary edema occurring after therapy with dexamethasone and terbutaline for premature labor. *Am. J. Obstet. Gynecol.*, **132**, 341–2

67 Eskes, T., Kornman, J., Bots, R., Hein, P., Gimbrère, J. and Vonk, J. (1980). Maternal morbidity due to beta-adrenergic therapy. Pre-existing cardiomyopathy aggravated by fenoterol. *Eur. J. Obstet. Gynecol. Reprod. Biol.*, **10**, 41–6

68 Benedetti, T., Hardgrove, J. and Rosene, K. (1982). Maternal pulmonary edema during premature labor inhibition. *Obstet. Gynecol.*, **59**, 33–7

69 Gropietsch, G., Fenske, M., Girndt, J., Uhlich, E. and Kuhn, W. (1980). The renin-angiotensin-aldosterone system antidiuretic hormone levels and water balance under tocolytic therapy with fenoterol and verapamil. *Int. J. Gynaecol. Obstet.*, **17**, 590–5

70 Fenske, M., Gropietsch, G. and Kuhn, W. (1981). Histologischer Nachweis der Lungenödementstehung unter Fenoterol ohne Auftreten von Herzmuskelnekrosen. *Arch. Gynäkol.*, **232**, 502–3

71 Wiest, W., Weidinger, H., Schleich, A., Hoffmann, W. and Schröder, D. (1978). Der Einfluss von Betasympathicomimetica und sogenamter calciumantagonischer Hemmstoffe auf den menschlichen fetalen Herzmuskel. In Schmidt, E., Dudenhausen, J. and Saling, E. (eds.) *Perinatale Medizin.* Vol. 7, pp. 149–51. (Stuttgart: G. Thieme Verlag)

72 Rona, G., Chappel, C. I., Balazs, T. and Gaudry, R. (1959). An infarctlike myocardial lesion and other toxic manifestations produced by isoproterenol in the rat. *Arch. Pathol.*, **67**, 443–55

73 Dhainaut, J., Boutonnet, F., Weber, S. and De Georges, M. (1978). Responsibilité des bêta-2-mimétiques au cours de la grossesse dans la génèse d'une cardiomyopathie du postpartum. *Nouv. Presse Méd.*, **7**, 4058

74 Gerris, J., Thiery, M., Bogaert, M. and De Schaepdryver, A. (1980). A randomized trial of two β-mimetic drugs for acute intrapartum tocolysis: results in a prospective comparative study with ritodrine and fenoterol. *Eur. J. Clin. Pharmacol.*, **18**, 443–8

75 Weidinger, H. and Mohr, D. (1973). Blutglukose und Immunreaktives Insulin unter dem Einfluss von Th 1165-a und Isotopin® bei Schwangeren mit und ohne tokolytische Therapie. *Z. Geburtsh. Perinatol.*, **177**, 244–51

76 Lang, H., Bellman, O., Hinckers, H. and Schlebusch, H. (1977). Carbohydrate and lipid metabolism during longtime treatment with beta-adrenergic drugs. In Weidinger, H. (ed.) *Labour Inhibition - Betamimetic Drugs in Obstetrics.* pp. 111–20. (Stuttgart: Gustav Fischer Verlag)

77 Lippert, R., De Grandi, P. and Fridrich, R. (1976). Actions of the uterine relaxant fenoterol on uteroplacental hemodynamics in human subjects. *Am. J. Obstet. Gynecol.*, **125**, 1093–8

78 Kloeck, F. and Chantraine, H. (1975). Möglichkeiten und Grenzen der intrauterinen Reanimation. *Z. Geburtsh. Perinatol.*, **179**, 401–19

79 Eskes, T. and De Haan, J. (1972). The influence of β-mimetic cathecholamines upon the fetal circulation. *Z. Geburtsh. Perinatol.*, **176**, 97–107

80 Beitzke, A., Winter, R., Zach, M. and Grubbauer, H. (1979). Kongenitales Vorhofflattern mit Hydrops fetalis durch mütterliche Tokolytikamedikation. *Klin Pädiatr.*, **191**, 410–17

81 Vogt, J., Schmidt-Redemann, B. and Urbanck, R. (1979). Neugeborenen-Kardiotoxizität nach Tokolyse mit Fenoterol hydrobromid? *Pädiatr. Pädol.*, **14**, 355–62

82 Brosset, P., Ronayette, D., Pierre, M., Le Lorier, B. and Bouquier, J. (1982). Cardiac complications of ritodrine in mother and baby. *Lancet*, **2**, 1468

83 Oddoy, A., Joschko, K., Frenzke, G. and Goldmann, M. (1981). Fetale Myokardschädigung durch Beta-Mimetika? *Zbl. Gynäkol.*, **103**, 1429–34

84 Mund-Hoym, S. and Goecke, H. (1981). Tierexperimentelle Untersuchungen zur Frage lichtmikroskopisch nachweisbarer Veränderungen am fetalen Myokard nach Gabe von Fenoterol. *Arch. Gynäkol.*, **232**, 505–6

85 Esteban-Altarriba, J. (1980). Fetal lung maturation. In Barden, T. P. (ed.) *Tocolytic Agents.* pp. 41–8. (New York: Biomedical Information Corporation)

86 Schutte, M., Treffers, P., Koppe, J. and Breur, W. (1980). The influence

of bethamethasone and orciprenaline on the incidence of RDS in the newborn after preterm labour. *Br. J. Obstet. Gynaecol.*, **87**, 127–31

87 Goeschen, K., Dudenhausen, J., Kynast, G. and Saling, E. (1981). Einfluss der Langzeittokolyse auf die Lungenreifung von Zwillingen. *Arch. Gynäkol.*, **232**, 516–17

88 Van Assche, F. A. and Aerts, L. (1983). The effect of beta-sympathico-mimetic on the fetal endocrine pancreas. *Eur. J. Obstet. Gynecol. Reprod. Biol.*, **15**, 406–8

89 Epstein, M., Nicolls, E. and Stubblefield, P. (1979). Neonatal hypogly-cemia after β-sympathicomimetic tocolytic therapy. *J. Pediatr.*, **94**, 449–53

90 Essed, G. G. M. (1983). Neonatal effects of beta-adrenergic drugs. *Eur. J. Obstet. Gynecol. Reprod. Biol.*, **15**, 397–400

91 Blouin, D., Murray, M. and Beard, R. (1976). The effect of oral ritodrine on maternal and fetal carbohydrate metabolism. *Br. J. Obstet. Gynaecol.*, **83**, 711–15

92 Karlsson, K., Krantz, M. and Hamberger, L. (1980). Comparison of various betamimetics on preterm labor, survival and development of the child. *J. Perinat. Med.*, **8**, 19–26

93 Gamissans, O., Carreras, M., Duran, P., Cararach, V., Calaf, J., Abril, V. and Esteban-Altirriba, J. (1972). The treatment of foetal acidosis with β-mimetic drugs. In Baumgarten, K. and Wesselius-de Casparis, A. (eds.) *Proceedings of the International Symposium on the Treatment of Fetal Risks*, Vienna. pp. 145–8.

94 Kloeck, F., Chantraine, H., Etzbrodt, A., Liedtke, H. and Schulte, H. (1973). Der Einfluss des β-stimulators Th-1165a auf mütterliche und fetale Stoffwechselparameter. In Dudenhausen, J. and Saling, E. (eds.) *Perina-tale Medizin*. pp. 384–6. (Stuttgart: G. Thieme Verlag)

95 Meinen, K. and Breinl, H. (1981). Zur Myokardtoxizität von Fenoterol bei Langzeittokolysen. *Arch. Gynäkol.*, **232**, 504–5

96 Kubli, F. (1983). Tocolytics: maternal hazards. *Eur. J. Obstet. Gynecol. Reprod. Biol.*, **15**, 404–5

97 Irmer, M., Trolp, R., Schmidt-Redemann, B. and Hillemanns, H. (1981). Auswirkungen einer Tokolysetherapie mit Fenoterol-Verapamil und Fen-oterol/Metoprolol auf das mütterliche und kindliche Herz. *Arch. Gynäkol.*, **232**, 519–20

98 Schneider, J., Wiest, W., Hettenbach, A., Hiltmann, W. and Hohlweg-Majert, P. (1981). Der Einfluss der kombinierten tokolytischen Therapie mit Fenoterol und Metoprolol auf den maternalen Stoffwechsel. *Arch. Gynäkol.*, **232**, 510–11

99 Ross, M. G., Nicolls, E., Stubblefield, P. G. and Kitzmiller, J. L. (1983). Intravenous terbutaline and simultaneous β_1-blockade for advanced premature labor. *Am. J. Obstet. Gynecol.*, **147**, 897–902

100 Ruckhaeberle, K. E., Vogtmann, Ch. and Viehweg, B. (1981). Risiko und Nutzen intravenöser Tokolyse bei drohender Frügeburt mit vorzeitigem Blasensprung. *Zbl. Gynäkol.*, **103**, 1417–28

101 Hettenbach, A., Wiest, W. and Hiltmann, W. D. (1982). Tocolysis in cases of premature rupture of the membranes. *J. Perinat. Med*, **10** (suppl. 2), 62–3

102 Berg, G., Andersson, R. and Rydén, G. (1983). *In vitro* study of phospho-diesterase inhibiting drugs: complement to β-sympathomimetic drug therapy in premature labor? *Am. J. Obstet. Gynecol.*, **145**, 802–6

103 Liu, D. and Blackwell, R. (1978). The value of a scoring system in predicting outcome of preterm labour and comparing the efficacy of treatment with aminophylline and salbutamol. *Br. J. Obstet. Gynaecol.*, **85**, 418–24

104 Berg, G., Andersson, R. and Rydén, G. (1982). Effects of selective β-adrenergic agonists on spontaneous contractions, cAMP levels and phosphodiesterase activity in myometrial strips from pregnant women treated with terbutaline. *Gynecol. Obstet. Invest.*, **14**, 56–64

105 Buckley, B. M., Braithwaite, R. A. and Vale, J. A. (1983). Theophylline poisoning. *Lancet*, **2**, 618

106 Whyte, K. F. and Addis, G. J. (1983). Toxicity of salbutamol and theophylline together. *Lancet*, **2**, 618–19

107 Ron, M., Hochner-Celnikier, D., Menczel, J., Palti, Z. and Kidroni, G. (1984). Maternal–fetal transfer of aminophyllin. *Acta Obstet. Gynecol. Scand.*, **63**, 217–28

108 Rubens, R., Carlier, A., Thiery, M. and Vermeulen, A. (1977). Pregnancy complicated by insulinoma. *Br. J. Obstet. Gynaecol.*, **84**, 543–7

109 Lammers, W., Nelemans, F., Bouwman, T., Van Noordwijk, J., Offerhaus, L., Rosinga, W. and Vermeulen, A. (1980). *Algemene Pharmacotherapie*. 4th Edn. (Alphen aan den Rijn: Stafleu)

110 Johansson, S., Andersson, R. and Wikberg, J. (1977). Mechanical and metabolic effects of diazoxide on rat uterus. *Acta Pharmacol. Toxicol.*, **41**, 328–36

111 MacLean, A., Doig, G., Chatfield, W. and Aickin, D. (1981). Small-dose diazoxide administration in pregnancy. *Aust. NZ J. Obstet. Gynaecol.*, **21**, 7–10

112 Boulos, B. M., Davis, L. E., Almond, C. H. and Jackson, R. L. (1971). Placental transfer of diazoxide and its hazardous effect on the newborn. *J. Clin. Pharmacol.*, **11**, 206–10

113 Smith, M., Aynsley-Green, A. and Redman, C. (1982). Neonatal hyperglycaemia after prolonged maternal treatment with diazoxide. *Br. Med. J.*, **284**, 1234

114 Milner, R. and Chouksey, S. (1972). Effects of fetal exposure to diazoxide in man. *Arch. Dis. Child.*, **47**, 537–43

115 Arnold, S. A. and Adamsons, K. (1982). A new indicator to predict the success of tocolytic therapy and the outcome of the neonate. In Thiery, M., Santerre, J. and Derom, R. (eds.) *Abstracts 8th European Congress of Perinatal Medicine* (Angleur: Nelissen), abstract 165

116 Adamsons, K. and Arnold, S. A. (1983). Diazoxide in the management of preterm labor. In Zuspan, F. and Christion, D. (eds.) *Controversies in Obstetrics and Gynecology*. pp. 91–7. (New York: Saunders)

117 Csapo, A. I. (1956). Progesterone block. *Am. J. Anat.*, **98**, 273–91

118 Fuchs, F. and Stakemann, G. (1960). Treatment of threatened premature labor with large doses of progesterone. *Am. J. Obstet. Gynecol.*, **79**, 172–6

119 Hendricks, C. H., Brenner, W. E. and Gabel, R. A. (1961). The effect of

progesterone administered intra-amniotically in the late human pregnancy. In Barnes, A. C. (ed.) *Progesterone*. pp. 109–21 (Augusta: Brook Lodge Press)

120 Kumar, D., Zourlas, P. and Barnes, A. G. (1963). *In vitro* and *in vivo* effects of magnesium sulfate on human uterine contractility. *Am. J. Obstet. Gynecol.*, **86**, 1036–40

121 Brenner, W. E. and Hendricks, C. H. (1962). Effect of medoxyprogesterone acetate upon the duration and characteristics of human gestation and labor. *Am. J. Obstet. Gynecol.*, **83**, 1094–8

122 Øvlisen, B. and Iversen, J. (1963). Treatment of threatened premature labor with α-methyl-17α-acetoxyprogesterone. *Am. J. Obstet. Gynecol.*, **86, 291–5**

123 Bengtsson, L. P. (1962). Experiments on the suppressive effect of synthetic gestagen on the activity of the pregnant human uterus. *Acta Obstet. Gynecol. Scand.*, **41**, 124–43

124 Ferre, F., Uzan, M., Janssens, Y., Tanguy, G., Jolivet, A., Bremhler, M., Sureau, C. and Cédard, L. (1984). Oral administration of micronized natural progesterone in late human pregnancy. *Am. J. Obstet. Gynecol.*, **148**, 26–34

125 Johnson, J. W., Lee, P. A., Zachary, A. S., Calhoun, S. and Migeon, C. J. (1979). High risk prematurity-progesterone treatment and steroid studies. *Obstet. Gynecol.*, **54**, 412–18

126 Colasson, F., Kamina, P., Courtois, P. and de Tourris, H. (1970). Les tocolytiques. *J. Gynécol. Obstét. Biol. Reprod.*, **9**, 377–88

127 Novy, M. J. (1978). Effects of indomethacin on labor, fetal oxygenation, and fetal development in Rhesus monkeys. In Coceani, F. and Olley, P. M. (eds.) *Advances in Prostaglandins and Thromboxane Research*, Vol. 4, *Prostaglandins and Perinatal Medicine*. pp. 285–300. (New York: Raven Press)

128 Amy, J. J., Volckaert, M., Foulon, W., Verhoeven, N. and De Brucker, P. (1984). The use of prostaglandin synthesis inhibitors in obstetrics. In Bottari, S., Thomas, J. P., Vokaer, A. and Vokaer, R. (eds.), *Uterine Contractility, Clinical and Therapeutic Aspects*. pp. 417–424. (New York: Masson)

129 Berman, W. Jr., Friedman, Z. and Vidyasagar, D. (1980). Pharmacokinetics of inhibitors of prostaglandin synthesis in the perinatal period. In Heymann, M. A. (ed.) *Prostaglandins in the Perinatal Period*. pp. 67–72. (New York: Grune and Stratton)

130 Helleberg, L. (1981). Clinical pharmacokinetics of indomethacin. *Clin. Pharmacokinet.*, **6**, 245–58

131 Aiken, J. W. (1972). Aspirin and indomethacin prolong parturition in rats: evidence that prostaglandins contribute to expulsion of foetus. *Nature (London)*, **240**, 21–5

132 Vane, J. R. and Williams, K. I. (1973). The contribution of prostaglandin production to contractions of the isolated uterus of the rat. *Br. J. Pharmacol.*, **48**, 629–39

133 Smith, I. D., Temple, D. M. and Shearman, R. P. (1975). The antagonism by anti-inflammatory analgesics of prostaglandin $F_{2\alpha}$-induced contrac-

tions of human and rabbit myometrium *in vitro*. *Prostaglandins*, **10**, 41–57

134 Chester, R., Dukes, M., Slater, S. R. and Walpole, A. L. (1972). Delay of parturition in the rat by anti-inflammatory agents which inhibit the biosynthesis of prostaglandins. *Nature (London)*, **240**, 37–8

135 Challis, J. R. G., Davies, I. J. and Ryan, K. J. (1975). The effects of dexamethasone and indomethacin on the outcome of pregnancy in the rabbit. *J. Endocrinol.*, **64**, 363–70

136 Lau, J. F., Saksena, S. K. and Chang, M. C. (1976). Effects of indomethacin and prostaglandin $F_{2\alpha}$ on parturition in the hamster. *Prostaglandins*, **11**, 859–69

137 Waltman, R., Tricomi, V. and Palav, A. (1973). Aspirin and indomethacin: effect on instillation–abortion time of mid-trimester hypertonic saline-induced abortion. *Prostaglandins*, **3**, 47–58

138 Reiss, U., Atad, J., Rubinstein, I. and Zuckerman, H. (1976). The effect of indomethacin in labour at term. *Int. J. Gynaecol. Obstet.*, **14**, 369–74

139 Lewis, R. B. and Schulman, J. D. (1973). Influence of acetylsalicylic acid, an inhibitor of prostaglandin synthesis, on the duration of human gestation and labour. *Lancet*, **2**, 1159–61

140 Collins, E. and Turner, G. (1975). Maternal effects of regular salicylate ingestion in pregnancy. *Lancet*, **2**, 335–7

141 Zuckerman, H., Reiss, U. and Rubinstein, I. (1974). Inhibition of human premature labor by indomethacin. *Obstet. Gynecol.*, **44**, 787–92

142 Wiqvist, N., Lundström, V. and Gréen, L. (1975). Premature labor and indomethacin. *Prostaglandins*, **10**, 515–26

143 Zuckerman, H., Shalev, E., Gilad, G. and Katzuni, E. (1984). Further study of the inhibition of premature labor by indomethacin. Part I. *J. Perinat. Med.*, **12**, 19–23

144 Gamissans, O., Canas, E., Cararach, V., Ribas, J., Puerto, B. and Edo, A. (1978). A study of indomethacin combined with ritodrine in threatened pre-term labor. *Eur. J. Obstet. Gynecol. Reprod. Biol.*, **8**, 123–8

145 Spearing, G. (1979). Alcohol, indomethacin and salbutamol. A comparative trial on their use in preterm labor. *Obstet. Gynecol.*, **53**, 171–4

146 Niebyl, J. R., Blake, D. A., White, R. D., Kumor, K. M., Dubin, N. H., Robinson, J. C. and Egner, P. G. (1980). The inhibition of premature labor with indomethacin. *Am. J. Obstet. Gynecol.*, **136**, 1014–19

147 Zuckerman, H., Shalev, E., Gilad, G. and Katzuni, E. (1984). Further study of the inhibition of premature labor by indomethacin. Part II. Double blind study. *J. Perinat. Med.*, **12**, 25–9

148 Wiqvist, N. (1979). The use of inhibitors of prostaglandin synthesis in obstetrics. In Keirse, M. J. N. C., Anderson, A. B. M. and Bennebroek Gravenhorst, J. (eds.) *Human Parturition*. pp. 189–200. (The Hague: Martinus Nijhoff)

149 Niebyl, J. R. (1981). Preterm parturition. Prostaglandin synthetase inhibitors. *Sem. Perinatol.*, **5**, 274–87

150 Stuart, M. J., Gross, S. J., Elrad, H. and Graeber, J. E. (1982). Effects of acetylsalicylic acid ingestion on maternal and neonatal hemostasis. *N. Engl. J. Med.*, **307**, 909–12

151 Reimann, I. W. and Fröhlich, J. C. (1981). Risiken einer Tokolysebe-handlung durch Kombination von Betaadrenergica und Prostaglandinsyn-thesehemmstoffen. Z. Geburtsh. Perinatol., 185, 305–12

152 Clyman, R. I. (1980). Ontogeny of the ductus arteriosus response to pros-taglandins and inhibitors of their synthesis. In Heymann, M. A. (ed.) Prostaglandins in the Perinatal Period. pp. 115–24. (New York: Grune and Stratton)

153 Sharpe, G. L., Thalme, B. and Larsson, R. S. (1974). Studies on closure of the ductus arteriosus. XI. Ductal closure in utero by a prostaglandin synthetase inhibitor. Prostaglandins, 8, 363–8

154 Heymann, M. A. and Rudolph, A. M. (1978). Effects of prostaglandins and blockers of prostaglandin synthesis on the ductus arteriosus: animal and human studies. In Coceani, F. and Olley, P. M. (eds.) Advances in Prostaglandin and Thromboxane Research, Vol. 4, Prostaglandins and Perinatal Medicine. pp. 363–72. (New York: Raven Press)

155 Rudolph, A. M. (1978). Discussion. In Coceani, F. and Olley, P. M. (eds.). Advances in Prostaglandin and Thromboxane Research, Vol. 4, Prosta-glandins and Perinatal Medicine. pp. 257–9. (New York: Raven Press)

156 Momma, K. and Takeuchi, H. (1983). Constriction of fetal ductus arter-iosus by non-steroidal anti-inflammatory drugs. Prostaglandins, 26, 631–43

157 Wiqvist, N. (1981). Preterm labour: other possibilities including drugs not to use. In Elder, M. G. and Hendricks, C. H. (eds.) Preterm Labor. pp. 148–75. (London: Butterworths)

158 Levin, D. L. (1980). Effects of inhibition of prostaglandin synthesis on fetal development, oxygenation, and the fetal circulation. In Heymann, M. A. (ed.) Prostaglandins in the Perinatal Period. pp. 35–44. (New York: Grune and Stratton)

159 Rudolph, A. M. (1981). The effects of nonsteroidal anti-inflammatory compounds on fetal circulation and pulmonary function. Obstet. Gyne-col., 58, 63–67

160 Kitterman, J. A. and Liggins, G. C. (1980). Fetal breathing movements and inhibitors of prostaglandin synthesis. In Heymann, M. A. (ed.) Pros-taglandins in the Perinatal Period. pp. 97–100. (New York: Grune and Stratton)

161 Sharpe, G. L., Krons, H. and Altschuler, G. (1977). Perinatal use of in-domethacin. Lancet, 2, 87

162 Van Kets, H., Thiery, M., Derom, R., Van Egmond, H. and Baele, G. (1979). Perinatal hazards of chronic antenatal tocolysis with indometha-cin. Prostaglandins, 18, 893–907

163 Röpke, F., Woraschk, H. and Seifert, B. (1977). The administration of β-sympathomimetic agents in obstetric emergencies. In Weidinger, H. (ed.) Labour Inhibition – Betamimetic Drugs in Obstetrics. pp. 235–41. (Stutt-gart: Gustav Fischer Verlag)

164 Zahn, V., Bittner, S. and Zach, H. (1977). Notfalltokolyse. Geburtsh. Frauenheilk., 37, 207–15

165 Thiery, M. and Gerris, J. (1980). Het indicatiegebied van de acute toco-lyse. T. Geneesk., 36, 415–20

166 Saling, E. and Müller-Holve, W. (1975). Die aussere Wendung des Feten aus Beckenendlage in Schädellage unter Tokolyse. *Geburtsh. Frauenheilk.*, **35**, 151-3

167 Pluta, M., Giffei, J.M. and Saling, E. (1981). Die aussere Wendung des Feten aus Beckenendlage bei Patientinnen mit Zustand nach Abdominaler Schnittenbindung. *Z. Geburtsh. Perinatol.*, **183**, 121-3

168 Van Dorsten, J.P., Schifrin, B.S. and Wallace, R.L. (1981). Randomized control trial of external cephalic version with tocolysis in late pregnancy. *Am. J. Obstet. Gynecol.*, **141**, 417-24

169 Lehman, R.E. (1983). Umbilican cord prolapse following external cephalic version with tocolysis. *Am. J. Obstet. Gynecol.*, **146**, 983-4

170 Fougner, A.C. and Wilson, S.J. (1984). The use of magnesium sulfate in the management of acute fetal distress. *Am. J. Obstet. Gynecol*, **149**, 587-8

171 Reece, E.A., Chervenak, F.A., Romero, R. and Hobbins, J.C. (1984). Magnesium sulfate in the management of acute intrapartum fetal distress. *Am. J. Obstet. Gynecol.*, **148**, 104-6

172 Conradt, A., Weidinger, H. and Rüth, H. (1978). Einsatz von Partusisten® bei einer Drillingsgeburt mit inkarzeriertem Feten. *Geburtsh. Frauenheilk.*, **38**, 1088-90

173 Thiery, M., Gerris, J. and Baele, G. (1978). Management of acute puerperal inversion of the uterus with betamimetic drugs and prostaglandins. *Eur. J. Obstet. Gynecol. Reprod. Biol.*, **8**, 359-61

13
Postpartum haemorrhage

M. K. TOPPOZADA

INTRODUCTION

Atonic postpartum haemorrhage is a serious obstetrical emergency and a major cause of maternal mortality. It occurs following normal labours as well as after complicated, abnormal deliveries. By today's standards, deaths due to postpartum haemorrhage should be considered preventable. Most cases of atonic postpartum haemorrhage occur within the first hour after delivery. Many physicians consider the first postpartum hour as the fourth stage of labour and recognize the necessity to guard against the occurrence of this complication.

It is important to understand the mechanisms involved in the normal cessation of bleeding after delivery. The two main factors involved in this process are vascular constriction and thrombosis. Vascular constriction results either from intrinsic factors (vessel wall contraction) or may be secondary to extrinsic pressure from the outside. Extrinsic forces which squeeze the bleeding vessels and obstruct blood flow are themselves the result of two different physiological processes; the first is intermittent myometrial contraction and relaxation superimposed on the second, i.e. sustained myometrial retraction. The contributions of each of these factors depend on a number of considerations, such as the type of tissue involved (artery, vein or sinus), the site of the vessel (muscular wall or decidua) and the patient's parity.

The greater part of the postpartum myometrium is composed of criss-cross layers of muscle fibres arranged in a trellis-like fashion. The blood vessels feeding the placental site traverse the interstices between the interlacing muscular bundles in a tortuous course. Contraction and retraction of these bundles squeeze the penetrating vascular channels and obstruct blood flow much like what occurs when a tourniquet is applied. Structural changes occurring during pregnancy in the terminal segments of the spiral arterioles supplying the placenta (replacement of the elastic lamina and smooth muscle of the media by a fibrin matrix) permit postpartum vascular collapse and occlusion by myometrial compression[1].

In contrast, the mechanism of thrombosis as a postpartum haemostatic factor is less well understood. Two components are involved in the formation of a thrombus: these are the platelet functions *per se* and the clotting mechanism. Thus, thrombus formation occurs in two phases. The first is the phase of platelet adhesion and aggregation (primary haemostasis), and the second is the phase of fibrin formation and consolidation of the plug (secondary haemostasis). The process of thrombosis probably occurs secondary to a combination of slowing of blood flow, vascular damage and changes in blood coagulation parameters. As such it appears to be a more chronic process than is muscular occlusion. Non-traumatic postpartum haemorrhage occasionally may result from defective blood clotting mechanisms; this fact should always be kept in mind by obstetricians.

PROSTAGLANDINS AND POSTPARTUM HAEMOSTASIS

If prostaglandins (PGs) play a fundamental role in the establishment of postpartum haemostasis, then this activity is achieved either through the process of myometrial contraction and retraction, vasoconstriction *per se* and/or thrombosis. The postpartum uterus continues to contract, albeit with a reduced frequency; at the same time there is a significant increase in the intensity of contractions, so that the activity level is far above the values achieved in the late first stage of labour. These pressures are sufficient to abolish myometrial blood flow by mechanical compression of the vessels at the placental bed[2-4]. Another, but less coordinated, contractility pattern may occur at prolonged intervals; this consists of small peaks combined into a contraction complex. In this case, however, the uterine tonus appears not to exceed that which exists during labour.

During the periods of myometrial relaxation which occur between contractions, mechanisms other than active compression must come into play to maintain uterine haemostasis, since uterine contractions can be inhibited by calcium antagonists such as nifedipine without clinical signs of atony[3-5]. These antagonists, however, do not reduce the resting uterine tone. Thus, a short phase of high amplitude contractions in the postpartum period may be sufficient to maintain decreased uterine size and assist in the production of clinical haemostasis. In this regard, the role of retraction is difficult to assess, but probably it exerts the crucial function of passive compression and occlusion of the spiral and radial arteries. It is possible that this is sufficient to maintain haemostasis until the slower coagulation mechanism becomes fully established.

The *in vivo* sensitivity of the postpartum myometrium to PGs has not been investigated until recently. An infusion of $PGF_{2\alpha}$ at a rate of 35 μg/min induces an initial hypertonus with superimposed irregular contractions with gradually increasing amplitudes. Uterine activity is sustained throughout the infusion period in contrast to the response to oxytocin infusion. This difference suggests a different mechanism of action[5].

It is not clear to what extent primary vasoconstriction acts in the physiological postpartum haemostatic balance. The stimulus for vaso-constriction may either be direct injury of the vessel wall or the release of vasoactive substances due to trauma, from platelet aggregations or following the initiation of clotting.

The vascular response to PGs depends on several factors. Among these are the species of the tissue being evaluated, its hormonal status, and the compound being tested. All can affect the muscular layer of the vessel directly or modify its adrenergic vasoconstrictor response[6]. Although available evidence suggests that endogenous PGs and throm-boxanes regulate vessel tone through a number of direct and indirect mechanisms, little is known about their possible role in the control of human postpartum uterine blood flow. In the non-pregnant primate, semiphysiological doses of $PGF_{2\alpha}$ dramatically reduce endometrial blood flow, but whether myometrial contractions or vasoconstriction are more important in inducing this response is not clear[7]. Moreover, vasoconstriction can be achieved *in vitro* by $PGF_{2\alpha}$ in preparations of human non-pregnant and pregnant uterine arteries[4,8]. The extent to which PGs, their intermediates and metabolites, and vasoactive sub-stances all interact in the puerperium to control uterine haemodyn-amics requires further study.

The interaction between platelets and the vascular endothelium pro-vides a sensitive system for controlling the haemostatic balance in the body where arachidonic acid metabolites seem to play a key role. The enzyme cyclo-oxygenase converts arachidonic acid to its endoperoxide intermediates (PGG_2 and PGH_2) which then are transformed by specific enzymes to prostacyclin (PGI_2), thromboxane A_2 (TxA_2), $PGF_{2\alpha}$, $PGE_{2\alpha}$ and PGD_2. In general, PGI_2, PGD_2 and, to a lesser extent, the E series are very potent inhibitors of platelet aggregation and act as vasodilators; as such, they have been designated 'prohae-morrhagic' or 'antithrombotic' PGs. On the other hand, the opposing group from the same family (arachidonic acid, the endoperoxides and TxA_2) possesses powerful vasoconstrictor and platelet aggregation properties and can justifiably be referred to as 'antihaemorrhagic' or 'prothrombotic' PGs.

The mechanism which directs arachidonic acid metabolism towards a pro- or antihaemorrhagic pathway is poorly understood; however, it includes substrate availability, chemical mediators and tissue selectivity. The latter is clearly illustrated by the predominance of TxA_2 generation by platelets which are rich in TX synthetase, while the vascular intima with abundant prostacyclin synthetase selectively promotes PGI_2 synthesis[9]. Specific enzyme inhibitors, if available, may enable clinicians to manipulate endogenous release in favour of either a pro- or an antithrombotic pathway. Such activity would have a major impact on the management of haemorrhagic tendencies or occlusive arterial disease.

Among the classical prostaglandins, PGE_1 inhibits platelet adhesiveness and interferes with clumping induced by specific aggregating agents such as thrombin, collagen, ADP and epinephrine. In contrast, PGE_2 has an initial weak PGE_1-like effect on the first phase of aggregation which is followed by a stimulatory action on the second or ADP-induced phase. On the other hand, $PGF_{2\alpha}$ has little effect on platelet aggregation[10]. The endoperoxide intermediates and TxA_2 induce rapid and irreversible aggregation of human platelets, while PGI_2 is the most potent inhibitor of platelet adhesiveness. PGI_2 is 20–30 times more potent than PGE_1 and 10–15 times more potent than PGD_2 in preventing human platelet aggregation induced by arachidonic acid, ADP or collagen. It can also reverse platelet aggregation induced by some aggregating agents. Moreover, alterations in platelet sensitivity to antiaggregatory PGs can serve as an index for a disturbance in haemostatic balance[11]. Also, it has been suggested that a plasma factor of still undefined nature regulates PGI_2 synthesis[12].

The compounds PGE_1, PGE_2 and $PGF_{2\alpha}$ appear to exert insignificant effects on coagulation factors and fibrinolytic activity[10]. Increased euglobulin lysis time and antithrombin activity in response to PGE_2 infusions have been reported with the implication of an antithrombotic effect rather than the reverse[13]. Moreover, induction of abortion by $PGF_{2\alpha}$ or labour with PGE_2 are associated with an activation of the coagulation system which was attributed to the increased uterine activity or to the process of fetal expulsion per se[14,15].

These observations all suggest that the effects of PGs on blood coagulation and thrombosis are mainly mediated through their influence on platelet function and, to a lesser extent, by their effect on the factors responsible for fibrin formation and fibrinolysis.

The local endogenous release of PGs after delivery may also provide important information related to haemostasis. However, the presently available data are not easily interpreted, since all human measurements of postpartum PG levels must be extrapolated from samples of peripheral venous blood which may not reflect the true uterine situation. Most reported concentrations of $PGF_{2\alpha}$ or its major metabolites have shown a peak near the end of labour[16,17], although one study demonstrated the highest level during placental separation and 30 min postpartum[18]. This finding may reflect placental separation, haemostasis and uterine involution.

The plasma concentration of PGI_2 and its major metabolite 6-keto-$PGF_{1\alpha}$ have been shown to increase towards term, rise significantly during labour and drop once again below prelabour values within 2 h after delivery[19-21]. The output of PGI_2 is reduced by uterine relaxants, and very small amounts of PGI_2 markedly potentiate the stimulant effect of oxytocin[22]. The low postpartum levels of PGI_2 and its interaction with oxytocin with regard to the myometrial contractility indicate that PGI_2 may have a physiological role in preventing postpartum haemorrhage. An increased endogenous production of PGI_2 is likely to have the opposite effect through dilating peripheral vessels and inhibiting platelet aggregation.

MANAGEMENT OF ATONIC POSTPARTUM HAEMORRHAGE

Diagnosis

Primary postpartum haemorrhage is commonly defined as bleeding in excess of 500 ml during the first 24 h after birth of the baby. All too frequently, the amount of blood loss is underestimated, since accurate quantitative measurement during parturition is performed only infrequently[23-25]. Accordingly, another definition may be far more suitable; it considers postpartum haemorrhage as any bleeding following labour, irrespective of its amount, which adversely affects the general condition of the mother. Obviously, this quantity would vary from case to case, depending on many pre-existing conditions and factors. Primary postpartum haemorrhage is divided into third stage haemorrhage and true (postplacental) postpartum haemorrhage. Secondary postpartum haemorrhage is bleeding after the first 24 h until the end of the puerperium.

It is imperative to rapidly identify the cause of bleeding, which may include uterine atony, trauma, retained placental remnants, uterine inversion or coagulation defects. More than one aetiological factor may be present, and this possibility should always be kept in mind. Although the differentiation between causes is essentially a clinical exercise, adequate surgical exploration of the vagina, cervix and uterine cavity under appropriate anaesthesia is a must to avoid any miscalculation with its serious consequences. Diagnostic aids such as ultrasound and computerized axial tomography may also be helpful in selected cases if they are close at hand and transfer of the patient to obtain these tests will not result in prolonged delays which, in themselves, may assume fatal importance.

Although atonic postpartum bleeding may occur without apparent cause, several factors predispose to its occurrence. Uterine overdistension (hydramnios, twins, large baby), grand multiparity, prolonged labour, ante- or intrapartum haemorrhage, and surgical intervention (Caesarean section, forceps, vacuum extraction and manual separation of placenta) are among the factors that should alert the physician to the possibility of postpartum atony. A past history of atonic haemorrhage or Caesarean section and young or old reproductive age are also relevant factors. Anaemia, hypovolaemia or severe toxaemia and heavy narcosis or use of certain anaesthetics are also important predisposing factors to haemorrhage.

Prevention and prophylaxis

The statement 'prevention is better than cure' applies for the prophylaxis against postpartum uterine atony and starts before the onset of labour. A recognition of risk factors and the treatment of anaemia and toxaemia during pregnancy, as well as skilled management of labour during the third stage, are all likely to reduce the incidence of atonic haemorrhage significantly.

In the third stage of labour, rough uterine manipulations and cord traction should be avoided, and the urinary bladder should be kept empty, as a full bladder predisposes to uterine atony[26]. Gentle uterine massage after placental expulsion may stimulate contractions and empty retained clots which may otherwise promote atony.

The routine parenteral administration of oxytocic agents after labour to reduce the occurrence of postpartum uterine atony and shorten the third stage is a common practice in many hospitals. However, the time of injection (before or after placental expulsion), the route of administration (i.v. or i.m.), and the type and dose of oxytocic agent used remain controversial. Ergot derivatives are powerful oxytocics, but the full effect after i.m. injection of ergometrine may not take place for several minutes[27,28]. To hasten this response, the addition of oxytocin (now available in combined form: Syntometrine®, Sandoz) or the enzyme hyaluronidase or changing the route of administration to i.v. have all been attempted with satisfaction. The mean time to maximum response after i.v. administration of ergometrine was found to be 41 s; in contrast, the i.m. route with added hyaluronidase required 5 min, and i.m. administration of ergometrine required 7 min to establish its full effect. Intramuscular Syntometrine, on the other hand, was effective within 2·5 min[27].

The effect of i.v. injection of oxytocin begins almost immediately but it lasts for only few minutes, and at most for an hour, after i.m. administration. In contrast, the contractions induced by ergonovine or methyl ergonovine tartrate persist for as long as 6 h after i.m. injection or oral ingestion[28]. It is generally believed that the administration of an oxytocic agent is most effective if administered at crowning or at the time of the delivery of the anterior shoulder. In a study carried out in Alexandria, Egypt (Toppozada *et al.*, unpublished data) the effect of postplacental i.v. injection of 0·2 mg methyl ergometrine maleate was compared to a placebo in two equal groups, each consisting of 40 normal deliveries. Objective measurement of blood loss was performed using the alkaline haematin spectrophotometric method. Though the mean amount of postpartum blood loss at 4 h was less in the treatment group than in the controls (153·7 ± 54·6 ml v. 164·8 ± 77·5 ml), this difference was not statistically or clinically significant. Nonetheless, two cases in the placebo group (their respective blood losses were not included in the mean values) developed atonic postpartum haemorrhage. These findings support the widespread recommendation for prophylactic administration of i.v. ergometrine (Table 13.1).

The efficacy of administering oxytocin or ergot alkaloids to prevent atony and excessive postpartum blood loss has also been evaluated[29]. Postplacental i.m. oxytocin decreased the incidence of atony from 22% to 5%, while the use of ergot alkaloids only decreased it to 16%. Another study compared the use of these drugs after delivery of the anterior shoulder and after placental separation[30]. It was shown that the mean duration of the third stage was not different, but that placental retention was more common in patients treated prior to expulsion.

It is important to remember that the use of oxytocic agents in all subjects in order to avoid a few cases of atonic postpartum haemorrhage is not without certain risks and side-effects. For example, the rapid i.v. administration of 5–10 units of synthetic oxytocin may cause a significant but transient drop in the blood pressure; electrocardi-

Table 13.1   Mean (± SD) amount of blood loss for 4 h after normal delivery measured by the alkaline haematin spectrophotometric method

Group I [Methergin®]		Group II (placebo)	
No. of cases	Mean blood loss (ml)	No. of cases	Mean blood loss (ml)
40	153·670	38*	164·767
	± 54·644		± 77·510

* Two cases not included due to atonic postpartum haemorrhage

ographic changes such as arrhythmias have also been reported[31,32]. Moreover, high dose oxytocin infusion may also lead to water intoxication if large fluid volumes are used[33]. Similarly, ergot preparations may cause hypertension in one third and nausea and vomiting in about one fifth of treated subjects, respectively[29,34].

In the late 1960s and early 1970s, PGs were determined to have sustained uterine stimulating effects in the early puerperium. The clinical evaluation of these agents followed shortly thereafter in Stockholm. The effect on blood loss and placental separation of i.v. doses of 1 mg PGE_1, 0·2 mg ergometrine, 3 IU oxytocin or a combination of PGE_1 and ergometrine was compared in a group of 188 women after delivery of the fetus[35]. Although PGE_1 induced sustained uterine contractions, it did not reduce blood loss or shorten the time for placental separation.

On the other hand, after successful induction of labour with $PGF_{2\alpha}$ in 21 women at term, continuation of the infusion during the third stage of labour resulted in a reduced amount of bleeding when compared to the use of oxytocin[36]. Another study compared the amount of blood loss after delivery in two groups of normal obstetric patients. Ten women were given i.m. syntometrine and five received i.m. 0·25 mg sulprostone at the moment of crowning; all blood lost at the time of delivery and during the next 24 h was measured by the alkaline haematin method[37]. The results indicated that the use of sulprostone was associated with a reduction in immediate postpartum blood loss compared to syntometrine (mean of 194·6 ml vs. 252·5 ml). Moreover, in the subsequent 24 h, the blood loss in the sulprostone group was also less (94·0 ml vs. 101·1 ml). Of interest, one woman in the syntometrine group required blood transfusion. Another study compared the effects of transabdominal intramyometrial injection of 1 mg $PGF_{2\alpha}$, the use of i.v. 0·2 mg ergometrine or no treatment after cord clamping in 140 cases. The three groups of patients were selected at random. In this study, $PGF_{2\alpha}$ significantly reduced the duration of the third stage of labour, blood loss, the incidence of subinvolution and subfebrility[38].

In another study conducted at Alexandria, Egypt, 90 patients thought to be at high risk for developing postpartum haemorrhage were divided into three equal groups according to a table of random numbers[39]. After delivery of the placenta, the first group was given an i.m. injection of 0·2 mg methyl ergometrine maleate, the second group received 0·25 mg of 15-methyl-$PGF_{2\alpha}$ and 0·5 mg sulprostone was administered to group three. The patients were evaluated by collecting vulvar pads for estimation of blood loss up to 4 h after labour. The amount of blood loss, as measured by weighing the pads, was 220, 173 and 274 g in each of the three groups, respectively. Significant differences were present between the first and third and first and second

Post-partum blood loss (gm)

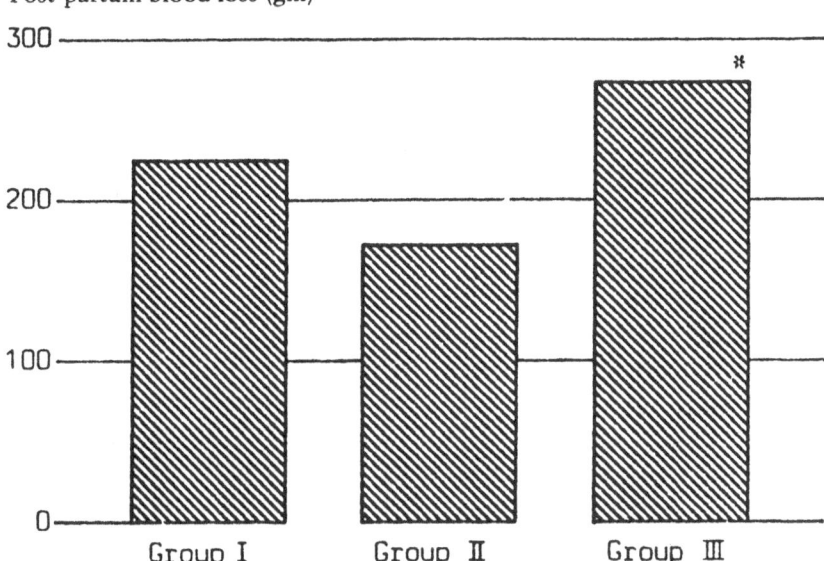

*Significantly different from Group II at 5% level

Figure 13.1 Amount of postpartum blood loss calculated by the weight difference of vulval pads collecting all blood lost for 4 h after placenta expulsion in three groups (30 each) of women at high risk of developing atonic haemorrhage. I = Methergin®; II = Prostin 15-M®; III = Sulprostone®. No difference was found between groups I and II and I and III, but groups II and III were significantly different[39]

groups, respectively (Figure 13.1). None of the patients in the three groups developed severe atonic postpartum haemorrhage or had significant side-effects related to any of the three test drugs.

The effect of postplacental i.m. administration of three different doses (62·5, 125 and 250 μg) of 15-methyl-PGF$_{2α}$ on postpartum blood loss was evaluated following 90 normal deliveries and found to have no significant difference[40]. In the same study, the author reported that the results of an ongoing study comparing 125 mg of 15-methyl-PGF$_{2α}$ with oxytocin and ergonovine in patients at high risk of developing atonic postpartum haemorrhage indicated that the prostaglandin analogue had superior effect.

Therapy

The methods used for the treatment of atonic haemorrhage can be collectively grouped under two main categories: (1) non-surgical

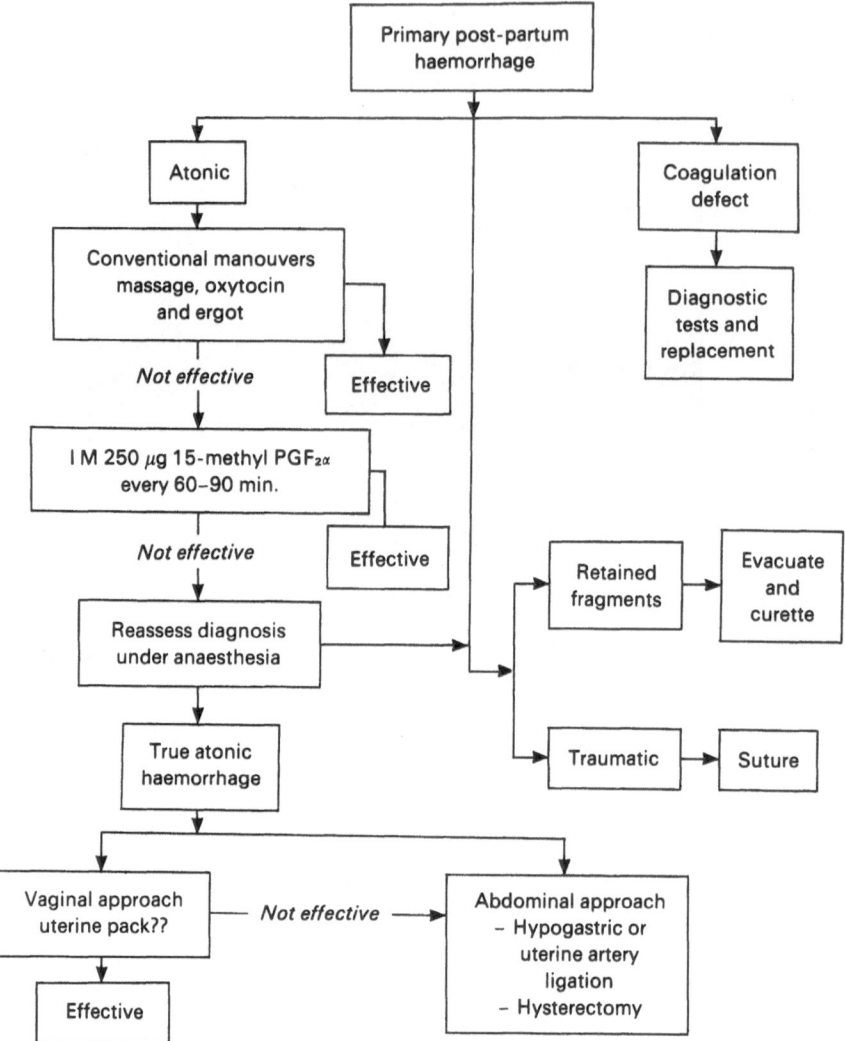

Figure 13.2 The figure summarizes recommended events of treatment in patients with postpartum haemorrhage

methods, and (2) surgical methods. The application of these methods is summarized in Figure 13.2.

Non-surgical methods
(1) Procedures for accurate diagnosis, management of shock and close monitoring of patients constitute essential requisites that should always be available in the obstetric ward. Proper fluid replacement, especially blood, should be started without delay to avoid critical cardiovascular collapse.

(2) Classical conventional manoeuvres for the treatment of atonic haemorrhage include gentle abdominal uterine massage and elevation of the uterus by a combination of vaginal and abdominal compression to enforce acute anteflexion while providing simultaneous fundal massage and digital compression of the paracervical areas. In addition, the uterus can also be compressed abdominally against the sacral promontory and the abdominal aorta as a temporary measure in very serious cases to reduce further blood loss.

(3) The administration of oxytocic agents constitutes the main pharmacological treatment for bleeding due to atony. Ergometrine and oxytocin have long been lifesaving drugs in this regard. Each agent can be administered as i.m. or i.v. injections or by i.v. infusion, alone or in combination with the other agent. In addition, the intramyometrial administration of oxytocin, ergonovine or methyl ergonovine has been efficacious at times when therapy by other routes has failed[41]. The spastic and prolonged effects of ergot derivatives make them particularly useful in postpartum haemorrhage secondary to uterine relaxation. When repeated doses are administered, long lasting and cumulative effects may be expected[42]. Recently, PGs have been used in similar situations with promising results (vide infra).

Bilateral transcatheter internal iliac artery vasopressin infusion (0·2 U/min) under angiographic control can also be used to control major postpartum haemorrhage. The therapeutic effect is achieved by inducing diffuse vasoconstriction while minimizing the development of collaterals supplying the bleeding site[43].

(4) Electrical stimulation to initiate myometrial activity has been reported as being successful, but this therapy has never attained widespread use[41].

(5) Controlled reduction of arterial blood pressure by continuous infusion of sodium nitroprusside has also been suggested as a beneficial supplemental treatment in atonic postpartum haemorrhage[44]. This approach has been reported in a case after bilateral hypogastric artery ligation.

(6) There are reports of angiographic arterial embolization of the internal iliac arteries, or, more specifically, the major source of collateral flow to the bleeding vessels, using gel foam[45,46].

(7) A non-specific life-saving measure is the use of the military aviator's 'anti-g suit' (MAST) to control intractable bleeding by external counterpressure[47]. The principle is the same for aortic compression; the lowered pressure within distal vessels and

diminished blood flow allows the patient's clotting system to effect haemostasis.

Surgical measures

Vaginal approach

This is usually the first line of therapy, since removal of retained clots and placental fragments, as well as the repair of traumatic cervical or vaginal lacerations, are essential therapeutic procedures. Curettage is usually a subsequent step which commonly arrests bleeding even if minimal or no retained tissues are recovered. Curettage probably acts via stimulation of local endogenous PG release from operative trauma.

(1) Hot intrauterine lavage may be effective in stimulating uterine contractions when oxytocic agents have failed. The heat of the solution (40-55 °C or 105-120 °F) is more important than the antiseptic constituents that are sometimes added. The solution is usually introduced by a catheter (metal or rubber) at a pressure of 110-140 mmHg. The complications of the procedure include air embolism, transplacental absorption of the fluid, transtubal leakage and infection. This procedure has almost disappeared from obstetric practice[48].

(2) The value of uterine pack or tamponade remains controversial. The procedure has its advocates as well as its opponents, and opinions are usually based on personal and limited experiences of success or failure to control bleeding in the past. The question of effectiveness is probably related to whether or not the tamponade is carried out properly. Inadequate or incomplete packing may further distend an already lax and atonic uterus, open up placental bed vessels, and leave an empty space above where blood can accumulate, thus turning a conspicuous blood loss into a concealed one. On the other hand, a complete, tight pack establishes an intrauterine pressure above that of the intravascular compartment, usually with gratifying results. The arrest of bleeding by the pack may only be temporary; however, it can provide sufficient time to improve the critical condition of the patient and to prepare her for further and more definitive surgical management. Packing is best carried out by hand, but a ring forceps or a special tube may also be utilized[49].

The vagina has also to be tightly packed for 24 h. After packing, supplemental oxytocin infusion and antibiotic therapy are usually needed. A packing-like technique for controlling atonic bleeding has been tried using a balloon similar to that used to control bleeding from oesophageal varices; although this operation has been reported to be effective, it has never gained popularity[50].

Abdominal approach
(1) Bilateral hypogastric (internal iliac) artery ligation is frequently used to control pelvic haemorrhage including uterine atony[51]. Though this procedure is followed by continued blood flow in all pelvic arteries through an extensive collateral circulation, the arteries distal to the ligatures demonstrate an 85% decrease in pulse pressure, a 24% reduction in the mean arterial blood pressure, and a 48% decrease in blood flow[52]. Selective ligation of the anterior hypogastric division is favoured by some surgeons, but this may be too time consuming in an emergency. Ligation of the uterine arteries is also effective[53]; ligation procedures such as these do not seem to jeopardize future pregnancies[41]. The additional ligation of both ovarian arteries has also been reported to be without compromise of either ovarian or uterine function[54].

(2) Total or subtotal abdominal hysterectomy with preservation of the ovaries may be necessary as a final life-saving measure in patients whose condition is considered critical.

USE OF PROSTAGLANDINS IN THE TREATMENT OF ATONIC POSTPARTUM HAEMORRHAGE

The potential for using PGs in the therapy of postpartum haemorrhage secondary to uterine atony was suggested more than 15 years ago by Bygdeman and his associates soon after recognition of the uterotonic activity of these compounds[55]. Unfortunately, this practical therapeutic suggestion was not tested in clinical practice until more than a decade later[56].

The first study on the successful control of atonic postpartum haemorrhage with PGs utilized $PGF_{2\alpha}$. Takagi *et al.*[56] reported that the intramyometrial (transabdominal or transvaginal) injection of 0·5–1·0 mg of $PGF_{2\alpha}$ induced powerful and sustained uterine contractions within a few minutes and arrested bleeding in patients with atonic haemorrhage when conventional measures and other oxytocic agents had failed. In contrast, the systemic administration (i.m. or i.v. infusion) of $PGF_{2\alpha}$ was not as effective[56]. Three years later, Jacobs and Arias[57] successfully treated three similar cases by injecting 1 mg $PGF_{2\alpha}$ directly into the myometrium through the abdominal wall. They considered this approach to be highly effective, safe, rapid and life-saving. The safety of local injection of up to 5 mg of $PGF_{2\alpha}$ had been documented previously in monkeys and shown to be without risk of myometrial damage or necrosis.

The transabdominal intramyometrial administration of PGF$_{2\alpha}$ has also proved to be effective in the control of delayed postpartum haemorrhage. In a woman who developed secondary postpartum haemorrhage 20 days after delivery and was unresponsive to curettage and uterine packing, the intramyometrial injection of 1·5 ml of PGF$_{2\alpha}$ caused intense uterine cramping within 3 min and the bleeding abruptly ceased[58]. PGE$_2$ vaginal suppositories (20 mg) have also been used with success to control postpartum uterine atony after Caesarean section, though side-effects in the form of fever, chills, tachycardia and hypotension were encountered[59]. However, since ongoing bleeding is likely to affect the degree of absorption of the compound, the therapeutic effect of suppositories is difficult to predict.

Corson and Bolognese[60] were the first to report that the 15-methyl analogue of PGF$_{2\alpha}$ was a potentially useful drug for treatment of postpartum atonic bleeding. In a single case, treatment with 0·25 mg injected intramuscularly arrested bleeding after several intramyometrial injections of 1 mg PGF$_{2\alpha}$ and i.v. ergotrate and oxytocin had failed. Zahradnik and co-workers[61] also suggested that the effects of i.m. 15-methyl-PGF$_{2\alpha}$ were superior to that of intramyometrial PGF$_{2\alpha}$ in producing haemostasis, probably through induction of sustained tetanic uterine contractions.

The data described above represent case reports introducing a potential therapeutic modality. Two large studies were published in which the results of i.m. 15-methyl-PGF$_{2\alpha}$ were almost identical[62,63]. The selection criteria necessitated that the subjects had haemorrhage secondary to uterine atony and be unresponsive to conventional treatment with oxytocin, ergot derivatives and uterine massage. Retained placental fragments and genital tract lacerations were ruled out prior to admission to the study, but coagulation tests were not done in either study.

Table 13.2   Patient characteristics in the two studies using i.m. 15-methyl PGF$_{2\alpha}$ for intractable atonic postpartum haemorrhage

		Labour			Blood loss (litres) before i.m. PG	Time to onset of i.m. PG (min)
	No. of cases	Spon- taneous	Forceps or vacuum	CS		
Toppozada et al., 1981[62]	16	7	3	6	1·80	100·0
Hayashi et al., 1981[63]	20	6	6	8	1·56	62·6

CS = Caesarean section

Table 13.3 Result of treatment with i.m. 15-methyl $PGF_{2\alpha}$ in two studies of intractable postpartum haemorrhage

	No. of cases	Success	Failure	Mean no. of PG inj. ($\times 250\ \mu g$)	A	F	H	GI	T
Toppozada et al., 1981[62]	16	15	1	1·75	6	5	3	6	—
Hayashi et al., 1981[63]	20	18	2	1·55	7	—	6	2	12

A = Absent; F = Flushing; H = Hypertension; GI = Gastrointestinal; T = Mild temperature rise

The dose of 15-methyl-$PGF_{2\alpha}$ was the same (0·25 mg), and the interval between injections (1·5 h) was similar in both studies. The success rate in arresting haemorrhage was extremely high (15 out of 16 and 18 out of 20 cases, respectively), and the incidence of side-effects was exceedingly low. The three failures in the combined investigations were associated with intrauterine infection. Relatively few cases developed a hypertensive response after PG treatment (three in the first and six in the second study), but these were generally self-limiting and sometimes related to pre-existing pregnancy toxaemia. The main academic criticism of these studies was the lack of 'control' data, but it can be argued that each subject served as her own control because she had already failed to respond with other oxytocic agents (Tables 13.2 and 13.3). The similarity of findings in these two studies provides a sense of assurance regarding the validity of the data.

In summary, the use of 15-methyl-$PGF_{2\alpha}$ via the i.m. route in a dose of 0·25 mg (repeated every 1–1·5 h whenever necessary) appears to be a highly effective and safe treatment in the management of intractable atonic postpartum bleeding. Special care is warranted in hypertensive patients; individuals with concomitant uterine sepsis are likely to be unresponsive. The mechanism of haemostasis appears to be through the induction of tetanic contractions of the myometrium, but an additional vasoactive effect cannot be ruled out.

Because of the high degree of efficacy resulting from use of i.m. 15-methyl-$PGF_{2\alpha}$ in atonic bleeding, when such treatment fails to arrest bleeding the clinician should be suspicious of a diagnostic error and consider that trauma, retained products or coagulation defects may be the cause of the bleeding instead of atony[62]. Other PG analogues such as 16-phenoxy-ω-tetranor PGE_2 methyl sulphonylamide (Sulprostone®) suitable for i.m. administration may also be found to be of therapeutic value for the control of atonic bleeding, but no supporting evidence from scientific studies has been presented in this regard to date.

References

1 Sheppard, B. L. and Bonnar, J. (1974). The ultrastructure of the arterial supply of the human placenta in early and late pregnancy. *Br. J. Obstet. Gynaecol.*, **81**, 497–511

2 Hendricks, C. H., Eskes, T. K. A. and Saameli, K. (1962). Uterine contractility at delivery and in the puerperium. *Am. J. Obstet. Gynecol.*, **83**, 890–8

3 Forman, A., Gandrup, P., Andersson, K. E. and Ulmsten, U. (1982). Effects of nifedipine on spontaneous and methyl-ergometrine-induced activity post-partum. *Am. J. Obstet. Gynecol.*, **144**, 442

4 Andersson, K. E., Forman, A. and Ulmsten, U. (1983). Pharmacology of labor. *Clin. Obstet. Gynecol.*, **26**, 56–77

5 Forman, A., Gandrup, P., Andersson, K. E. and Ulmsten, U. (1983). Effects of nifedipine on oxytocin and prostaglandin $F_{2\alpha}$ induced activity in the post-partum uterus. *Am. J. Obstet. Gynecol.*, **144**, 665

6 Clark, K. E., Farley, D. B., Van Orden, D. E. and Brody, M. J. (1977). Role of endogenous prostaglandins in regulation of uterine blood flow and adrenergic neurotransmission. *Am. J. Obstet. Gynecol.*, **127**, 455

7 Einer-Jenssen, N. (1973). Decreased endometrial blood flow and plasma progesterone level after instillation of $PGF_{2\alpha}$ into the lumen of the uteri of rhesus monkey. *Prostaglandins*, **4**, 517

8 Wilhelmsson, L., Lindblom, B., Wikland, M. and Wiqvist, N. (1981). Effects of prostaglandins on the isolated uterine artery of non-pregnant women. *Prostaglandins*, **22**, 223

9 Moncada, S. and Vane, J. R. (1978). Unstable metabolites of arachidonic acid and their role in haemostasis and thrombosis. *Br. Med. Bull.*, **34**, 129–35

10 Toppozada, M. K. (1979). Prostaglandins and their synthesis inhibitors in dysfunctional uterine bleeding. In Karim, S. M. M. (ed.) *Practical Applications of Prostaglandins and their Synthesis Inhibitors.* pp. 237–66. (Lancaster: MTP Press)

11 Hoche, C., Kefalides, A., Dadak, C. and Sinzinger, H. (1983). Platelet sensitivity to prostacyclin and prostaglandin in pregnancy and puerperium. In Lewis, P., Moncada, S. and O'Grady, J. (eds.) *Prostacyclin and Pregnancy.* pp. 189–94. (New York: Raven Press)

12 MacIntyre, D. E., Pearson, J. D. and Gordon, J. L. (1978). Localisation and stimulation of prostacyclin production in vascular cells. *Nature (London)*, **271**, 549–51

13 Howie, P. W., Calder, A. A., Forbes, C. D. and Prentice, C. R. M. (1973). Effects of intravenous prostaglandin E_2 on platelet function, coagulation and fibrinolysis. *J. Clin. Pathol.*, **26**, 354–62

14 Badraoui, M. H. H., Bonnar, J., Hillier, K. and Embrey, M. P. (1973). Coagulation changes during termination of pregnancy by prostaglandin and by vacuum aspiration. *Br. Med. J.*, **1**, 19–23

15 Badraoui, M. H. H., Bonnar, J., Hillier, K. and Embrey, M. P. (1984). Coagulation changes during induction of labour at 38 weeks by prostaglandin E_2 compared to syntocinon drip. In Toppozada, M. (ed.) *Prostaglandins in Human Reproduction.* pp. 45–9. (Alexandria: University Press)

16 Karim, S. M. M. (1968). Appearance of prostaglandin $F_{2\alpha}$ in human blood during labour. *Br. Med. J.*, **4**, 618
17 Gréen, K., Bygdeman, M., Toppozada, M. and Wiqvist, N. (1974). The role of $PGF_{2\alpha}$ in human parturition. *Am. J. Obstet. Gynecol.*, **120**, 25-31
18 Zuckerman, H., Reiss, U., Atad, J., Lampert, I., Ben Ezra, S. and Sklan, D. (1978). Prostaglandin $F_{2\alpha}$ in human blood during labor. *Obstet. Gynecol.*, **51**, 311-14
19 Ylikorkala, O., Makarainen, L. and Viinikka, L. (1981). Prostacyclin production increase during human parturition. *Br. J. Obstet. Gynaecol.*, **88**, 513-16
20 Kimball, F. A. (1983). Role of prostacyclin and other prostaglandins in pregnancy. In Lewis, P. J., Moncada, S. and O'Grady, J. (eds.) *Prostacyclin and Pregnancy.* pp. 1-13. (New York: Raven Press)
21 Seed, M. P., Williams, K. I. and Bamford, D. S. (1983). Influence of gestation on prostacyclin synthesis by human pregnant myometrium. In Lewis, P. J., Moncada, S. and O'Grady, J. (eds.) *Prostacyclin and Pregnancy.* pp. 31-6. (New York: Raven Press)
22 Williams, K. I., El-Tahir, K. E. H. and Seed, M. P. (1983). Influence of stimulant and relaxant drugs on myometrial prostacyclin formation. In Lewis, P. J., Moncada, S. and O'Grady, J. (eds.) *Prostacyclin and Pregnancy.* pp. 141-6. (New York: Raven Press)
23 Pritchard, J. A., Balwin, R. M., Dickey, J. C. and Wiggins, K. M. (1962). Blood volume changes in pregnancy and the puerperium. II. Red blood cells loss and changes in apparent blood volume during the following: vaginal delivery, Cesarean section, and Cesarean section plus total hysterectomy. *Am. J. Obstet. Gynecol.*, **84**, 1271-81
24 Newton, M. (1966). Post-partum hemorrhage. *Am. J. Obstet. Gynecol.*, **94**, 711
25 De Leeuw, N. K. M., Lowenestein, L., Tucker, E. C. and Dayal, S. (1968). Correlation of red cell loss at delivery with changes in red cell mass. *Am. J. Obstet. Gynecol.*, **100**, 1092-1101
26 Toppozada, H. K., Gaafar, A. A. and El-Sahwi, S. (1967). The urinary bladder and uterine cavity. *Am. J. Obstet. Gynecol.*, **98**, 904-12
27 Embrey, M. P. and Garrett, W. J. (1958). Ergometrine with hyaluronidase: speed of action. *Br. Med. J.*, **2**, 138-9
28 Hendricks, C. H. (1968). Uterine contractility changes in the early puerperium. *Clin. Obstet. Gynecol.*, **2**, 125-44
29 Friedman, E. A. (1957). Comparative clinical evaluation of post-partum oxytocics. *Am. J. Obstet. Gynecol.*, **73**, 1306
30 Sorbe, B. (1978). Active pharmacologic management of the third stage of labor. *Obstet. Gynecol.*, **52**, 694
31 Hendricks, C. H. and Brenner, W. E. (1970). Cardiovascular effects of oxytocic drugs used post-partum. *Am. J. Obstet. Gynecol.*, **108**, 751
32 Woodbury, R. A., Hamilton, W. F., Volpitto, P. P., Abreu, B. E. and Harper, H. T. (1944). Cardiac and blood pressure effects of pitocin (oxytocin) in man. *J. Pharmacol. Exp. Ther.*, **81**, 95
33 Liggins, G. C. (1963). Antidiuretic effects of oxytocin, morphine and pethidine in pregnancy and labour. *Aust. N.Z. J. Obstet. Gynecol.*, **3**, 81

34 Brazeau, P. (1965). Oxytocics. In Goodman, I. S. and Gillman, A. (eds.) *The Pharmacological Basis of Therapeutics*. p. 893. (New York: Macmillan)

35 Roth-Brandel, U., Bygdeman, M. and Wiqvist, N. (1970). A comparative study on the influence of prostaglandin E_1, oxytocin and ergometrine on the pregnant human uterus. *Acta Obstet. Gynecol. Scand.*, **49** (Suppl. 5), 1–13

36 Persianinov, L. S., Manuilova, I. A. and Chernukha, E. A. (1973). The result of using $PGF_{2\alpha}$ for induction and stimulation of labour. In Bergström, S. and Bernhard, S. (eds.) *Advances in Biosciences*. Vol. 9, p. 585. (New York: Pergamon Press)

37 Karim, S. M. M. (1979). Prostaglandins in obstetrics and gynecology. In Friebel, K., Schneider, A. and Wurfel, H. (eds.) *International Sulprostone Symposium*, Vienna, November 1978. pp. 7–28. (Berlin: Medical Scientific Department, Schering AG)

38 Kerekes, L. and Domokos, N. (1979). The effect of prostaglandin $F_{2\alpha}$ on third stage labor. *Prostaglandins*, **18**, 161–6

39 Toppozada, M. K., Souka, A. R., Ibrahim, M. A., El-Damarawy, H. and Sadek, W. (1985). Effects of intramuscular prostaglandins or ergometrine on post-partum blood loss. *Alex. Med. J.*, **25**, 215–23

40 Nelson, G. H. (1980). Prostaglandins in reproduction. In Goldstein, D. P. and Leventhal, J. M. (eds.) *Obstetrics and Gynecology*, p. 14 (Year Book Medical Publishers)

41 Kelly, J. V. (1976). Post-partum hemorrhage. *Clin. Obstet. Gynecol.*, **19**, 595–606

42 Cibils, L. A. and Hendricks, C. H. (1969). Effect of ergot derivative and spartene sulphate upon the human uterus. *J. Reprod. Med.*, **2**, 147

43 Sacks, B. A., Palestrant, A. M. and Cohen, W. R. (1982). Internal iliac artery vasopressin infusion for post-partum hemorrhage. *Am. J. Obstet. Gynecol.*, **143**, 601–3

44 Jackson, S. H., Liebermann, M. C. and Smith, D. E. (1981). Sodium nitroprusside-induced hypertension as a supplemental therapeutic modality in post-partum hemorrhage. *Obstet. Gynecol.*, **58**, 649–51

45 Brown, B. J., Heaston, D. K., Poulson, A. M., Gabert, H. A., Mineau, D. E. and Miller, F. J. (1979). Uncontrollable post-partum bleeding. A new approach to hemostasis through angiographic arterial embolization. *Obstet. Gynecol.*, **54**, 361–4

46 Pais, S. O., Glickman, M., Schwartz, P., Pringoud, E. and Berkowitz, R. (1980). Embolization of pelvic arteries for control of post-partum hemorrhage. *Obstet. Gynecol.*, **55**, 754–8

47 Pelligra, R. and Sandberg, E. C. (1979). Control of intractable abdominal bleeding by external counterpressure. *J. Am. Med. Assoc.*, **241**, 708

48 Fribourg, S. R., Rothman, L. A. and Rovinsky, J. J. (1973). Intrauterine lavage for control of uterine atony. *Obstet. Gynecol.*, **41**, 876–83

49 Hester, J. D. (1975). Post-partum hemorrhage and re-evaluation of uterine packing. *Obstet. Gynecol.*, **45**, 501

50 Holtz, S. R. (1951). The control of post-partum hemorrhage by the intra-uterine balloon. *Am. J. Obstet. Gynecol.*, **62**, 450

51 Fahmy, K. (1969). Internal iliac artery ligation and its efficiency in controlling pelvic hemorrhage. *Am. J. Obstet. Gynecol.*, 95, 320

52 Burchell, R. C. (1968). Physiology of internal iliac artery ligation. *J. Obstet. Gynaecol. Br. Commonw.*, 75, 642

53 O'Leary, J. L. and O'Leary, J. A. (1966). Uterine artery ligation in control of intractable post-partum hemorrhage. *Am. J. Obstet. Gynecol.*, 94, 920

54 Lucas, W. E. (1980). Post-partum hemorrhage. *Clin. Obstet. Gynecol.*, 23, 637–46

55 Bygdeman, M., Kwon, S. U., Mukherjee, T. and Wiqvist, N. (1968). Effect of infusion of PGE_1 and PGE_2 on the motility of the pregnant human uterus. *Am. J. Obstet. Gynecol.*, 102, 317–26

56 Takagi, S., Yoshida, T., Togo, Y., Tochigi, H., Abe, M., Sakata, H., Fuju, T. K., Takahashi, H. and Tochigi, B. (1976). The effects of intramyometrial injection of prostaglandin $F_{2\alpha}$ on severe post-partum hemorrhage. *Prostaglandins*, 12, 565–72

57 Jacobs, M. M. and Arias, F. (1979). Intramyometrial prostaglandin $F_{2\alpha}$ in the treatment of severe post-partum hemorrhage. *Obstet. Gynecol.*, 55, 665–6

58 Andrinopoulos, G. C. and Mendenhall, H. W. (1983). Prostaglandin $F_{2\alpha}$ in the management of delayed post-partum hemorrhage. *Am. J. Obstet. Gynecol.*, 146, 217–18

59 Hertz, R. H., Sokal, R. J. and Dieker, L. J. (1980). Treatment of post-partum uterine atony with prostaglandin E_2 vaginal suppositories. *Obstet. Gynecol.*, 56, 120–30

60 Corson, S. T. and Bolognese, R. J. (1977). Post-partum uterine atony treated with prostaglandins. *Am. J. Obstet. Gynecol.*, 129, 918

61 Zahradnik, H. P., Steiner, H., Hillemanns, H. G., Breckwoldt, M. and Ardelt, W. (1977). Prostaglandin $F_{2\alpha}$ und 15-methyl-prostaglandin $F_{2\alpha}$. Anwendung bein Massiven Uterine Blutungen (The use of prostaglandin $F_{2\alpha}$ and 15-methyl-$PGF_{2\alpha}$ in massive uterine bleeding). *Geburtsh. Frauenheilk.*, 37, 493–5

62 Toppozada, M. K., El-Bossaty, M., El-Rahman, H. A. and Shams, A. H. (1981). Control of intractable atonic post-partum hemorrhage by 15-methyl prostaglandin $F_{2\alpha}$. *Obstet. Gynecol.*, 58, 327–30

63 Hayashi, R. H., Castillo, M. S. and Noah, M. I. (1981). Management of severe post-partum haemorrhage due to uterine atony using an analogue of prostaglandin F. *J. Obstet. Gynecol.*, 58, 426–430

14
Abnormal intrauterine pregnancy

S. S. RATNAM and R. N. V. PRASAD

INTRODUCTION

The use of prostaglandins to treat abnormal pregnancy is a fairly recent therapeutic advance. In 1970 Karim[1] reported successfully terminating abnormal intrauterine pregnancies with PGE_2. Subsequently, several prostaglandins administered by various routes have also been used. Thiery[2] and Karim[3] have provided comprehensive reviews. This chapter will present a practical guide for clinicians on the use of prostaglandins and their analogues in the management of abnormal intrauterine pregnancy. Such conditions include molar and anencephalic pregnancies, missed abortion and intrauterine fetal death.

PHYSIOLOGICAL BASIS OF THE USE OF PROSTAGLANDINS IN ABNORMAL PREGNANCY

The prostaglandins have a well-established role in the termination of pregnancy and induction of labour. In contrast to oxytocin which elicits a response only in advanced pregnancy, prostaglandins and their synthetic analogues cause the pregnant uterus to contract at all stages of gestation. In abnormal pregnancies, an altered physiological state causes prostaglandins to be either more effective (e.g. in missed abortion, intrauterine fetal death and molar pregnancy) or less effective (e.g. in anencephalic pregnancy) than when used for termination of normal pregnancy of a corresponding gestational age. Although the exact mechanism of action of prostaglandins in initiating termination of abnormal pregnancy is unknown, it is likely that the fetal membranes are involved to a large extent, since these tissues have been clearly shown to rapidly metabolize prostaglandins and this capacity may play an important role in the maintenance of a normal pregnancy[4]. It is easier to induce labour with exogenous prostaglandins when the fetal membranes are absent (as in molar pregnancy) or are in some way inactive or damaged (as in missed abortion, intrauterine fetal death). On the other hand, in anencephalic pregnancy the abnormal fetus is alive and the membranes are normal. Hence, any exogenous prostaglandins are rapidly metabolized before they can act locally on the myometrium. Moreover, the fetal adrenal and hypothalamic deficiency assists in delaying labour and delivery in that during induced labour in anencephalic pregnancy the amniotic fluid prostaglandins do not show the progressive rise seen in normal and induced labour[4]. For these reasons, labour is more difficult to induce with prostaglandins in anencephalic pregnancy compared to normal pregnancy.

These considerations notwithstanding, there is little doubt that prostaglandins are generally better than other methods (oxytocin, surgical methods) in the management of abnormal intrauterine pregnancy since their efficacy is more predictable than that of oxytocin and the risk of complications is lower than surgical termination of a late second or third trimester pregnancy.

GENERAL CONSIDERATIONS

Intrauterine fetal death and missed abortion

Intrauterine fetal demise occurring before the 28th week of pregnancy is defined as missed abortion, whilst that occurring after 28 weeks is regarded as intrauterine fetal death. The management of either condition with prostaglandins is similar, and they can be dealt with together.

The majority of pregnancies complicated by fetal death *in utero* end in spontaneous expulsion of the products of conception within a few weeks. It is not uncommon, however, for the dead fetus to be retained *in utero* for many weeks or months. The longest recorded duration has been $2\frac{1}{4}$ years[5]. In addition to the psychological stress placed upon the mother by the knowledge that her pregnancy is non-viable, retention of a dead fetus predisposes to coagulation disorders which may result in disseminated intravascular coagulopathy. In fact, some degree of derangement of coagulation occurs in approximately one in three women who have retained a dead fetus *in utero* for more than 1 month[6]. The occurrence of overt hypofibrinogenaemia with disseminated intravascular coagulation and significant haemorrhage is less common, however. Even so, the possibility of this serious complication and the mental stress which the woman must face while awaiting spontaneous expulsion of a dead fetus are the basis of the clinician's decision to terminate these pregnancies at the earliest possible time after confirmation of intrauterine death.

Molar pregnancy

Evacuation of the uterus is required as soon as the diagnosis of molar pregnancy is made. Delay in evacuation may result in spontaneous abortion with risk of serious haemorrhage, and increased incidence of invasive mole and choriocarcinoma. Both oxytocin and prostaglandins can be administered to induce abortion in these cases, but the mainstay of treatment still remains suction evacuation (see below for further discussion).

Anencephalic pregnancy

Once the diagnosis of anencephalic pregnancy is made, it is best terminated as soon as possible. Not only is the mother spared the psychological trauma of carrying an abnormal pregnancy, but expectant treatment is particularly unkind as these pregnancies have a tendency to become prolonged. One reported case was terminated after 1 year and 24 days[7]. Discomfort from the associated polyhydramnios and the possibility of dystocia and postmaturity are further indications to terminate the pregnancy early.

Although prostaglandins and their analogues have been used for termination of anencephalic pregnancy and are more predictable and superior to conventional oxytocin therapy, they are less efficacious in anencephalic pregnancy when compared to their use in intrauterine death, missed abortion or molar pregnancy (see Figure 14.1).

MEDICAL METHODS USED FOR TERMINATION OF ABNORMAL PREGNANCY

Hormones and drugs have long been used for the medical termination of anencephalic pregnancy and fetal death *in utero*. They were given in the hope of inducing uterine contractions, but most often proved ineffective and unsafe. Oestrogen therapy, pioneered by Robinson *et al.*[8] in 1935 was shown to be ineffective by Martin and Menzies[9] some 20 years later, as it took an average of over 7 days for expulsion of the fetus in the occasional successful case.

Intra-amniotic hypertonic saline has also been used, but hypernatraemia and cerebral damage have occurred[10,11]. Intra-amniotic hypertonic dextrose carries an increased risk of infection with a dead fetus

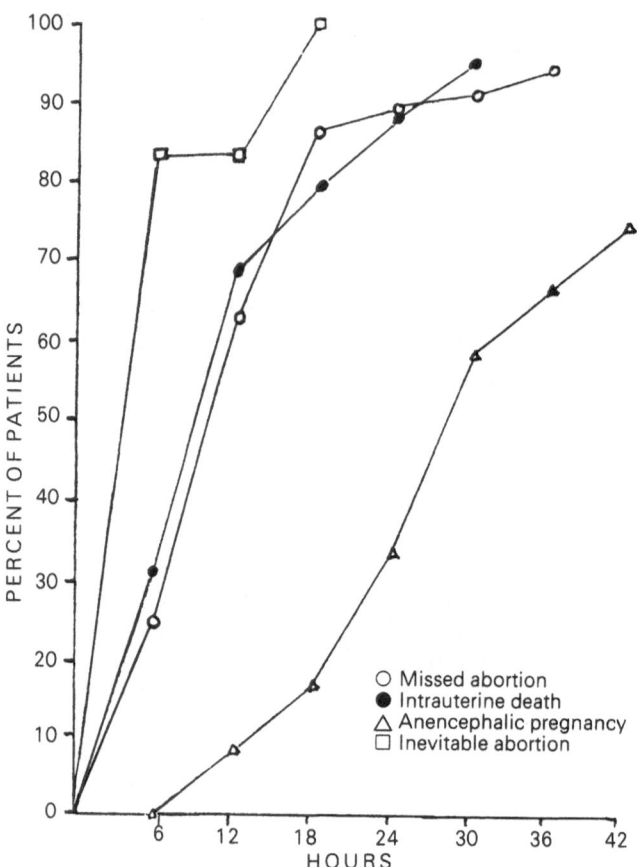

Figure 14.1 Cumulative abortion rates in percentage over induction–abortion interval in hours in 100 patients after intramuscular administration of 16,16-dimethyl *p*-benzaldehyde semicarbazone ester 150 μg at 6 hourly intervals (Tan, K. C.[24])

in utero. Moreover, both hypertonic dextrose and saline increase the risk of disseminated intravascular coagulation.

Intravenous oxytocin in high doses is often not successful when used alone for termination of anencephalic pregnancy and intrauterine fetal death. Moreover, the high doses required predispose to water intoxication, electrolyte disturbances and convulsions[12]. Amniotomy combined with intravenous oxytocin increases the success rate[13], but, in the failures, amniotomy in the presence of a dead fetus increases the risk of intrauterine sepsis and amniotic fluid embolism[14]. Unfortunately, the failure to initiate uterine contractions reasonably soon after amniotomy leaves the obstetrician no choice but to evacuate the uterus abdominally by Caesarean section or hysterotomy.

Prostaglandins and their analogues, in addition to being safe and effective in the treatment of abnormal pregnancy, may also exert a protective role against the development of disseminated intravascular coagulation. In cases of missed abortion and intrauterine fetal death, Briel *et al.*[15] have shown that intravenous PGE_2 infusion caused a significant decrease in platelet aggregation.

THE USE OF PROSTAGLANDINS FOR ABNORMAL PREGNANCY

Both natural prostaglandins and their synthetic analogues given via various routes have been used with varying degrees of success in the treatment of abnormal intrauterine pregnancy. In general, the prostaglandin analogues have proven better than the natural prostaglandins E_1, E_2 and $F_{2\alpha}$, as the former group of compounds are more effective with less side-effects. Moreover, natural prostaglandins require either an intravenous, intra-amniotic or extra-amniotic route of administration, whereas the newer analogues can be administered by the intramuscular route with comfort to the patient. On the other hand, the intramuscular administration of PGE_2 and $PGF_{2\alpha}$ causes intense pain, which sometimes lasts for several days.

Karim was the first to use prostaglandins in the management of intrauterine death. He used PGE_2 infused intravenously at a rate of $5\,\mu g/min$ in six patients with missed abortion, and at a rate ranging between $0.05\,\mu g$ and $2\,\mu g/min$ in 15 patients with intrauterine fetal death[1]. The mean induction–abortion interval was 8 h for missed abortion and slightly more than 12 h for intrauterine death. Many workers have subsequently also used PGE_2 for termination of abnormal pregnancy. Karim[3] has written a comprehensive review of the subject.

CLINICAL MANAGEMENT OF ABNORMAL INTRAUTERINE PREGNANCY

General considerations

The diagnosis of abnormal uterine pregnancy must be made unequivocally by a combination of clinical examination, hormonal assays, X-rays and/or ultrasound examination.

This is important for the following reasons:

(1) Unnecessary mental stress to the mother in cases of misdiagnosis is avoided.
(2) Prostaglandin analogues in the doses used for termination of abnormal pregnancy may not prove effective in a viable pregnancy. Moreover, should the patient elect to continue with her pregnancy if treatment with prostaglandin analogues fails, the safety of these drugs with regard to long-term fetal and neonatal effects has not been established.
(3) There are the obvious medicolegal implications of misdiagnosis.

Once diagnosis has been confirmed the clinical management follows the scheme shown on the flow chart in Figure 14.2. Selection of cases to be treated with prostaglandins is important in order to prevent serious complications. In abnormal pregnancy, the integrity of the fetal membranes should be preserved for as long as possible in order to prevent amnionitis. Once sepsis occurs, the clinician's hand is forced and hysterotomy or Caesarean section may be required if abortion or delivery is not imminent.

Missed abortion and intrauterine fetal death
The management of missed abortion depends on the uterine size. When the uterine size is smaller than 12 weeks size, it is preferable to evacuate the pregnancy surgically with dilatation and curettage or vacuum aspiration than to use prostaglandins. Surgical evacuation is quicker and more likely to be complete in contrast to prostaglandin therapy which may require several hours for an abortion which is frequently incomplete and requires subsequent curettage. Primary surgical evacuation in missed abortion is occasionally complicated by excessive uterine bleeding, and sometimes it may be difficult to achieve cervical dilatation prior to curettage. It may therefore be advisable to pretreat selected patients such as nulliparous women and others with clinical evidence of a firm cervix, with prostaglandins to soften and dilate the cervix much in the same manner as prostaglandins are used for the preoperative cervical dilatation prior to vacuum aspiration of the nulliparous first trimester pregnancy[16]. A single intramuscular dose of 150 μg of 16,16-dimethyl-PGE$_2$ given 3 h prior to evacuation suffices[17]. This treatment also reduces the amount of bleeding during curettage because of the contractile action of prostaglandins on the uterus.

In missed abortion, when the uterus is larger than 12 weeks size, treatment is best achieved with prostaglandins. Some gynaecologists still prefer dilatation and evacuation up to 16 weeks uterine size, but this procedure may be difficult because of the risk of laceration associated with removal of the bony fetal parts and excessive bleeding

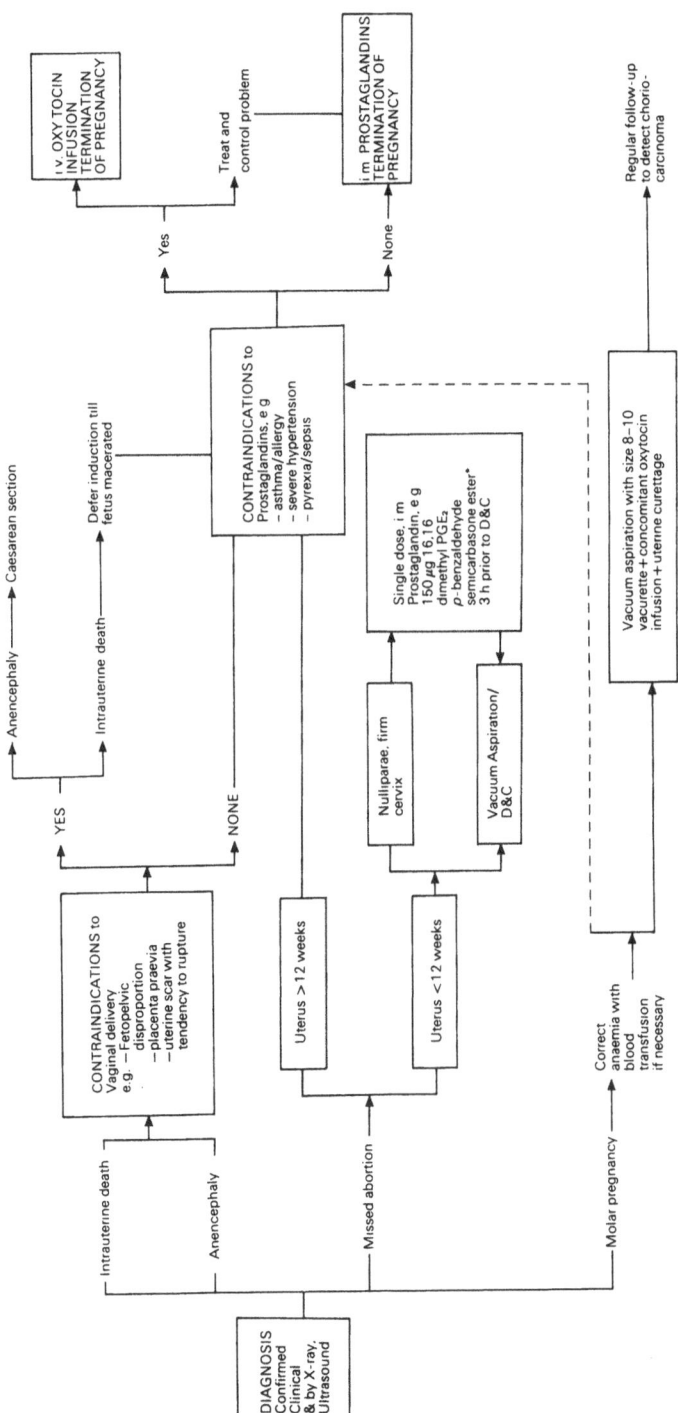

Figure 14.2   Clinical decision tree for the management of abnormal intrauterine pregnancy

* Editorial note: Because of its instability it is questionable whether this compound will become commercially available in the near future. Fortunately, sulprostone and 15-methyl PGF₂ₐ can be used with equal success (see Table 14.3) in the doses listed in Table 14.2

because of prolonged procedure times. The management of cases of late missed abortion with prostaglandins is similar to the treatment of intrauterine fetal deaths.

Before a patient with intrauterine fetal death is treated with prostaglandins, it must be ascertained that she has no contraindications to vaginal delivery or to prostaglandin therapy. In patients with a uterine scar prone to rupture (e.g. previous hysterotomy or classical Caesarean section scar), when fetopelvic disproportion is present or when there is a placenta praevia, repeat Caesarean section or hysterotomy is the preferable course of treatment in most instances. In the presence of minor fetopelvic disproportion or asymptotic placenta praevia, expectant treatment for about two weeks may enable vaginal delivery of the macerated foetus with prostaglandin treatment.

Contraindications to initiating prostaglandin treatment include active bronchial asthma and similar allergic conditions, severe hypertension complicating pregnancy or evidence of sepsis or pyrexia. These medical conditions should be adequately treated and controlled before initiating prostaglandin therapy. If medical therapy is unsuccessful, prostaglandins should not be started, and any aggravation of incipient medical problems should be regarded as an immediate indication to stop all further treatment with prostaglandins. In such patients it is advisable to await spontaneous expulsion or to institute treatment with conventional intravenous oxytocin infusion.

Anencephalic pregnancy

The comments on selection of patients for prostaglandin therapy in the preceding section apply equally to women with anencephalic pregnancy. In anencephalic pregnancy diagnosed near term, however, expectant treatment in cases complicated by major degrees of placenta praevia is of no use, as these babies are alive *in utero* and there is no hope of maceration with time, in contrast to what one can expect with intrauterine death. These infants are therefore best delivered by Caesarean section. Similarly, if there are medical contraindications to prostaglandin treatment, anencephalic pregnancy can then be treated by conventional oxytocin therapy.

Molar pregnancy

The preferred method of clinical management of hydatiform mole is debatable. Bygdeman[18] prefers using prostaglandins for the management of intact molar pregnancy, especially after 12 weeks. When the uterus is smaller than 12 weeks size, however, vacuum aspiration without prostaglandins can be performed. Satoh[19] uses dilatation and curettage for molar pregnancies of all gestational sizes, noting that:

(1) Abortion with prostaglandins in molar pregnancy requires time, and patients may bleed excessively while waiting for abortion to complete; and

(2) curettage is necessary anyway after abortion for molar pregnancy with prostaglandins; hence primary D & C may be a more rational approach, as it is quicker and more convenient.

In Singapore, at the National University Department of Obstetrics and Gynaecology, prostaglandin analogues have been used for termination of molar pregnancy and have not been found to offer any advantage to primary suction curettage. Although it has been suggested that prostaglandin-induced abortion may be associated with a high incidence of metastatic trophoblastic disease later on, there is no firm proof to support this contention[20]. In the younger patient, therefore, we consider suction currettage to be the treatment of choice, regardless of uterine size[21]. In the older, multiparous patient who has completed her family, hysterectomy may be performed as a primary therapy, as this management is thought to be associated with a lower incidence of malignant sequelae[21,22].

Vacuum aspiration for molar pregnancy is performed in conjunction with 30 Units of oxytocin in a 5% dextrose solution run in at 30 drops/min while suction curettage is in progress. Suction should be applied with a small bore curette (8–10 mm vacurette), as too rapid an evacuation can result in massive haemorrhage, disseminated intravascular coagulation and death[21]. After suction, the uterus should be curetted with standard metal, uterine curette.

Although suction curettage is the method of choice, prostaglandins have also been used to abort intact molar pregnancy. These agents are effective, but the patient requires curettage to complete the abortion process. Before using prostaglandins, however, the physician should determine that the patient is in optimum health and that she has no contraindications to use of this medication. Treatment with prostaglandin analogues is identical to that of other abnormal pregnancies summarized in Figure 14.2.

Whatever the treatment, careful long-term follow-up to detect the occurrence of choriocarcinoma is required and remains an important aspect in the treatment of hydatidiform mole.

Choice of prostaglandin and route of administration

Prostaglandin analogues have superseded natural prostaglandins in the treatment of abnormal pregnancy because of their enhanced efficacy and lower incidence of side-effects. The choice of the analogue and the dose and route of administration depend on the availability of the

Table 14.1 Recommended dose schedule of prostaglandins by route for anencephalic pregnancy and intrauterine death. (Modified after Karim[3], 1979)

Prostaglandin	Route of administration					
	Intravenous infusion*	Extra-amniotic infusion*	Extra-amniotic bolus‡	Intra-amniotic†	Intra-muscular	Vaginal
PGE_2	0·25–2·5 µg/min	0·5–25·0 µg/min	25–100 µg/2 h	2·5–5 mg/12–24 h	NS	ND
$PGF_{2\alpha}$	2·5–20 µg/min	2·5–5·0 µg/min	125–300 µg/2 h	25 mg/12 h	NS	ND
15-methyl $PGF_{2\alpha}$	ND	ND	ND	ND	50–100 µg/3 h	ND
2a,2b-dihomo-15-methyl $PGF_{2\alpha}$	ND	ND	ND	ND	0·25–0·5 mg/8 h	ND
Sulprostone	ND	ND	ND	ND	0·25–0·5 mg/6 h	ND
16,16-dimethyl PGE_2 p-benzaldehyde semicarbasone ester	ND_	ND	ND	ND	150 µg/6 h	ND

* = preferably given by infusion pump; † = not a preferred route; ‡ = preferably start with a small 'test' bolus dose; ND = no data; NS = not suitable

Table 14.2 Recommended dose scheduled by route for missed abortion and hydatidiform mole. (Modified after Karim[3], 1979)

Prostaglandin	Intravenous infusion*	Extra-amniotic infusion*	Extra-amniotic bolus‡	Intra-amniotic†	Intra-muscular	Vaginal
			Route of administration			
PGE₂	0·5–4·0 µg/min	1–4 µg/min	50–200 µg/2 h	2·5 mg/12–24 h	NS	20 µg/3–6 h
PGF₂α	5–40 µg/min	5–10 µg/min	250–600 µg/2 h	25 mg/12–24 h	NS	ND
15-methyl PGF₂α	ND	ND	ND	0·5 mg/12–24 h	0·1–0·2 mg 3 hourly	ND
2a,2b-dihomo-15-methyl PGF₂α	ND	ND	ND	ND	0·5 mg 8 hourly	ND
Suprostone	0·5–2·0 µg/min	ND	ND	ND	0·5–0 mg 6 hourly	ND
16,16-dimethyl PGE₂ p-benzaldehyde semicarbasone ester	ND	ND	ND	ND	150 µg/ 6 hourly	ND

* = preferably given by infusion pump; † = not applicable to molar pregnancy; not a preferred route for missed abortion; ‡ = preferably to start with a small 'test' dose; ND = no data; NS = not suitable

commercial medication – apart from any question of efficacy, side-effects and ease of administration. Tables 14.1 and 14.2 show the recommended dose schedule of various prostaglandin analogues used for termination of abnormal pregnancy[3]. These data should only serve as a guide and may require modification as more information, analogues and commercial products become available. Natural prostaglandins have been included in these Tables as analogues are not available for general use in certain countries.

In most instances, it is preferable to use intramuscular administration of prostaglandin analogues such as sulprostone or 16,16-dimethyl PGE$_2$ p-benzaldehyde semicarbasone ester, as they are effective and easy to administer. Intravenous PGE$_2$ and PGF$_{2\alpha}$, although almost 100% effective, have a high incidence of side-effects. These side-effects can be reduced by the use of an infusion pump, however. The extra-amniotic route requires a skilled technique for its administration and is often accompanied by side-effects, especially if large bolus doses are given. In addition, the accidental rupture of membranes, increased risk of infection and premature expulsion of the extra-amniotic catheter remain problems. Intra-amniotic instillation is not preferred, as infection can be introduced in the presence of a dead fetus. Moreover, intra-amniotic injection may also cause leakage of thromboplastin-rich material into the maternal circulation, and the altered fetal membranes present in intrauterine fetal death may permit an easier passage of the injected prostaglandins into the maternal circulation. Severe reactions have been observed when intra-amniotic injection has been attempted in cases of intrauterine death[23]. *Of particular importance is that this route cannot be used with molar pregnancy, as no intra-amniotic space exists!* The vaginal route, although simple to use, is accompanied by a high incidence of gastrointestinal side-effects, and early rupture of membranes or uterine bleeding may interfere with administration and absorption of prostaglandins.

Under these circumstances, prostaglandin analogues given by the intramuscular route are presently the best mode of therapy for abnormal intrauterine pregnancy in terms of their efficacy, comparative lack of side-effects and ease of administration. The analogues 2a,2b-dihomo-15(S)-15-methyl-PGF$_{2\alpha}$ methyl ester, 15-methyl-PGF$_{2\alpha}$, 16-phenoxy-ω-tetranor PGE$_2$ methyl sulphonylamide and 16,16-dimethyl-PGE$_2$ p-benzaldehyde semicarbasone ester injected intramuscularly have proven very effective for termination of pregnancy in intrauterine fetal death and missed abortion. A summary of the clinical results with these analogues is shown in Table 14.3.

Table 14.3 Results of termination of abnormal intrauterine pregnancy with intramuscular administration of prostaglandin analogues. (Karim[23], 1983)

No. and type of case	Success	Prostaglandin dose	Mean expulsion interval (average no. injections)	Side-effects	Reference
63 MP: 9 MA: 30 IUD: 19 AP: 5	61 (96·8%) within 24 h	16 phenoxy-ω-tetranor PGE$_2$ methyl-sulphonylamide (sulprostone) 0·5 mg 6 hourly	9·5 h (2·0)	GI 21% S 63%	Karim[25]
97 MA & IUD	96 (98·9%) within 30·5 h	15(S) 15-methyl PGF$_2$, 0·125–0·25 mg, 2 hourly	1·2–30·5 h (range) (Median 5·0)	GI 89%	Wallenberg[26]
631 MP: 82 MA: 233 IUD: 282 AP: 34	600 (95·1%) within 32 h	2a,2b-dihomo-15(S)-15-methyl PGE$_2$ methyl ester, 0·5 mg, 8 hourly	11·3 h (1·8)	GI 48% S 11·9%	Karim et al.[27]
100 MP: 1 MA: 37 IUD: 44 AP: 12 IA: 6	93 (93%) within 32 h	16,16-dimethyl PGE$_2$ p-benzaldehyde semicarbasone ester, 150 μg, 6 hourly	11·58 h (2·39)	GI 32% S 34% P 39%	Tan, K. C.[24]

MP = molar pregnancy: MA = missed abortion; IUD = intrauterine death; AP = anencephalic pregnancy; IA = inevitable abortion; S = shivering; GI = gastrointestinal; P = pyrexia >100 °F

Clinical management of prostaglandin treatment of abnormal intrauterine pregnancy

Prostaglandin therapy of all abnormal intrauterine pregnancies can be managed using the intramuscular route of administration (a summary is shown in Figure 14.3). Treatment in this manner is the best available regime at present. The analogues used in Singapore include sulprostone (commercially available as Nalador®), 16,16-dimethyl-PGE$_2$ p-benzaldehyde semicarbasone and 2a,2b-dihomo-15(S)-15-methyl PGF$_{2\alpha}$. They are all administered intramuscularly at the specified intervals until abortion/delivery occurs (see Figure 14.3). The clinical management protocol that follows applies to prostaglandins given by other routes as well.

Blood pressure, temperature and vital signs are monitored closely until after the pregnancy has been completely terminated. Side-effects such as bronchoconstriction, nausea, diarrhoea, vomiting and 'shivering' are frequent complications of treatment. The appearance of bronchoconstriction requires immediate discontinuation of prostaglandin therapy and prompt administration of bronchodilators. Gastrointestinal symptoms are usually mild and can be treated with the appropriate drugs, such as stemetil for nausea and vomiting and loperamide for diarrhoea; in such instances prostaglandin treatment can be continued. The occurrence of a blood pressure elevation of more than 150/95 mmHg, fever more than 38 °C (a common occurrence with sulprostone and 16,16-dimethyl-PGE$_2$ p-benzaldehyde semicarbasone ester) and 'shivering' require the omission of only one dose of the drug for prompt return to normal status. Prostaglandin treatment can then be resumed until abortion or delivery occurs.

With the standard regimen, abortion or delivery should occur within 30 h in cases of molar pregnancy, missed abortion and intrauterine death. Vaginal examinations to assess progress should be kept at a minimum since more frequent pelvic examinations may be followed by ascending infection. In most instances abortion or delivery would have occurred within 24 h. Delivery of an anencephalic fetus generally occurs much later, however, usually within 40 h of the start of therapy. In order to avoid the possibility of amnionitis artificial rupture of the fetal membranes should be delayed for as long as possible. In molar pregnancy and missed abortion, uterine curettage is required to empty the uterus.

If the pregnancy has not been terminated within the stipulated time in any case of abnormal pregnancy, then further management depends on the state of cervical dilatation and the status of the fetal membranes. If the cervix is effaced and more than 50% dilated, prostaglandin treatment can be continued for another 6–8 h, provided rapid progress and delivery occurs during that time.

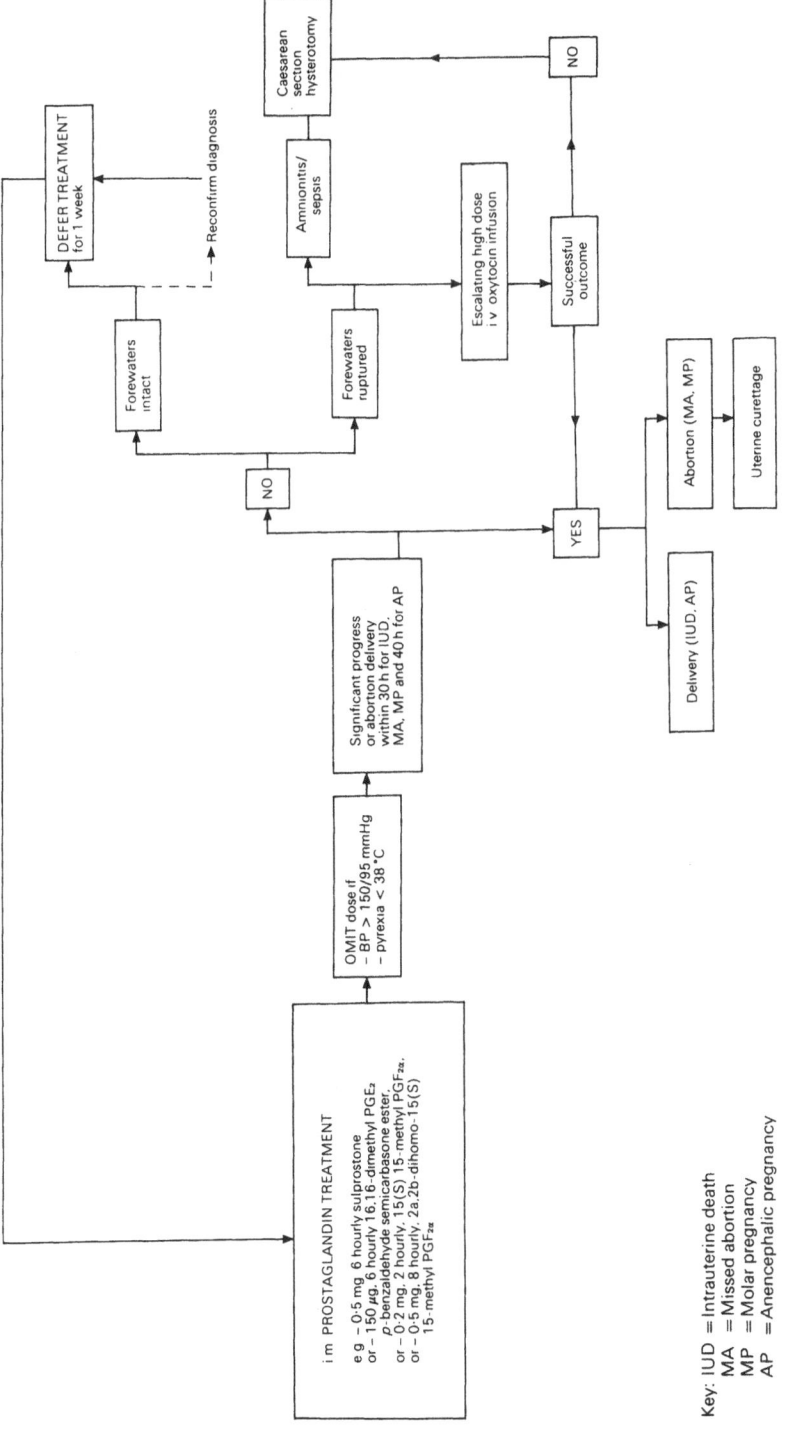

Figure 14.3 Clinical decision tree for subjects with abnormal pregnancy managed with prostaglandins

In the absence of any evidence of cervical dilatation, further management depends on the states of the fetal membranes. If they are intact and there is no evidence of amnionitis, then it is preferable to leave the patient alone and defer further treatment with prostaglandins for about 1 week. While waiting, it would be useful to repeat the ultrasound to confirm the diagnosis of abnormal intrauterine pregnancy, as on rare occasion a viable pregnancy or a uterine fibroid have been unwittingly treated with prostaglandins without success[17].

When the fetal membranes have spontaneously ruptured or in the presence of amnionitis, treatment has to be prompt and vigorous if delivery or abortion is not forthcoming within the stipulated time. Such patients will require escalating high dose, intravenous oxytocin administered via an infusion pump. In conjunction with this treatment, diuretics may be prescribed to prevent water intoxication. Oxytocin infusion should be begun with a small dose, as prior prostaglandin treatment is likely to have sensitized the uterus to oxytocin. In the absence of uterine tetany, the dose of oxytocin can be doubled every half an hour until adequate uterine contractions are achieved. In the absence of any progress and/or if amnionitis supervenes, Caesarean section (in case of late intrauterine death and anencephaly) or hysterotomy (with late missed abortion) must be performed. Fortunately, the need for such operative intervention occurs only rarely, but it may be required to prevent potentially fetal septicaemia.

CONCLUSION

When judiciously used, prostaglandin analogues at present provide the best means to terminate abnormal pregnancy without delay and with confidence. They represent a significant advance in obstetric and gynaecological therapy[3]. Prostaglandins offer an attractive alternative to oxytocin infusion therapy and surgical evacuation procedures. It is possible that prostaglandin therapy will be used in the treatment of other problems in the future. For example, Prasad[17] has used 16,16-dimethyl-PGE$_2$ p-benzaldehyde semicarbasone ester to successfully terminate abnormal uterine pregnancies in five patients, two of whom had hydrocephalic pregnancy (at 24 weeks and 36 weeks) and three who developed hydrops fetalis in the third trimester of pregnancy. With the more routine use of ultrasound, fetoscopy, amniocentesis and chorionic villi biopsy, it is conceivable that abnormal pregnancies can be diagnosed much earlier and termination with prostaglandin therapy will continue to offer specific advantages.

ACKNOWLEDGEMENT

The authors wish to thank Ms Surjit Kaur for her expert secretarial assistance.

References

1 Karim, S. M. M. (1970). Use of prostaglandin E₂ in the management of missed abortion, missed labour and hydatidiform mole. *Br. Med. J.*, **3**, 196

2 Thiery, M., De Hemptinne D., Yo Le Sian, A. and Amy, J. J. (1975). The use of prostaglandins for the termination of 'pregnancy' in cases of hydatidiform mole, intrauterine death and anencephaly. *T. Geneesk*, **9**, 441

3 Karim, S. M. M., Ng. S. C. and Ratnam, S. S. (1979). Termination of abnormal intrauterine pregnancy with prostaglandins. In Karim, S. M. M. (ed.) *Advances in Prostaglandin Research. Practical Applications of Prostaglandins and their Synthesis Inhibitors.* pp. 319–74. (Lancaster: MTP Press)

4 Karim, S. M. M. and Hilier, K. (1975). Physiological roles and pharmacological actions of prostaglandins in relation to human reproduction. In Karim, S. M. M. (ed.) *Prostaglandins and Human Reproduction.* pp. 57–8. (Lancaster: MTP Press)

5 El-Sherbini, R. M. (1963). Retention of an 8 month foetus *in utero* for two-and-a-half years. *J. Obstet. Gynaecol. Br. Commonw.*, **70**, 154

6 Hardisty, R. M. and Ingram, G. I. C. (1965). In *Bleeding Disorders. Investigations and Management.* (Oxford: Blackwell Scientific Publications)

7 Higgins, L. G. (1954). Prolonged pregnancy (partus serotinus). *Lancet*, **2**, 1154

8 Robinson, A. L., Datnow, H. M. and Jeffcoate, T. N. A. (1935). Induction of abortion and labour by means of oestrin. *Br. Med. J.*, **1**, 749

9 Martin, R. H. and Menzies, D. N. (1955). Oestrogen therapy in missed abortion and labour. *J. Obstet. Gynaecol. Br. Emp.*, **62**, 256

10 Cameron, J. M. and Dayan, A. D. (1966). Association of brain damage with therapeutic abortion induced by amniotic fluid replacement. Report of 2 cases. *Br. Med. J.*, **1**, 1010

11 Wagatsuma, T. (1965). Intraamniotic saline for therapeutic abortion. *Am. J. Obstet. Gynecol.*, **93**, 743

12 Liggins, G. C. (1962). The treatment of missed abortion by high dosage syntocinon intravenous infusion. *J. Obstet. Gynaecol. Br. Commonw.*, **69**, 277

13 Ursell, W. (1972). Induction of labour following fetal death. *J. Obstet. Gynaecol. Br. Commonw.*, **79**, 260

14 Peterson, E. P. and Taylor, H. B. (1970). Amniotic fluid embolism. An analysis of 40 cases. *Obstet. Gynecol. (NY)*, **35**, 787

15 Briel, R. C., Kunz, S., Kidess, E. and Dieter, B. (1976). Studies on platelet function during application of prostaglandin F₂ in missed abortion and

missed labour. Presented at the *8th World Congress of Gynecology and Obstetrics*, 7–22 Oct. (Abstr), p. 37 (Amsterdam: Excerpta Medica)

16 Karim, S. M. M. and Prasad, R. N. V. (1979). Preoperative cervical dilatation with prostaglandins. In Karim, S. M. M. (ed.) *Advances in Prostaglandin Research. Practical Applications of Prostaglandins and their Synthesis Inhibitors.* pp. 283–99. (Lancaster: MTP Press)

17 Prasad, R. N. V. (1984). Unpublished observations

18 Bygdeman, M. (1983). In Karim, S. M. M. (ed.) *Proceedings of a Symposium on 'Cervagem – A New Prostaglandin in Obstetrics and Gynaecology'*, Singapore, 31st July, 1982. p. 108. (Lancaster: MTP Press)

19 Satoh, K. (1983). In Karim, S. M. M. (ed.) *Proceedings of a Symposium on 'Cervagem – A New Prostaglandin in Obstetrics and Gynaecology'*, Singapore, 31st July, 1982. p. 108, (Lancaster: MTP Press)

20 Ratnam, S. S. and Chew, S. C. (1979). The modern management of trophoblastic disease. In Stallworthy, J. and Bourne, G. (eds) *Recent Advances in Obstetrics and Gynaecology*, Vol. 13, p. 239. (London: Churchill Livingstone)

21 Ratnam, S. S. and Ilancheran, A. (1982). Disease of the trophoblast. In Philpott, R. H. (ed.) *Clinics in Obstetrics and Gynaecology – Obstetric Problems in the Developing World.* Vol. 9, No. 3, December 1982, pp. 550–1. (London: Saunders Co Ltd.)

22 Tow, W. S. H. (1966). The classification of malignant growth of the chorion. *J. Obstet. Gynaecol. Br. Commonw.*, **73**, 1000

23 Karim, S. M. M. (1983). Clinical applications of prostaglandins in obstetrics and gynaecology. In Karim, S. M. M. (ed.) *Proceedings of a Symposium on 'Cervagem – A New Prostaglandin in Obstetrics and Gynaecology.'* Singapore, 31st July, 1982. pp. 27–9. (Lancaster: MTP Press)

24 Tan, K. C., Karim, S. M. M., Kottegoda, S. R., Prasad, R. N. V. and Ratnam, S. S. (1984). Termination of abnormal pregnancy with intramuscular 16,16 dimethyl PGE_2 *p*-benzaldehyde semicarbasone ester. Personal communication

25 Karim, S. M. M., Lim, A. L., Prasad, R. N. V., Yeo, K. C., Ng, S. C., Salmon, Y. M., Choo, H. T. and Ratnam, S. S. (1979). Termination of abnormal intrauterine pregnancy with intramuscular administration of sulprostone. *Singapore J. Obstet. Gynecol.*, **10**, 33

26 Waleenberg, H. S. G., Keirse, M. J. N. C., Freie, H. M. P. and Blacquiere, J. F. (1980). Intramuscular administration of 15(s) 15 methyl prostaglandin $F_{2\alpha}$ for induction of labour in patients with fetal death. *Br. J. Obstet. Gynaecol.*, **87**, 203

27 Karim, S. M. M., Ratnam, S. S., Hutabarat, H., Hanafiah, J., Simanjuntak, P., Teoh, S. K., Ong, S. K., Sen, D. K. and Sinnathuray, T. N. A. (1982). Termination of pregnancy in cases of intrauterine fetal death missed abortion, molar and anencephalic pregnancy with intramuscular administration of 2a, 2b-dihomo 15(s) 15 methyl $PGF_{2\alpha}$ methyl ester – a multicentre study. *Ann. Acad. Med. Singapore*, **11**, 508

15
Induced abortion

N. H. LAUERSEN

INTRODUCTION

Prostaglandins have been available in numerous countries for investigations as well as routine performance of abortion procedures for more than a decade[1] (Table 15.1). This chapter will provide the clinician with an adequate information base for the decision of *when* to use a prostaglandin, *what* prostaglandin to use, and *how* to use that prostaglandin in the induction of first and second trimester abortion. Each subsection of this chapter initially presents a survey of the development of prostaglandins for this particular aspect of pregnancy termination. This survey includes discussion of and, where possible, resolution of the 'controversies' surrounding the use of PG in termination of a particular stage of pregnancy.

The techniques for routine utilization of a prostaglandin will be detailed, when there is a routine use, and there will be an exposition of the expected range of efficacy and side-effects of the agent. Finally, the status of prostaglandin research will be discussed to indicate which particular prostaglandin or mode of prostaglandin administration may become the 'routine' technique of tomorrow, and the direction of future prostaglandin research and development.

PROSTAGLANDINS AS PRIMARY ABORTIFACIENTS – FIRST TRIMESTER

The first trimester of pregnancy is generally regarded as the period up through 12 weeks from the beginning of the last menstrual period. The vast majority of induced abortions are performed during the first

Table 15.1 Prostaglandins – application and availability for first and second trimester abortion

Name	Manufacturer	Registered	Clinical use	Route
Prostaglandin $F_{2\alpha}$ ($PGF_{2\alpha}$) 'Prostin F_2 Alpha' 'Prostarmon-F' 'Prostaglan'	Upjohn ONO ONO	50 countries worldwide Japan, Korea, Taiwan Brazil	Primary abortifacient – 2nd trimester	Intra-amniotic
Prostaglandin E_2 (PGE_2) 'Prostin E_2'	Upjohn	15 countries	Primary abortifacient – 2nd trimester Intrauterine fetal demise Hydatidiform mole	Vaginal suppository Vaginal suppository
15(S)-15-methyl prostaglandin $F_{2\alpha}$ (THAM (tromethamine salt)) 'Prostin 15 M'	Upjohn	USA, India	Primary abortifacient – 1st and 2nd trimester Intrauterine death Failed induced 2nd trimester abortion	Intramuscular Intra-amniotic Intramuscular Intramuscular
16-phenoxy-ω-tetranor-PGE_2 methyl sulphonylamide Sulprostone 'Nalador'	Schering AG	Germany, several European countries	Primary abortifacient – 1st and 2nd trimester Preoperative cervical priming 1st and 2nd trimester	Intramuscular Intramuscular
16,16-dimethyl-trans-Δ^2-PGE_1 methyl ester ONO-802 Cervagem	ONO May & Baker	Japan England, several European countries	Primary abortifacient – 1st trimester Preoperative cervical priming 1st and 2nd trimester	Vaginal suppository
9-deoxo-16,16-dimethyl-9-methylene PGE_2 potassium salt	Upjohn		Preoperative cervical priming 1st and 2nd trimester	Vaginal suppository

trimester. Suction curettage, the most prevalent technique for first trimester abortion, was developed concurrently in China and Eastern Europe[2-6] to minimize the complications of infection and blood loss encountered with dilatation and sharp curettage. Suction curettage, however, is not without immediate and future risk, including the hazard of anaesthesia and the potential of cervical damage resulting from mechanical dilatation of the cervix. A desire to minimize these risks, coupled with the development of highly sensitive pregnancy tests, led to the development of the early suction abortion which has been variously termed menstrual regulation, menstrual extraction and mini-abortion[7-9]. Later experience with this technique, however, demonstrated the routine need for local anaesthesia to minimize vasovagal symptoms, and, in some patients, mechanical cervical dilatation could not be avoided. In addition, even with local anaesthesia, approximately 20% of the patients experienced episodes of syncope and sweating during the procedure, 4% had developed endometritis when seen at follow-up, and another 4% required a repeat extraction or suction curettage for incomplete abortion[10]. The development of the prostaglandins offered the hope of an effective pharmacological alternative to surgical interruption of first trimester pregnancy that would avoid the potential problems associated with anaesthesia, mechanical dilatation of the cervix and uterine instrumentation.

The prostaglandins have the ability to stimulate uterine activity and to interrupt pregnancy throughout all stages of gestation. In the application of the prostaglandins to the termination of first trimester pregnancy, prostaglandin administration must be compared to well-established and widely used methods of surgical interruption such as suction curettage. The standards of safety, ease of administration, cost effectiveness and patient acceptability by which prostaglandins are evaluated must therefore be very stringent in the first trimester of pregnancy.

Natural prostaglandins

In the early 1970s, pioneering prostaglandin investigators Karim and Filshie[11,12] and Bygdeman and Wiqvist[13-16] demonstrated the ability of intravenous infusions of the naturally occurring prostaglandins, prostaglandin $F_{2\alpha}$ and E_2, to terminate first trimester pregnancy. Their work was subsequently confirmed by researchers such as Kinoshita et al.[17], Haspels and Luigies[18] and Wentz and Jones[19]. The variation in the abortion success rate from 43 to 100%, the practical consideration of the inconvenience of a prolonged intravenous infusion, and the problem of adverse systemic response in the form of prolonged painful uterine contractions and gastrointestinal side-effects[20] precluded the routine application of this technique for early pregnancy interruption.

The side-effects of the prostaglandins, as is the case with the side-effects of most drugs, are dose related. Administration of the prostaglandin directly to the target organ, the myometrium, would enhance the myometrial effects while reducing the systemic absorption of prostaglandin, and therefore minimizing side-effects. Extra-amniotic administration of prostaglandin $F_{2\alpha}$ and E_2 yielded more consistently acceptable success rates (generally 90–100%)[18,21-23,39-42] than did intravenous infusion in the first trimester, and side-effects were considerably reduced. Karim suggested that this method of administration appeared to be a valuable clinical tool in the termination of pregnancy during the late first trimester and early second trimester (13–15 weeks of gestation), the period when suction curettage may not be indicated and when the uterus is still too small for intra-amniotic injection[20]. Successful interruption in the late first trimester, however, usually required either multiple administrations or continuous infusion of the prostaglandin through an indwelling catheter. The potential for infection inherent in the use of the indwelling catheter limited the general acceptance of the technique.

The application of prostaglandin for 'menstrual induction' in very early pregnancy in lieu of a 'menstrual extraction' initially appeared to be a most promising area for investigation for two reasons. First, prostaglandin has a luteolytic effect in several subprimate species, and second, prostaglandins act as early abortifacients through myometrial stimulation[20]. Intravenous infusion of prostaglandin $F_{2\alpha}$ in a limited series of nine patients resulted in a termination of pregnancy in five[19]. Intrauterine administration was more consistently successful[24-28] and, in general, abortion could be achieved with a single intrauterine injection, eliminating the need for and potential hazard of the indwelling catheter. The success rate during the early first trimester ranged from 65%[27] to 100%[25,26]. Since the technique of intrauterine injection was not adaptable to self-medication, hospitalization was required.

Vaginal administration of the natural prostaglandins avoided the problem of the indwelling catheter and offered the possibility of eventual self-medication. As study of this technique progressed[29-38] the success rates in the early first trimester (37–90%) and late first trimester (44–80%) were not clinically acceptable. Absorption from the vagina appeared to be variable, and vaginal bleeding induced by the prostaglandin administration could wash the drug from the vagina so that the patient did not receive a dose sufficient to induce abortion.

The inherent characteristics of the natural prostaglandins, i.e. rapid inactivation necessitating continuous or at least frequent administration and gastrointestinal side-effects, precluded the widespread acceptance of these agents for first trimester abortion following their approval in a number of countries.

Prostaglandin analogues

The rapid inactivation of the natural prostaglandins, most probably due to dehydrogenation at carbon-15, resulted in prostaglandin metabolites that possessed greatly reduced biological activity. Bundy and co-workers[43] synthesized prostaglandin analogues which were modified at carbon-15 and resisted enzymatic degradation by prostaglandin 15-dehydrogenase. These modified prostaglandins were 15(S)-15-methyl prostaglandin E_2 and $F_{2\alpha}$. Karim and Sharma[44] reported that the 15-methyl-prostaglandin E_2 methyl ester and 15-methyl-prostaglandin $F_{2\alpha}$ free acid were more potent and possessed more sustained uterine stimulating activity than the naturally occurring parent compounds. The abortifacient effectiveness of the 15-methyl analogues was initially established in the midtrimester of pregnancy.

Intramuscular administration of 15-methyl-prostaglandin E_2 methyl ester was an extremely potent and effective abortifacient in the midtrimester with a success rate of 97% and a mean abortion time of less than 10 h in one study of 30 patients[45]. Gastrointestinal side-effects were minimal, but pyrexia was a consistent problem. In addition, the drug was extremely unstable and had to be carefully stored at temperatures below 0 °C and kept refrigerated even while being utilized. This instability and the pyrexic effects limited its clinical usefulness.

The 15-methyl analogue of prostaglandin $F_{2\alpha}$ proved to be an effective midtrimester abortifacient and the more clinically valuable agent. Intramuscular injections of 15-methyl-prostaglandin $F_{2\alpha}$ (THAM) successfully induced abortion in midtrimester and late first trimester pregnancy[46-48]. The intramuscular route offered the advantage of ease of administration and the ability to tailor the dose to the individual patient's response. When administered in the early first trimester of pregnancy, intramuscular administration of 15-methyl-prostaglandin $F_{2\alpha}$ (THAM) was a fairly effective abortifacient[49-51]. In a study of the luteolytic and abortifacient effects of intramuscular injections of 15-methyl-prostaglandin $F_{2\alpha}$ (THAM) in a series of nine patients at 5-6 weeks of amenorrhoea, a drop in progesterone and oestradiol was demonstrated 4 h subsequent to prostaglandin administration[49]. It was not certain that the abortifacient effectiveness of the prostaglandin was due to luteolysis since the prostaglandin immediately initiated strong persistent uterine activity which may have played the major role in the abortion process. Intramuscular administration of 15-methyl-prostaglandin $F_{2\alpha}$ (THAM) was characterized by a fairly high frequency of gastrointestinal side-effects. Despite this, the ease of administration, predictability of effect and low doses required have made 15-methyl-prostaglandin $F_{2\alpha}$ (THAM) administered by the intramuscular route an acceptable form of first trimester abortion in a number of countries[1].

Throughout the latter half of the 1970s, research focused on vaginal administration of the methyl ester of 15-methyl-prostaglandin for induction of abortion in the early first trimester, the period up to 8 weeks from the last menstrual period. Vaginal administration of 15-methyl-prostaglandin $F_{2\alpha}$ methyl ester via either a single or multiple vaginal suppositories or a vaginal Silastic device impregnated with the drug offered the advantage of ease of administration (possible self-medication in an out-patient setting) with reduced gastrointestinal side-effects. Multiple vaginal suppositories[50,52-59] evoked fairly consistent abortifacient responses (65–100%). Side-effects, although reduced from the systemic administration, were not eliminated. The use of multiple suppositories on an out-patient basis, however, required self-medication and a degree of patient compliance and motivation that may not be consistently found in a number of clinical situations. A single long acting suppository, usually 3·0 mg 15-methyl-prostaglandin $F_{2\alpha}$ methyl ester, did not require patient compliance for effectiveness[57,59-65]. The abortion rate ranged from 70 to 100%.

The development of a Silastic device impregnated with 15-methyl-prostaglandin $F_{2\alpha}$ methyl ester appeared to approach the 'ideal' method for first trimester abortion[66]. In 1976 and 1977, a series of studies examined various size devices containing varying concentrations of drug[67-77]. Eventually, it became clear that consistent abortifacient effectiveness with this device remained elusive, and these studies were terminated in 1977.

Vacuum aspiration for first trimester abortion has a success rate of 95–98%[1]. Any other first trimester abortion technique must therefore approach that success rate to gain widespread acceptance as a routine procedure. Preliminary studies evaluating the 15-methyl-prostaglandin analogues demonstrated their potential as abortifacients in early pregnancy. Despite the failure of any specific prostaglandin or technique of prostaglandin administration to clearly emerge as a 'routine' or primary first trimester abortifacient technique, investigation of newer, 'third generation' analogues continued with promising results.

Intramuscular injections of sulprostone (16-phenoxy-ω-tetranor prostaglandin E_2 methyl sulphonylamide)[65,78-81] proved an effective first trimester abortifacient with a success rate of 95–100% and an incidence of gastrointestinal side-effects around 30%[1]. Vaginal suppositories containing 16,16-dimethyl-prostaglandin E_2[57,82], 16,16-dimethyl-prostaglandin E_1 methyl ester[81,83] and 9-methylene-prostaglandin E_2[81] (Table 15.2) were shown to produce consistently high abortion rates in initial studies in the first trimester of pregnancy.

Table 15.2 First trimester abortion induced by prostaglandin analogues

Reference	No. of women	Period of gestation	Dose schedule	Success rate (%) effective (complete)
16-phenoxy-PGE$_2$ methyl sulphonylamide				
Karim et al., 1977[78]	240	5-6 weeks	50 μg extraovular	95
Csapo and Pulkkinen, 1979[65]	10	≤6 weeks	500 μg i.m. q 4 h	100
Csapo et al., 1980[79]	90	7 weeks	500 μg i.m. q 4 h × 2	96
Fleischer et al., 1982[80]	19	≤7 weeks	4-10 mg vag. supp. q 1 h	95
Bygdeman et al., 1983[81]	34	≤7 weeks	500 μg i.m. q 4 h × 3	94
16,16-dimethyl-PGE$_2$ vaginal suppository				
Lundström et al., 1977[82]	50	≤8 weeks	0·8-1·0 mg q 3 h × 4	87-94
Mackenzie et al., 1978[57]	34	≤4 weeks	1·0 mg q 6-8 h × 2 q 3 h × 4	100
16,16-dimethyl-PGE$_1$ methyl ester				
Karim et al., 1977[83]	50	≤6 weeks	1·0 mg vag. supp. q 1 h × 5	92
Ninagawa et al., 1976[84]	24	≤9 weeks	20 μg × 2 intrauterine 40 μg injection 40 μg × 2	100
Bygdeman et al., 1983[81]	63	≤7 weeks	1·0 mg vag. supp. q 3 h × 5	92
9-methylene-PGE$_2$ vaginal suppositories				
Bygdeman et al., 1983[81]	101	≤7 weeks	75 mg + 30 mg - 6 h 60 mg + 45 mg - 6 h	92-98

Conclusions

It may be concluded that prostaglandins can be used as a non-surgical procedure for termination of very early pregnancy. Both vaginal and intramuscular administration of the latest generation of PG analogues can be equally effective as vacuum aspiration if the treatment is restricted to the first 3 weeks following the first missed menstrual period. Gastrointestinal side-effects and uterine pain are still problems, although of significantly less importance than if natural prostaglandins are used. The development of new prostaglandin analogues as well as more effective prostaglandin delivery systems may further improve the treatment and make pharmacological alternatives to surgical methods of first trimester abortion more widely acceptable.

CERVICAL PRIMING WITH PROSTAGLANDINS PRIOR TO SURGICAL INTERRUPTION OF PREGNANCY

The importance of cervical integrity to the successful maintenance of pregnancy and its contribution to parturition beyond the role of a passive muscle sphincter are widely recognized[85]. Surgical interruption of pregnancy threatens cervical integrity by requiring mechanical dilatation of the cervix for effective evacuation of the products of conception. In first trimester suction curettage, the reported clinical incidence of cervical tears or laceration ranges from 0·07% to 4·8%. Histological examination of the cervix, however, following forced dilatation for diagnostic curettage in non-pregnant patients reveals an incidence of small cervical tears of 39%[86]. Forceful dilatation of the cervix to 10 mm to accommodate suction evacuation at 10 weeks results in the internal os remaining open to 4 mm at postabortion follow-up[87]. Dilatation and evacuation (D&E) is being performed more frequently for early second trimester interruption; this technique requires even more extensive dilatation of the cervix, i.e. as much as 14–16 mm. Indeed, cervical injury is the most frequently reported complication of D&E, with a rate of 1·16 per 100 abortions[83].

Preoperative cervical priming with laminaria tents has been suggested to minimize the potential for cervical damage[87,88]. The utilization of laminaria, however, is not without risk and cost. Insertion can be painful, and the patient may experience vasovagal symptoms of syncope and sweating. There is an increased risk of infection[88], and insertion of the laminaria can produce a false cervical passage[89]. Furthermore, the laminaria require time to produce desired cervical changes and therefore are usually placed the night prior to the procedure. If the patient does not return to the abortion facility within 24 h for the procedure, the risk of infection and/or bleeding is significant.

Prostaglandins induce cervical changes through both the initiation of uterine contractions and via a direct effect on the smooth muscle and collagen of the cervix. Prostaglandin E_2 relaxes the smooth muscle tissue of the pregnant cervix *in vitro* in very low doses, and while prostaglandin $F_{2\alpha}$ exerts a strong stimulus to the human myometrium *in vitro*, it does not influence the spontaneous activity of isolated cervical smooth muscle[90]. In addition, prostaglandins influence the connective tissue which makes up 80–90% of the cervix. The mechanism by which prostaglandins affect connective tissue to produce cervical ripening is under investigation[91,92]. That prostaglandins can be used to induce cervical dilatation *in vivo* was first demonstrated by Wiqvist *et al.*[93]. These authors administered $PGF_{2\alpha}$ or PGE_2 extra-amniotically for 2–8 h prior to vacuum aspiration and the cervix had dilated to such a degree that uterine evacuation could be effected with minimal or no instrumental dilatation. The practical application of prostaglandins for routine cervical priming prior to surgical interruption of pregnancy, however, requires more than the ability to induce the desired cervical changes. The cervical effects must be induced with an acceptable interval from administration to procedure and with a minimum of side-effects. For the indication of cervical priming, prostaglandin-induced uterine contractions of sufficient strength to cause partial expulsion of the products of conception might be considered an undesirable side-effect.

Natural prostaglandins

In 1973, Brenner and co-investigators[94] demonstrated that vaginal administration of 50 mg of naturally occurring prostaglandin $F_{2\alpha}$ 3 h prior to suction curettage induced cervical softening and dilatation in 40 nulliparous patients. The same research team, in 1975, reported on the development of an Electronic Force Monitor, which provided precise measurements of the force required to overcome cervical resistance during instrumental dilatation of the cervix. Through the use of the Electronic Force Monitor, Dingfelder and co-workers[95] were able to demonstrate that vaginal administration of natural prostaglandin suppositories (50 mg $PGF_{2\alpha}$ or 20 mg PGE_2) for 3 h in nulliparous patients in the first trimester of pregnancy dramatically reduced the force needed to produce the degree of cervical dilatation required for complete evacuation of the products of conception. Gastrointestinal side-effects, however, were common with the use of both prostaglandins. The 50 mg prostaglandin $F_{2\alpha}$ suppository produced vomiting in 40% of patients and diarrhoea in 45%, and prostaglandin E_2 resulted in vomiting in 60% and diarrhoea in 15% of the patients.

Administration of the naturally occurring prostaglandins for cervical priming was pursued through the development of delivery systems that permitted the dose of the drug to be reduced, therefore minimizing side-effects, without eliminating the desired cervical effects. In West Germany, a prostaglandin $F_{2\alpha}$ intracervical gel, which consists of 3 ml of 5% Tylose and 5 mg prostaglandin $F_{2\alpha}$, was developed and shown to be a safe and practical method to prime and dilate the cervix in more than 2500 interruptions of pregnancy at the Department of Obstetrics and Gynecology, University of Göttingen. Subsequently, this technique has been successfully applied in more than 10 000 surgical interruptions in West Germany[96]. In a limited study of 100 patients in the 7–12th week of gestation, Rath et al.[96] provided an objective demonstration of the dilatory effects of the intracervical gel through the use of a specially designed tonometer to measure cervical resistance. 35 patients, 20 of whom were in the 10–12th week of gestation, spontaneously expelled the fetus and/or the placenta during the night prior to the operation. 99 patients were dilated to at least Hegar 8, and 63 to Hegar 10 at 6–8 h following administration. In contrast to the findings of Caspi et al.[87] that the cervix remains dilated at follow-up in cases of instrumental dilatation, cervical resistance in 92% of patients was the same at the 5–6 weeks postabortion follow-up visit as it was prior to prostaglandin $F_{2\alpha}$ gel application.

Promising results have also been reported by Wingerup and co-workers[85] with the intracervical administration of a low dose of prostaglandin E_2 in a viscous gel. These investigators developed a gel based on a cross-linked starch polymer which contained 0·5 mg of prostaglandin E_2. The gel was tested against placebo gel in nulliparous patients prior to medical termination of pregnancy in the late first trimester. The gel was administered on the night prior to the scheduled termination and the patients permitted to leave the hospital 1 h after administration. Prior to vacuum aspiration the following day, the placebo patients had a mean cervical dilatation of 5·2 mm, and the cervix had the same consistency as it had in the pretreatment period. In contrast, the patients treated with prostaglandin E_2 had a cervical dilatation of 9·6 mm, and the cervix was significantly softer than in the pretreatment period. A subsequent three centre study of 120 patients in the 9–12th week of gestation, confirmed these results. The frequency of gastrointestinal side-effects was very low. Only 3·3% of patients experienced nausea, and only 1·7% had either vomiting or diarrhoea. In addition, the intracervical administration of the prostaglandin E_2 gel did not induce abortion prior to surgical evacuation in any of the patients.

Prostaglandin analogues

The synthesis of the 15-methyl analogues of prostaglandin with sustained biological activity provided agents for investigation using the indication of presurgical priming. Extra-amniotic administration of 25 μg of 15-methyl-prostaglandin E₂ methyl ester was investigated by Cheng et al.[97] and Choo and co-workers[98]. In the study by Cheng et al., the drug was administered to 75 nulliparous patients in the 8–13th week of gestation 3 h prior to uterine evacuation. At the time of the procedure, 36 women required no further dilatation and, when dilatation was required, it was performed with ease. Gastrointestinal side-effects were within an acceptable range; 12% of patients experienced vomiting and 3% diarrhoea. An extensive investigation of the effectiveness of a single extra-amniotic dose of 25 μg 15-methyl-prostaglandin E₂ methyl ester administered 12–14 h prior to surgical termination in 1785 women at 7–14 weeks gestation was carried out by Choo and co-workers[98]. The prostaglandin administration resulted in cervical dilatation of 8 mm or more in 96% of patients, and these patients did not require instrumental dilatation of the cervix for uterine evacuation. In 40% of the patients, however, the prostaglandin induced partial or complete expulsion of the products of conception prior to the scheduled operative procedure. The incidence of gastrointestinal side-effects was similar to that found by Cheng et al. (vomiting in 4% of patients), but there was a higher incidence of diarrhoea (7·5%) with the more prolonged administration. Blood loss was minimal, and 89% of patients had a blood loss of less than 30 ml. Despite these initially promising results, the essential instability of the 15-methyl-prostaglandin E₂ methyl ester compound appears to preclude the widespread application of this effective drug[45].

As previously noted, the 15-methyl analogue of prostaglandin F₂α is more stable than the prostaglandin E₂ analogue, and 15-methyl-prostaglandin F₂α (THAM) administered by intramuscular or extra-amniotic injection can induce cervical dilatation and softening. Toppozada and co-workers[99] demonstrated that either 0·4 mg of 15-methyl-prostaglandin F₂α administered extra amniotically or 0·3 mg and 0·8 mg administered intramuscularly every 4 h for 18 h induced preoperative cervical dilatation of at least 7 mm in all 67 patients studied.

Vaginal administration via suppositories of the methyl ester of 15-methyl-prostaglandin F₂α was extensively tested from 1975 to 1984 (Table 15.3). Multiple administrations of 1·0 and 1·5 mg 15-methyl-prostaglandin F₂α suppositories over an 18 h period produced cervical dilatation of at least 10 mm in 56 of 58 patients[100]. The development of a long acting 15-methyl-prostaglandin F₂α suppository afforded the

Table 15.3 Preoperative cervical dilatation with 16-methyl-prostaglandin $F_{2\alpha}$ methyl ester administered by vaginal suppository(ries)

Reference	No. of women	Dose schedule	Time to procedure	Dilatation 7-9 mm	≤10 mm	Blood loss ≤100 ml
15-methyl-PGF$_{2\alpha}$ methyl ester						
Borell et al., 1976[100]	58	Multiple 1·0-1·5 mg × 4	18 h	2	56	58
Ganguli et al., 1977[101]	20	Single 2·0 mg	6 h	8	7	mean = 62 ml
	20	Single 2·5 mg	12 h	5	15	mean = 50 ml
Lauersen et al., 1979[102]	10	Single 1·0 mg	12 h	mean = 8·4 mm		mean = 108 ml
Lauersen and Wilson, 1980[103]	13	Single 1·0 mg	12 h	mean = 8·2 mm		NR
Niloff and Stubblefield, 1982[105]	15	Single 0·5 mg	3-6 h	mean = 7·7 mm		NR
	20	Single 1·0 mg		mean = 9·2 mm		NR
Lauersen et al., 1982[89]	20	Single 0·5 mg	1-12 h	mean change +3·6		NR
	20	Single 1·0 mg		mean change +3·8		
Fayemi et al., 1983[106]	19	Single 1·0 mg	3 h	external os = 10 mm internal os = 8·05 mm		NR
Kent et al., 1983[104]	60	Single 1·0 mg	12 h	mean = 9·8 mm		NR
Fylling and Jerve, 1983[107]	37	Single 1·0 mg	3-4 h	mean = 6·9 mm		NR
Lauersen and Graves, 1984[108]	23	Single 0·5 mg	1-2 h	mean increase = 1·3 mm		mean = 50·7 ml

NR = not recorded

convenience of one-time administration, and research focused on manipulating the factors of dosage and time of administration to achieve the desired cervical effects with minimal side-effects. A single suppository containing either 2·0 or 2·5 mg 15-methyl-prostaglandin $F_{2\alpha}$ methyl ester administered 6 and 12 h, respectively, prior to surgical termination produced mean cervical dilatation of 9 and 11 mm, respectively. This degree of dilatation was significantly different from the untreated control group who had a mean dilatation of 4·8 mm. Ganguli and co-workers[101] reported the occurrence of gastrointestinal side-effects with the 2·0 and 2·5 mg suppository, but did not specify the incidence. Gastrointestinal side-effects with prostaglandins are dose-related; therefore, subsequent studies of single 15-methyl-prostaglandin $F_{2\alpha}$ methyl ester focused on lower dose (1·0 mg and 0·5 mg) vaginal suppositories (Table 15.3).

Administration of a single 1·0 mg 15-methyl-prostaglandin $F_{2\alpha}$ vaginal suppository for 12 h prior to surgical termination consistently resulted in a mean cervical dilatation of at least 8 mm[102-104]. In order to provide for this extended administration to procedure interval, an abortion candidate must either be hospitalized overnight or, if the patient is sent home, she must be carefully informed of the probability of labour-like uterine contractions, vaginal bleeding and the possibility of gastrointestinal side-effects. In addition, with this prolonged prostaglandin administration period, there is a chance of expulsion of the products of conception prior to the surgical evacuation. 12 h of administration may not be required to achieve the desired cervical changes. In a study of 80 patients, in the late first trimester and early second trimester, Lauersen and co-workers[89] tested administration times of 1–6 h and 12 h. Administration of the prostaglandin 4–5 h prior to the surgical evacuation appeared to achieve the maximal cervical effect, and administration of a single long-acting vaginal suppository for a longer period seemed to result in a somewhat lesser degree of cervical changes. Other studies have substantiated the ability of a single 15-methyl-prostaglandin $F_{2\alpha}$ vaginal suppository to induce cervical dilatation and softening when administered from 1 to 6 h prior to surgical evacuation[105-108].

As the administration to procedure period was decreased, the amount of 15-methyl-prostaglandin $F_{2\alpha}$ methyl ester contained in the single vaginal suppository was also decreased. A single suppository containing only 0·5 mg of the drug was investigated by Niloff and Stubblefield[105], Lauersen et al.[89] and Lauersen and Graves[108], who reported cervical dilatation at evacuation of a mean of 7·7 mm and cervical changes up to an increase of 3·6 mm (Table 15.3). Episodes of vomiting were generally minimal with the 0·5 mg suppository and ranged from 0%[89] to 7%[105]. The reported incidence of diarrhoea was more variable; Niloff and Stubblefield[105] observed diarrhoea in 40% of their patients, whereas the Lauersen studies[89,108] did not report diarrhoea in any of their patients. These initially promising results have resulted in continued interest and investigation of the most appropriate technique for prostaglandin administration for cervical dilatation.

The Norwegian experience with the prostaglandins for therapeutic abortion led Fylling and Jerve[107] to the opinion that the main application of prostaglandins in the future will be for the indication of preoperative cervical dilatation. These authors compared the effects of single and multiple 15-methyl-prostaglandin $F_{2\alpha}$ methyl ester vaginal suppositories with 16-phenoxy-PGE_2 methyl sulphonylamide administered either intramurally in the cervix, intramuscularly or subcutaneously on the evening prior to surgical evacuation in a series of 180 patients. The overnight treatment resulted in abortion in 30% of the patients and was determined to be impractical. The administration to procedure time was reduced to 3–4 h, and the effects of a 1·0 mg 15-methyl-prostaglandin $F_{2\alpha}$ vaginal suppository was compared to 250 μg of 16-phenoxy-PGE_2 methyl sulphonylamide administered intramuscularly in 37 and 20 patients, respectively. The 15-methyl-prostaglandin resulted in a mean cervical dilatation of 6·9 mm with an incidence of 0·7 episodes of gastrointestinal side-effects per patient. The 16-phenoxy-PGE_2 methyl sulphonylamide produced a mean cervical dilatation of 7·5 mm; this effect was reported without gastrointestinal side-effects and with minimal discomfort to the patient. Using a double-blind methodology, Fehrmann and Praetorius[109] studied the effect of 500 μg of 16-phenoxy-PGE_2 methyl sulphonylamide administered intramuscularly 3 h prior to vacuum aspiration in 25 primigravidae and 19 previously pregnant patients compared to 29 primigravidae controls and 18 previously pregnant controls. The 500 μg of 16-phenoxy-PGE_2 methyl sulphonylamide produced a significant degree of cervical dilatation; ten patients (22%) required analgesia for pain and complained of nausea and vomiting. Because of the incidence of side-effects, the authors concluded that cervical priming with 16-phenoxy-PGE_2 methyl sulphonylamide should be limited to specific clinical situations such as the young primigravidae in the late first trimester.

The stability of 16,16-dimethyl prostaglandin E_1 permits the administration of the prostaglandin analogue via a vaginal suppository. Welch and Elder[110] examined the effects of a 1 mg 16,16-dimethyl-prostaglandin E_1 suppository administered to 25 primigravidae patients, 7–12 weeks of gestation, at 3 h prior to surgical termination. The results from the prostaglandin patients were compared to a control group of 18 patients, and allocation to the study groups was on a random basis. The mean degree of cervical dilatation at 3 h post-treatment was 6·36 mm in the prostaglandin group compared to 3·06 in the control group; the mean force required to insert the maximum size of dilator was 0·86 kg in the prostaglandin group compared to 1·16 kg in the control group. These differences were highly statistically significant in both instances. Side-effects were within an acceptable range: 11 patients experienced some degree of preoperative pain; three patients experienced nausea but only one vomited, and 14 of the treated patients (56%) were free of any prostaglandin side-effects. These results encouraged Welch and Elder to suggest routine preoperative priming with this prostaglandin analogue.

In addition to 16,16-dimethyl-prostaglandin E_1, 9-methylene-prostaglandin E_2 possesses sufficient stability to allow local administration through a vaginal suppository. Lauersen and co-workers[89] compared the effects of 9-methylene-prostaglandin E_2, 15-methyl-prostaglandin $F_{2\alpha}$ and laminaria administered prior to dilatation and evacuation in the late first trimester and in the second trimester. A total of 100 women with a mean gestational age of 15 weeks (range 10–20 weeks) were included in the investigation. 20 women acted as a control group and received no preoperative therapy; 20 women received 4–6 laminaria tents approximately 12 h prior to the dilatation and evacuation procedure. Prostaglandin was administered to 80 women who received one of two forms: 15-methyl-prostaglandin $F_{2\alpha}$ at a dose level of 0·5 or 1·0 mg of 9-methylene-prostaglandin E_2 at a dose level of either 30 or 60 mg. Each dose schedule of both prostaglandins was administered to 20 patients in the form of a vaginal suppository in a 800 mg waxy base. Administration occurred from 1 to 12 h prior to the scheduled dilatation and evacuation. The mean cervical change for all prostaglandin treated patients was an increase of 4·2 mm and, as previously noted, the maximum cervical change occurred at 4–5 h following prostaglandin administration. The cervical dilatory effects of the prostaglandins were dose-related within the prostaglandin analogue, and the maximum mean change of 4·7 mm was seen with 60 mg of 9-methylene-prostaglandin E_2. No further mechanical dilatation was required in 26% of the prostaglandin patients. The laminaria, however, were more consistent in their ability to achieve cervical dilatation; only two patients (10%) required further dilatation at the time of the operation. Gastrointestinal side-effects were more frequent with the prostaglandins, but the overall morbidity rate including blood loss, cervical lacerations and incomplete abortions was higher with laminaria or in the control group (Table 15.4).

On making the clinical choice between prostaglandin and laminaria for cervical priming, the rapid effectiveness of prostaglandin in combination with its low morbidity should be considered. Subsequent studies have substantiated the rapid effectiveness of 9-methylene-prostaglandin E_2 administered 3 h prior to surgical evacuation in 97 patients

Table 15.4 Side-effects and complications of preoperative cervical priming

Side-effects and complications	Control n = 20	Laminaria n = 20	15-methyl-PGF$_{2\alpha}$		9-methylene PGE$_2$	
			0·5 mg n = 20	1·0 mg n = 20	30 mg n = 20	60 mg n = 20
Vomiting	3 (15%)	2 (10%)	0	0	3 (15%)	6 (30%)
Diarrhoea	0	0	0	2 (10%)	1 (5%)	8 (40%)
Vasovagal symptoms	0	3 (15%)	0	0	0	0
Blood loss > 200 ml	3 (15%)	5 (25%)	0	0	0	0
Cervical laceration	1 (5%)	0	0	2 (10%)	0	1 (5%)
Incomplete abortion	1 (5%)	1 (5%)	0	0	0	0

in the 9–12 weeks of gestation[111]. The mean cervical dilatation at the time of the procedure was 7·3 mm, and although 91% of the patients required further dilatation, this was easily performed. A 12 h pretreatment with the prostaglandin produced more significant cervical dilatation, but the convenience of the 3 h treatment made it more acceptable in the first trimester. It was suggested that the more prolonged 12 h treatment be reserved for the early second trimester.

Lauersen and Graves[108] demonstrated that cervical priming can also be achieved with an administration to procedure time of 1–2 h. In a study of 23 women in the first trimester of pregnancy, 30 mg of 9-methylene-prostaglandin E_2 administered by vaginal suppository 1–2 h prior to suction curettage resulted in an increase in cervical dilatation of 3·3 mm; only 17% of patients requiring further dilatation at the time of the procedure. Side-effects were minimal and only one patient experienced gastrointestinal side-effects.

Administration of a prostaglandin to facilitate surgical termination of pregnancy not only minimizes the potential of cervical trauma, but also appears to significantly reduce blood loss. In a recent study of the effectiveness of a vaginal suppository containing 30 mg of 9-methylene-prostaglandin E_2 administered 3 h prior to dilatation and evacuation in 51 patients at 9–14 weeks of gestation, Sidhu and Kent[112] observed that the blood loss in the 51 treated patients was significantly less than the blood loss in the control group of 44 patients. The prostaglandin patients experienced a mean blood loss of 69 ml and the control group had a mean blood loss of 151 ml ($p < 0.001$). In addition, the mean dilatation at procedure was 8·72 mm for the prostaglandin patients as compared with 4·14 mm for the control group; again this was a statistically significant difference. Although the mechanism through which prostaglandins inhibit blood loss remains to be elucidated, the ability to minimize blood loss is not exclusive to 9-methylene-prostaglandin E_2. Lauersen and co-workers[89,102] also reported that blood loss was effectively reduced through the use of 15-methyl-prostaglandin $F_{2\alpha}$ compared to the untreated control group or to the use of laminaria. Fehrmann and Praetorius[109] previously observed a similar beneficial effect with 16-phenoxy-PGE_2 methyl sulphonylamide; the mean blood loss in the control group was 140 ml, while in the patients treated with 16-phenoxy-PGE_2 methyl sulphonylamide, the mean blood loss was 59 ml; this difference was statistically significant ($p < 0.005$).

Conclusion

In summary, the preoperative cervical priming by prostaglandins can reduce both immediate complications and long-term sequelae of surgical termination of pregnancy by minimizing cervical trauma and reducing abortion-related blood loss. The Medical Advisory Committee of the International Planned Parenthood Federation has also recently recommended pretreatment with either laminaria or prostaglandin prior to vacuum aspiration in first trimester abortion. Prostaglandin therapy is more practical than laminaria; the introduction of the tent has to be performed by the medical staff whereas prostaglandin treatment can be taken care of by the attending nurse. In addition, prostaglandin treatment results in less blood loss at surgery.

PROSTAGLANDINS AND MIDTRIMESTER ABORTION

Abortion-related mortality and morbidity increase significantly as gestation increases. During a 5 year period in the United States, midtrimester abortion accounted for 11% of legal abortions, but was associated with two thirds of major complications and over one half of abortion-related fatalities. In addition, there is a strong inverse correlation between gestational age and patient age[113]. Midtrimester abortion is often performed on very young patients who are at the beginning of their fertile years and who present clinical problems because of physical immaturity. The development of safe and effective techniques for abortion in the second trimester of pregnancy, when conventional suction evacuation is contraindicated, is a major clinical challenge. At this time the uterus is notoriously quiescent and unresponsive to stimuli[114], and this quiescent uterus and large fetus are the basis of the problem of midtrimester abortion.

Natural prostaglandins

In 1970, Karim and Filshie[11] and Roth-Brandel and co-workers[13] first reported that intravenous infusion of either prostaglandin E_1 or prostaglandin $F_{2\alpha}$ could effectively induce abortion in the midtrimester of pregnancy. These findings were substantiated by the investigations of Wiqvist and Bygdeman[14], Hendricks et al.[115] and Kaufman et al.[116]. Intravenous administration of effective doses of the naturally occurring prostaglandins, however, was associated with a high incidence of gastrointestinal side-effects, characteristically nausea, vomiting and diarrhoea. Furthermore, local irritation at the infusion site was observed in up to 60% of patients[115]. These side-effects severely curtailed the clinical utility of intravenous administration.

The rapid enzymatic degradation of the naturally occurring prostaglandins requires that the compounds be administered either continuously or very frequently to maintain their uterine stimulatory effect. In

1971, Karim and Sharma[117] and Bygdeman and associates[118] indepen-
dently reported on induction of midtrimester abortion by intra-
amniotic administration of prostaglandin E_2 and $F_{2\alpha}$. All patients in
both studies aborted successfully, and the majority of patients were
free of gastrointestinal side-effects. Uterine response to intra-amniotic
instillation of prostaglandin was frequently characterized by an immediate elevation of basal tonus followed by the development of high
frequency, low amplitude contractions. In some patients there was a
slow stimulatory response, and the rise in basal tonus was not always
present. It was hypothesized that the prostaglandins reached the myometrium by a slow diffusion through the fetal membranes[119].

Clinical investigations of the effectiveness of intra-amniotic instillation of prostaglandin $F_{2\alpha}$ led to the November 1973 approval of prostaglandin $F_{2\alpha}$ by the United States Food and Drug Administration as
an abortifacient. At present, drug regulatory authorities in approximately 50 countries have approved prostaglandin $F_{2\alpha}$ for this indication[120].

Although the pioneering prostaglandin investigations examined
single and multiple dose intra-amniotic administration, the focus
eventually shifted to the single dose administration to avoid the hazard
of repeat amniocentesis. Numerous studies using this technique have
been published since 1971[66,121-138]. The dose of prostaglandin $F_{2\alpha}$ varied
from a low of 15 mg[125] to a high of 100 mg[126], but the most frequent
dosage was 40 or 50 mg[121,123-125,128-139]. Although authors varied in their
definition of abortifacient success, the mean instillation to abortion
interval in most studies was below 24 h. A study-to-study comparison
of results is complicated by the presence of oxytocin augmentation in
some studies[66,140] but not in others[141]. The overall clinical results of
prostaglandin administration have been sufficiently acceptable so that
intra-amniotic prostaglandin $F_{2\alpha}$ has routinely replaced intra-amniotic
saline in midtrimester abortion in many institutions. In 1977, at The
New York Hospital–Cornell University Medical Center, there were
438 prostaglandin-induced abortions and only four abortions induced
by intra-amniotic instillation of saline. These saline abortions were
performed in patients with gestations beyond 20 weeks, when therapeutic abortion was indicated because of genetic defects of the fetus.

Intra-amniotic prostaglandin vs. intra-amniotic saline
Prior to the development of prostaglandins, intra-amniotic hypertonic
saline administration was the primary method for induction of midtrimester abortion. The utilization of prostaglandin for this indication
and by the almost identical technique of intra-amniotic instillation
inevitably led to a series of investigations which compared the effects
and efficacy of these two abortifacients. At least 12 such studies, per-

formed between 1973 and 1978 on a total of approximately 4000 prostaglandin patients and 12 000 saline patients, have been published[139-150].
A cursory examination of the results of these studies reveals one
consistent finding: mean times from instillation to abortion were
shorter when abortions were induced by intra-amniotic instillation of
prostaglandin $F_{2\alpha}$. In addition, even though various investigators may
have used different definitions for abortifacient success, they almost
consistently reported a higher success rate with prostaglandin compared to the saline administration[139,140,143-149]. The extensive comparative study of the World Health Organization (WHO)[145], reported
results in favour of prostaglandins, although the study acknowledged
that there was a higher frequency of minor side-effects such as vomiting and diarrhoea. This study was criticized by Cates et al.[151] of the
Center for Disease Control in the United States. He re-analysed the
WHO results and found prostaglandins to be faster but more hazardous than saline. Subsequently these authors have been chief advocates
of saline over prostaglandins for second trimester abortion[152,153]. In
turn, their position and analysis has been criticized and challenged.
Speroff, in an editorial comment on the controversy emphasized the
importance of defining safety. 'It is not totally clear whether safety
means no deaths or serious complications with significant morbidity,
or whether it refers to an overall incidence of complications'[154].
Furthermore, the deaths reported with prostaglandin administration
may or may not be related to the drug; the saline abortion deaths were
directly related to the saline.

Intrauterine (extraovular) instillation
Wiqvist and Bygdeman[21] observed that prostaglandin side-effects could
be significantly reduced if the drug was injected close to the target
organ, i.e. the myometrium, by administration into the extra-amniotic
space. Similar effectiveness in conjunction with reduced side-effects
was reported by Embrey and Hillier[155] in their investigation of intrauterine instillation of prostaglandin $F_{2\alpha}$ and E_2. Abortion could be
induced by this technique in both the first and second trimester of
pregnancy. The technique eliminates the need for amniocentesis and
can be effectively used in the 'grey zone', i.e. 12–16 weeks, when
conventional surgical interruption of pregnancy is not recommended,
and amniocentesis is difficult or impossible due to the small amount of
amniotic fluid.
    Lauersen and Wilson[156,157] reported that continuous extraovular ad-
ministration of prostaglandin $F_{2\alpha}$ at a rate of 4 mg/h was a very effective abortifacient with a 97% success rate and mean abortion time of
16 h. The incidence of side-effects with the extraovular administration

of prostaglandin $F_{2\alpha}$ is lower than the incidence with intra-amniotic instillation of this drug[21,155-157].

Continuous extraovular administration of prostaglandin eliminates the potential hazards of amniocentesis, but introduces the potential complication of endometritis. Lauersen and Wilson[157] reported seven cases of endometritis in 76 patients; six of these seven patients had abortion procedures lasting *in toto* beyond 24 h. Since the incidence of endometritis may be related to the presence of a foreign body in the uterus, it is important that the Foley catheter be removed within 24 h if abortion has not occurred[157,158].

Vaginal administration
In 1971, Karim and Sharma[30] reported successful induction of abortion in gestations from 7 to 23 weeks with intravaginal administration of prostaglandin $F_{2\alpha}$ and E_2 suppositories. Bolognese and Corson[159] confirmed these results with prostaglandin E_2 suppositories; they had a 98% success rate in 62 midtrimester patients with a mean abortion time of 12 h. Lauersen and co-workers[160] reported similar results in 71 patients with a success rate of 99% and a similar mean abortion time of 12 h. The characteristic uterine response to vaginal prostaglandin E_2 is a gradual development of uterine activity and a low intrauterine tonus. Fever is the most frequently encountered side-effect, although gastrointestinal side-effects, particularly nausea and vomiting, are common. Prostaglandin E_2 has a mild hypotensive effect, and the patient's vital signs should be monitored frequently throughout the abortion procedure. Comparisons of intravaginal prostaglandin E_2 and intra-amniotic prostaglandin $F_{2\alpha}$[161,162] have demonstrated that they are equivalent midtrimester abortifacients. Although the prostaglandin E_2 has a shorter abortion time, its administration is associated with a higher incidence of side-effects.

Prostaglandin E_2 vaginal suppositories have also proven to be an effective technique for induction of labour in patients with either missed abortion or fetal death *in utero*[163-165]. This procedure eliminates the waiting for spontaneous labour in patients with IUFD or missed abortion, reduces the possible hazard of coagulopathy with hypofibrinogenaemia, and relieves the patient of the psychological burden of carrying a dead fetus.

Prostaglandin analogues

The synthesis of prostaglandins, modified at carbon-15 to resist enzymatic degradation, provided potent new agents for induction of midtrimester abortion.

Intramuscular administration of PGF$_{2\alpha}$
Serial intramuscular injections of 15-methyl-PGE$_2$ methyl ester appeared to be an effective abortifacient technique, with extremely rapid abortion times, i.e. a mean of $9\frac{1}{2}$h[41]. On a weight-for-weight basis, 15-methyl-prostaglandin E$_2$ appeared to be most potent of the prostaglandins, and abortion could be induced with as little as 25 μg of the drug. Studies of intramuscular 15-methyl-PGE$_2$ revealed that more than 50% of the patients experienced no gastrointestinal side-effects; however, almost three quarters of the patients had a temperature elevation and two thirds experienced chills and shaking[41,166-168]. These observations show that while the F$_{2\alpha}$ prostaglandins have a more specific effect on gastrointestinal motility, the E$_2$ compounds have an effect on temperature control. The clinical application of this extremely effective abortifacient was precluded by the inherent instability of 15-methyl-PGE$_2$ and its definite pyretic effects. The 15-methyl analogue of prostaglandin F$_{2\alpha}$ is a far more stable compound with minimal pyretic effects. A series of intramuscular injections of 15-methyl-prostaglandin F$_{2\alpha}$ (THAM) is a very efficient abortifacient. Its success rate ranges from 85% to 98% in studies encompassing more than 900 patients[66,169-172]; the mean abortion time ranged from 14·0 to 18·2 h. Uterine response to an initial intramuscular injection of 15-methyl-prostaglandin F$_{2\alpha}$ is rapid and characterized by the appearance of low-amplitude, high-frequency contractions and a rapid rise in baseline intrauterine tonus. Subsequent injections produce further increments in uterine activity[47]. Lauersen[66] observed that 121 patients with gestations of 9–16 weeks aborted significantly faster (mean 12·8 h) than 98 patients with gestations of 17 weeks (mean 16·6 h) ($p < 0·01$). Intramuscular injections of 15-methyl-prostaglandin F$_{2\alpha}$ were particularly effective in the so-called 'grey zone'. In contrast to the natural prostaglandins, the intramuscular injection of the 15-methyl analogue causes little discomfort and no local reaction at the injection site. A commonly used dose schedule includes an initial dose of 100–200 μg 15-methyl-prostaglandin F$_{2\alpha}$ followed by 250 μg every 2–4 h for 24 h. The most prevalent side-effects are gastrointestinal. Premedication and routine administration of antiemetic and antidiarrhoeal agents can limit the severity and frequency of gastrointestinal disturbances. The WHO study[172] of intramuscular administration of 15-methyl-prostaglandin F$_2$ concluded that these gastrointestinal side-effects limited the value of this method as a primary abortifacient but recommended the treatment in patients where the initial procedure has failed.

Intrauterine administration of 15-methyl-PGF$_{2\alpha}$
Once the effectiveness of intramuscular administration of 15-methyl-prostaglandin F$_{2\alpha}$ (THAM) had been demonstrated, the effectiveness

of this analogue administered via the intra- and extra-amniotic route was investigated. The WHO Prostaglandin Task Force[173,174] and Tejuja and co-workers[175] demonstrated that both extra-amniotic and intra-amniotic administrations of 15-methyl-prostaglandin $F_{2\alpha}$ (THAM) are effective abortifacient techniques. In a total of 1349 women[174,175], intra-amniotic administration of 2·5 mg 15-methyl-prostaglandin $F_{2\alpha}$ (THAM) successfully induced abortion in 88–95% of patients. The mean abortion time ranged from 18 to 21 h. Similarly, extra-amniotic administration of 0·92[173] to 1·0 mg[175] of the drug resulted in a 78–80% success rate, with a mean abortion time of 15·4–16·2 h in 2229 women. At present intra-amniotic administration of this analogue seems to be the best single agent procedure for termination of second trimester pregnancy which requires only one application.

Intramuscular administration of 16-phenoxy-PGE$_2$ methyl sulphonylamide

16-phenoxy-PGE$_2$ methyl sulphonylamide, initially administered via an intravenous infusion by Schmidt-Gollwitzer et al.[176] effectively induced midtrimester abortion in 114 of 116 (98·3%) women with a mean abortion time of 13 h. Abortion was complete in 106 women (91·4%), and gastrointestinal side-effects were within acceptable levels. 47% of patients experienced no side-effects. Those promising results were later confirmed by Karim et al.[177] and the WHO Task Force[178] in studies of serial intramuscular injections of 0·5 mg or 1·0 mg 16-phenoxy-PGE$_2$ methyl sulphonylamide, every 4 h or 8 h, respectively, in the midtrimester of pregnancy. Karim and co-workers[177] reported a success rate of 92% in 60 patients, with a mean abortion time of 18–20 h. Nausea and vomiting were the most common side-effects in 15–30% of patients, but diarrhoea was minimal. The success rate in the WHO study of 295 patients[178], as judged by abortion within 30 h, was 84%, with a mean abortion time of 15–16 h. Intramuscular administration produced complete abortion in 26–32% of patients. Gastrointestinal side-effects were less frequent than those experienced with the 15-methyl analogue of prostaglandin $F_{2\alpha}$ and the WHO Task Force judged that the success rate was similar.

The efficacy of the treatment can be further increased and the frequency of gastrointestinal side-effects further reduced if the patients are pretreated with a laminaria tent. Karim et al.[179] and Bygdeman and Christensen[180] used pretreatment with one medium-size laminaria tent for 12 h followed by intramuscular injections of 16-phenoxy-PGE$_2$ methyl sulphonylamide 0·5 mg every 4 h. The success rate was almost 100% within 24 h of prostaglandin therapy. The mean duration of labour was short, or approximately 10 h, and the frequency of gastrointestinal side-effects low or only marginally higher than that reported for hypertonic saline. The frequency of cervical laceration was also comparable to that observed with hypertonic saline, at least if augmented with an intravenous infusion of oxytocin. These data have recently been confirmed in a multicentre study of almost 600 patients performed by the WHO[181].

15-methyl-$PGF_{2\alpha}$ and failed abortion

If induction of midtrimester abortion fails to effect expulsion of the products of conception, the clinician is faced with the dilemma of deciding between reinstillation of hypertonic saline or prostaglandin or surgical termination. Both choices are associated with potential hazards and prolonged hospitalization of the patient.

Lauersen and Wilson[182] observed in their study of the effectiveness of serial intramuscular injections of 15-methyl-prostaglandin $F_{2\alpha}$ (THAM) that if the placenta was not expelled spontaneously along with the fetus, continued intramuscular administration of this prostaglandin analogue would facilitate the completion of the abortion. In addition, intramuscular administration of 15-methyl-prostaglandin $F_{2\alpha}$ resulted in placental expulsions when abortion induced by intravaginal prostaglandin E_2 suppositories was prolonged by vaginal bleeding which diluted the effectiveness of the vaginally administered prostaglandin.

In a study of the effects of intramuscular injection of 15-methyl-prostaglandin $F_{2\alpha}$ in failed abortions, Lauersen and Wilson[182] reported that $250\,\mu g$ or $500\,\mu g$ every 2 h successfully induced abortion in 36 of 38 patients who had failed to abort with other methods of induction. These results were substantiated with a later study of 27 patients who failed to abort with vaginal administration of 15-methyl-prostaglandin $F_{2\alpha}$ methyl ester[183].

Induced midtrimester abortion vs. dilatation and evacuation

Cervical dilatation and suction evacuation (D&E) is the most common technique for early, 13–15 weeks, midtrimester abortion in the United States. This technique has been espoused by Grimes and Cates[184] for more advanced gestations. These authors stated that D&E appears to be the safest available method of abortion up to 20 weeks of gestation. Hern[185] has extended the use of D&E up to the 25th week of menstrual age, albeit with pretreatment using multiple laminaria and, after 20 weeks of gestation, an adjunctive urea amnioinfusion on the day of the procedure. Studies from the United States indicate that patients and the nursing staff find D&E more acceptable and less arduous than amnioinfusion-induced abortion[186]. This may, however, not be true in many other countries. Dilatation and evacuation requires highly skilled and experienced surgeons. In his testimony before the Food and Drug Administration, Bernard Greenberg, Dean of the School of Public Health, University of North Carolina stated, 'To assume that the D and E procedure is the safer procedure in the hands of all operators is a serious error in wishful thinking'[187]. The complications of cervical laceration and uterine perforation associated with D&E affect a patient's subsequent fertility and cannot be ignored[83]. It is probable, however, that the prostaglandins combined with D&E will provide a safer, more acceptable early midtrimester abortion technique in those locations where the surgical skill and the facilities for D&E are available. Cervical priming prior to D&E minimizes trauma to the cervix and may help preserve the future fertility of the patient[89].

MANAGEMENT OF SECOND TRIMESTER PROSTAGLANDIN ABORTION

Failed prostaglandin abortion

In general, the prostaglandins are extremely effective abortifacients. When midtrimester prostaglandin-induced abortion fails to occur within the expected time for the agent and the technique employed, the presence of uterine malformation should be considered. In a study of 529 midtrimester abortions induced by prostaglandins, Lauersen et al.[188] reported a failure rate of 1·9%. Two of the ten patients who failed to abort had uterine malformations; one patient had the pregnancy in a blind uterine horn, and the second patient was pregnant in one horn of a uterus didelphys. In addition, five patients had distorted uterine cavities due to myomata uteri. When abortion induced by prostaglandin fails to occur, evaluation with ultrasonography is indicated, along with a repeat test to confirm the pregnancy.

Side-effects of prostaglandin-induced abortion

Side-effects, such as nausea, vomiting, and diarrhoea, are characteristic of prostaglandin administration and are due to prostaglandins' stimulatory effect on the gastrointestinal tract. These side-effects can usually be minimized by premedication with antiemetic and antidiarrhoeal agents. Brenner and co-workers[189] reported decreased frequency of vomiting during prostaglandin-induced abortions after premedication with prochlorperazine. Wentz et al.[190] observed that 11 of 12 patients experienced multiple episodes of diarrhoea when not premedicated with antidiarrhoeal agents; in contrast, only three of eight patients placed on a regimen of diphlenoxylate hydrochloride with atrophine sulphate (Lomotil®) had any sign of diarrhoea. Similar results were reported by Brenner[191] who found that premedication with Lomotil reduced the mean number of diarrhoeal episodes from a mean of four to a mean of 0·7 episodes per patient.

Complications of prostaglandin abortion

Cervical fistulae have been reported subsequent to prostaglandin-induced abortions; this complication characteristically occurs when prostaglandin is administered to a nulliparous patient and it is augmented by oxytocin infusion[192-194]. Cervical fistulae have also occurred when midtrimester abortion was induced by intra-amniotic instillation of saline[195]. Preprostaglandin use of laminaria tents has been suggested as a method of minimizing cervical laceration and achieving shorter abortion times[196-200]. Duenhoelter and co-workers observed that laminaria inserted at the time of prostaglandin administration significantly reduced abortion times in nulliparous patients as compared to prostaglandin used alone; on the other hand, no significant differences were observed for multiparous patients. Although the use of laminaria tents minimized the risk of cervical trauma, complications such as bleeding, infection and failure to complete the abortion occurred in 28 of the 116 patients in their series[201].

The multiparous rather than the nulliparous patient is at risk for the rare but serious complication of uterine rupture associated with prostaglandin-induced abortion[202,203]. Concomitant intravenous oxytocin infusion has been associated with the reported cases of rupture, but laminaria which may reduce the risk of cervical injury apparently are not equally effective in minimizing the risk of uterine rupture. In two reported cases of uterine rupture during abortion induced with intra-amniotic administration of prostaglandin $F_{2\alpha}$, both multiparous patients had been previously primed with laminaria[202]. In the case of uterine rupture following intravaginal administration of prostaglandin E_2 for intrauterine fetal demise[203], the patient had a scarred uterus; her previous pregnancy had resulted in premature delivery at 34 weeks via low-segment Caesarean section for uncontrolled hypertension. Caution must be used when prostaglandin-induced uterine activity is augmented by an additional stimulant in the case of a multiparous patient or a patient with a history of a previous operative delivery. A similar caution is valid in the case of saline-induced abortion. Grimes et al.[204] reported on the deaths of two multiparous patients with no history of previous uterine surgery following intra-amniotic instillation of saline augmented by intravenous oxytocin administration for induction of abortion. In one patient, who survived the rupture by 3·5 days, the effects of the rupture were complicated by the systemic absorption of hypertonic saline, and her course was further complicated by disseminated intravascular coagulation and acute renal tubular necrosis.

Retained placenta can be a serious problem with induced midtrimester abortion. Rates of retained placenta after abortion by intra-amniotic instillation of hypertonic saline ranged from 7 to 60%. In analysis of 5000 saline-induced abortions, Kerenyi et al.[205] observed retained placenta in 13% of patients 4 h after abortion of the fetus. In an initial evaluation of the efficacy of intra-amniotic instillation of prostaglandin $F_{2\alpha}$ for induction of abortion[130], the placenta was spontaneously expelled in 80% of the patients and removed digitally or by sponge forceps in the other patients. Anderson and Steage[133] reported that 77% of 500 patients completely aborted following intra-amniotic instillation of 40 mg prostaglandin $F_{2\alpha}$. However, in an analysis of the routine use of intra-amniotic instillation of prostaglandin $F_{2\alpha}$ in 200 patients[66], only 25% of the patients spontaneously expelled the placenta. This lower rate of expulsion might possibly be due to the concomitant infusion of oxytocin in this series and uterine contractions resulting in a placenta trapped by the cervix. In 61% of the patients, the placenta could be removed easily either digitally or by sponge forceps. Heavy bleeding was most often associated with prolonged retention of the placenta. Spontaneous expulsion usually occurs within 2 h of abortion. If the placenta has not been spontaneously expelled within 2 h, attempts should be made to remove it either digitally or surgically, but routine curettage following midtrimester abortions is not advised.

An intravenous infusion of prostaglandin $F_{2\alpha}$ at the rate of 50–200 μg/min does not produce significant changes in cardiac output, central venous pressure, blood pressure or heart rate[206]. If intra-amniotically administered prostaglandin $F_{2\alpha}$ should inadvertently pass into the general circulation, side-effects such as bronchoconstriction, gastrointestinal disturbances, hypotension, bradycardia and cardiac arrythmias may develop; these effects should dissipate spontaneously, however, due to the rapid metabolism of the prostaglandin. Abortion induced by prostaglandins alone is not characterized by the coagulation changes observed with saline procedures. MacKenzie and co-workers[207] compared the coagulation changes in patients undergoing abortion by intra-amniotic instillation of prostaglandin E_2 alone and in combination with a hypertonic solution of either urea or glucose. They observed no changes suggestive of intravascular coagulation with the prostaglandin alone. There was, however, a rise in quantity of fibrin degradation products and a fall in plasma-fibrinogen and platelet count in patients treated with prostaglandin and hypertonic urea and, to a lesser extent, in patients receiving prostaglandin and hypertonic glucose. Phillips *et al.*[208] reported an increase in platelets, fibrinogen, factor V and factor VIII when abortion was induced by intra-amniotic instillation of prostaglandin $F_{2\alpha}$.

The possibility of delivery of a living fetus following prostaglandin-induced abortion is of concern. Whereas the fetus aborted after saline-induced abortions is usually macerated and the fetal tissues without viability, prostaglandin is not directly lethal to the fetus[209]. This difference notwithstanding, there have been live births following a hypertonic saline infusion. Stroh and Hinman[210] reported on 27 live births in a series of 9327 intra-amniotic instillations of saline or a rate of 2·9 live births/1000 procedures. The survival rate with prostaglandins appears to be higher. If a fetus survives a saline-induced abortion, central nervous system damage from salt poisoning represents a serious problem. The cause of fetal death when abortion is induced with intra-amniotic installation of prostaglandin is not known. The fetal heart rate is usually lost 2 h after intra-amniotic instillation of 40 mg of prostaglandin $F_{2\alpha}$[66], but fetal tissue following prostaglandin-induced abortion was viable and 86% of the expelled tissue was successfully cultured[209]. It appears that fetal death in the prostaglandin-induced abortions is due to hypoxia following the uterine hypertonus. In order to avoid the delivery of a live fetus, some institutions have as a policy decided to perform midtrimester abortion with intra-amniotic instillation of saline rather than prostaglandin after the 20th week of gestation.

Intra-amniotic instillation of prostaglandin $F_{2\alpha}$ in combination with urea is an economical means of pregnancy interruption that minimizes the possibility of live births. Initial studies demonstrated that the intra-amniotic dose of prostaglandin $F_{2\alpha}$ could be reduced by half, i.e. to 20 mg, when combined with 80 mg of urea and still result in abortion within acceptable mean abortion times[211,212]. The dose of prostaglandin was reduced even further to 10 mg and 5 mg in subsequent and expanded studies with urea, and abortion times remained within satisfactory limits (mean of 15·1 and 16·7 h, respectively)[213].

A statistical analysis comparing 2805 urea-prostaglandin patients with 4778 saline patients for induction of midtrimester abortion revealed a significantly lower rate of serious complications and a significantly shorter induction-to-abortion time (mean 14·2 h) with the urea-prostaglandin[214]. The results of this study suggested that urea-prostaglandin was superior to hypertonic saline as an abortifacient. The superiority of urea-prostaglandin to prostaglandin alone, however, requires analysis. The prostaglandin dose can be lowered when the drug is used in combination with urea; therefore, cost and systemic prostaglandin side-effects are reduced. In addition, the possibility of live birth is minimized. Intra-amniotic administration of urea, however, is not without its own attendant side-effects and potential complications, such as temporary but characteristic biochemical changes, coagulation problems and the possibility of intravascular injection of the hyperosmolar urea[213].

Approximately 34 000 abortions in the United States were induced by prostaglandin $F_{2\alpha}$ administration between November 1973 and 1975. During this period there were six maternal deaths associated with these prostaglandin abortion procedures. The mortality rate was 17·8 deaths/100 000 procedures. During the same time approximately 153 000 saline-induced abortions were also performed in the United States. These were associated with 35 maternal deaths, for a mortality rate of 22·9 deaths/100 000 procedures. Although only 15 000 hysterotomies were performed for pregnancy interruption during the same time span, there were seven deaths; this resulted in a mortality rate of 45 deaths/100 000 procedures. Using these statistics, induction of abortion by prostaglandin appears to have a slightly lower mortality rate than induction of abortion by saline and a much lower mortality rate than hysterotomy. Cates and co-workers[215] analysed in detail the six deaths associated with administration of prostaglandin $F_{2\alpha}$; none appeared to be directly related to prostaglandin but rather to complications arising from the abortions. The first patient, a chronic alcoholic, died from aspiration of haematemesis. The second had severe congenital heart disease and underwent cardiac arrest 3 min after the passage of the fetus. The third, who had chronic hypertension and severe pre-eclampsia, died of intracranial haemorrhage shortly after a hypertensive crisis. The fourth patient died of respiratory arrest shortly after an intravenous narcotic injection in association with intravenous phenothiazine. The fifth patient died from an over-whelming infection not directly associated with the use of prostaglandin. The sixth and last succumbed from microscopic pulmonary emboli and water intoxication following extraovular administration of prostaglandin $F_{2\alpha}$ with concomitant oxytocin infusion. The authors[215] felt that the last two deaths were not associated with the prostaglandin administration and estimated a death-to-case rate of intra-amniotic prostaglandin $F_{2\alpha}$ of 10·5 deaths/100 000 abortions.

Adachi et al.[216] subsequently reported an additional death with intra-amniotic prostaglandin $F_{2\alpha}$ in combination with saline; the cause of death in this patient appeared to be saline-related haemolysis or disseminated intravascular coagulation (DIC). In 1979, Cates and Jordaan[217] reported on the sudden collapse and eventual death of two apparently healthy women following induction of abortion with intra-amniotic instillation of 40 mg of prostaglandin $F_{2\alpha}$. The first patient, a 35-year-old, grand multipara, collapsed soon after the administration of the prostaglandin in her physician's office. On admission to the hospital, she was comatose and without palpable cardiac activity. Her neurological status never improved and she eventually died 5 months after her sudden collapse. Autopsy attributed the death to a pulmonary embolus and severe anoxic brain damage. The second patient was a 26-year-old multipara who complained of chest pain, headache, abdominal cramps and difficulty in breathing within 3 min of the prostaglandin administration. The patient's condition deteriorated rapidly, and despite eventual restoration of cardiac function, she became and remained comatose until her death 4 weeks later. Autopsy revealed extensive cortical necrosis in the brain and two myocardial infarcts. Two additional deaths in India were reported in a multicentre study of prostaglandins for induction of abortion[218]. In the first case, intra-amniotic instillation of prostaglandin $F_{2\alpha}$ in a grand multipara resulted in uterine rupture, shock and death during hysterectomy. The second patient was also a grand multipara; she received intra-amniotic administration of 15-methyl-prostaglandin $F_{2\alpha}$, experienced shock on the 12th postoperative day from delayed postabortal bleeding, and died during hysterectomy.

Conclusion

Midtrimester abortion is a potentially hazardous undertaking and, although the prostaglandins have helped reduce the risk compared to other methods, the hazard has not been completely eliminated. The optimum chance for successful outcome is provided by the informed and alert clinician who appreciates the potential risks of the procedure and who is prepared to deal with those risks. The natural prostaglandins are preferably administered either extra- or intra-amniotically. In comparison with these procedures, treatment with prostaglandin analogues offers further advantages, the most important being the possibility of using non-invasive routes of administration. Analogues are now available which are highly effective in stimulating uterine contractility, associated with a low frequency of side-effects, suitable for both vaginal and intramuscular administration and applicable for termination of pregnancy during both the early and late parts of the second trimester.

References

1 Population Information Program (1980). Prostaglandins, the use of PGs in human reproduction. *Population Rep.*, 8, G-79–119
2 T'sai, K. I. (1958). Application of electric vacuum suction in artificial abortions. 30 cases. *China J. Obstet. Gynecol.*, 6, 445
3 Wu-Yuan-Tai and Wu-Hsein-Chen (1958). Suction curettage for artificial

abortions. Preliminary report of 300 cases. *China J. Obstet. Gynecol.*, 6, 447

4   Vojta, M. A. (1960). A review of technics of artificial interruption of pregnancy and prevention of complication. *Cesk. Gynekol.*, 25, 717

5   Vladov, E., Ivanov, I. and Angelov, A. (1965). Schwangerschaftsunter-brechung mit Vakuum-aspiration. *Gynaecologia (Basel)*, 159, 54

6   Kerslake, D. and Casey, D. (1967). Abortion induced by means of the uterine aspirator. *Obstet. Gynecol.*, 30, 35

7   Goldsmith, E. and Margolis, A. J. (1971). Aspiration without cervical dilatation. *Am. J. Obstet. Gynecol.*, 110, 580–2

8   Karman, H. and Potts, M. (1972). Very early abortion using syringe as a vacuum source. *Lancet*, 1, 1051–2

9   Margolis, A. J. and Goldsmith, E. (1972). Early abortion without cervical dilatation, pump or syringe aspiration. *J. Reprod. Med.*, 9, 237

10 Landesman, R., Kaye, R. E. and Wilson, K. H. (1973). Menstrual extraction: review of 400 procedures at Women's Services, New York, N.Y. *Contraception*, 8, 527–39

11 Karim, S. M. M. and Filshie, G. M. (1970). Therapeutic abortion using prostaglandin $F_{2\alpha}$. *Lancet*, 1, 157–9

12 Karim, S. M. M. (1971). Prostaglandins as abortifacients. *N. Engl. J. Med.*, 285, 1534–5

13 Roth-Brandel, U., Bygdeman, M., Wiqvist, N. and Bergström, S. (1970). Prostaglandins for induction of therapeutic abortion. *Lancet*, 1, 190

14 Wiqvist, N. and Bygdeman, M. (1970). Induction of therapeutic abortion with intravenous prostaglandin $F_{2\alpha}$. *Lancet*, 2, 889

15 Bygdeman, M. and Wiqvist, N. (1971). Early abortion in the human. *Ann. N.Y. Acad. Sci.*, 180, 473–82

16 Wiqvist, N., Bygdeman, M. and Toppozada, M. (1971). Induction of abortion by the intravenous administration of prostaglandin $F_{2\alpha}$. A critical evaluation. *Acta Obstet. Gynecol. Scand.*, 50, 381–9

17 Kinoshita, K., Wagatsuma, T., Hogaki, M. and Sakamoto, S. (1971). The induction of abortion by prostaglandin $F_{2\alpha}$. *Am. J. Obstet. Gynecol.*, 111, 855–8

18 Haspels, A. A. and Luigies, J. H. H. (1972). Induction of abortion by intravenous and intrauterine administration of prostaglandin $F_{2\alpha}$. In Southern, E. M. (ed.) *The Prostaglandins*. pp. 433–41. (Mount Kisco N.Y.: Futura)

19 Wentz, A. C. and Jones, G. S. (1973). Intravenous prostaglandin $F_{2\alpha}$ for induction of menses. *Fertil. Steril.*, 24, 569–77

20 Karim, S. M. M. and Amy, J. J. (1975). Interruption of pregnancy with prostaglandins. In Karim, S. M. M. (ed.) *Prostaglandin and Reproduction*. pp. 77–148. (Lancaster: MTP Press)

21 Wiqvist, N. and Bygdeman, M. (1970). Therapeutic abortion by local administration of prostaglandin. *Lancet*, 2, 716–17

22 Roberts, G., Cassie, R. and Turnbull, A. C. (1971). Therapeutic abortion by intrauterine instillation of prostaglandin E_2. *J. Obstet. Gynaecol. Br. Commonw.*, 78, 834–7

23 Embrey, M. P. and Hillier, K. (1972). Therapeutic abortion by extra-

amniotic administration of prostaglandins. In Southern, E. M. (ed.) *The Prostaglandins*. pp. 381-90. (Mount Kisco, N.Y.: Futura)

24 Csapo, A. I., Kivikoski, A. and Wiest, W. G. (1972). Massive initial prostaglandin impact in postconceptional therapy. *Prostaglandins*, 2, 125-34

25 Mocsary, P. and Csapo, A. I. (1973). Delayed menstruation induced by prostaglandin in pregnant patients. *Lancet*, 2, 683

26 Karim, S. M. M. (1973). Intrauterine prostaglandins for outpatient termination of very early pregnancy. *Lancet*, 2, 794

27 Lichtman, A. S., Brenner, P. and Mishell, D. R. (1974). Intrauterine administration of prostaglandin $F_{2\alpha}$ as an outpatient procedure for termination of early pregnancy. *Contraception*, 9, 403-8

28 Ylikorkala, O., Jouppila, P., Ylöstalo, P. and Jarvinen, P. A. (1974). Intrauterine injection of prostaglandin $F_{2\alpha}$ for termination of early pregnancy in out-patient. *Prostaglandins*, 7, 57-70

29 Bygdeman, M., Beguin, F., Toppozada, M., Wide, L. and Wiqvist, N. (1972). Postconceptional fertility control by prostaglandin $F_{2\alpha}$. In Bergström, S., Green, K. and Samuelsson, B. (eds.) *Prostaglandins in Fertility Control*. pp. 175-81. (Stockhholm: WHO Research and Training Centre on Human Reproduction)

30 Karim, S. M. M. and Sharma, S. D. (1971). Therapeutic abortion and induction of labour by intravaginal administration of prostaglandin E_2 and $F_{2\alpha}$. *J. Obstet. Gynaecol. Br. Commonw.*, 78, 294-9

31 Karim, S. M. M. (1971). Once-a-month vaginal administration of prostaglandin E_2 and $F_{2\alpha}$ for fertility control. *Contraception*, 3, 173-83

32 Corlett, R. C., Sribyatta, B., Mishell, A., Ballard, C., Nakamura, R. M. and Thorneycroft, I. (1972). Termination of early gestations with vaginal prostaglandin $F_{2\alpha}$ tablets. *Prostaglandins*, 2, 453

33 Freid, N. D., Tredway, D. R. and Mishell, D. R. (1973). Termination of early pregnancy with prostaglandin E_2 vaginal suppositories. *Contraception*, 8, 255

34 Sato, T., Ami, K. and Matsumoto, S. (1973). The induction of abortion and menstruation by intravaginal administration of prostaglandin $F_{2\alpha}$. *Am. J. Obstet. Gynecol.*, 116, 287-9

35 Jones, J. R., Perez, R. J. and Bienart, W. (1974). Intravaginal PGE_2 in early abortion. *Prostaglandins*, 7, 149-63

36 Cheng, M. C. E., Tan, P. M. and Ratnam, S. S. (1973). Prostaglandin vaginal pessary as a fertility regulating agent. In *Proceedings of 8th Singapore-Malaysia Congress of Medicine*, 8, 111-14

37 Bolognese, R. J. and Corson, S. L. (1973). Abortion of early pregnancy by the intravaginal administration of prostaglandin $F_{2\alpha}$. *Am. J. Obstet. Gynecol.*, 117, 246-50

38 Tredway, D. R. and Mishell, D. R. (1973). Therapeutic abortion of early human gestation with vaginal suppositories of prostaglandin $F_{2\alpha}$. *Am. J. Obstet. Gynecol.*, 116, 795-8

39 Midwinter, A., Shepherd, A. and Bowen, M. (1973). Continuous extra-amniotic prostaglandin E_2 for therapeutic termination and the effectiveness of various infusion rates and dosages. *J. Obstet. Gynaecol. Br. Commonw.*, 80, 371-3

40 Hingorani, V., Dua, A. and Bhuyan, U. N. (1974). Induction of abortion by intrauterine administration of prostaglandin $F_{2\alpha}$ and histopathological studies of gestational sac. *Contraception*, **10**, 13-23

41 Zoltan, I., Csillag, M., Zsolnai, B., Zubek, L., Moksony, I. and Matanyi, S. (1974). The termination of first trimester pregnancy by extraovular 'prostaglandin impact'. *Prostaglandins*, **6**, 211-16

42 Fylling, P. and Refsdal, A. (1974). Therapeutic abortion by a single extra-amniotic instillation of prostaglandin $F_{2\alpha}$. *Arch. Gynäk.*, **217**, 119-25

43 Bundy, G., Lincoln, F., Nelson, N., Pike, J. and Schneider, W. (1971). Novel prostaglandin syntheses. *Ann. N.Y. Acad. Sci.*, **190**, 76-90

44 Karim, S. M. M. and Sharma, S. D. (1972). Termination of second trimester pregnancy with 15 methyl analogues and prostaglandin E_2 and $F_{2\alpha}$. *J. Obstet. Gynaecol. Br. Commonw.*, **79**, 737-43

45 Lauersen, N. H., Secher, N. J. and Wilson, K. H. (1975). Midtrimester abortion induced by serial intramuscular injections of 15(S)-15-methyl-prostaglandin E_2 methyl ester. *Am. J. Obstet. Gynecol.*, **123**, 665-70

46 Lauersen, N. H. and Wilson, K. H. (1975). Midtrimester abortion induced by serial intramuscular injection of 15(S)-15-methyl-prostaglandin $F_{2\alpha}$. *Am. J. Obstet. Gynecol.*, **121**, 273-6

47 Lauersen, N. H. and Wilson, K. H. (1976). Termination of midtrimester pregnancy by serial intramuscular injection of 15(S)-15-methyl-prostaglandin $F_{2\alpha}$. *Am. J. Obstet. Gynecol.*, **124**, 169-76

48 Lauersen, N. H. and Wilson, K. H. (1975). Serial intramuscular injection of 15(S)-15-methyl-prostaglandin $F_{2\alpha}$ in the induction of midtrimester abortion. *Prostaglandins*, **10**, 1029

49 Lauersen, N. H. and Wilson, K. H. (1976). Luteolytic and abortifacient effects of serial intramuscular injections of 15(S)-15-methyl-prostaglandin $F_{2\alpha}$ in early pregnancy. *Am. J. Obstet. Gynecol.*, **124**, 425-9

50 Fylling, P. and Jerve, F. (1977). 15(S)15-methyl prostaglandin $F_{2\alpha}$ for termination of very early human pregnancy: a comparative study of a single intramuscular injection and vaginal suppositories. *Prostaglandins*, **14**, 785-90

51 Seifert, B., Liedtke, M. P., Brockmann, J., Beissert, M., Gstottner, H., Alexander, H. and Herter, U. (1978). Investigations of hormones during early abortion induced by prostaglandin $F_{2\alpha}$ and 15(S)-methyl-$PGF_{2\alpha}$. *Acta Biologica Medica Germanica*, **37**, 955-7

52 Bygdeman, M., Martin, J. N. Jr., Leader, A., Lundström, V., Ramadan, M., Eneroth, P. and Green, K. (1976). Early pregnancy interruption by 15(S)15-methyl prostaglandin $F_{2\alpha}$ methyl ester. *Obstet. Gynecol.*, **48**, 221-4

53 Ylikorkala, O., Jarvinen, P. A., Puukka, M. and Viinikka, L. (1976). Abortifacient efficiency of 15(S)15-methyl-prostaglandin $F_{2\alpha}$ methyl ester administered vaginally during early pregnancy. *Prostaglandins*, **12**, 609-24

54 Leader, A., Bygdeman, M., Eneroth, P., Lundstrom, V. and Martin, J. N., Jr. (1976). Induced abortion in the 8-9th week of pregnancy with vaginally administered 15-methyl $PGF_{2\alpha}$ methyl ester. *Prostaglandins*, **12**, 631-7

55 Zoremthangi, B., Agarwal, N., Puri, C. P., Laumas, K. R. and Hingorani, V. (1976). Evaluation of 15(S) 15-methyl-PGF$_{2\alpha}$ methyl ester suppositories with a two dose schedule for termination of early pregnancy. *Contraception*, **14**, 519-27

56 Hamberger, L., Nilsson, L., Bjorn-Rasmussen, E., Atterfelt, P. and Wiqvist, N. (1978). Early abortion by vaginal prostaglandin suppositories: blood loss in relation to elimination of serum chorionic gonadotrophin, progesterone and estradiol-17β. *Contraception*, **17**, 183-94

57 Mackenzie, I. Z., Embrey, M. P., Davies, A. J. and Guillebaud, J. (1978). Very early abortion by prostaglandins. *Lancet*, **1**, 1223-6

58 Ganguli, A. C., Krishna, U. R., Raote, V. B. and Purandare, V. N. (1978). Prostaglandin suppositories in first trimester termination of pregnancy. *Clinician*, **42**, 400-4

59 Lauersen, N. H. and Wilson, K. H. (1980). Early pregnancy interruption with two 15-ME-PGF$_{2\alpha}$ suppositories. *Contraception*, **21**, 273-82

60 Kinra, G., Agarwal, N., Jagannath, K. T. and Hingorani, V. (1978). Evaluation of a single dose schedule of 15(S)15-methyl PGF$_{2\alpha}$ methyl ester suppository for the termination of 10-14 weeks of pregnancy. *Contraception*, **17**, 455-64

61 Green, K., Bygdeman, M. and Bremme, K. (1978). Interruption of early first trimester pregnancy by single vaginal administration of 15-methyl-PGF$_{2\alpha}$-methyl ester. *Contraception*, **18**, 551-60

62 Mandelin, M. and Kajanoja, P. (1978). Induction of second trimester abortion: comparison between vaginal 15-methyl-PGF$_{2\alpha}$ methyl ester and intra-amniotic PFG$_{2\alpha}$. *Prostaglandins*, **15**, 995-1001

63 Dutt, T. P., Blair, M. and Elder, M. G. (1979). Ultrasonic assessment of uterine emptying in first-trimester abortions induced by intravaginal 15-methyl prostaglandin F$_{2\alpha}$ methyl ester. *Am. J. Obstet. Gynecol.*, **133**, 484-8

64 Csapo, A. I. and Pulkkinen, M. G. (1979). The mechanism of prostaglandin action on the pregnant human uterus. *Prostaglandins*, **17**, 283-99

65 Csapo, A. I. and Pulkkinen, M. G. (1979). The mechanism of prostaglandin action on the early pregnant human uterus. *Prostaglandins*, **18**, 479-90

66 Lauersen, N. H. (1979). Investigation of prostaglandins for abortion. *Acta Obstet. Gynecol. Scand. Suppl.*, **81**, 1-36

67 Spilman, C. H., Beuving, D. C. and Forbes, A. D. (1976). Evaluation of vaginal delivery systems containing 15(S)-15-methyl PGF$_{2\alpha}$ methyl ester. *Prostaglandins*, **12**, 1-26

68 Bygdeman, M., Green, K., Lundström, V., Ramadan, M., Fotiou, S. and Bergström, S. (1976). Induction of abortion by vaginal administration of 15(S)15-methyl prostaglandin F$_{2\alpha}$ methyl ester: a comparison of two delivery systems. *Prostaglandins*, **12**, 27-52

69 Dillon, T. F., Mootabar, H., Phillips, L. L. and Risk, A. (1976). The efficacy of intravaginal 15-methyl prostaglandin F$_{2\alpha}$ methyl ester in first and second trimester abortion. *Prostaglandins*, **12**, 81-98

70 Hendricks, C. H., Dingfelder, J. R. and Gruber, W. S. (1977). Clinical observations with a prostaglandin-containing Silastic vaginal device for pregnancy termination. *Prostaglandins*, **12**, 99-122

71 Robins, J. (1976). The use of a Silastic vaginal device containing 15(S)-15-methyl prostaglandin F₂ alpha methyl ester for early first trimester pregnancy termination. *Prostaglandins*, **12**, 123–34

72 Duenhoelter, J. H. and Santos-Ramos, R. (1976). First experiences with a 15(S)-15-methyl prostaglandin F₂ₐ vaginal device for termination of early first trimester pregnancy with serial sonographic observations. *Prostaglandins*, **12**, 135–45

73 Lauersen, N. H. and Wilson, K. H. (1976). Hormone release and abortifacient effectiveness of a newly developed silastic device containing 15-ME-PGF₂ₐ methyl ester in concentration of 0·5% and 1·0%. *Prostaglandins*, **12**, 63–79

74 Lauersen, N. H. and Wilson, K. H. (1976). The abortifacient effects of a newly developed vaginal silastic device impregnated with an 0·5% concentration of 15(S)-15-methyl-prostaglandin F₂ₐ methyl ester in the first trimester. *Contraception*, **13**, 697–705

75 Lauersen, N. H. and Wilson, K. H. (1976). The abortifacient effectiveness and plasma prostaglandin concentration with 15(S)-15-methyl prostaglandin F₂ₐ methyl ester containing vaginal silastic devices. *Fertil. Steril.*, **27**, 1366–73

76 Robins, J. (1977). Early first-trimester abortion induction by Silastic vaginal devices for continuous release of 15(S)-15-methyl prostaglandin F₂ₐ methyl ester. *Fertil. Steril.*, **28**, 1048–55

77 Corson, S. L. and Bolognese, R. J. (1977). Abortion of early pregnancy on an outpatient basis using Silastic 15(S)-15-methyl prostaglandin F₂ₐ vaginal devices. *Fertil. Steril.*, **28**, 1056–62

78 Karim, S. M. M., Rao, B., Ratnam, S. S., Prasad, R. N. V., Wong, Y. M. and Ilancheran, A. (1977). Termination of early pregnancy (menstrual induction) with 16-phenoxy-ω-tetranor-PGE₂ methylsulfonylamide. *Contraception*, **16**, 377–81

79 Csapo, A. I., Peskin, E. G., Sauvage, J. P., Pulkkinen, M. O., Lampe, L., Godeny, S., Laajoki, V. and Kivikoski, A. (1980). Menstrual induction in preference to abortion (Letter). *Lancet*, **1**, 90–1

80 Fleischer, A., Schulman, H., Blattner, P., Jagani, N. and Fayemi, A. (1982). Early pregnancy-abortion model using sulprostone. *Prostaglandins*, **23**, 643–55

81 Bygdeman, M., Christensen, N. J., Green, K., Zheng, S. and Lundström, V. (1983). Termination of early pregnancy: future development. *Acta Obstet. Gynecol. Scand. Suppl.*, **113**, 125–9

82 Lundström, V., Bygdeman, M., Fotiou, S., Green, K. and Kinoshita, K. (1977). Abortion in early pregnancy by vaginal administration of 16,16-dimethyl-PGE₂ in comparison with vacuum aspiration. *Contraception*, **16**, 167–73

83 Karim, S. M. M., Ratnam, S. S. and Ilancheran, A. (1977). Menstrual induction with vaginal administration of 16,16 dimethyl trans-Δ²-PGE₁ methyl ester (ONO-802). *Prostaglandins*, **14**, 615–16

84 Ninagawa, T., Ohta, M., Hiroshima, T., Tomita, Y., Ito, K., Imoto, N. and Matsukawa, R. (1976). Application of prostaglandins in fertility control: Japanese experience. In Karim, S. M. M. (ed.) *Obstetrics and*

Gynaecological Uses of Prostaglandin. pp. 55-66. (Baltimore: University Park Press)

85 Wingerup, L., Ekman, G. and Ulmsten, U. (1983). Local application of prostaglandin E$_2$ in gel. *Acta Obstet. Gynecol. Scand. Suppl.*, **113**, 131-6

86 Hulka, J. F. and Higgins, G. (1961). Trauma to the internal cervical os during dilatation for diagnostic curettage. *Am. J. Obstet. Gynecol.*, **82**, 913-19

87 Caspi, E., Schneider, D., Sadovsky, G., Weinraub, Z. and Bukovsky, I. (1983). Diameter of cervical internal os after induction of early abortion by laminaria or rigid dilatation. *Am. J. Obstet. Gynecol.*, **146**, 106-8

88 Stubblefield, P. G. (1981). Laminaria and other adjunctive methods. In Berger, G. S., Brenner, W. E. and Keith, L. G. (eds.) *Second Trimester Abortion.* pp. 136-61. (Boston: John Wright PSG Inc.)

89 Lauersen, N. H., Den, T., Iliescu, C., Wilson, K. H. and Graves, Z. R. (1982). Cervical priming prior to dilatation and evacuation: a comparison of methods. *Am. J. Obstet. Gynecol.*, **144**, 890-4

90 Bryman, I., Sahni, S., Norström, A. and Lindblom, B. (1984). Influence of prostaglandins on contractility of the isolated human cervical muscle. *Obstet. Gynecol.*, **63**, 280-4

91 Uldbjerg, N., Ekman, G., Herltoft, P., Malmstrom, A., Ulmsten, U. and Wingerup, L. (1983). Human cervical connective tissue and its reaction to prostaglandin E$_2$. *Acta. Obstet. Gynecol. Scand. Suppl.*, **113**, 163-6

92 Wilhelmsson, L., Norström, A., Hamberger, L., Wikland, M., Lindblom, B. and Wiqvist, N. (1983). Interaction between PGs and catecholamines on cervical collagen synthesis. *Acta Obstet. Gynecol. Scand. Suppl.*, **113**, 171-2

93 Wiqvist, N., Beguin, F., Bygdeman, M., Fernström, I. and Toppozada, M. (1972). Induction of abortion by extra-amniotic prostaglandin administration. *Prostaglandins*, **1**, 37-53

94 Brenner, W. E., Dingfelder, J. R., Staurovsky, L. G. and Hendricks, C. (1973). Vaginally administered PGF$_{2\alpha}$ for cervical dilatation in nulliparas prior to suction curettage. *Prostaglandins*, **4**, 819-36

95 Dingfelder, J. R., Brenner, W. E., Hendricks, C. H. and Staurovsky, L. G. (1975). Reduction of cervical resistance by prostaglandin suppositories prior to dilatation for induced abortion. *Am. J. Obstet. Gynecol.*, **122**, 25-30

96 Rath, W., Kühnle, H., Theobald, P. and Kuhn, W. (1982). Objective demonstration of cervical softening with a prostaglandin F$_{2\alpha}$ gel during first trimester abortion. *Int. J. Gynaecol. Obstet.*, **20**, 195-9

97 Cheng, M. C. E., Ahmed, R., Selvadurai, V. and Ratnam, S. S. (1976). The use of prostaglandins for pre-operative cervical dilatation in first trimester nulliparous pregnancy. In Karim, S. M. M. (ed.) *Obstetrics and Gynaecological Uses of Prostaglandins.* pp. 187-90. (Baltimore: University Park Press)

98 Choo, H. T., Chong, P., Yam, K. L. and Karim, S. M. M. (1976). Cervical dilatation with prostaglandin analogue prior to evacuation of the first trimester pregnant uterus. In Karim, S. M. M. (ed.) *Obstetric and Gynae-*

cological Use of Prostaglandins. pp. 191–8. (Baltimore: University Park Press)

99 Toppozada, M., Bygdeman, M., Papageorgiou, C. and Wiqvist, N. (1973). Administration of 15-methyl prostaglandin $F_{2\alpha}$ as a preoperative means of cervical dilatation. *Prostaglandins*, **4**, 371–9

100 Borell, U., Bygdeman, M., Leader, A., Lundström, V. and Martin, J. N. (1976). Successful first trimester abortion following the use of 15(S) 15-methyl-prostaglandin $F_{2\alpha}$ methyl ester vaginal suppositories. *Contraception*, **13**, 87–94

101 Ganguli, A. C., Green, K. and Bygdeman, M. (1977). Preoperative dilatation of the cervix by single vaginal administration of 15-ME-PGF$_{2\alpha}$ methyl ester. *Prostaglandins*, **14**, 779–84

102 Lauersen, N. H., Seidman, S. and Wilson, K. H. (1979). Cervical priming prior to first-trimester suction abortion with a single 15-methyl-prostaglandin $F_{2\alpha}$ vaginal suppository. *Am. J. Obstet. Gynecol.*, **135**, 1116–18

103 Lauersen, N. H. and Wilson, K. H. (1980). The role of a long-acting vaginal suppository of 15-ME-PGF$_{2\alpha}$ in first and second trimester abortion. In Samuelsson, B., Ramwell, P. W. and Paoletti, R. (eds.) *Advances in Prostaglandin and Thromboxane Research.* Vol. 8, pp. 1435–41. (New York: Raven Press)

104 Kent, D. R., Goldstein, A. I. and Milokvich, D. (1983). Pre-operative cervical dilatation with a single long-acting prostaglandin analog suppository. *J. Reprod. Med.*, **28**, 778–80

105 Niloff, J. M. and Stubblefield, P. G. (1982). Low-dose vaginal 15-methyl prostaglandin F_2 for cervical dilatation prior to vacuum curettage abortion. *Am. J. Obstet. Gynecol.*, **142**, 596–7

106 Fayemi, A., Schulman, H., Mitchell, J. and Fleischer, A. (1983). Internal and external cervical os dilatation with vaginal 15-methyl prostaglandin $F_{2\alpha}$. *Am. J. Obstet. Gynecol.*, **146**, 219–20

107 Fylling, P. and Jerve, F. (1983). Experience with prostaglandins for therapeutic abortion in Norway. *Acta Obstet. Gynecol. Scand. Suppl.*, **113**, 113–16

108 Lauersen, N. H. and Graves, Z. R. (1984). Preabortion cervical dilatation with a low-dose prostaglandin suppository. *J. Reprod. Med.*, **29**, 133–5

109 Fehrmann, H. and Praetorius, B. (1983). The cervix-ripening effect of the prostaglandin E analogue Sulprostone, before vacuum aspiration in first trimester pregnancy. *Acta Obstet. Gynecol. Suppl.*, **113**, 141–4

110 Welch, C. and Elder, M. G. (1982). Cervical dilatation with 16,16-dimethyl-trans-Δ^2 PGE$_1$ methyl ester vaginal pessaries before surgical termination of first trimester pregnancies. *Br. J. Obstet. Gynaecol.*, **89**, 849–52

111 Moberg, P. J., Bygdemann, M., Carnsjö, L. G., Frankman, O. and Green, K. (1983). Preabortion treatment with a singel (sic) vaginal suppository containing 9-deoxo-16, 16-dimethyl-9-methylene PGE$_2$ in the late first and early second trimester pregnancies. *Acta Obstet. Gynecol. Scand. Suppl.*, **113**, 137–40

112 Sidhu, M. S. and Kent, D. R. (1984). Effects of prostaglandin E$_2$ analogue suppository on blood loss in suction abortion. *Obstet. Gynecol.*, **64**, 128–30

113 Tietze, C. (1981). Second-trimester abortion: a global view. In Berger, G. S., Brenner, W. E. and Keith, L. G. (eds.) *Second Trimester Abortion.* pp. 1-11. (Boston: John Wright PSG Inc.)

114 Hendricks, C. H. (1981). Physiology. In Berger, G. S., Brenner, W. E. and Keith, L. G. (eds.) *Second Trimester Abortion.* pp. 69-78. (Boston: John Wright PSG Inc.)

115 Hendricks, C. H., Brenner, W. E., Ekbladh, L., Brotanek, V. and Fishburne, J. I., Jr. (1971). Efficacy and tolerance of intravenous prostaglandins $F_{2\alpha}$ and E_2. *Am. J. Obstet. Gynecol.*, 111, 564-78

116 Kaufman, R. G., Freeman, R. K. and Mishell, D. R. (1971). Abortifacient activity of intravenously administered prostaglandins. *Contraception*, 3, 121-32

117 Karim, S. M. M. and Sharma, S. D. (1971). Second trimester abortion with single intra-amniotic injection of prostaglandins E_2 or $F_{2\alpha}$. *Lancet*, 2, 47-8

118 Bygdeman, M., Toppozada, M. and Wiqvist, N. (1971). Induction of midtrimester abortion by intra-amniotic administration of prostaglandin $F_{2\alpha}$. *Acta Physiol. Scand.*, 82, 415-16

119 Toppozada, M., Bygdeman, M. and Wiqvist, N. (1971). Induction of abortion by intra-amniotic administration of prostaglandin $F_{2\alpha}$. *Contraception*, 4, 293-303

120 Bygdeman, M. A. (1981). Prostaglandin procedures. In Berger, G. S., Brenner, W. E. and Keith, L. G. (eds.) *Second Trimester Abortion.* pp. 89-106. (Boston: John Wright PSG Inc.)

121 Anderson, G. G., Hobbins, J. C., Rajkovic, V., Speroff, L. and Caldwell, B. V. (1972). Midtrimester abortion using intra-amniotic prostaglandin $F_{2\alpha}$. *Prostaglandins*, 1, 147-55

122 Karim, S. M. M., Sharma, S. D. and Filshie, G. M. (1972). Termination of second trimester pregnancy with intra-amniotic administration of prostaglandins E_2 and $F_{2\alpha}$. In Southern, E. M. (ed.) *The Prostaglandins - Clinical Applications in Human Reproduction.* pp. 403-16. (Mount Kisco, N.Y.: Futura)

123 Bergstrom, S. (1973). *Prostaglandins in Fertility Control.* Vol. 3. (Stockholm: WHO Research and Training Centre on Human Reproduction)

124 Brenner, W. E., Dingfelder, J. R., Hendricks, C. H. and Staurovsky, L. (1973). Induction of therapeutic abortion with a single dose of intra-amniotically administered prostaglandin $F_{2\alpha}$. *Prostaglandins*, 4, 485-98

125 Brenner, W. E., Hendricks, C. H., Fishburne, J. I., Jr., Staurovsky, L., Braaksma, J. and Taft, R. (1973). Intraamniotic prostaglandin $F_{2\alpha}$ dose-twenty-four-hour abortifacient response. *J. Pharm. Sci.*, 62, 1278-82

126 Craft, I. (1973). Intra-amniotic prostaglandin E_2 and $F_{2\alpha}$ for induction of abortion: a dose-response study. *J. Obstet. Gynecol. Br. Commonw.*, 80, 46-7

127 Roberts, G., Gomersall, R., Adams, M. and Turnbull, A. C. (1973). Therapeutic abortion by intraamniotic injection of prostaglandins. *Adv. Biosci.*, 9, 555-60

128 Wiqvist, N., Bygdeman, M. and Toppozada, M. (1973). Intra-amniotic

prostaglandin administration – a challenge to the currently used methods for induction of midtrimester abortion. *Contraception*, **8**, 113–31

129 Corlett, R. C., Jr. and Ballard, C. A. (1974). The induction of midtrimester abortion with intraamniotic prostaglandin $F_{2\alpha}$. A single-dose technique. *Am. J. Obstet. Gynecol.*, **118**, 353–7

130 Lauiersen, N. H. and Wilson, K. H. (1974). Midtrimester abortion induced with a single intra-amniotic instillation of prostaglandin $F_{2\alpha}$. *Am. J. Obstet. Gynecol.*, **118**, 210–17

131 Brenner, W. E. (1975). The current status of prostaglandins as abortifacients. *Am. J. Obstet. Gynecol.*, **123**, 306–28

132 Kajanoja, P., Jungner, G., Seppälä, M., Karjalainen, O. and Widholm, O. (1975). Prostaglandin induction of midtrimester abortions: three years' experience of 626 cases. *Acta Obstet. Gynecol. Scand. Suppl.*, **37**, 51

133 Anderson, G. G. and Steage, J. F. (1975). Clinical experience using intraamniotic prostaglandin $F_{2\alpha}$ for midtrimester abortion in 600 patients. *Obstet. Gynecol.*, **46**, 591–5

134 Bhatt, R. V., Patel, N. F., Pathak, Chauhan, L. and Bhiwandiwala, P. (1976). Prostaglandins in midtrimester termination of pregnancy – a comparative study. In Karim, S. M. M. (ed.) *Obstetric and Gynecological Uses of Prostaglandins*. pp. 171–4. (Baltimore: University Park Press)

135 Ragab, M. I. and Edelman, D. A. (1976). Midtrimester abortion: a comparison of intraamniotic prostaglandin $F_{2\alpha}$ and hypertonic saline. *Int. J. Gynaecol. Obstet.*, **14**, 393–6

136 WHO Task Force on the Use of Prostaglandins for the Regulation of Fertility (1977). Prostaglandins and abortion. III. Comparison of single intra-amniotic injections of 15-methyl prostaglandin $F_{2\alpha}$ and prostaglandin $F_{2\alpha}$ for termination of second-trimester pregnancy: an international multicenter study. *Am. J. Obstet. Gynecol.*, **129**, 601–5

137 Kajanoja, P. (1983). Induction of abortion by prostaglandins in the second trimester of pregnancy. *Acta Obstet. Gynecol. Scand. Suppl.*, **113**, 145–51

138 Lange, A. P. (1983). Prostaglandins as abortifacients in Denmark. *Acta Obstet. Gynecol. Scand. Suppl.*, **113**, 117–24

139 Corson, S. L., Bolognese, R. J. and Merola, J. (1973). Intraamniotic prostaglandin $F_{2\alpha}$ to induce midtrimester abortion. *Am. J. Obstet. Gynecol.*, **117**, 27–34

140 Bostofte, E., Stakemann, G. and Stocklund, K. E. (1975). A comparison of termination of pregnancies in the 2nd trimester induced by intra-amniotic injection of hypertonic saline, prostaglandin $F_{2\alpha}$ or both drugs. *Acta. Obstet. Gynecol. Scand. Suppl.*, **37**, 47–50

141 Edelman, D. A., Brenner, W. E., Mehta, A. C., Philips, F. S., Bhatt, R. V. and Bhiwandiwala, P. (1976). A comparative study of intra-amniotic saline and two prostaglandin $F_{2\alpha}$ dose schedules for midtrimester abortion. *Am. J. Obstet. Gynecol.*, **125**, 188–95

142 Lauersen, N. H., Wilson, K. H., Beling, C. G. and Fuchs, F. (1974). Comparison of prostaglandin $F_{2\alpha}$ and hypertonic saline for induction of midtrimester abortion. *Am. J. Obstet. Gynecol.*, **120**, 875–89

143 Nielsen, K. R., Gregersen, E., Larsen, J. F. and Olsen, C. E. (1974). Pros-

taglandin $F_{2\alpha}$ and oxytocin compared with hypertonic saline and oxytocin for the induction of second trimester abortion. *Acta Obstet. Gynecol. Scand. Suppl.*, 37, 57–60

144 Risk, A., Mootabar, H., Porta, P. J. and Dillon, T. F. (1975). Second trimester abortions: review of four procedures. *N.Y. State J. Med.*, 75, 1022–7

145 World Health Organization. Task Force on the Use of Prostaglandins For the Regulation of Fertility (1976). Comparison of intra-amniotic prostaglandin $F_{2\alpha}$ and hypertonic saline for induction of second-trimester abortion. *Br. Med. J.*, 1, 1373–6

146 Berger, G. S. and Edelman, D. A. (1977). A clinical comparison of pros-taglandin $F_{2\alpha}$ and intra-amniotic saline for induction of midtrimester abortion. *Ann. Chir. Gynaecol. Fen.*, 66, 55–8

147 Grimes, D. A., Schulz, K. F., Cates, W. Jr. and Tyler, C. W. Jr. (1977). Midtrimester abortion by intraamniotic prostaglandin $F_{2\alpha}$: safer than saline? *Obstet. Gynecol.*, 49, 612–16

148 Stroup, P. E. (1977). Medical termination of midtrimester pregnancy in a community hospital. *J. Med. Soc. N.J.*, 74, 747–52

149 Bygdeman, M. (1978). Comparison of prostaglandin and hypertonic saline for termination of pregnancy. *Obstet. Gynecol.*, 52, 424–9

150 Robins, J. (1978). A clinical comparison of intra-amniotic prostaglandin $F_{2\alpha}$ and intra-amniotic hypertonic saline for midtrimester pregnancy ter-mination. *Adv. Plan. Parenth.*, 13, 27–34

151 Cates, W., Jr., Grimes, D. A., Schulz, K. F., Ory, H. W. and Tyler, C. W., Jr. (1978). World Health Organization studies of prostaglandins versus saline as abortifacients. *Obstet. Gynecol.*, 52, 493–8

152 Grimes, D. A. and Cates, W., Jr. (1979). The comparative efficacy and safety of intraamniotic prostaglandin $F_{2\alpha}$ and hypertonic saline for second-trimester abortion. *J. Reprod. Med.*, 22, 248–54

153 Grimes, D. A. and Cates, W., Jr. (1980). The brief for hypertonic saline. *Contemp. OB/GYN*, 15, 29–38

154 Speroff, L. (1980). Editorial comment. *Contemp. OB/GYN*, 15, 43

155 Embrey, M. P. and Hillier, K. (1971). Therapeutic abortion by intrauter-ine instillation of prostaglandins. *Br. Med. J.*, 1, 588–90

156 Lauersen, N. H. and Wilson, K. H. (1973). Continuous prostaglandin $F_{2\alpha}$ infusion for middle-trimester abortion. *Lancet*, 1, 1195

157 Lauersen, N. H. and Wilson, K. H. (1974). Continuous extraovular ad-ministration of prostaglandin $F_{2\alpha}$ for midtrimester abortion. *Am. J. Ob-stet. Gynecol.*, 120, 273–80

158 Toppozada, M., Bygdeman, M. and Wiqvist, N. (1974). Intra-uterine prostaglandin administration – A promising approach for inducing second trimester abortion. *OB/GYN Digest*, **December**, 14–25

159 Bolognese, R. J. and Corson, S. L. (1974). Prostaglandin E_2 vaginal sup-pository as an early second trimester abortifacient. *Obstet. Gynecol.*, 43, 104–8

160 Lauersen, N. H., Secher, N. J. and Wilson, K. H. (1975). Mid-trimester abortion induced by intravaginal administration of prostaglandin E_2 sup-positories. *Am. J. Obstet. Gynecol.*, 122, 947–54

161 Labed, J. P., Rubin, A. and Millman, A. E. (1980). Comparison between intraamniotic $PGF_{2\alpha}$ and vaginal PGE_2 for second trimester abortion. *Obstet. Gynecol.*, **56**, 90–6

162 Rettenmaier, M. A. and Hanson, F. W. (1983). A comparative study of two types of prostaglandins for abortion during the second trimester. *Surg. Gynecol. Obstet.*, **156**, 585–8

163 Lauersen, N. H. and Wilson, K. H. (1977). Induction of labor in patients with missed abortion and fetal death *in utero* with prostaglandin E_2 suppositories. *Am. J. Obstet. Gynecol.*, **127**, 609–11

164 Southern, E. M., Gutknecht, G. D., Mohberg, N. R. and Edelman, D. A. (1978). Vaginal prostaglandin E_2 in the management of fetal intrauterine death. *Br. J. Obstet. Gynecol.*, **85**, 437–41

165 Christensen, N. J. and Bygdeman, M. (1983). The use of prostaglandins for termination of abnormal pregnancy. *Acta Obstet. Gynecol. Scand. Suppl.*, **113**, 153–7

166 Bieniarz, J., Hunter, G., Scommegna, A. and Altamirano, Z. (1974). Efficacy and acceptability of 15(S)-methyl-prostaglandin E_2 methyl ester for midtrimester pregnancy termination. *Am. J. Obstet. Gynecol.*, **120**, 840–3

167 Brenner, E. E., Dingfelder, J. K., Staurovsky, L. G., Kumarasamy, T. and Grimes, D. A. (1974). Intramuscular administration of 15(S)-15-methyl prostaglandin E_2 methyl ester for induction of abortion. *Am. J. Obstet. Gynecol.*, **120**, 833–6

168 Brenner, W. R., Dingfelder, J. R., Staurovsky, L. G., Kumarasamy, T. and Grimes, D. A. (1975). Intramuscular administration of 15(S)-15-methyl prostaglandin E_2 methyl ester for induction of abortion. A comparison of two dose schedules. *Fertil Steril.*, **26**, 369–79

169 Brenner, W. E., Gruber, W., Staurovsky, L. G. and Dingfelder, J. R. (1976). Induction of abortion with intramuscular administration of 15(S)-15 methyl PGE_2 and 15(S)-15 methyl $PGF_{2\alpha}$. In Karim, S. M. M. (ed.) *Obstetric and Gynaecological Uses of Prostaglandins.* pp. 175–86. (Baltimore: University Park Press)

170 Dhall, G. I., Devi, P. K., Gulati, S. and Dhall, K. (1976). Induction of midtrimester abortion with serial intramuscular injections of 15(S)-15 methyl $PGF_{2\alpha}$. In Karim, S. M. M. (ed.) *Obstetric and Gynaecological Uses of Prostaglandins.* (Baltimore: University Park Press)

171 Lange, A. P. and Secher, N. J. (1977). Midtrimester and missed abortion treated with intramuscular 15(S)-15 methyl $PGF_{2\alpha}$. *Prostaglandins*, **14**, 389–95

172 World Health Organization. Task Force on the Use of Prostaglandins for the Regulation of Fertility (1977). Prostaglandins and abortion. 1. Intramuscular administration of 15-methyl prostaglandin $F_{2\alpha}$ for induction of abortion in weeks 10 to 20 of pregnancy. *Am. J. Obstet. Gynecol.*, **129**, 593–6

173 World Health Organization. Task Force on the Use of Prostaglandins for the Regulation of Fertility (1977). Prostaglandins and abortion. 2. Single extra-amniotic administration of 0·92 mg of 15-methyl prostaglandin $F_{2\alpha}$ in Hyskon for termination of pregnancies in weeks 10 to 20 of gestation:

an international multicenter study. *Am. J. Obstet. Gynecol.*, **129**, 597–600

174 World Health Organization. Task Force on the Use of Prostaglandins for the Regulation of Fertility (1977). Prostaglandins and abortion. 3. Comparison of single intra-amniotic injections of 15-methyl prostaglandin $F_{2\alpha}$ and prostaglandin $F_{2\alpha}$ for termination of second-trimester pregnancy: an international multicenter study. *Am. J. Obstet. Gynecol.*, **129**, 601–6

175 Tejula, S., Choudhury, S. D. and Manchanda, P. K. (1978). Use of intra- and extra-amniotic prostaglandins for the termination of pregnancies: report of multicentric trial in India. *Contraception*, **18**, 641–52

176 Schmidt-Gollwitzer, K., Schuessler, B., Elger, W. and Schmidt-Gollwitzer, M. (1980). Improvement in artificial second trimester abortion with a new tissue-selective prostaglandin E_2 derivative. *Am. J. Obstet. Gynecol.*, **137**, 867–8

177 Karim, S. M. M., Choo, H. T., Lim, A. L., Yeo, K. C. and Ratnam, S. S. (1978). Termination of second trimester pregnancy with intramuscular administration of 16 phenoxy-ω-17,18,19,20 tetranor PGE_2 methyl sulfonylamide. *Prostaglandins*, **15**, 1063–8

178 World Health Organization. Task Force on the Use of Prostaglandins for Fertility Regulation (1982). Termination of second trimester pregnancy by intramuscular injection of 16-phenoxy-17,18,19,20-tetranor-PGE_2 methyl sulphonylamide. *Int. J. Gynaecol. Obstet.*, **20**, 383–6

179 Karim, S. M. M., Ratnam, S. S. and Lin, A. L. (1982). Termination of second trimester pregnancy with laminaria and intramuscular 16-phenoxy-ω-17,18,19,20-tetranor PGE_2 methyl sulfonylamide. *Prostaglandins*, **23**, 257–63

180 Bygdeman, M. and Christensen, N. J. (1983). Termination of second trimester pregnancy by laminaria and intramuscular injections of 15-methyl-$PGF_{2\alpha}$ or 16-phenoxy-ω-17,18,19,20-tetranor PGE_2 methyl sulfonylamide. A randomized study. *Acta Obstet. Gynecol. Scand.*, **62**, 535–7

181 World Health Organization Task Force on Prostaglandins for Fertility Regulation (1985). Termination of second trimester pregnancy with laminaria and intramuscular 15-methyl-$PGF_{2\alpha}$ or 16-phenoxy-ω-17,17,19,20-tetranor PGE_2 methyl sulfonylamide. A randomized multicentre study. *Int. J. Gynaecol. Obstet.* (in press)

182 Lauersen, N. H. and Wilson, K. H. (1977). The effects of intramuscular injections of 15(S)-15-methyl-prostaglandin $F_{2\alpha}$ in failed abortions. *Fertil. Steril.*, **28**, 1044–7

183 Lauersen, N. H., Den, T., Scher, J., Iliescu, C. and Wilson, K. H. (1982). A new abortion technique: intravaginal and intramuscular prostaglandin. *Obstet. Gynecol.*, **58**, 96–100

184 Grimes, D. A. and Cates, W., Jr. (1981). Dilatation and evacuation. In Berger, G. S., Brenner, W. E. and Keith, L. G. (eds.) *Second Trimester Abortion*. pp. 119–33. (Boston: John Wright PSG Inc.)

185 Hern, W. M. (1984). Serial multiple laminaria and adjunctive urea in late outpatient dilatation and evacuation abortion. *Obstet. Gynecol.*, **63**, 543–9

186 Kaltreider, N. B., Goldsmith, S. and Margolis, A. (1979). The impact of midtrimester abortion techniques on patients and staff. *Am. J. Obstet. Gynecol.*, **135**, 235-8

187 Greenberg, B. G. (1978). Testimony presented to Food and Drug Administration, Obstetrics and Gynecology Advisory Committee, January 30, 1978, Rockville, Maryland

188 Lauersen, N. H., Wilson, K. H., Zervoudakis, I. A. and Saary, Z. (1976). Management of failed prostaglandin abortions. *Obstet. Gynecol.*, **47**, 473-8

189 Brenner, W. E., Hendricks, C. H., Braaksma, J. T., Fishburne, J. I., Kroncke, F. G., Jr. and Staurovsky, L. (1972). Intra-amniotic administration of prostaglandin $F_{2\alpha}$ to induce therapeutic abortion. *Am. J. Obstet. Gynecol.*, **114**, 781-7

190 Wentz, A. C., Austin, K. and King, T. M. (1973). Abortifacient efficacy of intravaginal prostaglandin $F_{2\alpha}$. *Am. J. Obstet. Gynecol.*, **115**, 27-32

191 Brenner, W. E. (1976). Fertility control - symposium on prostaglandins. *Fertil. Steril.*, **27**, 1380-6

192 Wentz, A. C., Thompson, B. H. and King, T. M. (1973). Posterior cervical rupture following prostaglandin-induced midtrimester abortion. *Am. J. Obstet. Gynecol.*, **115**, 1107-10

193 Lowensohn, R. and Ballard, C. A. (1974). Cervicovaginal fistula: an apparent increased incidence with prostaglandin $F_{2\alpha}$. *Am. J. Obstet. Gynecol.*, **119**, 1057-61

194 Duenhoelter, J. H. and Gant, N. F. (1975). Complications following prostaglandin $F_{2\alpha}$-induced midtrimester abortion. *Obstet. Gynecol.*, **46**, 247-50

195 Goodlin, R., Newell, J., O'Hare, J. and Sturz, H. (1972). Cervical fistula. *Obstet. Gynecol.*, **40**, 82-4

196 Engel, T., Greer, B., Kochenour, N. and Droegemueller, W. (1973). Midtrimester abortion using prostaglandin $F_{2\alpha}$, oxytocin, and laminaria. *Fertil. Steril.*, **24**, 265-8

197 Stubblefield, P. G., Naftolin, F., Frigoletto, F. D. and Ryan, K. J. (1974). Pretreatment with laminaria tents before midtrimester abortion with intra-amniotic prostaglandin $F_{2\alpha}$. *Am. J. Obstet. Gynecol.*, **118**, 284-5

198 Stubblefield, P. G., Naftolin, F., Frigoletto, F. and Ryan, K. J. (1975). Laminaria augmentation of intra-amniotic $PGF_{2\alpha}$ for midtrimester pregnancy termination. *Prostaglandins*, **10**, 413-22

199 Golbus, M. S., Margolis, A. J., Sweet, R. L. and Laros, R. K. (1976). Experience with 276 intra-amniotic prostaglandin $F_{2\alpha}$ induced midtrimester abortions. *Prostaglandins*, **11**, 841-54

200 Droegemueller, W., Weinstein, L. and Milzer, G. (1980). Low-dose prostaglandins for second-trimester abortion. *Contemp. OB/GYN*, **15**, 19-23

201 Duenhoelter, J. H., Gant, N. F. and Jimenez, J. M. (1976). Concurrent use of prostaglandin $F_{2\alpha}$ and laminaria tents for induction of midtrimester abortion. *Obstet. Gynecol.*, **47**, 469-72

202 Propping, D., Stubblefield, P. G., Golub, J. and Zuckerman, J. (1977). Uterine rupture following midtrimester abortion by laminaria, prostaglandin $F_{2\alpha}$, and oxytocin. *Am. J. Obstet. Gynecol.*, **128**, 689-90

203 Sandler, R. Z., Knutzen, V. K., Milano, C. M. and Gleicher, N. (1979). Uterine rupture with the use of vaginal prostaglandin E_2 suppositories. *Am. J. Obstet. Gynecol.*, **134**, 348-9

204 Grimes, D. A., Cates, W., Jr., Petitti, D. B. and Pakter, J. (1978). Fatal uterine rupture during oxytocin-augmented saline-induced abortion. *Am. J. Obstet. Gynecol.*, **130**, 591-3

205 Kerenyi, T., Mandelman, N. and Sherman, D. H. (1973). Five thousand consecutive saline inductions. *Am. J. Obstet. Gynecol.*, **116**, 593-600

206 Fishburne, J. I., Jr., Brenner, W. E., Braaksma, J. T., Staurovsky, L. G., Mueller, R. A., Hoffer, J. L. and Hendricks, C. H. (1972). Cardiovascular and respiratory responses to intravenous infusion of prostaglandin $F_{2\alpha}$ in the pregnant woman. *Am. J. Obstet. Gynecol.*, **114**, 765-72

207 MacKenzie, I. Z., Sayers, L., Bonnar, J. and Hillier, K. (1975). Coagulation changes during second-trimester abortion induced by intra-amniotic prostaglandin E_2 and hypertonic solutions. *Lancet*, **2**, 1066-9

208 Phillips, L. L., Mohajer-Shojai, E. and Dillon, T. C. F. (1974). Coagulation studies during second-trimester abortions induced by $PGF_{2\alpha}$. *Am. J. Obstet. Gynecol.*, **119**, 577-82

209 Golbus, M. S. and Erickson, R. P. (1974). Mid-trimester abortion induced by intra-amniotic prostaglandin $F_{2\alpha}$: fetal tissue viability. *Am. J. Obstet. Gynecol.*, **119**, 268-70

210 Stroh, G. and Hinman, A. R. (1976). Reported live births following induced abortion: two and one-half years' experience in Upstate New York. *Am. J. Obstet. Gynecol.*, **126**, 83-9

211 King, T. M., Átienza, M. F., Burkman, R. T., Burnett, L. S. and Bell, W. R. (1974). The synergistic activity of intra-amniotic prostaglandin $F_{2\alpha}$ and urea in the mid-trimester elective abortion. *Am. J. Obstet. Gynecol.*, **120**, 704-18

212 Wellman, L. and Jacobsson, A. (1976). Intra-amniotic prostaglandin $F_{2\alpha}$ and urea for midtrimester abortion. *Fertil. Steril.*, **27**, 1374

213 Burkman, R. T., King, T. M. and Atienza, M. F. (1981). Hyperosmolar urea. In Berger, G. S., Brenner, W. E. and Keith, L. G. (eds.) *Second Trimester Abortion*. pp. 107-18. (Boston: John Wright PSG Inc.)

214 Binkin, N. J., Schulz, K. F., Grimes, D. A. and Cates, W., Jr. (1983). Urea-prostaglandin versus hypertonic saline for instillation abortion. *Am. J. Obstet. Gynecol.*, **146**, 947-52

215 Cates, W., Jr., Grimes, D. A., Haber, R. J. and Tyler, C. W., Jr. (1977). Abortion deaths associated with the use of prostaglandin $F_{2\alpha}$. *Am. J. Obstet. Gynecol.*, **127**, 219-22

216 Adachi, A., Wilson, L. and Herzig, N. (1977). Prostaglandin $F_{2\alpha}$, hypertonic saline and oxytocin in midtrimester abortion. *N.Y. State J. Med.*, **77**, 46-9

217 Cates, W., Jr. and Jordaan, H. V. F. (1979). Sudden collapse and death of women obtaining abortions induced with prostaglandin $F_{2\alpha}$. *Am. J. Obstet. Gynecol.*, **133**, 398-400

218 Tejuja, S., Choudhury, S. D. and Manchanda, P. K. (1978). Use of intra- and extra-amniotic prostaglandins for termination of pregnancies. Reports of multicentric trial in India. *Contraception*, **18**, 641-52

16
Dysmenorrhoea and other menstrual disorders

P. T. RUSSELL and O. M. OWENS

INTRODUCTION

Primary dysmenorrhoea is defined as menstrual pain not associated with recognizable pelvic pathology. Ideally, the term should be reserved for pain severe enough to limit daily activity or require medical attention. Depending on the aetiology, dysmenorrhoea may be classified as primary or secondary. Primary dysmenorrhoea is due to factors intrinsic to the uterus and occurs only in ovulatory cycles. Pain normally occurs a few hours before commencement of menstrual flow, but can also occur simultaneously with or a few hours after the beginning of menstruation. Associated symptoms may include nausea, vomiting, headache, malaise and depression. The diagnosis of primary dysmenorrhoea is based on the patient's medical history and normal findings at pelvic examination.

In contrast, the apparent cause of menstrual pain in secondary dysmenorrhoea is related to some form of pelvic pathology. Most commonly noted are endometriosis, intrauterine myomas or polyps. Intrauterine devices (IUDs) can also be related to menstrual pain. The menstrual pain of secondary dysmenorrhoea typically begins a few to several days before the onset of menstrual flow. This pain is more consistent and continuous throughout the menstrual period and may even continue 2–3 days after cessation of flow.

The recognition that a substance in menstrual blood[1], later defined as prostaglandin, stimulated contractions in smooth muscle, focused attention on the role of these agents in normal menstruation and menstrual related problems. The serendipitous observations that a non-steroidal anti-inflammatory drug, phenylbutazone, ameliorated the symptoms of dysmenorrhoea in patients who were treated for arthritis led to clinical trials evaluating the efficacy of these drugs as therapeutic agents for dysmenorrhoea. As knowledge increased regarding the chemical nature of the prostaglandins, their actions in physiological systems, and the sites of action of the non-steroidal anti-inflammatory drugs, a rational basis for the aetiology of dysmenorrhoea was proposed with prospects for effective treatment.

NORMAL MENSTRUATION

The initiation of menstrual bleeding derives from a decline in levels of circulating ovarian steroids, spasms of the spiral uterine arteries, ischaemic necrosis of the superficial layers of cells, and finally, endometrial desquamation. Increased contractile activity of the endometrium occurs with the menstrual bleeding. It is quite possible that prostaglandins play an active role in more than one of these events[2, 3].

Prostaglandins are found in high concentrations in the endometrium as menstruation approaches, and they are also found in abundance in menstrual blood[4-6]. Prostaglandins can initiate myometrial contractions both *in vivo* and *in vitro* which could account for the increased uterine contractions that occur at menstruation. Prostaglandin $F_{2\alpha}$ has been identified as the main prostaglandin found in menstrual fluid, and infusion of prostaglandin $F_{2\alpha}$ during the luteal phase of the menstrual cycle results in premature bleeding with focal endometrial necrosis[7]. Measurements of menstrual prostaglandins show that the total amount of menstrual fluid and the total amount of prostaglandin released per menstruation is fairly constant for a given woman[8]. Approximately 95% of the prostaglandins are released during the first 48 h; by 72 h nearly 99% of the total menstrual prostaglandins have been released. All of the drugs effective for inhibiting prostaglandin production (i.e. non-steroidal anti-inflammatory agents) potentially can decrease menstrual flow and delay the onset of menses.

PHYSIOLOGICAL BASIS OF DYSMENORRHOEA

The painful cramps associated with ovulatory cycles experienced by many women are thought to be prostaglandin related[9-12]. Pickles and co-workers[13] first suggested that prostaglandins might be associated with primary dysmenorrhoea based on their observation that endometrial and menstrual fluid concentrations of prostaglandin $F_{2\alpha}$ ($PGF_{2\alpha}$) were higher in women with a diagnosis of primary dysmenorrhoea than in women with pain-free menses. Once menstruation begins, large amounts of $PGF_{2\alpha}$ appear in the menstrual fluid. In all probability this derives from the endometrium and represents what remains after absorption and uterine metabolism have occurred. Under these circumstances, the myometrium is probably exposed to high local concentrations of $PGF_{2\alpha}$ during menstruation, and these concentrations could give rise to the contractions observed at this time. A good temporal correlation can be demonstrated between the quantity of prostaglandin released and the clinical symptoms experienced by the woman over a 24-h period during her menses[12]. As $PGF_{2\alpha}$ is the main prostaglandin found in menstrual fluid and as it is a potent myometrial stimulant, dysmenorrhoea could be due to painful contractions induced by this substance.

Uterine activity during the proliferative phase of the cycle is characterized by contractions of short duration, relatively small amplitude and stable resting tone. In contrast, during the secretory phase the primary contractions become slower, but they are now superimposed with secondary contractions of faster but smaller character. Contractile amplitude increases at this time until it reaches its highest level just before menstruation. Frequency, however, is now lower compared to previous times in the cycle. The contractile pattern in the late luteal phase resembles uterine activity seen during the late prelabour period in term pregnancy. At the onset of menstruation in women with pain-free menses, enhancement of regular uterine activity is noted with larger contractions. The resting tone during normal menstruation is similar to that observed in the secretory phase[14].

In women with dysmenorrhoea, uterine activity is abnormal during menstruation[14,15]. Primary dysmenorrhoea is accompanied by a spastic hypercontractility pattern, characterized either by an extremely high tone between contractions, excessive amplitudes of contractions or asynchronous propagation of the waves (Table 16.1).

Blood flow measurements of the endometrium show a local decrease during well-delineated contractions[14]. Minimal blood flow usually occurs later than the peak of the contraction which, in most instances, coincides with greatest pain sensation experienced by the woman with dysmenorrhoea. Furthermore, the most pronounced decrease in blood flow occurs coincidentally with the most intensive contractions of high

Table 16.1 Intrauterine pressure and frequency of contractions (from Lundstrom[14])

	Tone (mmHg)	Amplitude (mgHg)	Frequency (n/10 min)
Normal menstruation	10–30	50–150	3–10
Primary dysmenorrhea	30–75	100–350	6–12

Table 16.2   Menstrual fluid volume in normal controls and dysmenorrhoeic subjects (from Dawood[12])

	Total menstrual fluid per cycle (g)
Normal controls (2 subjects, 4 cycles)	33.4 ± 1.5
Dysmenorrhoeic subjects (8 patients, 16 cycles)	60.1 ± 10.7

amplitude and long duration or is accompanied by frequent contractions without intermittent periods of relaxation. These observations suggest that ischaemia, due to hypercontractility, causes primary dysmenorrhoea.

Besides the abnormal myometrial contractile patterns identified in dysmenorrhoea and the increased prostaglandin content of menstrual blood, the total volume of menstrual fluid is significantly higher in dysmenorrhoeic subjects than in normal controls (Table 16.2). This observation seems particularly important, as painful menses appear well correlated to circumstances, congenital or otherwise, where menstrual fluid remains in the uterine cavity for excessive periods of time.

The pain of primary dysmenorrhoea is believed to consist of both physiological and psychological components[16,17], although the problems of measuring pain make this an area of controversy. Pain is greatest about 12 h after the onset of flow when the endometrium is shedding rapidly. The pain can result from three sources[12]: (1) ischaemia following the decrease in uterine blood flow after abnormal uterine contractions, (2) direct production by the prostaglandin endoperoxides if they are generated in excessive quantities, and (3) sensitization of the pain fibres by the prostaglandins to mechanical or other chemical stimuli. The fact that the most intense sensation of pain occurs coincidentally with low uterine blood flow[18] favours the explanation of the ischaemia for its primary cause.

HYPOTHESES ON THE AETIOLOGY OF PRIMARY DYSMENORRHOEA THAT INVOLVE PROSTAGLANDINS

Current hypotheses[9,11,19] to explain the aetiology of primary dysmenorrhoea have four basic features in common.

(1) Prostaglandins are responsible for the abnormal myometrial activity at the time of menses.
(2) There is an increased and abnormal uterine activity during menstruation.
(3) This activity causes a reduced uterine blood flow.
(4) Reduced uterine blood flow causes pelvic pain.

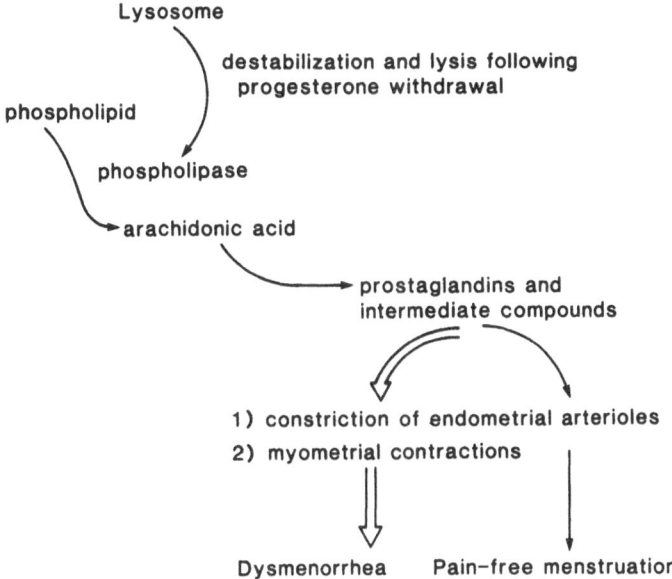

Figure 16.1 Scheme explaining primary dysmenorrhoea (after Rosenwaks and Seeger-Jones[11])

Figure 16.1 presents one hypothesis to explain the aetiology of primary dysmenorrhoea. This hypothesis is attractive because it shares key features in common with mechanisms proposed for menstruation and the onset of labour. Both the process of menstruation and the onset of labour are thought to begin by the destabilization of the lysosomal membranes of the endometrial cells under the influence of progesterone withdrawal. Progesterone depletion, subsequent to luteolysis, sets in motion the process for menstruation (see Figure 16.1). Destabilized lysosomes release their contents including a host of enzymes whose functions seem to be degradative in nature. A prominent member of the lysosomal enzyme family is phospholipase A_2 which hydrolyses fatty acids from the 2 position of the phospholipids. This position is occupied by unsaturated fatty acids, most frequently arachidonic acid. The secretory or progesterone dominated endometrium is rich in phospholipid with an abundant supply of arachidonic acid. Arachidonic acid, once set free by the action of the phospholipase A_2, is the substrate for the enzymes which lead to prostaglandin formation and for the formation of the other members of the arachidonic acid cascade. With the increased prostaglandin formation and diffusion, contractile activity is stimulated in adjacent myometrium.

A reasonable explanation why some women experience dysmenorrhoea and while others do not could be related to the significantly increased volume of menstrual fluid that occurs with dysmenorrhoea (Table 16.2). Some women, being very responsive to the influences of cyclic oestrogens and progesterone, could possibly develop a thickened endometrium that does not clear the uterus at the time of sloughing as easily as when there is less endometrial tissue. Endometrial concentrations of prostaglandins are higher during the secretory than during the proliferative phase of the cycle[12]. Thus, an exaggerated prostaglandin influence on the small blood vessels of the endometrium and a tocolytic effect on the myometrium could cause the myometrial contractile patterns observed in dysmenorrhoea. This hypothesis suggests that women with heavier monthly flows should be more prone to have primary dysmenorrhoea.

POSSIBILITIES FOR PROSTAGLANDIN INVOLVEMENT IN OTHER MENSTRUAL DISORDERS

Support for a primary role for prostaglandins in the genesis of menstrual pain whose aetiology is pelvic pathology is less convincing than it is for primary dysmenorrhoea. Prostaglandins are components of the inflammatory process[20] and they are generated with tissue trauma[12]; thus their presence may be but coincidental in secondary dysmenorrhoea.

Pelvic pain around the time of menstruation can occur in the presence of pelvic inflammatory disease, endometriosis, intrauterine myomas, or polyps[20-24]. Though the pattern of pain onset, maximum intensity, and time of disappearance differs between these conditions, prostaglandins are usually believed to have some role in each. Their contribution probably is related to the process of tissue inflammation, through irritation of tissue, trauma associated with the stretching of pelvic adhesions or through irritation caused by polyps or myomas associated with the uterine contractions during menstruation. Though increased prostaglandin $F_{2\alpha}$ levels have been identified in peritoneal fluid of women with endometriosis[25], pain may not always be experienced nor are prostaglandin antagonists always effective in relieving pain when it occurs. Obviously, a causal relationship between prostaglandin and pain has yet to be established.

The passage of large blood clots or a membranous cast may be associated with dysmenorrhoea. The intensity of the pelvic pain may mimic labour-like cramps. Prostaglandin levels have been shown to be elevated with abnormal uterine bleeding, but again a causal relationship has yet to be established.

One of the most common causes of secondary dysmenorrhoea is the intrauterine device (IUD). The cause of the pain with the IUD in place is not known, but it may be related to endometrial trauma or leuko-cytic infiltrations adjacent to the IUD[26,27]. Both circumstances could lead to increased prostaglandin synthesis and endometrial content, which, in turn, could result in enhanced myometrial activity[28,29]. IUD-induced hyperactivity of the myometrium might contribute to a spasmodic form of dysmenorrhoea, such as may occur with endometrial polyps. It might also be that the pelvic pain associated with IUD is secondary to pelvic inflammatory disease of variable severity[30,31].

All types of IUDs increase the incidence and the severity of dysmenorrhoea with the exception of the progesterone-containing IUD which is reputed to relieve the symptoms of dysmenorrhoea[32]. The reason for this may be related to inflammatory suppression by progesterone or to a decrease in menstrual blood loss. Copper-containing IUDs also have the potential to exacerbate the severity of dysmenorrhoea. Copper ions, leaching from the IUDs, would direct endometrial prostaglandin biosynthesis to favour $PGF_{2\alpha}$ formation at the expense of PGE_2[33].

DRUGS FOR DYSMENORRHOEA

Non-steroidal anti-inflammatory drugs

The use of non-steroidal anti-inflammatory drugs (NSAIDs) in the treatment of dysmenorrhoea was reported before prostaglandins were identified in the endometrium and before it was recognized that these drugs inhibit prostaglandin synthesis[34]. The successful amelioration of the symptoms of dysmenorrhoea in patients who were treated with phenylbutazone for arthritis led to the early trials of this drug for relief of menstrual pain[35]. Recognition of the inhibitory action of NSAIDs on prostaglandin synthesis[20,36,37] led to the use of flufenamic acid specifically for the treatment of dysmenorrhoea[38]. The non-steroidal drugs, although they have diverse chemical structures[39], all share anti-inflammatory, analgesic and antipyretic actions.

The relationship of the non-steroidal anti-inflammatory drugs to prostaglandin production is complex because of the nature of the enzyme system which forms prostaglandins from polyunsaturated fatty acid precursors. The system for prostaglandin synthesis (prostaglandin synthetase) exists in many molecular forms within the mammalian body[40], and the synthetase system exhibits varying degrees of sensitivity to the inhibitors[41]. The biosynthetic scheme for the production of prostaglandins and related compounds is described in detail in Chapter 1.

The sites of action of the anti-inflammatory agents are shown in an abbreviated representation of this scheme (Figure 16.2). Arachidonic

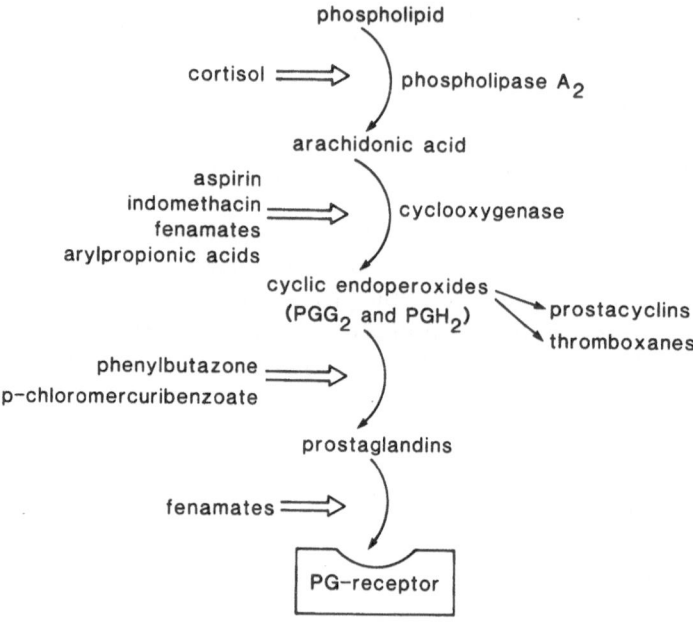

Figure 16.2   Biosynthesis of prostaglandins, prostacyclins and thromboxanes from phospholipids. Scheme indicates sites of drug inhibition

acid or some other precursor fatty acid is first converted to the endoperoxides (PGG$_2$ and PGH$_2$). These are unstable and highly reactive intermediate compounds that can then be further transformed into the prostaglandins, prostacyclins, and thromboxanes depending upon the tissue of the body and the enzymatic capacity of that tissue. For the purposes of this discussion, it is important to recognize that virtually all of the compounds belonging to the arachidonic acid cascade have some action on smooth muscle or the vasculature system; this can make a difference where the inhibitor has its action. If the inhibition takes place prior to endoperoxide formation, all of the subsequent compounds of this cascade should not be formed. If, on the other hand, inhibition occurs after endoperoxide formation, varied effects could occur depending on the site and nature of the inhibition. Aspirin, indomethacin, fenamate compounds and arylproprionic acids all act prior to cyclic endoperoxide formation; in contrast, phenylbutazone and p-chloromercuribenzoate inhibit the breakdown of the endoperoxides to the prostaglandins. Cortisol appears to inhibit at the phospholipase step. The fenamates also inhibit myometrial prostaglandin binding to its receptor (Figure 16.2).

The primary site of action of the non-steroidal anti-inflammatory drugs used for dysmenorrhoea is at the cyclo-oxygenase step, prior to

Table 16.3 Prostaglandin synthetase inhibitor dosage regimens, as prescribed in the USA (from ref. 43)

PGSIs	Dosage regimens
Aspirin	650 mg q.4 h
Indomethacin	25 mg t.i.d.
Mefenamic acid	500 mg, then 250 mg q.6 h
Ibuprofen	400 mg q.i.d.
Naproxen	500 mg, then 250 mg q.4–6 h
Naproxen sodium	550 mg, then 275 mg q.4–6 h
Piroxicam	20 mg/day

endoperoxide formation (see Figure 16.2). This means that all classes of products that are formed by way of the endoperoxides should be affected. It is interesting to note, however, that prostaglandin production is not totally eliminated by these drugs[42]. Drugs which antagonize the action of prostaglandins by direct blocking action at the receptor site should provide more effective and faster relief of dysmenorrhoea.

Because the non-steroidal anti-inflammatory drugs block the initial steps in the arachidonic acid cascade, these drugs are indiscriminant in their action, suppressing prostaglandin biosynthesis as well as the synthesis of the thromboxanes and prostacyclins. The NSAIDs are not specific for the uterus, and the ability of the various drugs to reach and act within the uterus can be an important variable.

Four different chemical groups of the non-steroidal anti-inflammatory drugs, all of which inhibit prostaglandin synthetase, have been evaluated in the treatment of dysmenorrhoea. Though each of these four groups of compounds expresses its action by interfering with prostaglandin formation, the precise mechanisms by which these agents exert their actions differ in certain respects. Table 16.3 lists the

Table 16.4 Relief obtained with prostaglandin synthetase inhibitors in primary dysmenorrhoea (double-blind crossover studies only) (from refs. 43–45)

	Moderate or complete relief (%)
Aspirin	21–30
Ibuprofen	61–100
Indomethacin	65–75
Mefenamic acid	43–100
Naproxen	61–88
Naproxen sodium	70–86

usual dosage regimens for many of the drugs, while Table 16.4 gives typically reported responses to the relief of pain in primary dysmenorrhoea.

Acetylsalicylic acid (aspirin)

Acetylsalicylic acid expresses its action by direct acetylation of the cyclo-oxygenase protein at the active site[46]. This is the only compound of the non-steroidal anti-inflammatory drugs that has an active acetyl group, and thus acts by this mechanism. These compounds generally have been found to be of relatively little benefit in the treatment of dysmenorrhoea in doses up to 650 mg every 4-6 h, as needed. In one study of adolescent girls, however, aspirin provided greater relief than placebo, but whether this effect was because the girls began their medication 2-3 days before the onset of menses or because adolescents might be more responsive to therapy was unclear[39].

Indomethacin

Indomethacin is an extensively studied indoleacetic acid derivative[39,47] that has proven effective in the treatment of dysmenorrhoea (Table 16.4). It is potentially toxic, however, showing a relatively high incidence of side-effects (Table 16.5). Nearly 20-30% of patients taking

Table 16.5 Prostaglandin synthetase-related adverse effects by drugs for dysmenorrhoea[43]

PGSIs	Incidence of reported side-effects* (%)
Aspirin	0–10
Indomethacin	0–56
Mefenamic acid	6–10
Ibuprofen	0–12
Naproxen	0–21
Naproxen sodium	6–15

* Double-blind crossover studies only

25 mg three times a day experience symptoms of headache, drowsiness and/or gastrointestinal symptoms. When the dosage is at 50 mg, this may increase to 50%[8]. Because of the high incidence of dose-related side-effects, indomethacin cannot be recommended for routine dysmenorrhoeic therapy.

Fenamates

The fenamates[8,39,48] are anthranilic acid derivatives and have proven effective for the treatment of dysmenorrhoea (Table 16.4). Flufenamic acid, mefenamic acid and tolfenamic acid are commonly utilized members of the fenamate group of compounds. Flufenamic acid was used early as a prostaglandin synthesis inhibitor specifically for the treatment of dysmenorrhoea. Flufenamic acid has been very effective at doses of 200 mg, three times a day. Side-effects are generally minor. Mefenamic acid, at doses of 500 mg three times a day or 250 mg four times a day, also provides significant relief in dysmenorrhoea. Tolfenamic acid, in dosages of 133 mg three times a day beginning 2 days before the onset of menses, has given results comparable to indomethacin, i.e. affording relief in 88% of treated cycles with no reported side-effects.

Besides inhibiting prostaglandin formation, these compounds also antagonize prostaglandin activity on the myometrium, presumably by competition at the receptor site. Similar side-effects are reported for these drugs as are for all of the prostaglandin synthetase inhibitors, but mefanamic acid additionally seems more prone to cause an allergic-type of diarrhoea. In susceptible patients, its use is therefore limited.

Arylpropionic acids

The arylpropionic acids, ibuprofen and naproxen, are probably the most commonly used prostaglandin synthetase inhibitors (PGSI) for treatment of dysmenorrhoea. Ketoprofen is also a member of this group. Naproxen sodium was the first PGSI specifically approved for the treatment of dysmenorrhoea. Both naproxen and naproxen sodium should be equally effective therapeutic agents once they are in the bloodstream, since the acid (naproxen) has a pK_a of 4.2 and, therefore, can be expected to nearly completely dissociate to the salt form (naproxen sodium) in the blood[49].

High initial doses are often prescribed both for naproxen and naproxen sodium (Table 16.3). These drugs may need to be taken every 6 h in contrast to the regimen for these drugs in conditions other than dysmenorrhoea. At least one report suggests that dosages on an as-needed basis may also provide adequate relief[43]. Higher doses than those indicated in Table 16.3 are not recommended due to the increased incidence of gastrointestinal side-effects. However, if the first medication cycle does not provide relief, the starting dose for the next cycle can be increased by 50 or 100%[43].

The only demonstrable difference between naproxen and its sodium salt as they are administered is purely kinetic[49]. When naproxen sodium comes in contact with stomach acid, it acidifies and precipitates as very fine particles of naproxen, whereas naproxen *per se* remains in the same particle size as when it was administered. Consequently, the

very fine particles of naproxen produced in the stomach when naproxen sodium is administered provide a larger surface area for absorption. The result is a more rapid absorption and higher initial
plasma levels. Naproxen sodium reaches peak concentrations at about
40 min after administration; in contrast, peak concentration is not
reached until about 2 h after administration of naproxen[49]. More rapid
relief of pain should be consistent with the high, earlier blood levels
from naproxen sodium. Both naproxen and naproxen sodium have
proven to be highly effective in dysmenorrhoea (Table 16.4).

Ibuprofen, 400 mg t.i.d., has also been effective in dysmenorrhoea
and is relatively free of side-effects. When a prophylactic regimen of
ibuprofen was used (400 mg, four times per day), all the subjects reported good to excellent relief. A comparison of prophylactic treatment
with treatment beginning on day 1 of menses showed no difference in
the effectiveness of the therapy. All reports of this drug seem very
favourable.

Ketoprofen administered in a dose of 50 mg three times per day has
been found to be equally effective compared to indomethacin administered at a dose of 27 mg, three times per day, starting on day 1 of
menses for a maximum of 4 days. Nearly 90% of the patients were
afforded relief; only mild side-effects were recorded[50].

Calcium antagonists

As previously mentioned, prostaglandin action on the myometrium is
inextricably linked to calcium translocation and binding. This provides
the rationale for the use of calcium antagonists for the treatment of
dysmenorrhoea, as these agents prevent the transfer of calcium across
the cellular membrane and thereby inhibit uterine contractions exac-
erbated or dependent upon prostaglandins.

Both electrical and mechanical activities of the myometrium are
affected by calcium and by $PGF_{2\alpha}$. $PGF_{2\alpha}$ acts at the cellular level to
increase calcium ion transport across the myometrial cellular
membrane and, possibly, to reduce protein binding of calcium within
the cell. These changes directly or indirectly alter cellular events
effected through cAMP mechanisms or alter cellular metabolism by
specific changes in enzyme activity. The net result of this $PGF_{2\alpha}$ action
is to cause an increase in free calcium and trigger contractions of the
myofibrils. If the pain of dysmenorrhoea is caused by uterine contractions not only the drugs that inhibit prostaglandin synthesis but also
other tocolytic agents should alleviate it.

The calcium antagonist, nifedipine, has proven effective for the treatment of primary dysmenorrhoea because of its ability to inhibit myometrial hypercontractility[51,52]. Oral doses of between 20 and 40 mg

suppress all uterine contractions and pain within 10–30 min[53]. Uterine contractility recordings demonstrate that nifedipine is more rapid in its effect than are the prostaglandin synthetase inhibitors[54].

The use of this type of tocolytic agent for dysmenorrhoea is limited because of side-effects which often are severe. Flushes and headache are typical and may be related to a reduction in calcium-dependent blood vessel tonus or to alterations of other calcium-dependent mechanisms of the vascular system.

MANAGEMENT*

The causes of pelvic pain at the time of menstruation are varied. Thus, an accurate and thorough history and complete examination are proper first steps to management (Figure 16.3). Particular attention is directed to the characteristics of the pain and to its location[17,22,24,55]. Under certain circumstances, the history and physical examination may indicate that laparoscopy would be most useful early in treatment. For example, a patient with infertility presenting with dysmenorrhoea may best have laparoscopy before being given a trial of antiprostaglandin medication, since the cause of her dysmenorrhoea may be related to the cause of her infertility. The physician must always bear in mind that treatment of dysmenorrhoea should be highly individualized, taking into consideration the patient's history and physical findings, the severity of her symptoms, and the potential risks and side-effects associated with treatment.

A diagnosis of primary dysmenorrhoea suggests two possible avenues of treatment, i.e. combination birth control pills or prostaglandin synthetase inhibitors. The choice of treatment is based on the patient's need or desire for contraception. Patients with primary dysmenorrhoea who want contraception should be given a trial of combination oral contraceptives. These contraceptives are effective in the therapy of primary dysmenorrhoea, providing relief for more than 90% of the women treated. They alter uterine motility, and thus may decrease the sensitivity of the myometrium to prostaglandins. Oral contraceptives also decrease total endometrial mass, the volume of the menstrual fluid, and prostaglandin production and accumulation in the endometrium[9]. The combination type oral contraceptive should be tried for 2–4 months and, if successful in eliminating pelvic pain, this therapy should be continued. If the oral contraceptives are not successful, then an appropriate PGSI should be prescribed supplementally. If relief is incomplete, the PGSI dose may be increased to the maximum level. If

---

*Editor's note: generic terms are used in this section

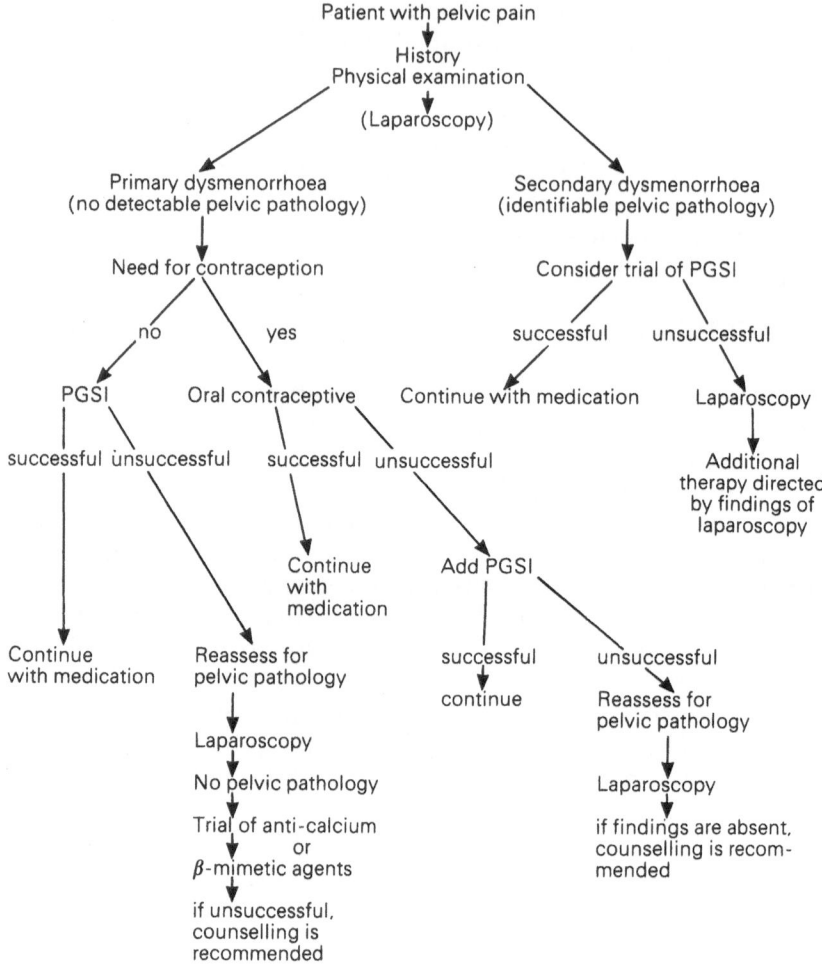

Figure 16.3   Scheme for the management of the patient with pelvic pain

there is still no improvement, a change to a different type of PGSI may prove beneficial.

Without a specific preference or need on the part of the patient for birth control, prostaglandin synthetase inhibitors are the drugs of choice. These agents have few side-effects and only need to be taken for 2–3 days of a menstrual cycle. For sexually inactive women, this brief exposure to medication is of more practical benefit than taking drugs 3 weeks of every month. If a patient does not respond to the PGSIs, then it is appropriate to consider a trial of the anti-calcium agents. If these latter agents fail to elicit a favourable response, laparoscopy should be initiated to rule out any pelvic pathology. If laparoscopy demonstrates the absence of pelvic pathology, then a re-evaluation of the patient's psychosocial situation may be of use[17].

Secondary dysmenorrhoea is treated by ascertaining its aetiology and directing an appropriate remedy to the specific problem. Although the role of prostaglandins in secondary dysmenorrhoea is still equivocal, a trial of PGSIs is warranted before proceeding to more aggressive treatment. If PGSIs are successful, then therapy with these agents should be continued. However, if PGSIs are not successful, laparoscopy is warranted. The findings at laparoscopy should dictate the appropriate surgical or medical treatment[22].

The criteria for choosing the appropriate PGSI in the treatment of primary dysmenorrhoea depends on a number of factors[16]. The drug of choice should be effective in inhibiting endometrial prostaglandin synthesis prior to cyclic endoperoxide formation. This not only decreases $PGF_{2\alpha}$ production, but it also eliminates possible participation of the thromboxanes and prostacyclins from contributing to the dysmenorrhoeic symptoms. If possible, the drug should also block myometrial prostaglandin receptors, be rapidly absorbed and quickly achieve therapeutic blood levels, have minimal, tolerable or inconsequential side-effects and be safe for long-term use. Of the drugs approved in the United States for the treatment of dysmenorrhoea, the following best satisfy these criteria: ibuprofen, naproxen sodium and mefenamic acid.

The vast majority (70–80%) of dysmenorrhoeic women have decreased symptoms with almost all of the PGSIs[44,56]. Aspirin (acetylsalicylic acid) is not very effective in controlling menstrual pain, and propoxyphene appears to be less effective than the other PGSIs[43]. These exceptions notwithstanding, all of the other PGSIs are nearly equally effective. Choices for prescribed use are made, therefore, on the other considerations such as patient tolerance and rate of uptake. Comparative studies have shown that indomethacin has higher rates of side-effects than ibuprofen, although both drugs are equally efficacious in relieving pain[57]. Ibuprofen and naproxen do not differ in efficacy, rates of side-effects or in average time of onset or relief of symptoms (about 1 h)[58], but naproxen sodium has a quicker onset of action than does naproxen[49]. Dosage schedules should be individualized and physicians need to become familiar with more than one drug that offers relief in dysmenorrhoea in order to provide patients the best chance for effective and appropriate treatment.

It is not necessary to institute pretreatment with PGSIs before the menstrual flow to achieve relief of primary dysmenorrhoea. With ibuprofen, for example, there is no significant difference in menstrual fluid prostaglandin levels whether the drug is begun 3 days before the onset of menses or at its onset. The absorption of orally administered ibuprofen is sufficiently rapid so that 400 mg gives peak blood levels within 20-30 min after ingestion. Obviously the faster the absorption, the faster the relief from pain. Also, naproxen sodium in doses of 275 mg q.i.d. for 3-5 days is effective even when taken at the onset of pain, since its action is also very rapid, i.e. within 1 h[49]. Pretreatment is contraindicated because it carries the risk of drug exposure in early pregnancy.

Medication should be prescribed for the first 2-3 days of menstruation. The rationale for this is to suppress prostaglandin production at the time of its maximal release during the first 48 h of menstrual flow[59]. If the relief from pain is not complete within hours during the first cycle with the PGSIs, the starting dose can be boosted 50-100% at the next cycle. After the initial increase, however, the maintenance dose should be kept essentially the same as before[16]. Doses higher than those indicated in Table 16.3 are not recommended because of the potential for gastrointestinal side-effects. PGSIs should be used for several cycles before judged to be ineffective. If a second PGSI is described after an ineffective response to the initial drug, the replacement should be of a different structural class. If the patient continues to fail to respond to the PGSIs, secondary dysmenorrhoea should be suspected and necessary measures for diagnosis be implemented.

Similar side-effects have been reported for all PGSIs, i.e. gastrointestinal discomfort, nausea, diarrhoea, headache and visual disturbances[43]. All PGSIs potentially can decrease menstrual flow and delay its onset if taken as pretreatment. The side-effects may be identical to or resemble the symptoms of dysmenorrhoea and make it difficult to adduce whether complaints are related to the condition or to the drug. Gastrointestinal symptoms may also reflect a lack of drug efficacy rather than the occurrence of side-effects. The intravaginal administration of the PGSI may assist in decreasing the incidence of side-effects if they are a problem[59]. Women who have a history of asthma or ulcers of the gastrointestinal tract should not be given prostaglandin synthetase inhibitors (or they should be followed carefully during therapy) because these conditions may be aggravated. The side-effects of the PGSIs and antagonists, however, are generally relatively mild and are usually very well tolerated.

Since prostaglandin synthetase inhibitors have been effective even in women who have not obtained relief from menstrual pain by using oral contraceptives, they have become the preferred method of treatment of primary dysmenorrhoea. This treatment is rational, effective with few side-effects and only required for a few days during each cycle. This should be the first type of therapy used in the patient with suspected primary dysmenorrhoea.

FUTURE DIRECTIONS

Primary dysmenorrhoea is a condition that requires better definition. Our knowledge of its pathophysiology is in good part based on circumstantial evidence, and in the future we will need to increase our basic understanding of this condition. Drug design has become sufficiently sophisticated so that it should be possible to block specific pathways or receptor sites. To benefit from this level of sophistication, however, requires that we understand the precise nature of dysmenorrhoea. The specific role of endoperoxides, thromboxanes or prostacyclins in the genesis of menstrual pain requires clarification. Current therapy is non-selective both biochemically and physiologically, and its action is not restricted to the uterus.

As the biochemistry of the mechanism of pain is further exposed, it may be possible to treat the pain more specifically, particularly in those patients where pain of undetermined cause persists. Continued progress in understanding the way neuropeptides and opiate receptors function in the pain process may help explain the inexplicable failures of current modes of therapy.

References

1 Pickles, V. R. (1957). A plain-muscle stimulant in the myometrium. *Nature (London)*, **180**, 1198

2 Williams, K. I. and Vane, J. R. (1975). Inhibition of uterine motility: The possible role of the prostaglandins and aspirin-like drugs. *Pharmacol. Therap.*, **1**, 89

3 Henzl, M. R., Smith, R. E., Boost, G. and Tyler, E. T. (1972). Lysosomal concept of menstrual bleeding in humans. *J. Clin. Endocrinol. Metab.*, **34**, 860

4 Pickles, V. R. (1967). Prostaglandins in human endometrium. *Int. J. Fertil.*, **12**, 335

5 Karim, S. M. M. (1972). Physiological role of prostaglandins in the control of parturition and menstruation. *J. Reprod. Fertil. Suppl.*, **16**, 105

6 Singh, E. J., Baccarini, I. M. and Zuspan, F. P., (1975). Levels of prostaglandin $F_{2\alpha}$ and $E_{2\alpha}$ in human endometrium during the menstrual cycle. *Am. J. Obstet. Gynecol.*, **121**, 1003

7 Turksoy, R. N. and Safaii, H. S. (1975). Immediate effect of prostaglandin $F_{2\alpha}$ during the luteal phase of the menstrual cycle. *Fertil. Steril.*, **26**, 634

8 Chan, W. Y. (1983). Prostaglandins and nonsteroidal anti-inflammatory drugs in dysmenorrhoea. *Ann. Rev. Pharmacol.*, **23**, 131–49

9 Dawood, M. Y. (1983). Dysmenorrhoea. *Clin. Obstet. Gynecol.*, **26**, 719–27

10 Ylikorkala, O. and Dawood, M. Y. (1978). New concepts in dysmenorrhoea. *Am. J. Obstet. Gynecol.*, **130**, 833

11 Rosenwaks, Z. and Seegar-Jones, G. (1980). Menstrual pain. Its origins and pathogenesis. *J. Reprod. Med.*, 25, 207

12 Dawood, M. Y. (1981). Hormones, prostaglandins and dysmenorrhoea. In Dawood, M. Y. (ed.) *Dysmenorrhoea.* pp. 21–52. (Baltimore: Williams and Wilkins)

13 Pickles, V. R., Hall, W. J., Best, F. A. and Smith, G. N. (1965). Prostaglandins in endometrium and menstrual fluid from normal and dysmenorrhoeic subjects. *Br. J. Obstet. Gynecol.*, 72, 185–92

14 Lundstrom, V. (1981). Uterine activity during the normal cycle and dysmenorrhoea. In Dawood, M. Y. (ed.) *Dysmenorrhoea.* pp. 53–74. (Baltimore: Williams and Wilkins)

15 Csapo, A. I. (1980). A rationale for the treatment of dysmenorrhoea. *J. Reprod. Med.*, 25, 213–21

16 Dawood, M. Y. (1981). Overall approach to the management of dysmenorrhea. In Dawood, M.Y. (ed.) *Dysmenorrhea.* pp. 261–79. (Baltimore: Williams and Wilkins)

17 Cox, D. J. and Santirocco, L. L. (1981). Psychological and behavioural factors in dysmenorrhoea. In Dawood, M. Y. (ed.) *Dysmenorrhea.* pp. 75–93. (Baltimore: Williams and Wilkins)

18 Akerlund, M., Andersson, K. E. and Ingemasson, I. (1976). Effects of terbutaline on myometrial activity, uterine blood flow and lower abdomen pain in women with primary dysmenorrhoea. *Br. J. Obstet. Gynecol.*, 83, 673

19 Akerland, M. (1979). Pathophysiology of dysmenorrhoea. *Acta Obstet. Gynecol.*, 87, 27

20 Vane, J. R. (1976). The mode of action of aspirin and similar compounds. *J. Allergy Clin. Immunol.*, 58, 691

21 Jacobsen, J., Cavalli-Bjorkman, K., Lundstrom, V., Nilsson, B. and Norbeck, M. (1979). Prostaglandin synthetase inhibitors: a survey and personal clinical experience. *Acta Obstet. Gynecol. Scand. Suppl.*, 87, 73

22 Laros, R. K., Jr. (1981). Secondary dysmenorrhoea (excluding endometriosis). In Dawood, M. Y. (ed.) *Dysmenorrhea.* pp. 155–64. (Williams and Wilkins)

23 Lundstrom, V. (1978). Treatment of primary dysmenorrhea with prostaglandin synthetase inhibitors. A promising therapeutic alternative. *Acta Obstet. Gynecol. Scand.*, 57, 421

24 Lamb, E. J. (1981). Clinical features of dysmenorrhoea. In Dawood, M. Y. (ed.) *Dysmenorrhea.* pp. 107–29. (Williams and Wilkins)

25 Schneider, G. T. (1980). In discussion of Malinak, I. R., Buttram, V. C., Elias S. and Simpson, J. L. Heritable aspects of endometriosis. II. Clinical characteristics of familial endometriosis. *Am. J. Obstet. Gynecol.*, 137, 332

26 Chaudhuri, G. (1971). Intrauterine device: Possible role of prostaglandins. *Lancet*, 1, 480

27 Sagiroglu, N. and Sagiroglu, E. (1970). Biologic mode of action of the Lippes loop in intrauterine contraception. *Am. J. Obstet. Gynecol.*, 106, 506

28 Bengtsson, L. P. and Moawad, A. H. (1966). Lippes loop and myometrial activity. *Lancet*, 1, 146

29 Bengtsson, L. P. and Moawad, A. H. (1967). The effect of the Lippes loop on human myometrial activity. *Am. J. Obstet. Gynecol.*, **98**, 957

30 Dawood, M. Y. and Birnbaum, S. J. (1975). Unilateral tubo-ovarian abscess and intrauterine contraceptive device. *Obstet. Gynecol.*, **46**, 429

31 Taylor, E. S., McMillan, J. H., Greer, B. E., Droegemueller, W. and Thompson, H. E. (1975). The intrauterine device and tubo-ovarian abscess. *Am. J. Obstet. Gynecol.*, **123**, 338

32 Trobough, G. E. (1978). Pelvic pain and the IUD. *J. Reprod. Med.*, **20**, 167

33 Lee, R. E. and Lands, W. E. M. (1972). Co-factors in the biosynthesis of prostaglandin $F_{1\alpha}$ and $F_{2\alpha}$. *Biochim. Biophys. Acta*, **260**, 203

34 Strazza, J. A., Bloomfield, N. J. and Ressetar, M. (1953). Butazolidine in the treatment of arthritis. *J. Med. Soc. N. J.*, **50**, 333

35 Black, E. F. E. (1958). The treatment of dysmenorrhea with phenylbutazone. *Can. Med. Assoc. J.*, **79**, 752

36 Vane, J. R. (1971). Inhibition of prostaglandin synthesis as a mechanism of action for aspirin-like drugs. *Nature New Biol.*, **231**, 232

37 Flower, R. J. and Vane, J. R. (1974). Inhibition of prostaglandin biosynthesis. *Biochim. Pharmacol.*, **23**, 1439

38 Schwartz, A., Zor, U., Lindner, R. and Naor, S. (1974). Primary dysmenorrhea, alleviation by an inhibitor of prostaglandin synthesis and action. *Obstet. Gynecol*, **44**, 709

39 Chan, W. Y. (1981). Prostaglandin inhibitors and antagonists in dysmenorrhea therapy. In Dawood, M. Y. (ed.), *Dysmenorrhea*. pp. 209–245. (Baltimore: Williams and Wilkins)

40 Flower, R. J. (1979). Prostaglandins and related compounds. In Vane and Ferreira (eds.), *Handbook of Experimental Pharmacology* 50(I). pp. 374–422. (New York: Springer/Verlag)

41 Shen, T. Y. (1979). Prostaglandin synthetase inhibitors I. In *Handbook of Experimental Pharmacology*, Vol. 50 pp. 305–347

42 Samuelsson, B. (1973). Quantitative aspects on prostaglandin synthesis in man. In Bergstrom and Bernhard (eds.), *Advances in the Biosciences*, Vol. 9, pp. 7–14. (New York: Pergamon Press)

43 Wenzloff, N. J. and Shimp, L. (1984). Therapeutic management of primary dysmenorrhea. *Drug Intelligence and Clin. Pharm.*, **18**, 22

44 Dingfelder, J. R. (1981). Primary dysmenorrhea treatment with prostaglandin inhibitors: A review. *Am. J. Obstet. Gynecol.*, **140**, 874

45 Owen, P. R. (1984). Prostaglandin synthetase inhibitors in the treatment of primary dysmenorrhea. *Am. J. Obstet. Gynecol.*, **148**, 96

46 Roth, G. J., Stanford, N. and Majerus. P. W. (1975). Acetylation of prostaglandin synthetase by aspirin. *Proc. Natl. Acad. Sci.*, **73**, 3073

47 Shen, T.-Y. and Winter, C. A. (1977). Chemical and biological studies on indomethacin, sulindac and their analogs. In Harper and Simmonds (eds.), *Advances in Drug Research*. pp. 186–211. (New York: Academic Press)

48 Nickander, R., McMahon, F. G. and Ridolfo, A. S. (1979). Nonsteroidal anti-inflammatory agents. *Ann. Rev. Pharmacol. Toxicol.*, **19**, 469

49 Segre E. J. (1980). Naproxen sodium (Anaprox). Pharmacology, pharmacokinetics and drug interactions. *J. Reprod. Med.*, **25**, 222

50 Kauppila, A., Puolakka, J. and Ylikorkala, O. (1979). The relief of primary dysmenorrhea by ketoprofen and indomethacin. *Prostaglandins*, **18**, 647

51 Sandahl, B., Ulmsten, U. and Andersson, K.-E. (1979). Trial of calcium antagonist nifedipine in the treatment of primary dysmenorrhoea. *Arch. Gynecol.*, **227**, 147

52 Ulmsten, U., Andersson, K.-E. and Forman, A. (1978). Relaxing effects of nifedipine on the non-pregnant human uterus *in vitro* and *in vivo*. *Obstet, Gynecol.*, **52**, 436

53 Andersson, K.-E. and Ulmsten, U. (1978). Effects of nifedipine on myometrial activity and lower abdominal pain in women with primary dysmenorrhea. *Br. J. Obstet. Gynecol.*, **85**, 142

54 Fuchs, F. (1981). Uterine tocolytic agents in primary dysmenorrhea. In Dawood, M.Y. (ed.), *Dysmenorrhea*. pp. 247-60. (Baltimore: Williams and Wilkins)

55 Heinrichs, W.L. and Adamson, G.D. (1980). A practical approach to the patient with dysmenorrhea. *J. Reprod. Med.*, **25**, 236

56 MacKinnon, G.L. and Parker, W.A. (1982). Current concepts – the management of primary dysmenorrhea. *Can. Pharm. J.*, **1150**, 3

57 Halbert, D.R. and Demers, L.M. (1978). A clinical trial of indomethacin and ibuprofen in dysmenorrhea. *J. Reprod. Med.*, **21**, 219

58 Kajanoja, P. and Vesanto, T. (1979). Naproxen and indomethacin in the treatment of primary dysmenorrhea. *Acta Obstet. Gynecol, Scand.*, **87** (Suppl. 1), 87

59 Chan, W.Y., Dawood, M.Y. and Fuchs, F. (1979). Relief of dysmenorrhea with the prostaglandin synthetase inhibitor ibuprofen. Effect on prostaglandin levels in menstrual fluid. *Am. J. Obstet. Gynecol.*, **135**, 102

SECTION IV
FUTURE APPLICATIONS

17
Pregnancy-induced hypertension

F. BROUGHTON PIPKIN and E. M. SYMONDS

INTRODUCTION

The best approach to the management of pregnancy-induced hypertension (PIH) would be its prevention (Table 17.1). Unfortunately, this goal will continue to elude us until we understand the primary pathogenesis of this condition. Some hypotheses are beginning to emerge. Prominent among them is the increasing conviction that the prostaglandins play a major causative role in the vasoconstriction which results in the cardinal feature of PIH, namely raised blood pressure.

Table 17.1 Primary and secondary objectives of care in pregnancy-induced hypertension

Symptom	Primary aim	Secondary aim
Vasoconstriction-raised bp	Prevent onset –	Treat, with successful
Widespread DIC	impossible at present	outcome
Deteriorating renal function		

This chapter provides a brief review of the production, concentration and possible role of the prostaglandins in normal pregnancy, and then considers the same parameters in PIH. Subsequently, a more detailed consideration of the role of locally-produced and circulating prostaglandins as regulators of vascular reactivity includes their interrelationship with the renin–angiotensin system in the regulation of blood flow in specific vascular beds. Their effects on platelet activity and disseminated intravascular coagulation are considered next, before the formulation of a unifying hypothesis regarding their possible role in the various forms of PIH. This is followed by a discussion of the theoretical means by which therapeutic manipulation of prostaglandin concentrations could be brought about and the results of animal and preliminary human studies in this area. A final section considers some theoretical drawbacks to such manipulations in pregnancy and looks to the future of such treatment.

PROSTAGLANDINS IN NORMAL PREGNANCY

Prostaglandin E_2 (PGE_2) and prostaglandin F (PGF) are among the earliest hormones produced by the mammalian blastocyst. Both are present and can be produced in culture prior to implantation[1]. Arachidonic acid (AA), the essential fatty acid (EFA) precursor of the prostaglandins, forms some 20% of the lipids in the human amnion and chorion, compared with less than 1% in adult mesenteric adipose tissue. It appears that the AA in the amnion is primarily derived from EFAs in the amniotic fluid, while that in chorion may also have a direct maternal source through the decidua vera[2]. Short-term cultures of first trimester human placental villi can synthesize prostacyclin (PGI_2) measured as its metabolite, 6-keto-prostaglandin $F_{1\alpha}$ (6-keto-$PGF_{1\alpha}$), although production is lower than at term[3]. There is a suggestion that placental production of PGE and PGF is maximal in the second trimester[4], but this has yet to be confirmed. The amnion, chorion, decidua and placental tissue contain and are capable of synthesizing PGE, PGF, the 13,14-dihydro-15-keto PGF metabolite (PGFM) and PGI_2[5-7]. The fetal membranes produce mainly PGE and 6-keto-$PGF_{1\alpha}$, while the decidua produces mainly PGF and thromboxane. These differences between predominantly vasodilator and vasoconstrictor production undoubtedly have functional significance.

Human umbilical smooth muscle synthesizes immunoreactive PGE[8]. The generation of PGI_2 by human umbilical arteries (UA) and placental veins (PV) is substantially higher than by adult blood vessels[9]. Umbilical vessels also produce more PGI_2 than do either the amnion or placental vessels[10,11]. 6-keto-$PGF_{1\alpha}$ appears to be the only product of arachidonic acid metabolism in human UA microsomes[12]. The conversion of [^{14}C]AA to 6-keto-$PGF_{1\alpha}$ by the UA does not change with gestation age during the last trimester of pregnancy[13]. A recent report[14] identified Mead's acid (5,8,11-eicosatrienoic acid) in umbilical arteries. This is rarely found in adults, and since it is not a substrate for cyclo-oxygenase activity, it could limit the production of PGI_2 when present in high concentration by limiting the availability of the substrate, AA.

Thromboxane B_2 (TxB_2), the stable metabolite of thromboxane A_2 (TxA_2), and $PGF_{2\alpha}$ were only produced in smaller quantities (2·3%) by

Table 17.2 Summary of information regarding content of and synthetic capacity for various prostanoids and thromboxane in fetal and maternal tissues in normal pregnancy. Full references are quoted in the text

Tissue	Prostanoids/thromboxane	
Fetal placenta and membranes	PGE ⎫ PGF ⎭	production ?maximal in 2nd trimester or ↑ to term
	PGI_2	production rises to term
Umbilical smooth muscle	PGI_2	the primary prostanoid produced
	PGE	much less production
	TxB_2 ⎫ PGF_2 ⎭	minimal production
Decidua	PGF ⎫ Tx ⎭	predominant forms
	PGI_2	some production, rising to term

umbilical arteries[15]. Information on tissue production of prostanoids and thromboxanes in normal pregnancy is summarized in Table 17.2.

Umbilical cord serum contains less non-esterified fatty acids (NEFA) than does maternal, but there is a greater percentage of AA in the umbilical cord NEFA, phospholipids, triglycerides and cholesterol esters than in maternal serum[16]. PGE, PGF, PGFM and PGI_2 have all been measured in the umbilical circulation at birth and noted to be in higher concentrations than those in the maternal circulation[10,17,18]. There is some evidence that fetal plasma PGE increases towards term, while PGF and 6-keto-$PGF_{1\alpha}$ fall, the latter from very high second trimester levels, but there is a wide scatter of data[18]. After birth, plasma 6-keto-$PGF_{1\alpha}$ concentrations rise, while concentrations of PGE and PGF fall[7].

Concentrations of 6-keto-$PGF_{1\alpha}$ in amniotic fluid show a linear increase during the third trimester and are approximately five times higher than amniotic fluid TxB_2 concentrations which also rise as term approaches[19]. PGE_2 and $PGF_{2\alpha}$ are also present in high concentration in the amniotic fluid at term[20]. It is possible that the fetal kidney, in addition to the fetal membranes, decidua or placenta may be a source of these prostaglandins[20]. Biosynthesis of the primary PGs occurs in the renal medulla of the human fetus by at least 22 weeks gestation, and this production increases rapidly during the second trimester[21]. PGE production is consistently greater than that of PGF. This is reflected in the somewhat higher urinary PGE_2 and $PGF_{2\alpha}$ in the neonate delivered by Caesarean section and not subjected to the stress of labour[20]. Concentrations of PGE, PGF and 6-keto-$PGF_{1\alpha}$ all rise in amniotic fluid during labour[22], and similar rises are seen in the first voided urine of neonates following labour[20].

The onset of labour is associated with increases in both umbilical plasma and amniotic fluid concentrations of PGE_2, $PGF_{2\alpha}$, PGFM and 6-keto-$PGF_{1\alpha}$[20,23] compared with similar samples taken at elective Caesarean section. The production of PGE_2 by the umbilical artery and of 6-keto-$PGF_{1\alpha}$ by the amnion are also increased following vaginal delivery[24,25].

With regard to the mother, human myometrium and decidua can generate PGI_2 from at least the 15th week of gestation; this process increases towards term[5,26,27]. Uterine arteries are also capable[11] of the synthesis of PGI_2. The maternal plasma concentration of $PGF_{2\alpha}$ and PGFM rise significantly in early pregnancy but remain stable thereafter, only rising once again during labour[7,28,29]. In contrast, maternal concentrations of PGE do not rise significantly until the third trimester[28]. 6-keto-$PGF_{1\alpha}$ concentrations have been variously reported as increasing[30], remaining largely stable[7,29] and decreasing with increasing gestational age[31]. The latter study[31] was longitudinal, lending it more weight. The reported fall in concentrations of 6-keto-$PGF_{1\alpha}$ would be consistent with the fall in plasma stimulatory factor for PGI_2 reported in late pregnancy by Remuzzi et al.[32] Maternal plasma TxB_2 is increased during pregnancy[29,33], possibly increasing further towards term.

The maternal urinary excretion of PGE_2 rises in the first trimester[34,35] and appears to fall off towards term. The excretion of $PGF_{2\alpha}$ is also increased early in pregnancy and continues throughout[35], as is the excretion of PGI_2 metabolites[36].

A summary of information regarding plasma, amniotic and urinary concentrations of various prostaglandins and thromboxane is given in Table 17.3. At least in early pregnancy, there appears to be a preponderance of production of the primarily vasodilator, antiaggregatory prostaglandins in the fetoplacental and the maternal circulation. This makes sense from a teleological point of view. The fetoplacental circulation is characterized by vasodilation and low resistance; it is a high flow, low pressure system requiring a high cardiac output to maintain adequate tissue oxygenation in the face of a low PaO_2. The umbilical cord and placenta are without innervation, and their vascular tone must therefore be under hormonal control. Both PGE_1 and PGI_2 exert a biphasic effect on the human umbilical and chorionic plate vasculature; they cause relaxation at low doses and contractions at higher doses[37-39]. TxA_2, PGE_2 and $PGF_{2\alpha}$ are all constrictors of the umbilical vasculature and the perfused placental lobule[37,40]. Furthermore, prostaglandin synthetase inhibitors in vitro reduce the tone of human UA preparation. However, indirect evidence from human pregnancies suggests a direct association between the capacity of the

Table 17.3 Summary of information regarding plasma, amniotic fluid and urinary concentrations of various prostanoids and thromboxane in normal pregnancy. Full references are quoted in the text

Source		Prostanoids/thromboxane
Fetal plasma	PGE	>maternal, ↑ to term; ↓ after birth
	PGF	>maternal, ?↓ from 2nd trimester; ↓ after birth
	$6kF_{1\alpha}$	>maternal, ↓ from 2nd trimester; ↑ after birth
Maternal plasma	PGE	no significant ↑ until 3rd trimester
	$PGF_{2\alpha}$	↑ in 1st trimester; stable thereafter
	$6kF_{1\alpha}$	↑ in 1st trimester; stable thereafter
	TxB_2	↑ to term
Amniotic fluid	PGE	present in high concentration at term
	PGF	
	$6kF_{1\alpha}$	↑ over 3rd trimester; $\times 5 > TxB_2$
	TxB_2	↑ over 3rd trimester
Maternal urine	PGE_2	↑ rapidly in 1st trimester; ↓ towards term
	PGF_2	↑ rapidly in 1st trimester; ?stable to term
	$6kF_{1\alpha}$	↑ in 1st trimester; ?stable to term

umbilical arteries to synthesize PGI_2 and the umbilical blood flow measured ultrasonically shortly before delivery[41]. However, studies attempting to relate maternal placental flow as measured with a ^{133}Xe washout technique to simultaneously measured plasma concentrations of 6-keto-$PGF_{1\alpha}$ and TxB_2 failed to show any association[42]. This result should be interpreted with caution, however, since Tx certainly, and PGI_2 probably, act as local hormones, so that their circulating concentrations may be of little relevance in terms of their action within the placenta. Thus, the balance between the various prostaglandins may well determine vascular tone.

In the chronically prepared non-pregnant sheep, PGE_1 and PGE_2 both elicited increases in uterine blood flow (UBF) with PGE_2 exerting the greater effect[43]; in contrast, $PGF_{2\alpha}$ reduced UBF. In the pregnant sheep, indomethacin causes vasoconstriction in both the cotyledonary (placental) and extracotyledonary uterine vascular beds along with a decrease in uteroplacental blood flow, thus suggesting an overall vasodilator effect of the prostaglandins[44,45]. Indomethacin also reduces UBF in anaesthetized rabbits and dogs[46,47].

The high uteroplacental blood flow is partly a consequence of the replacement of musculoelastic tissue in the walls of the spiral arteries by fibrinoid material[48]. The arteries are thus 'damaged', and Wallenburg[49] has pointed out that although they might therefore be expected to be highly susceptible to platelet aggregation and thrombosis leading to placental infarction; this is an uncommon occurrence in normal pregnancy. It may be that PGI_2 production by the uterine arteries[11] normally prevents platelet aggregation and thrombosis in these 'damaged' spiral arteries. The antiaggregatory role of PGI_2 in pregnancy has yet to be fully determined, but may well be of considerable importance in the placental circulation. There is a small rise in the production of the pro-aggregatory TxA_2, which might explain the increased platelet reactivity and potential for thromboembolic complications in pregnancy, but it appears to be largely balanced by the increase in vasodilator prostaglandins in normal pregnancy.

It is likely that the prostaglandins, whether fetal or maternal in origin, are also concerned with the initiation of labour; this subject is discussed in detail elsewhere in this book.

The prostaglandins are not the only vasoactive hormones to be present in high concentrations in the maternal and fetoplacental circulations. In 1968, Symonds et al.[50] identified the presence of renin-secreting cells in human myometrium and chorion. Warren et al.[51] then showed that tissue cultures of human myometrium produced both active and inactive renin. The use of an antibody to pure human renin together with peroxidase:antiperoxidase techniques has now allowed the study of renin-secreting cells in the human uterus[52]. These studies have shown the presence of cells containing renin granules in clusters in the interarteriolar connective tissue within the myometrium adjacent to the endometrium, cells bearing a striking resemblance to those of the juxta-glomerular apparatus in the kidney. The clusters of renin-secreting cells around the spiral arterioles in the uterus appear analogous to the parallel structures around the afferent arterioles in the kidney and presumably may play some mediating role in the regulation of uterine blood flow.

The human chorion has also been shown to contain high concentrations of renin[53], and the proximity of the chorion and myometrium to the decidua gave rise to speculation that the cellular source of renin might arise within the decidua. However, Symonds et al.[54] showed, by karyotyping chorion from a male fetus, that there were independent sources of renin in the chorion and the myometrium, apparently with the same structure as those forms of renin produced in the kidney[55].

The mechanism of release of renin from genital tract sources is not clearly understood. Ferris et al.[56] showed in nephrectomized pregnant rabbits that reduction in uteroplacental blood flow resulted in the release of renin into the peripheral circulation, but the sensor mechanism for this process has not been identified. In a study on the generation of AI in the female genital tract, Craven et al.[57] showed that explants of human chorio-decidua generated AI in large quantities. This generation was not influenced[58] by perfusion with PGE_2, but was abolished by the addition of antirenin. These experiments indicate that both renin and its substrate are released from these tissues and that the vasoactive peptides can be generated *in situ*. Angiotensin converting enzyme can also be identified in the placenta and chorion[59]. Thus, the entire system could allow the local production of angiotensin II. If the renin–angiotensin system plays a significant role in the genital tract, then it is likely to do so at a local level and probably by its interaction at the cellular level with vasodilator prostaglandins.

Angiotensin II (AII), the most potent circulating vasoconstrictor yet described, shows regular plasma concentration changes during the menstrual cycle, peaking at around the time of ovulation[60]. Should conception occur, the concentration of AII continues to rise and does not stabilize until the third trimester (for reference, see Symonds, 1981[61]). In the non-pregnant adult, arterial concentrations of AII are only just below those capable of influencing arterial blood pressure[62]. Given these circumstances, it is pertinent to ask, why then are not all pregnant women hypertensive?

The answer appears to lie in an apparently specific diminution in pressor responsiveness to AII in pregnancy, noted some 20 years ago by Chesley *et al.*[62] and repeatedly confirmed since. Evidence from animal pregnancies that this resistance might be mediated via the vaso-dilator prostaglandins[64,65] was rapidly followed by studies in humans. The administration of indomethacin was associated with a decrease in the amount of AII required to evoke a specified pressor response[66,67]. Conversely, the administration of PGE_2 was associated with a diminution in pressor response to AII in pregnant, but not non-pregnant subjects[68]. A similar effect was noted with PGE_2[69] and PGI_2[70].

The vasodilator prostaglandins and the renin–angiotensin system appear to a large extent to be inter-regulatory in pregnancy as well as in the non-pregnant state. *In vitro* studies show that AII can stimulate PGE production in cultured human umbilical vein smooth muscle cells[8] and that prostaglandin synthetase inhibitors can inhibit AII-induced contraction of the bovine umbilical artery[71]. *In vivo* studies have included: (1) those on the effect of dietary deprivation of PG precursors on AII pressor responsiveness in pregnant rabbits[65]; (2) the stimulatory effect of infused AII or PGE concentrations in the uterine vein of the pregnant monkey[72,73]; (3) the effect of dietary supplementation in the pregnant woman with EFAS and co-factors on AII responsiveness[74]; (4) the stimulatory effect of PGE_2 on plasma renin concentration (PRC) in human pregnancy[75], and (5) the observation of positive associations between PRC and urinary PGE_2[35].

PROSTAGLANDINS IN HYPERTENSIVE PREGNANCY

Placental tissue from hypertensive pregnant women has been reported to have a lower PGE content and synthetic capacity than that from normotensive pregnant women[6,76,77]. Unfortunately, in none of these studies were the hypertensive women matched for gestational age with the controls and, in view of the changes in PG concentration and production in the third trimester (*vide supra*), the magnitude of the reported differences may be questioned. The ability of placental tissues to metabolize PGE_1 has also been reported to be less in PIH with the same provisos relating to gestation age[78]. This may be of significance in view of the biphasic effect of PGE_1, in that low concentrations act as vasodilators while high concentrations act as vasoconstrictors. The concentrations or production of PGF in placentae from hypertensive women have been variously described as lower[6], unchanged[77], or higher[76] than in normal pregnant women. The catabolic capacity of placentae from PIH pregnancies for $PGF_{2\alpha}$ appears to be diminished[77]. PGI_2 production in placental tissues has been reported to be diminished in PIH, while that of TxB_2 is increased[79] or unchanged[6].

There is general agreement that umbilical blood vessels both contain and synthesize less PGI_2 in hypertensive pregnancies than in normal pregnancies[13,80-83], the production of PGI_2 in umbilical arteries from hypertensive pregnancies being only equivalent to that in adult arterial specimens[84]. Downing et al.[81], on the basis of kinetic analysis of microsomal action on AA, suggested that there is a smaller quantity of enzyme complex present in umbilical arteries from hypertensive pregnancies rather than a decreased enzyme activity. Placental venous content and production of 6-keto-$PGE_{1\alpha}$ were lower than those in the umbilical arterial system and were also lower than in the corresponding vessels from normal pregnancies[80,82,85]. Subcutaneous and uterine maternal blood vessels also contain markedly less 6-keto-$PGF_{1\alpha}$ in hypertensive than normotensive pregnancies[85]. Indirect indices of factors influencing prostaglandin production can be difficult to interpret. Plasma concentrations of PGI_2-stimulating factor were raised in late pregnancy in six women with severe PIH compared with normals[32]. This could be a response to the lower PGI_2 production reported in PIH. On the other hand, serum concentrations of an endogenous inhibitor of PG synthesis have been found to be unchanged in PIH[86]. These data are summarized in Table 17.4.

Reliable radioimmunoassays for plasma 6-keto-$PGF_{1\alpha}$ and TxB_2 have become available only recently, and there is so far very little information concerning plasma concentrations of these two metabolites in PIH. One case report[87] showed a dramatic fall in maternal plasma concentrations of 6-keto-$PGF_{1\alpha}$ in association with rapidly

Table 17.4 Summary of information regarding content of, and/or synthetic capacity of fetal and maternal tissues for various prostanoids and thromboxane in PIH pregnancy. Full references are quoted in the text

Tissue	Prostanoids/thromboxane	
Fetal placenta	PGE	<normal pregnancy; PGE_1
and membranes	PGF	metabolism impaired data
		conflict; metabolism impaired
	$6kF_{1\alpha}$	<normal pregnancy
	TxB_2	>normal pregnancy
Umbilical and maternal	$6kF_{1\alpha}$	<normal pregnancy. ↓ umbilical
arterial smooth muscle		production also possible in EHT
		and IUGR

worsening PIH in the second trimester. Martensson and Wallenburg[88] showed a marked increase in uterine venous and umbilical venous TxB_2 concentrations in PIH by comparison with normal pregnancies, while 6-keto-$PGF_{1\alpha}$ concentrations were somewhat lower. The higher uterine than peripheral venous concentrations of both 6-keto-$PGF_{1\alpha}$ and TxB_2 in both PIH and normotensive women suggest a uteroplacental origin for at least some of the parent prostaglandins. The uterine venous:maternal peripheral venous concentration ratio for both metabolites was greater than 2 in the PIH group, but only just above 1 in the normal group; this observation may suggest an increased catabolism of both substances in PIH. The lower proportion of AA in the NEFA and triglyceride fractions of umbilical venous blood in PIH[16] could possibly contribute to a PGI_2:Tx imbalance, since the fetal circulation appears to be geared primarily to PGI_2 production.

The urinary excretion of PGEs is reduced in the third trimester in women with PIH or chronic hypertension[34,35], as is that of PGI_2 metabolites[36]. Indeed, one patient with eclampsia at 40 weeks of gestation had undetectable urinary PGE excretion on admission; however, levels climbed steeply after delivery[34]. Urinary excretion of $PGF_{2\alpha}$ does not differ in normal and hypertensive patients[35]. It is generally assumed that urinary PGE_2 reflects renal synthesis, since the majority of PGE_2 is metabolized in the lungs. These data suggest an overall impairment in vasodilator PG production in hypertensive pregnancy, rather than one confined to the fetoplacental unit.

Concentrations of 6-keto-$PGF_{1\alpha}$ are also reduced in the amniotic fluid in PIH in face of unchanged TxB_2 levels[19,89], again suggesting an impairment of production of the vasodilator form and a consequent tipping of the ratio towards vasoconstriction. These data are summarized in Table 17.5.

Table 17.5   Summary of information regarding plasma, amniotic fluid and urinary concentrations of various prostanoids and thromboxane in hypertensive pregnancy. Full references are quoted in the text

Source	Prostanoids/thromboxane	
Fetal plasma	$6kF_{1\alpha}$	<normal pregnancy
	TxB_2	>normal pregnancy
Maternal plasma	$6kF_{1\alpha}$	<normal pregnancy
	TxB_2	>normal pregnancy especially in uterine vein
Amniotic fluid	$6kF_{1\alpha}$	<normal pregnancy
	TxB_2	>normal pregnancy
Maternal urine	PGE	<normal pregnancy (but also in EHT)
	$6kF_{1\alpha}$	<normal pregnancy
	PGF	=normal pregnancy

PATHOGENESIS OF PREGNANCY-INDUCED HYPERTENSION

How may this information contribute to our understanding of the pathogenesis of PIH? In normal obstetric practice, the first symptom of PIH is that of hypertension. A blood pressure of 140/90 mmHg on at least two separate occasions after the 20th week of pregnancy in a woman previously normotensive should be classified as PIH. Arguments are also presented (see MacGillivray[90]) for using a rise of 15 or 20 mmHg diastolic, or 30 mmHg systolic pressure above non-pregnant values. These may well be useful cut-off points in populations having a lower basal blood pressure than is the case in a basically Caucasian population. They could, however, result in the overlooking of patients at risk, for example, in a black population with a higher incidence of hypertensive disease, since fetal mortality and morbidity begins to rise rapidly after a diastolic blood pressure of 90 mmHg has been reached[91].

One of the remarkable features of PIH is the rapidity with which the hypertension usually resolves after delivery. The classical treatment of severe PIH, that of urgent delivery, springs from this observation. And yet, weeks before the blood pressure reaches a point at which concern is expressed, other changes have already occurred (Table 17.6). Placental bed biopsy has shown that the normal secondary wave of trophoblast invasion which finally erodes the musculoelastic lining of the spiral arterioles in the second trimester is incomplete in hypertensive pregnancies[48,92]. This appears to be a consistent finding in PIH,

Table 17.6 Chronological changes of some measured variables in normal and hypertensive pregnancy

Outcome	Weeks gestation							
	12	16	20	24	28	32	36	40
(A) Normal	Spiral arteries widen BP at nadir		BP$_D$ starts to ↑		Vascular response to AII ↑			Birthweight, mortality, morbidity normal
(B) Mild/moderate PIH	Secondary trophoblast invasion fails Vascular response to AII ↑				25% distinguishable by MAP			
(C) Severe PIH	As (B)				50% distinguishable by MAP PV contracts ↑Plasma and blood viscosity, hct, fibrinogen Low birthweight; mortality + morbidity Problems of prematurity per se			

AII = angiotensin II; BP = blood pressure; BP$_D$ = diastolic blood pressure; MAP = mean arterial pressure; PV = plasma volume

but it has also been shown in cases of intrauterine growth retardation (IUGR) without PIH[93]. There is also some evidence to suggest an immunological deficit in PIH, with an absent or deficient maternal immune response to the fetus (for refs., see *Lancet*, 1980[94]), but the way in which this might directly impede trophoblast invasion of maternal tissues is not clear.

Another early change in PIH is that of the progressive loss of protection against the pressor effects of the raised circulating AII concentrations. Abdul-Karim and Assali[95] showed that a standard dose of AII given i.v. evoked a consistently greater pressor response postpartum than in the same women studied in the third trimester of pregnancy. This work was confirmed and extended by Chesley *et al.*[96] who showed, in addition, that the renal antidiuretic and antinatriuretic effects of AII were also less in pregnancy, this effect being more marked in the second trimester. The effect appeared to be specific for AII, and could not be demonstrated for noradrenaline[63,97]. Lumbers[97], who studied changing hand blood flow in response to AII in pregnant and non-pregnant patients, showed that there appeared to be a genuine change in reactivity of the vessels, not simply an alteration in baroreceptor or central activity. A longitudinal study of predominantly teenage black primigravidae showed that the blunting of response in pregnancies which remained normotensive was effectively complete by 14 weeks gestation, being maintained until 30 weeks after which it was progressively lost. Even so, at term, pressor responsiveness was still less than that of non-pregnant subjects[98]. However, patients who subsequently developed PIH showed a progressive loss of protection from 18 weeks which became statistically significant by the middle of the second trimester[98]. This finding provided confirmation of Chesley's observations[99] that women with PIH were more responsive to exogenous AII.

Various attempts have been made to identify the cause of the blunted pressor response to AII in normal pregnancy. Those of Gant's group were largely summarized in 1980[100]. Other authors have investigated the effects of L-DOPA, theophylline and sodium restriction on vascular responsiveness to AII in pregnancy[101-2]. The most consistent point of view implicates alterations in prostaglandin synthesis. O'Brien *et al.*[64] showed that the administration of PGE_2 blunted, while that of indomethacin augmented, the pressor response to AII in pregnant rabbits. Everett *et al.*[66] then showed that the administration of prostaglandin synthetase inhibitors in human pregnancy also augmented the pressor response to AII. On the other hand, the infusion of PGE_2, PGE_1 or PGI_2 was associated with a diminution in AII pressor response in second trimester human pregnancy[68,69,75]. Directly comparable experiments with other vasodilators have not been performed, but the effects of indomethacin and other prostaglandin synthetase inhibitors suggest that the blunting of response is PG-mediated. If so, then the subnormal production of vasodilator prostaglandins by vascular smooth muscle in PIH may be directly responsible for the vasoconstriction.

Another phenomenon which antedates the onset of clinically recognizable hypertension is that of a contracted plasma volume in patients who subsequently develop proteinuric PIH with IUGR[90]. However, hypertension alone is not associated with either a contracted plasma volume or a shift in the distribution of birth weight[90]. The level of hyperuricaemia reached as a consequence of the decreased uric acid clearance and its increased renal reabsorption in PIH is directly correlated with the contraction in plasma volume[104].

Suboptimal diameter of the spiral arteries of the placental bed is of little apparent functional significance in the early and middle part of the second trimester, since the uteroplacental blood flow is adequate to meet the requirements for growth of the fetus. Preliminary information suggests that an increase in resistance to flow in the arcuate arteries can be detected as early as 18 weeks gestation in pregnancies complicated by IUGR, using a Doppler technique[105]. Comparable measurements have yet to be made in pregnancies which subsequently become hypertensive, but it seems highly probable that a similar picture will emerge.

The narrowed spiral arteries present a progressively more damaging impediment to flow later in pregnancy, as fetal demands outstrip the ability of the narrowed vessels to deliver. Uteroplacental blood flow at normal term is of the order of 500 + ml/min. A reduction in diameter of the spiral vessels of only 10% would reduce this to 320 ml/min, since flow varies with the fourth power of the radius. Analysis of velocity waveforms of the uteroplacental circulation in established PIH does indeed show an appearance of increased pulsatility and lowered diastolic velocity characteristic of an increased resistance[105]. The measurement of actual uterine blood flow in humans presents many difficulties; such studies as exist confirm the suggestion of a decreased uteroplacental blood flow in PIH (for refs., see MacGillivray[90]).

Narrowing of the spiral arteries may not be unique to PIH, and not all the changes seen in such arteries in PIH are found in those from pregnancies with IUGR[106]. Scrutiny of the data relating to birthweights in mild to moderate PIH without proteinuria shows no excess incidence of low birthweight infants[90,91].

How then, if at all, may we link up the various early changes which occur in PIH to produce a unifying theory? It has been argued that mild PIH without proteinuria is a different entity from the severe, proteinuric, early-onset form. It seems to us, however, much more likely that they are opposite ends of a spectrum of disease. It may even be that the severe form is an example of a physiological response (i.e. mild PIH) breaking down.

Let us begin with an analogy. Renal artery stenosis, whether caused experimentally or naturally, is associated with an increase in renin release into the renal venous blood[107] with a consequent increase in systemic arterial AII concentrations, which themselves evoke an increase in systemic arterial pressure. The potentially constrictor effect of the raised AII levels in the renal circulation is partly countered by a secondary intrarenal release of PGE_2[108]. Administration of indomethacin prevents this release of PGE_2 and augments the renal vasoconstrictor action of the AII[108]. Thus, perfusion pressure to the kidney is maintained at a cost to the general circulation of a raised blood pressure.

It can be shown experimentally that reduction of blood flow to the uterus in anaesthetized pregnant dogs[109] or chronically cannulated pregnant monkeys[110] is associated with a marked increase of circulating AII and plasma renin activity (PRA). The monkeys developed both hypertension and proteinuria over the course of the study. Kokot and Cekanski[111] and Broughton Pipkin et al.[112] reported raised renin and AII concentrations in human uterine venous compared with peripheral venous blood at Caesarean section of hypertensive patients, while in normotensive patients the gradient was usually reversed.

Blockade of the RAS with the angiotensin-converting enzyme inhibitor, Captopril in pregnant rabbits resulted in a diminution in uterine blood flow (UBF) of some 30% and a marked fall in uterine venous PGE[112]. The AII receptor blocker, Saralasin, evoked comparable results[114]. Conversely, the administration of AII to pregnant dogs and monkeys was associated with a rise in UBF and in the uterine venous concentration of PGE[47,72]. Interestingly, Naden and Rosenfeld[115] showed AII to have a biphasic effect on uterine blood flow in the chronically cannulated pregnant ewe, in that it evoked a rise in UBF at lower, and a fall in UBF at higher, doses. It is possible that this may reflect the biphasic effect of PGE_1 on vascular smooth muscle (see above), since major stimulation of PGE_1 by AII might result in vasoconstrictor concentrations being produced.

There is thus evidence that AII and the vasodilator PGs can interact in the gravid uterus in the regulation of flow as they do in the kidney. Since PGs can themselves stimulate renin release in man[116], the stage is set for a 'benevolent circle'.

We can thus postulate that in the mild and moderate forms of PIH, the gradual retardation of the normal rise in UBF, as a consequence of the inadequate erosion of the spiral arteries, results in a stimulation of renin release and hence of increased AII concentration. This both increases systemic pressure and stimulates production of vasodilator

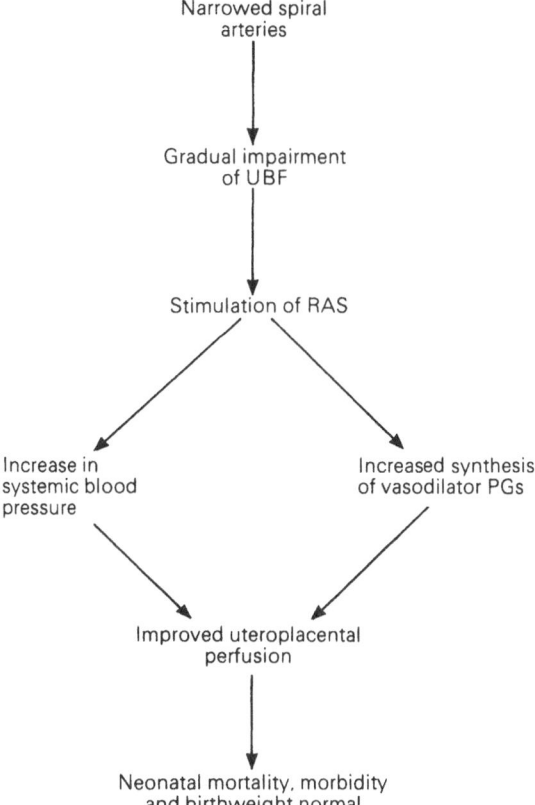

Figure 17.1   Schematic representation of proposed events in mild and moderate PIH

PGs, with an overall improvement in uteroplacental blood flow. There is then the ultimate production of an infant whose perinatal mortality and morbidity may even be marginally better than that of his normotensive cousins and whose birthweight is unimpaired[90]. On this hypothesis, the mild and moderate forms of PIH may be viewed as a physiological adaptation of the pathologically-narrowed spiral arteries. This is summarized in Figure 17.1.

The severe forms of the disease can be imagined as sequelae of the inability to synthesize the vasodilator prostaglandins in both fetal and maternal tissues in this condition. Human umbilical blood flow has been shown[41] to be positively associated with the umbilical arterial production of PGI_2, so that a fall in this capacity could be associated with a fall in umbilical blood flow. There are, of course, other consequences of a breakdown in vasodilator production, especially in terms of diminished renal function. The protection against the unopposed

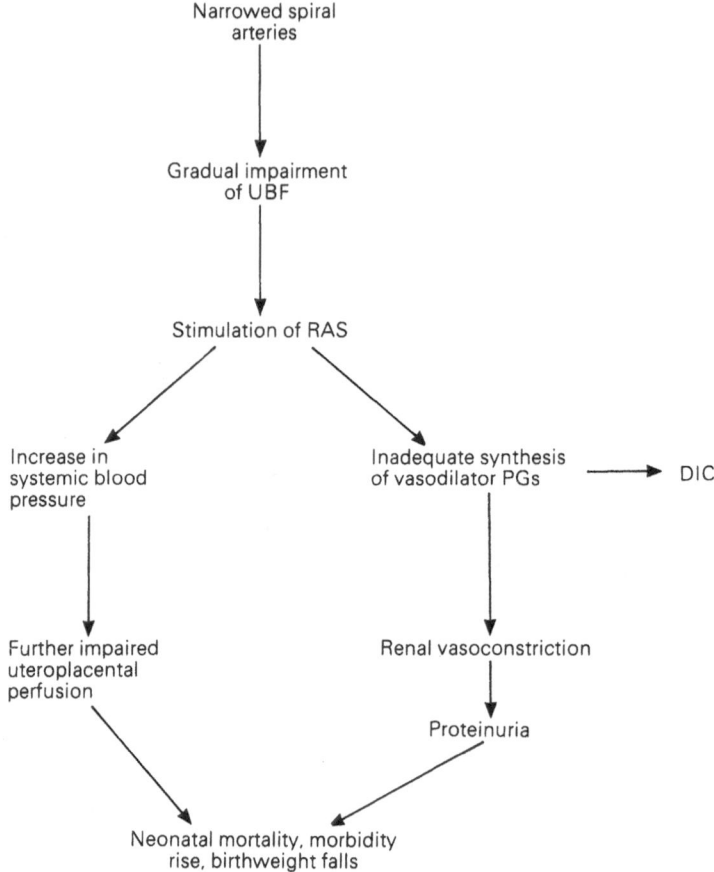

Figure 17.2   Schematic representation of proposed events in severe PIH

vascular effects of AII is lost, both systemically and in the utero-placental and possibly renal vasculature, and the positive feedback to renin production by the PGs is also lost. This may explain the observations of normal or low PRA in severe PIH[117]. However, with an enhanced vascular reactivity to AII, normal or even low plasma AII concentrations may be sufficient to cause intense vasoconstriction and hypertension. The renal effects of severe PIH, which appear to be secondary, could also be ascribed to an unopposed action of a vaso-constrictor such as AII which, in high doses, is diuretic and natri-uretic[118], and related perhaps to the contracted plasma volume in these patients[90]. The developing proteinuria of severe PIH has been reported as being 'vasoactive' in type, similar to that experimentally induced by AII[119].

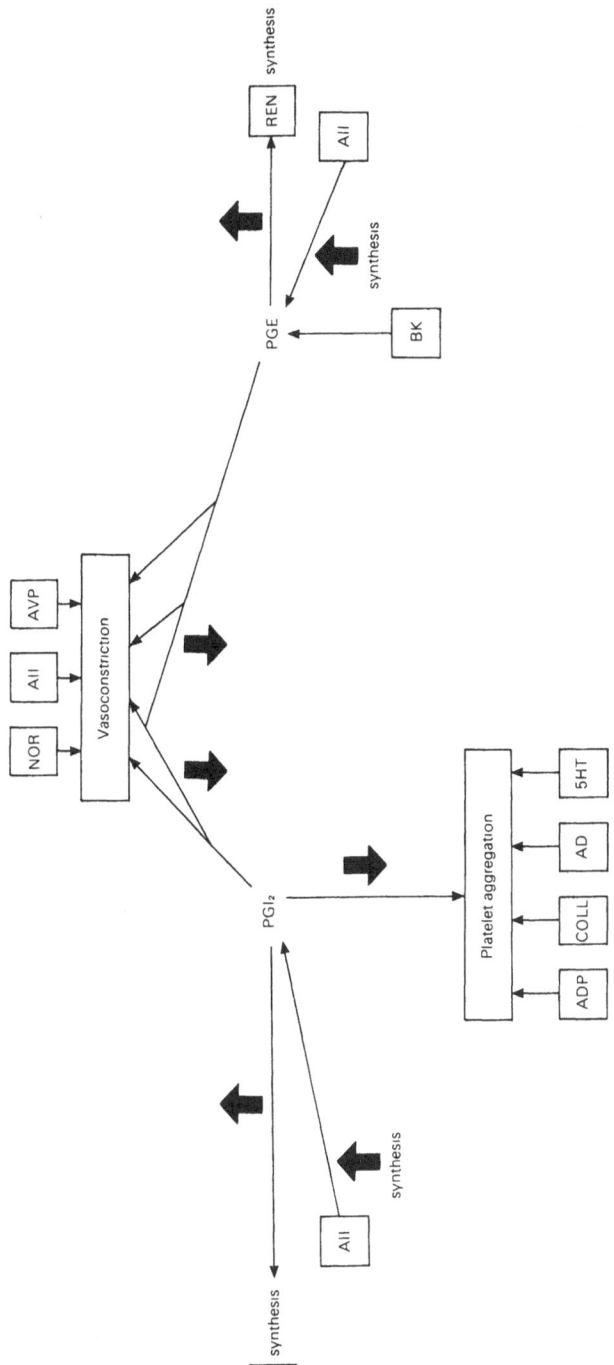

AII = angiotensin II; AD = adrenaline; ADP = adenosine diphosphate; AVP = arginine vasopressin; BK = bradykinin; 5HT = 5-hydroxytryptamine; NOR = nor-adrenaline; REN = renin

Figure 17.3   Outline interrelationships between prostacyclin and E series prostaglandins and some other vasoactive substances

The reduced production of PGI_2 in this condition may also be related to the imbalance in platelet-vascular homoeostasis, with platelet production tipped towards TxA_2. Furthermore, the very sensitivity of the platelets to the antiaggregatory properties of PGI_2 is reduced in PIH[120]. Certainly, there is evidence for *in vivo* platelet activation, with increased platelet consumption, which is a major component of disseminated intravascular coagulation. The platelets are also less responsive to aggregating agents while there are raised plasma concentrations of 5HT and β-thromboglobulin[49]. Interestingly, the administration of AII, together with thrombin, produced widespread disseminated intravascular coagulation in rabbits, although each alone failed to do so[121]. Thus, AII may also exert a permissive or contributory effect on disseminated intravascular coagulation. This hypothesis is summarized in Figure 17.2. An outline of the interrelationships between PGI_2 and E series prostaglandins and some other vasoactive substances is given in Figure 17.3.

TREATMENT OF PREGNANCY-INDUCED HYPERTENSION

It is evident that although an abnormality in the spiral arteries is a primary cause of PIH, it is the inability of fetal and maternal tissues to synthesize vasodilator prostaglandins which appears to convert a basically physiological response into a pathological one. One can thus regard PIH as a form of deficiency disease and attempt to treat it accordingly, either by attempting to stimulate the body to synthesize more prostaglandins, or by direct replacement therapy, either acute or chronic.

Human uterine and intrauterine tissues can convert AA into several lipoxygenase products known to be potent inhibitors of the biosynthesis of PGI_2 and suggested as being implicated in various diseases, including ischaemic heart disease[122]. Concentrations of these lipoxygenase products have yet to be measured in hypertensive pregnancy, but it seems at least possible that a lipoxygenase inhibitor could have a role in augmenting PGI_2 production in this condition.

Eicosapentaenoic acid (EPA) is the precursor of the prostaglandins with three double bonds (e.g. thromboxane A_3, TxA_3), and may interfere with the lipoxygenase pathways. Various studies of the effect of dietary supplementation with cod-liver oil, a substance which is very rich in EPA, have been carried out in man[124-126]. These have shown that the long-chain polyunsaturated fatty acids with three double bonds rapidly substitute a proportion of the AA in platelet membranes and compete for cyclo-oxygenase activity. TxA_3 is much less proaggregatory than TxA_2, with which it competes. Experiments have been performed *in vitro* with human tissues including rings of umbilical artery. These have shown that, following incubation with EPA, vascular tissues synthesize prostacyclin I_3 (PGI_3). While somewhat less active than PGI_2, this substance is nevertheless antiaggregatory and a vasodilator[125,127]. The production of PGI_3 in man following the ingestion of cod-liver oil is confirmed by the appearance of the 17-2,3-dinor-6-keto-$PGF_{1\alpha}$ metabolite in the urine[126]. There is a small fall in upright blood pressure after such dietary supplementation, but bleeding time increases and the platelet count falls[125].

Thus, the use of cod-liver oil, an easily obtained and cheap dietary supplement, might be of benefit in pregnancy by substituting the considerably less active TxA_3 for a proportion of the TxA_2 produced, without a major effect on the PGIs.

Another approach might be dietary supplementation with linoleic acid, a precursor of the prostaglandins with two double bonds. An infusion of safflower oil has been reported to increase urinary excretion of 6-keto-$PGF_{1\alpha}$ in man[128]. Furthermore, O'Brien and Broughton Pipkin[74] have reported that dietary supplementation with evening primrose oil (73% linoleic acid) and a vitamin/co-factor preparation for 7 days in second trimester human pregnancy is associated with a diminution in the pressor response to AII. These authors made no measurement of plasma or urinary prostaglandin concentration, but it is tempting to postulate that the fall in pressor response was a consequence of increased synthesis of I or E series PGs, as found following their direct i.v. administration.

Since there is some evidence for a relative excess of TxA_2 production in PIH, an alternative approach might be the use of inhibitors of Tx synthesis. Aspirin in low doses irreversibly inhibits platelet cyclo-oxygenase, and thus the production of TxA_2. Crandon and Isherwood[129] reported a lower incidence of PIH in an unselected group of patients of mixed parity who took unspecified amounts of aspirin frequently throughout pregnancy. However, salicylates cross freely into the fetal circulation, and the birthweights of infants born to mothers who had taken aspirin daily throughout pregnancy were significantly lower than those of carefully-matched controls[130]. Both stillbirth and perinatal mortality rates were also significantly raised. A commentary in *Lancet* (1980)[131] drew attention to the fetal and neonatal problems which appear to be associated with the maternal ingestion of non-steroidal anti-inflammatory drugs such as aspirin. These range from prolongation of gestation to primary pulmonary hypertension of the newborn and premature closure of the ductus arteriosus. Pregnancy should be regarded as a contraindication to the administration of these drugs.

UK-37,248 (Dazoxiben, Pfizer) is an orally-active thromboxane synthetase inhibitor rather than an inhibitor of cyclo-oxygenase. It has an apparently specific effect on TxA_2 production in man[132] and should, in theory, tip the TxA_2:PGI_2 balance towards PGI_2. UK-37,248 has been used in the treatment of four patients with very severe proteinuric PIH which developed in the second trimester[133]. Clinical details are not given for two patients; the other two both had evidence of underlying renal disease postpartum. Two infants survived and two died. Data of this kind are extremely difficult to evaluate, but the authors considered that there was a definite improvement in clinical and haematological condition in one patient which was synchronous with the start of therapy with Dazoxiben.

If, as Downing et al.[81] suggest, there is a smaller quantity of the enzyme complex required for PGI_2 synthesis in PIH, then neither dietary supplementation with precursors nor presumably blockade of TxA_2 synthesis will increase vasodilator prostaglandin synthesis. If this is so, then another therapeutic approach would be the direct administration of the prostaglandins themselves.

Toppozada et al.[134] reported that the acute administration of $1 \mu g$ $kg^{-1} min^{-1}$ PGA for 2 h to women with severe proteinuric PIH (mean BP 172/135 mmHg) was associated with a fall in systemic blood pressure and an improvement in renal function, especially in GFR. The same infusion was without significant effect on the blood pressure in normal patients. Labour was induced in half of the hypertensive and normotensive patients by the infusion; however, fetal heart rate abnormalities were noted. The same group[135] have also described a hypotensive, oxytocic effect of a dose of $0.5 \mu g$ $kg^{-1} min^{-1}$ given to seven patients, mostly nulliparae, with severe PIH at 36–40 weeks gestation, all of whom delivered within 24 h. The dose of PGA required to evoke a hypotensive effect was always associated with an increase in uterine activity.

In contrast with PGA, PGI_2 appears to have no oxytocic activity[26]. Fidler et al.[136] reported the first use of PGI_2 as a hypotensive agent in pregnancy. Severe proteinuric PIH developed in the second trimester in both patients reported. After conventional treatment had failed, a continuous infusion of PGI_2 at 8 ng $kg^{-1} min^{-1}$ succeeded in maintaining the diastolic blood pressure at or below 90 mmHg. In one patient, three prolonged periods of fetal bradycardia after 3 days of therapy prompted delivery at 27 weeks gestation and this fetus survived. However, the fetus of the second patient, and that of a patient similarly treated by another group[87] both died in utero without warning during the process of cardiotocographic monitoring. Obviously, the perinatal mortality rate will be high in patients with poor obstetric histories and early-onset severe PIH; nevertheless, the lack of warning of imminent IUFD in two of the three cases is worrying. Non-specific chest pain is a known side-effect of PGI_2 infusion in adults, and increasing severity of other symptoms such as headache, nausea, vomiting and both tachy- and bradycardia necessitate the 'tailing off' of dosage in long-term ($>6\,h$) infusions[137]. It seems possible that 'flooding' the fetal circulation with PGI_2 or its metabolites may be more than an already compromised fetus can tolerate.

As there is no naturally occurring animal model for PIH, there is an unavoidable element of experimentation in any attempted manipulation of prostaglandin, prostacyclin and thromboxane concentrations in patients with PIH. This problem is compounded since it is only in the very severe cases that such intervention is felt to be reasonable. Since the prognosis in such cases is already poor, any assessment of the effectiveness of therapy becomes very difficult indeed.

SUMMARY

In summary, if dietary deficiencies of precursors, trace elements or vitamins are partly to blame, balanced dietary supplementation may have a role to play in the prevention as well as the treatment of PIH, and these are unlikely to be associated with undesirable sequelae. Selective tipping of the balance between thromboxane and prostacyclin production towards production of the latter, whether by dietary supplementation with precursors of the trienoic prostaglandins or by thromboxane synthetase inhibitors, could also be valuable. At present, direct treatment with vasodilator prostaglandins or PGI_2 appears to be associated with too many undesirable side-effects to make it a reasonable proposition. However, the search is already on for PGI_2 analogues tailored to specific requirements in view of PGI_2's undoubted effectiveness in some forms of vascular disease outside pregnancy. It may not be too much to hope that a 'son of prostacyclin' may yet give us a physiologically appropriate means of treating, if not of preventing, PIH.

References

1 Shemesh, M. and Hansel, W. (1983). Hormone production by the early bovine embryo. *J. Steroid. Biochem.*, **19**, 979–83

2 Okita, J. R., Johnston, J. M. and MacDonald, P. C. (1983). Sources of prostaglandin precursor in human fetal membranes: arachidonic acid content of amnion and chorion laeve in diamniotic–dichorionic twin placentas. *Am. J. Obstet. Gynecol.*, **147**, 477–82

3 Jogee, M., Myatt, L., Moore, P. and Elder, M. G. (1983). Prostacyclin production by human placental cells in short-term culture. *Placenta*, **4**, 219–30

4 Siler-Khodr, T. M., Khodr, G. S., Valenzuela, G. and Harper, M. (1984).

Prostaglandin production by the human placenta throughout gestation. *Proc. Soc. Gynecol. Invest.*, San Francisco, p. 142

5 Myatt, L. and Elder, M. G. (1977). Inhibition of platelet aggregation by a placental substance with prostacyclin-like activity. *Nature (London)*, **268**, 159–60

6 Robinson, J. S., Redman, C. W. G., Clover, L. and Mitchell, M. D. (1979). The concentrations of the prostaglandins E and F, 13,14-dihydro-15-oxo prostaglandin F and thromboxane B1 in tissues obtained from women with and without pre-eclampsia. *Prostaglandins Med.*, **3**, 223–34

7 Mitchell, M. D. (1981). Prostacyclin during human pregnancy and parturition. In Lewis, P. J. and O'Grady, J. (eds.) *Clinical Pharmacology of Prostacyclin.* pp. 121–9. (New York: Raven)

8 Alexander, R. W. and Gimbrone, M. A. (1976). Stimulation of prostaglandin E synthesis in cultured human umbilical vein smooth muscle cells. *Proc. Natl. Acad. Sci. USA*, **73**, 1617–20

9 Remuzzi, G., Misiani, R., Muratore, D., Marchesi, D., Livio, M., Schieppati, A., Mecca, G., de Gaetano, G. and Donati, M. B. (1979). Prostacylin and human fetal circulation. *Prostaglandins*, **18**, 341–8

10 Mitchell, M. D., Jamieson, D. R. S., Sellers, S. M. and Turnbull, A. C. (1980). 6-keto-PGF$_1$: concentration in human umbilical plasma and production by umbilical vessels. *Adv. Prostaglandin Thromboxane Res.* **7**, 891–6

11 Kawano, M. and Mori, N. (1983). Prostacyclin producing activity of human umbilical, placental and uterine vessels. *Prostaglandins*, **26**, 645–62

12 Downing, I., Blair, I. and Lewis, P. J. (1981). 6 Oxo PGF$_1$ synthesis by human umbilical artery. In Lewis, P. J. and O'Grady, J. (eds.) *Clinical Pharmacology of Prostacyclin.* pp. 131–5. (New York: Raven)

13 Stuart, M. J., Sunderji, S. G., Yambo, T., Clark, D. A., Allen, J. B., Elrad, H. and Slott, J. H. (1981) Decreased prostacyclin production: a characteristic of chronic placental insufficiency syndromes. *Lancet*, i, 1126–8

14 Ongari, M. A., Ritter, J. M., Orchard, M. A., Waddell, K. A., Blair, I. A. and Lewis, P. J. (1984). Correlation of prostacyclin synthesis by human umbilical artery with status of essential fatty acid. *Am. J. Obstet. Gynecol.*, **149**, 455–60

15 Ritter, J. M., Ongari, M. A., Barrow, S. E., Orchard, M. A., Blair, I. A. and Lewis, P. J. (1982). Prostanoid synthesis by human umbilical artery. *Prostaglandins*, **24**, 881–6

16 Ogburn, P. L., Williams, P. P., Johnson, S. B. and Holman, R. T. (1984). Serum arachidonic acid levels in normal and pre-eclamptic pregnancies. *Am. J. Obstet. Gynecol.*, **148**, 5–9

17 Mitchell, M. D., Brunt, J., Bibby, J., Flint, A. P. F., Anderson, A. B. M. and Turnbull, A. C. (1978). Prostaglandins in the human umbilical circulation at birth. *Br. J. Obstet. Gynaecol.*, **85**, 114–18

18 MacKenzie, I. Z., MacLean, D. A. and Mitchell, M. D. (1980). Prostaglandins in the human fetal circulation in mid-trimester and term pregnancy. *Prostaglandins*, **20**, 649–54

19 Ylikorkala, O., Mäkilä, U.-M. and Viinikka, L. (1981). Amniotic fluid

prostacyclin and thromboxane in normal, pre-eclamptic and some other complicated pregnancies. *Am. J. Obstet. Gynecol.*, **141**, 487–90

20 Casey, M. L., Cutrer, S. I. and Mitchell, M. D. (1983). Origin of prostanoids in human amniotic fluid: the fetal kidney as a source of amniotic fluid prostanoids. *Am. J. Obstet. Gynecol.*, **147**, 547–51

21 Friedman, Z. V. I. and Demers, L. M. (1979). Prostaglandin synthetase in the human neonatal kidney. *Pediatr. Res.*, **14**, 190–3

22 Mitchell, M. D., Keirse, M. J. N. C., Brunt, J. D., Anderson, A. B. M. and Turnbull, A. C. (1979). Concentrations of the prostacyclin metabolite, 6-keto prostaglandin F_1, in amniotic fluid during late pregnancy and labour. *Br. J. Obstet. Gynaecol.*, **86**, 350–3

23 Bibby, J. G., Brunt, J. D., Hodgson, H., Mitchell, M. D., Anderson, A. B. M. and Turnbull, A. C. (1979). Prostaglandins in umbilical plasma at elective Caesarean section. *Br. J. Obstet. Gynaecol.*, **86**, 282–4

24 Willman, E. A., Rodeck, C. H., Collins, W. P. and Clayton, S. G. (1977) The relation between umbilical cord tissue prostaglandin E_2 levels, mode of onset of labour, fetal distress and method of delivery. *Br. J. Obstet. Gynaecol.*, **83**, 605–7

25 Mitchell, M. D., Bibby, J. G., Hicks, B. R. and Turnbull, A. C. (1978). Possible role for prostacyclin in human parturition. *Prostaglandins*, **16**, 931–7

26 Omini, C., Folco, G. C., Pasargiklian, R., Fano, M. and Berti, F. (1979). Prostacyclin (PGI_2) in pregnant human uterus. *Prostaglandins*, **17**, 113–20

27 Bamford, D. S., Jogee, M. and Williams, K. I. (1980). Prostacyclin formation by the pregnant human myometrium. *Br. J. Obstet. Gynaecol.*, **87**, 215–18

28 Whalen, J. B., Clancey, C. J., Farley, D. B. and van Orden, D. E. (1978). Plasma prostaglandins in pregnancy. *Obstet. Gynecol.*, **51**, 52–5

29 Koullapis, E. N., Nicolaides, K. H., Collins, W. P., Rodek, C. H. and Campbell, S. (1982). Plasma prostanoids in pregnancy-induced hypertension. *Br. J. Obstet. Gynaecol.*, **89**, 617–21

30 Lewis, P. J., Boylan, P., Friedman, L. A., Hensby, C. N. and Downing, I. (1980). Prostacyclin in pregnancy. *Br. Med. J.*, **1**, 1581–2

31 Spitz, B., Deckmyn, H., van Assche, F. A. and Vermylen, J. (1983). Prostacyclin production in whole blood throughout normal pregnancy. *Clin. Exp. Hyp.-Hyp. Preg.*, **B2**, 191–202

32 Remuzzi, G., Zoja, C., Marchesi, D., Schieppati, A., Mecca, G., Misiani, R., Donati, M. B. and de Gaetano, G. (1981). Plasmatic regulation of vascular prostacyclin in pregnancy. *Br. Med. J.*, **282**, 512–14

33 Ylikorkala, O. and Viinikka, L. (1980). Thromboxane A_2 in pregnancy and puerperium. *Br. Med. J.*, **281**, 1601–2

34 Moutquin, J. M. and Leblanc, N. (1982). A prospective study of urinary prostaglandins E in women with normal and hypertensive pregnancies. *Clin. Exp. Hyp.-Hyp. Preg.*, **B1**, 539–52

35 Pedersen, E. B., Christensen, N. J., Christensen, P., Johannesen, P., Kornerup, H. J., Kristensen, S., Lauritsen, J. G., Leyssac, P. P., Rasmussen, A. and Wohlert, M. (1983). Pre-eclampsia – a state of prostaglandin

deficiency? Urinary prostaglandin excretion, the renin–aldosterone system and circulating catecholamines in pre-eclampsia. *Hypertension*, 5, 105–11

36 Goodman, R. P., Killam, A. P., Brash, A. R. and Branch, R. A. (1982). Prostacyclin production during pregnancy: comparison of production during normal pregnancy and pregnancy complicated by hypertension. *Am. J. Obstet. Gynecol.*, 142, 817–22

37 Tuvemo, T. (1980). Role of prostaglandins, prostacyclin and thromboxanes in the control of the umbilical-placental circulation. *Sem. Perinatal.*, 4, 91–5

38 Pomerantz, K., Sintetos, A. and Ramwell, P. (1978). The effect of prostacyclin on the human umbilical artery. *Prostaglandins*, 15, 1035–44

39 Kitson, G. E. and Broughton Pipkin, F. (1981). Effects and interactions of prostaglandins E_1 and E_2 on human chorionic plate arteries. *Am. J. Obstet. Gynecol.*, 140, 683–8

40 Mak, K. K.-W., Gude, N. M. Walters, W. A. W. and Boura, A. L. A. (1984). Effects of vasoactive autacoids on the human umbilical-fetal placental vasculature. *Br. J. Obstet. Gynaecol.*, 91, 99–106

41 Mäkilä, U.-M., Jouppila, P., Kirkinen, P., Viinikka, L. and Ylikorkala, O. (1983). Relation between umbilical prostacyclin production and blood flow in the fetus. *Lancet*, 1, 728–9

42 Ylikorkala, O., Jouppila, P., Kirkinen, P. and Viinikka, L. (1983). Maternal prostacyclin, thromboxane and placental blood flow. *Am. J. Obstet. Gynecol.*, 145, 730–2

43 Resnik, R. and Brink, G. W. (1978). Effects of prostaglandins E_1, E_2 and F_2 on uterine blood flow in non-pregnant sheep. *Am. J. Physiol.*, 234, H557–61

44 McLaughlin, M. K., Brennan, S. C. and Chez, R. A. (1978). Effects of indomethacin on sheep uteroplacental circulations and sensitivity to angiotensin II. *Am. J. Obstet. Gynecol.*, 132, 430–5

45 Rankin, J. H. G., Berssenbrugge, A., Anderson, D. and Phernetton, T. (1979). Ovine placental vascular response to indomethacin. *Am. J. Physiol.*, 236, H61–4

46 Venuto, R. C., O'Dorisio, T., Stein, J. H. and Ferris, T. F. (1975). Uterine prostaglandin E secretion and uterine blood flow in the pregnant rabbit. *J. Clin. Invest.*, 55, 193–7

47 Terragno, N. A., Terragno, D. A., Pacholczyk, D. and McGiff, J. C. (1974). Prostaglandins and the regulation of uterine blood flow in pregnancy. *Nature (London)*, 249, 57–8

48 Brosens, I. A. (1977). Morphological changes in the utero-placental bed in pregnancy hypertension. In Symonds, E. M. (ed.) *Hypertensive states in pregnancy (Clinics in Obstetrics and Gynaecology)*. pp. 573–93. (London: W. B. Saunders)

49 Wallenburg, H. C. S. (1981). Prostaglandins and the maternal placental circulation: review and perspectives. *Biol. Res. Preg.*, 2, 15–22

50 Symonds, E. M., Stanley, M. A. and Skinner, S. L. (1968). Production of renin by *in vitro* cultures of human chorion and uterine muscle. *Nature (London)*, 217, 1152–3

51 Warren, A. Y., Craven, D. J. and Symonds, E. M. (1982). Production of

active and inactive renin by cultured explants from the human female genital tract. *Br. J. Obstet. Gynaecol.*, **89**, 628–32

52 Johnson, J., Johnson, I. R., Ronan, J. E. and Craven, D. J. (1984). The site of renin in the human uterus. *Histopathology*, **8**, 273–8

53 Skinner, S. L., Lumbers, E. R. and Symonds, E. M. (1968). Renin concentration in human foetal and maternal tissues. *Am. J. Obstet. Gynecol.*, **101**, 529–33

54 Symonds, E. M., Skinner, S. L., Stanley, M. A., Kirkland, J. A. and Ellis, R. C. (1970). An investigation of the cellular source of renin in human chorion. *J. Obstet. Gynaecol. Br. Commonw.*, **77**, 885–90

55 Craven, D. J., Warren, A. Y. and Symonds, E. M. (1982). Correlation of direct and indirect measurements of renin from human myometrium and chorion. *Biomed. Res.*, **3**, 330–2

56 Ferris, T. F., Stein, J. H. and Kauffman, J. (1972). Uterine blood flow and uterine renin secretion. *J. Clin. Invest.*, **51**, 2827–33

57 Craven, D. J., Warren, A. Y. and Symonds, E. M. (1983). Generation of angiotensin I by tissues of the human female genital tract. *Am. J. Obstet. Gynecol.*, **145**, 749–51

58 Cooke, G. F., Craven, D. J. and Symonds, E. M. (1985). Prostaglandin E$_2$ and the release of renin from human chorion. *Clin. Exp. Hyp.-Hyp. Preg.*, **B4**, 49–62

59 Warren, A. Y., Craven, D. J. and Symonds, E. M. (1984). Angiotensin converting enzyme in human fetal membranes, placenta, amniotic fluid and cord venous serum. *Clin. Exp. Hyp.-Hyp. Preg.*, **B3**, 51–60

60 Sundsfjord, J. A. and Aakvaag, A. (1970). Plasma angiotensin II and aldosterone excretion during the menstrual cycle. *Acta Endocrinol.*, **64**, 452–8

61 Symonds, E. M. (1981). The renin–angiotensin system in pregnancy. *Obstet. Gynaecol. Ann.*, **10**, 45–67

62 Chinn, R. H. and Düsterdieck, G. (1972). The response of blood pressure to infusion of angiotensin II: relation to plasma concentrations of renin and angiotensin II. *Clin. Sci.*, **42**, 489–504

63 Chesley, L. C., Talledo, E., Bohler, C. S. and Zuspan, F. P. (1965). Vascular reactivity to angiotensin II and norepinephrine in pregnant and non-pregnant women. *Am. J. Obstet. Gynecol.*, **91**, 837–42

64 O'Brien, P. M. S., Filshie, G. M. and Broughton Pipkin, F. (1977). The effect of prostaglandin E$_2$ on the cardiovascular response to angiotensin II in pregnant rabbits. *Prostaglandins*, **13**, 171–81

65 O'Brien, P. M. S. and Broughton Pipkin, F. (1979). The effects of deprivation of prostaglandin precursors on vascular sensitivity to angiotensin II and on the kidney in the pregnant rabbit. *Br. J. Pharmacol.*, **65**, 29–34

66 Everett, R. B., Worley, R. J., MacDonald, P. C. and Gant, N. F. (1978). Effect of prostaglandin synthetase inhibitors on pressor response to angiotensin II in human pregnancy. *J. Clin. Endocrinol. Metab.*, **46**, 1007–10

67 Jaspers, W. J. M., De Jong, P. A. and Mulder, A. W. (1981). Angiotensin II sensitivity and prostaglandin synthetase inhibition in pregnancy. *Eur. J. Obstet. Gynaecol. Reprod. Biol.*, **11**, 379–84

68 Broughton Pipkin, F., Hunter, J. C., Turner, S. R. and O'Brien, P. M. S. (1982). Prostaglandin E_2 attentuates the pressor response to angiotensin II in pregnant, but not non-pregnant, humans. *Am. J. Obstet. Gynecol.*, **142**, 168–76

69 Broughton Pipkin, F., O'Brien, P. M. S. and Sant-Cassia, L. J. (1982). The effect of prostaglandin E_1 on the pressor response to angiotensin II in second trimester human pregnancy and in the non-pregnant subject. *Clin. Exp. Hyp.-Hyp. Preg.*, **B1**, 493–504

70 Broughton Pipkin, F., Morrison, R. and O'Brien, P. M. S. (1984). Effects of prostacyclin on the pressor response to angiotensin II in human pregnancy. In Proceedings of the Meeting of the European Society for Clinical Investigation, Milan. *Eur. J. Clin. Invest.*, **14**, 3

71 Goodfriend, T. L. and Simpson, R. U. (1981). Angiotensin receptors in bovine umbilical artery and their inhibition by non-steroidal anti-inflammatory drugs. *Br. J. Pharmacol.*, **72**, 247–55

72 Franklin, G. O., Dowd, A. J., Caldwell, B. V. and Speroff, L. (1974). The effect of angiotensin II intravenous infusion on plasma renin activity and prostaglandins A, E and F levels in the uterine vein of the pregnant monkey. *Prostaglandins*, **6**, 271–80

73 Speroff, L., Haning, R. V., Ewaschuk, E. J., Alberino, S. L. and Kieliszek, F. Y. (1976). Uterine artery blood flow studies in the pregnant monkey. In M. D. Lindheimer, Katz, A. I. and Zuspan, F. P. (eds.) *Hypertension in Pregnancy.* pp. 315–26. (New York: John Wiley & Sons)

74 O'Brien, P. M. S. and Broughton Pipkin, F. (1983). The effect of essential fatty acid and specific vitamin supplements on vascular sensitivity in the mid-trimester of human pregnancy. *Clin. Exp. Hyp.-Hyp. Preg.*, **B2**, 247–54

75 Broughton Pipkin, F., Hunter, J. C., Turner, S. R. and O'Brien, P. M. S. (1984). The effect of prostaglandin E_2 upon the biochemical response to infused angiotensin II in human pregnancy. *Clin. Sci.*, **66**, 399–406

76 Demers, L. M. and Gabbe, S. G. (1976). Placental prostaglandin levels in pre-eclampsia. *Am. J. Obstet. Gynecol.*, **126**, 137–9

77 Valenzuela, G. and Bodkhe, R. R. (1980). Effect of pregnancy-induced hypertension upon placental prostaglandin metabolism: decreased prostaglandin F_2 catabolism with normal prostaglandin E_2 catabolism. *Am. J. Obstet. Gynecol.*, **136**, 255–6

78 Alam, N. A., Clary, P. and Russell, P. T. (1973). Depressed placental prostaglandin E_1 metabolism in toxaemia of pregnancy. *Prostaglandins*, **4**, 363–70

79 Walsh, S. W. and Fenner, P. C. (1984). Toxemia: an imbalance in placental prostacyclin (PGI) and thromboxane (TxA) production. *Clin. Exp. Hyp.-Hyp. Preg.*, **B3**, 159

80 Remuzzi, G., Marchesi, D., Zoja, C., Muratore, D., Mecca, G., Misiani, R., Rossi, E., Barbato, M., Capetta, P., Donati, M. B. and de Gaetano, G. (1980). Reduced umbilical and placental vascular prostacyclin in severe pre-eclampsia. *Prostaglandins*, **20**, 105–10

81 Downing, I., Shepherd, G. L. and Lewis, P. J. (1980). Reduced prostacyclin production in pre-eclampsia. *Lancet*, **2**, 1374

82 Carreras, L. O., DeFreyn, G., van Houte, E., Vermylen, J. and van Assche, A. (1981). Prostacyclin and pre-eclampsia. *Lancet*, 1, 442

83 Mäkilä, U.-M., Viinikka, L. and Ylikorkala, O. (1984). Evidence that prostacyclin deficiency is a specific feature in pre-eclampsia. *Am. J. Obstet. Gynecol.*, 148, 772-4

84 Misiani, R., Remuzzi, G. and Mecca, G. (1981). Prostacyclin generation by human umbilical and placental vessels in normal and hypertensive pregnancy. In Lewis, P. J. and O'Grady, J. (eds.) *Clinical Pharmacology of Prostacyclin*. pp. 137-40. (New York: Raven)

85 Bussolino, F., Bendetto, C., Massobrio, M. and Camussi, G. (1980). Maternal vascular prostacyclin activity in pre-eclampsia. *Lancet*, 2, 702

86 Redman, C. W. G., Brennecke, S. P. and Mitchell, M. D. (1981). Prostaglandins and pre-eclampsia. *Lancet*, 1, 731

87 Lewis, P. J., Shepherd, G. L., Ritter, J., Chan, S. M. T., Bolton, P. J., Jogee, M., Myatt, L. and Elder, M. G. (1981). Prostacyclin and pre-eclampsia. *Lancet*, 1, 559

88 Martensson, L. and Wallenburg, H. C. S. (1984). Uterine venous concentrations of 6-keto-PGF$_1$ (6-k) in normal pregnant (NP) and pregnancy-induced hypertensive (PIH) women. Presented at *Proc. Soc. Gynecol. Invest.*, San Francisco. Abstract 410, p. 243

89 Bodzenta, A., Thomson, J. M. and Poller, L. (1980). Prostacyclin activity in amniotic fluid in pre-eclampsia. *Lancet*, 2, 650

90 MacGillivray, I. (1984). *Pre-eclampsia: The Hypertensive Disease of Pregnancy*. (London: W. B. Saunders)

91 Page, E. W. and Christianson, R. (1976). Influence of blood pressure changes with and without proteinuria upon outcome of pregnancy. *Am. J. Obstet. Gynecol.*, 126, 821-9

92 Robertson, W. B., Brosens, I. and Dixon, H. G. (1967). The pathological response of the vessels of the placental bed to hypertensive pregnancy. *J. Pathol. Bacteriol.*, 93, 581-92

93 Sheppard, B. L. and Bonnar, J. (1980). Ultrastructural abnormalities of placental villi in placentas from pregnancies complicated by intrauterine fetal growth retardation: their relation to decidual spiral arterial lesions. *Placenta*, 1, 145-56

94 Editorial (1980). Genetic control of pre-eclampsia. *Lancet*, 1, 634-5

95 Abdul-Karim, R. and Assali, N. S. (1961). Pressor response to angiotonin in pregnant and non-pregnant women. *Am. J. Obstet. Gynecol.*, 82, 246-51

96 Chesley, L. C., Wynn, R. M. and Silverman, N. I. (1963). Renal effects of angiotensin II infusions in normotensive pregnant and non-pregnant women. *Circulation Res.*, 13, 232-8

97 Lumbers, E. R. (1970). Peripheral vascular reactivity to angiotensin and noradrenaline in pregnant and non-pregnant women. *Aust. J. Exp. Biol. Med. Sci.*, 48, 493-500

98 Gant, N. F., Daley, G. L., Chand, S., Whalley, P. J. and MacDonald, P. C. (1973). A study of angiotensin II pressor response throughout primigravid pregnancy. *J. Clin. Invest.*, 52, 2682-9

99 Chesley, L. C. (1966). Vascular reactivity in normal and toxemic pregnancy. *Clin. Obstet. Gynecol.*, **9**, 871-80

100 Gant, N. F., Worley, R. J., Everett, R. B. and MacDonald, P. C. (1980). Control of vascular responsiveness during human pregnancy. *Kidney Int.*, **18**, 253-8

101 Kaulhausen, H., Oney, T., Feldmann, R. and Leyendecker, G. (1981). Decrease of vascular angiotensin sensitivity by L-Dopa during human pregnancy. *Am. J. Obstet. Gynecol.*, **140**, 671-5

102 Kaulhausen, H., Oney, T. and Mey, F. (1983). Acute inhibition of vascular angiotensin sensitivity by theophylline in human pregnancy. *Clin. Exp. Hyp.-Hyp. Preg.*, **B2**, 415-20

103 Jaspers, W. J. M., De Jong, P. A. and Mulder A. W. (1983). Decrease of angiotensin sensitivity after bed rest and strongly sodium-restricted diet in pregnancy. *Am. J. Obstet. Gynecol.*, **145**, 792-6

104 Davison, J. M. (1983). The kidney in pregnancy: a review. *J. R. Soc. Med.*, **76**, 485-501

105 Griffin, D., Cohen-Overbeek, T. and Campbell, S. (1983). Fetal and utero-placental blood flow. In Campbell, S. (ed.) *Clinics in Obstetrics and Gynaecology*. pp. 565-602. (London: W. B. Saunders)

106 Fox, H. (1982). Do the morphological abnormalities found in pre-eclampsia have an immunological basis? In Sammour, M. B., Symonds, E. M., Zuspan, F. P. and El-Tomi, N. (eds.) *Pregnancy Hypertension*. pp. 249-53. (Cairo: Air Shams University Press)

107 Keeton, T. K. and Campbell, W. B. (1981). The pharmacologic alteration of renin release. *Pharmacol. Rev.*, **31**, 81-227

108 Aiken, J. W. and Vane, J. R. (1973). Intrarenal prostaglandin release attenuates the renal vasoconstrictor activity of angiotensin. *J. Pharmacol. Exp. Ther.*, **184**, 678-87

109 Bell, C. (1973). Vasoactive substances in the circulation of the pregnant dog during acute fetal ischemia. *Am. J. Obstet. Gynecol.*, **117**, 1088-92

110 Abitbol, M. M., Ober, W. B., Gallo, G. R., Driscoll, S. G. and Pirani, C. L. (1977). Experimental toxemia of pregnancy in the monkey, with a preliminary report on renin and aldosterone. *Am. J. Pathol.*, **86**, 573-90

111 Kokot, F. and Cekanski, A. (1972). Plasma renin activity in peripheral and uterine vein blood in pregnant and non-pregnant women. *J. Obstet. Gynaecol. Br. Commonw.*, **79**, 72-6

112 Broughton Pipkin, F., Craven, D. J. and Symonds, E. M. (1981). The uteroplacental renin–angiotensin system in normal and hypertensive pregnancy. *Contr. Nephrol.*, **25**, 49-52

113 Ferris, T. F. and Weir, E. K. (1983). Effect of Captopril on uterine blood flow and prostaglandin E synthesis in the pregnant rabbit. *J. Clin. Invest.*, **71**, 809-15

114 O'Brien, P. M. S. and Broughton Pipkin, F. (1982). The effect of essential fatty acid and specific vitamin supplements on vascular sensitivity in the mid-trimester of human pregnancy. *Clin. Exp. Hyp.-Hyp. Preg.*, **B1**, 286

115 Naden, R. P. and Rosenfeld, C. R. (1981). Effect of angiotensin II on uterine and systemic vasculature. *J. Clin. Invest.*, **68**, 468-74

116 Oates, J. A., Whorton, A. R., Gerkens, J. F., Branch, R. A., Hollifield,

J. W. and Frolich, J. C. (1979). The participation of prostaglandins in the control of renin release. *Fed. Proc.*, **38**, 72-4

117 Weir, R. J., Fraser, R., Lever, A. F., Morton, J. J., Brown, J. J., Kraszewski, A., McIlwaine, G. M., Robertson, J. I. S. and Tree, M. (1973). Plasma renin, renin substrate, angiotensin II and aldosterone in hypertensive disease of pregnancy. *Lancet*, **1**, 291-4

118 Pickering, G. W. (1968). *High Blood Pressure.* (London: J. & A. Churchill Ltd.)

119 Wood, S. M., Burnett, D. and Studd, J. (1976). Selectivity of proteinuria assessed by different methods. In Lindheimer, M. D., Katz, A. I. and Zuspan, F. P. (eds.) *Hypertension in Pregnancy.* pp. 75-83. (New York: John Wiley and Sons)

120 Briel, R. C., Kieback, D. G. and Lippert, T. H. (1984). Platelet sensitivity to a prostacyclin analogue in normal and pathological pregnancy. *Prostaglandins, Leukotrienes Med.*, **13**, 335-40

121 Whitaker, A. N., Bunce, I., Nicol, P. and Dowling, S. V. (1973). Interaction of angiotensin with disseminated intravascular coagulation. *Am. J. Pathol.*, **72**, 1-12

122 Saeed, S. A. and Mitchell, M. D. (1983). Lipoxygenase activity in human uterine and intrauterine tissues: new prospects for control of prostacyclin production in pre-eclampsia. *Clin. Exp. Hyp.-Hyp. Preg.*, **B2**, 103-11

123 Moncada, S., Higgs, E. A. and Vane, J. R. (1977) Human arterial and venous tissues generate prostacyclin (prostaglandin X), a potent inhibitor of platelet aggregation. *Lancet*, **1**, 18-21

124 Hay, C. R. M., Durber, A. P. and Saynor, R. (1982). Effect of fish oil on platelet kinetics in patients with ischaemic heart disease. *Lancet*, **1**, 1269-72

125 Lorenz, R., Spengler, U., Fischer, S., Duhm, J. and Weber, P. C. (1983). Platelet function, thromboxane formation and blood pressure control during supplementation of the western diet with codliver oil. *Circulation*, **67**, 504-11

126 Fischer, S. and Weber, P. C. (1984). Prostaglandin I₃ is formed *in vivo* in man after dietary eicosapentaenoic acid. *Nature (London)*, **307**, 165-8

127 Dyerberg, J. and Jorgensen, K. A. (1980). The effect of arachidonic- and eicosapentaenoic acid on the synthesis of prostacyclin-like material in human umbilical vasculature. *Artery*, **8**, 12-17

128 Epstein, M., Lifschitz, M. and Rappaport, K. (1982). Augmentation of prostaglandin production by linoleic acid in man. *Clin. Sci.*, **63**, 565-71

129 Crandon, A. J. and Isherwood, D. M. (1979). Effect of aspirin on incidence of pre-eclampsia. *Lancet*, **1**, 1356

130 Turner, G. and Collins, E. (1975). Fetal effects of regular salicylate injection in pregnancy. *Lancet*, **2**, 338-40

131 Editorial (1980). PG-Synthetase inhibitors in obstetrics and after. *Lancet*, **2**, 185-6

132 Tyler, H. M., Saxton, C. A. P. D. and Parry, M. J. (1981). Administration to man of UK 37, 248-01, a selective inhibitor of thromboxane synthetase. *Lancet*, **1**, 629-32

133 Van Assche, F. A., Spitz, B., Vermylen, J. and Deckmijn, H. (1984).

Preliminary observations on treatment of pregnancy-induced hypertension with a thromboxane synthetase inhibitor. *Am. J. Obstet. Gynecol.*, **148**, 216–18

134 Toppozada, M., Ghoneim, A., Habib, Y. A., El-Ziadi, L. and El-Damarawy, H. (1979). Effect of prostaglandin A_1 on renal hemodynamics in pregnancy toxemia. *Am. J. Obstet. Gynecol.*, **135**, 581–5

135 Toppozada, M. K., Shaala, S. A. and Moussa, H. A. (1983). Therapeutic use of PGA_1 infusions in severe pre-eclampsia – a major clinical potential. *Clin. Exp. Hyp.-Hyp. Preg.*, **B2**, 217–32

136 Fidler, J., Ellis, C., Bennett, M. J., DeSwiet, M. and Lewis, P. J. (1981). Prostacyclin and pre-eclamptic toxaemia. In Lewis, P. J. and O'Grady, J. (eds.) *Clinical Pharmacology of Prostacyclin.* pp. 141–3. (New York: Raven Press)

137 Data, J. L., Molony, B. A., Meinzinger, M. M. and Gorman, R. R. (1981). Intravenous infusion of prostacyclin sodium in man: clinical effects and influence on platelet adenosine diphosphate sensitivity and adenosine $3':5'$-cyclic monophosphate levels. *Circulation*, **64**, 4–12

18
Premenstrual syndrome

P. W. BUDOFF

INTRODUCTION

Premenstrual syndrome (PMS) is a group of discomforting symptoms experienced by many women 7-10 days before menstruation. Typically, each woman has her own personal complex of symptoms and these occur cyclically, month after month. Common symptoms include tenderness and swelling of breasts, headaches, weight gain, abdominal cramps, bloating, thirst, nausea, specific food or alcohol cravings, joint pain, dizziness, acne, hyperalgesia, irritability, fatigue and lethargy, anxiety, depression, hostility and aggression. In most cases, symptoms cease completely within 24 h after the onset of menses.

Numerous theories have been advanced to explain the diverse symptoms associated with PMS. Some of these relate the syndrome either to hormone imbalance, vitamin deficiency or psychosomatic aberrations. Treatments based on these theories, however, have failed to relieve all the symptoms of PMS. More recently, the presence of prostaglandins (PGs) has been proposed as causing the disparate symptoms of PMS. In many cases, the symptoms of PG excess resemble those of PMS. At the present time, the PG theory appears to offer a reasonable explanation for this syndrome. Because PG production is related to sex hormone production, the PG and hormonal theories of PMS may overlap somewhat.

HISTORY AND DEFINITIONS

PMS was first identified by Frank[1] who, in 1931, characterized cyclic emotional disturbances, weight gain, subcutaneous haemorrhages, œdema and epilepsy as premenstrual disease. 10 years later, Gray[2] discussed a premenstrual psychosexual disorder. However, it was not until Dalton's report in 1953 that PMS began to be systematically investigated as a specific syndrome. Dalton designated the time when PMS symptoms occur as the 'paramenstruum'[3]. PMS has been reported to occur in 21–54% of otherwise normal women[4].

At various times PMS has been called premenstrual tension (PMT) and the premenstrual tension syndrome (PMTS). Some authors have suggested that the syndrome might be more accurately called the 'cyclic syndrome', since it need not occur in menstruating women – prepubescent girls, amenorrhoeic women, postmenopausal women and hysterectomized women also can have PMS symptoms[5,6].

The psychological symptoms of PMS include irritability, short-temperedness, impatience, hostility and aggression, fatigue and lethargy, anxiety, sadness and crying spells, depression, suicidal feelings, difficulty concentrating, clumsiness and changes in libido. Dermatological changes associated with PMS include acne, premenstrual urticaria, herpetic outbreaks, pruritus, periocular pigmentation and changes in the dryness of scalp or hair. Migraine and other types of headaches, dizziness, palpitations, fainting, vertigo, hyperalgesia, asthma, rhinitis and, in epileptics, increased seizure activity are PMS-related neurological symptoms. Musculoskeletal problems include backache, joint pain, arthritis-like changes and swelling of fingers and legs. Gastrointestinal symptoms such as constipation (often followed by loose stool), abdominal cramping, food cravings, anorexia and nausea may also be associated with PMS, as may renal and mammary changes such as less frequent urination and breast tenderness and swelling, respectively.

Abraham has proposed that PMS is really several syndromes, each with its own aetiology[4]. He has divided the syndrome complex into four subgroups, based on symptoms and corresponding pathophysiological changes that he and others have observed: PMT-A (anxiety, irritability, tension); PMT-H (weight gain, abdominal bloating, mastalgia); PMT-C (increased appetite, fatigue, palpitations and headache); and PMT-D (depression, withdrawal, suicidal ideation). According to Abraham, PMT-A, -H, and -C occur in 40–80% of symptomatic patients, and a given patient is likely to belong to more than one subgroup[4].

AETIOLOGY OF PMS

Early theories postulated that psychological factors or environmental and evolutionary forces were responsible for the symptoms of PMS.

Sexual and marital maladjustment, guilt feelings and repulsive feelings toward the menses were proposed as causes. It has also been suggested that PMS results from the stresses and strains of modern urban life – a notion countered by the finding that women in remote villages in Third World countries also experience symptoms of PMS[7]. Another theory presents PMS as serving an evolutionary function in discouraging male sexual advances when conception is not possible[8].

More recent theories have focused on a wide variety of physiological processes, most prominently those involving female hormones. Oestrogen excess (because of its water- and sodium-retaining effect), progesterone deficiency, suppression of prolactin and production of α-melanocyte releasing hormone (MRH) have all been proposed as causes. Abraham suggests that PMT-A, which he relates to anxiety, is due to changes in oestrogen/progesterone ratio and PMT-D, which is linked to depression, may be associated with low oestrogen and high progesterone levels[4]. An association between dysmenorrhoea and PMS has also been suggested. Both have been related to PG levels, and there is some indication that nearly one quarter of patients with PMS also suffer from primary dysmenorrhoea[9].

Other theories relate PMS to deficiencies in vitamin B_6 and vitamin A, to psychosomatic factors, fluid retention, reactive hypoglycaemia, defects in fatty acid metabolism, neuroendocrine disorders, allergic reactions to endogenous hormones and local physiological processes at symptom sites. The role of endorphins in prolactin and vasopressin release has also been suggested as a cause. Most recently, an aberration in the cyclic functioning of the hypothalamic–pituitary axis has been related to PMS[10,11]. Labrum has suggested that abnormal fluctuations in brain levels of serotonin and γ-aminobutyric acid (GABA) and interrelated neuroendocrine processes may be a common cause of the many diverse PMS symptoms. Abraham has noted that PMT-C, which is characterized by food cravings, headache and palpitations, may be due to excessive intake of refined carbohydrates and a magnesium deficiency[4].

PROSTAGLANDINS AND PMS

There is considerable evidence linking PGs to PMS symptoms. These ubiquitous prostaglandins, produced in PMS-associated tissues such as the brain, breasts, gastrointestinal tract, kidney and reproductive tract, participate in many important regulatory functions by interacting with other hormones and with cyclic AMP. In most cases, the symptoms associated with PG excess are similar to those of PMS.

Reproductive tract

Changes in PG levels in the female reproductive tract have been shown to affect the cyclic regression of the corpus luteum[13], menstrual shedding of the endometrium[14] and the decidual reaction[15]. Studies have demonstrated that one PG, prostacyclin, can be found in the amnion, decidua and chorion[16]. Increased blood flow during pregnancy may be regulated by this PG. Prostacyclin may also influence the contractions induced by other PGs. These PGs may in turn influence physiological processes by reinforcing or, alternately, antagonizing the actions of hormones and/or neurotransmitters[13].

The effects of luteal hormone (LH) on ovulation may also be mediated by PGs, although the data are not clear on this point[17]. In the absence of LH, PGs can induce many ovarian functions, such as nuclear maturation division of the oocyte, shift from oestrogen to progesterone production and rupture of the follicle with extrusion of a fertilizable oocyte[18]. These findings are based on studies using indomethacin or aspirin, PG inhibitors that can block ovulation in rats[19] and rabbits[20]. Changes in PG levels have been associated with several disorders of the female reproductive tract, including dysmenorrhoea and endometriosis. Basic research as well as clinical reports confirm that PGs are implicated in primary dysmenorrhoea. Support for the theory that excess PGs cause dysmenorrhoea comes from three findings:

(1) Exogenous PGs cause symptoms similar to those of dysmenorrhoea. For instance, the primary action of $PGF_{2\alpha}$ is to constrict uterine blood vessels, causing anoxia and also myometrial contraction[21].

(2) There are higher PG levels in secretory endometrium than in proliferative; this is consistent with the fact that dysmenorrhoea generally occurs in ovulatory cycles[22].

(3) High levels of PGs are found in the endometrium, jet washings and menstrual fluid of women suffering from dysmenorrhoea[22].

Other studies show that progesterone levels influence the action of PGs on uterine tissue. The uterus is resistant to PG stimulation when progesterone levels are high. But as progesterone levels fall before the menses, high levels of PGs – due to aberrant metabolism or excess or imbalanced production – may trigger dysmenorrhoea[21].

In 1974, Schwartz et al. reported the first successful treatment of dysmenorrhoea with an anti-inflammatory drug, flufenamic acid[14]. Subsequent studies have also shown that PG inhibitors can lessen or eliminate the symptoms of dysmenorrhoea[23-26]. However, dysmenorrhoea is a complex of symptoms which may range far beyond menstrual pain. In a 1982 study, Budoff recognized 25 symptoms of what she termed dysmenorrhoea syndrome, of which 13 were primary: abdominal cramping, backache, headache, nausea, vomiting, diarrhoea, weakness and dizziness, leg pain, insomnia, fatigue/lethargy, anxiety/irritability, swelling and depression[27]. In this double-blind cross-over study, non-steroidal anti-inflammatory drugs were significantly more effective than placebo in relieving 12 of 13 primary dysmenorrhoea symptoms and six of 13 associated symptoms. Non-steroidal anti-inflammatory drugs have also been shown to reduce menorrhagia and alleviate uterine pain after insertion of an IUD[28-31].

Interestingly, women with dysmenorrhoea appear to be somewhat prone to PMS and vice versa. 58% (255 out of 442) of women with PMS also had menstrual pain in one study[32]. In another survey of 137 women with PMS, 31, or 22%, had primary dysmenorrhoea[9]. Hargrove and Abraham[9] also found that PMS was most frequently associated with endometriosis, a disorder that is related to elevated PG levels. Both disorders were alleviated with a non-steroidal anti-inflammatory drug.

Budoff[25], as well as Henzel et al.[26], has suggested that anti-prostaglandins are 'ideal drugs' for the treatment of dysmenorrhoea because their analgesic effects occur quickly, they have few troublesome side-effects and they need only be taken when pain is actually experienced. Approximately 80-85% of women with dysmenorrhoea obtain relief of uterine pain and cramping, nausea, vomiting and diarrhoea with the first prescribed PG inhibitor[23]. These drugs would appear to be good candidates for the treatment of PMS if, like dysmenorrhoea, PMS is wholly or partly caused by excess PGs.

Although PG levels rise during the luteal phase of the menstrual cycle, the symptoms associated with dysmenorrhoea and PMS could be caused by aberrations in PG metabolism, an imbalance among individual PGs or heightened PG production interacting with shifting hormonal levels.

Breast tissue

Both PGE_1[33] and PGE_2[34] have been found in normal human breast tissue. The latter has also been detected in rat breast tissue[35]. PGE_1 mediates prolactin's regulation of breast growth after puberty and during pregnancy. PGE_2's biological role remains more obscure[36].

It is widely accepted that PGs are involved in inflammatory processes, such as vasodilation[37,38]. Studies have also shown that PGs can cause hyperalgesia and pain[38]. It is possible that PGE_2 plays a role in PMS-related breast swelling and tenderness. During the menses, breast engorgement has been determined volumetrically to increase 30–40%, along with increased tenderness[39]. Thermographic examinations have shown that PGE_2 can cause painful vasodilator and thermic effects[36].

In 1976, Preese et al.[40,41], suggested that mastalgia is due to systemic factors, such as an endocrine abnormality. Supporting that hypothesis is a French study by Rolland et al.[36]. These researchers found that patients with benign breast disease had excess oestrogen levels compared to their progesterone levels. This finding suggests that oestrogen-directed biosynthesis of PGs, which occurs in the uterus, may also take place in the breast. Rolland et al.[36] also showed that patients with benign breast mastopathies and mastodynia had significantly higher plasma PGE_2 levels than normal women during both phases of the menstrual cycle.

Most importantly, mastodynia symptoms were relieved when Rolland administered PG inhibitors to ten patients with benign mastopathies[36]. In addition, breast thermogenic responses were positive. in seven patients. The inhibition of prostaglandin biosynthesis results in reduced breast temperatures in most patients through the inhibition of vasodilator and thermic phenomena. These results suggest that breast swelling and tenderness associated with PMS also may be alleviated by PG inhibitors.

Central nervous system

Considerable evidence suggests that PGs are produced at numerous neural and non-neural central nervous system (CNS) sites. Wolfe and Coceani have noted that PGs of the CNS interact with hormones and neurotransmitters to cause both physiological and pathological changes[42]. These authors also noted that PGs, including PGl_2 (prostacyclin) and thromboxane which are synthesized in aggregating platelets or in the parenchyma of organs, may be regulators of cerebrovascular homoeostasis.

Prostaglandins also appear to play a role in regulating body water content, food intake and body temperature. Their precise effects on thirst appear controversial, however. PGs have suppressed drinking in the rat[43] and increased drinking in another species[44,45]. PGs also stimulate the release of antidiuretic hormone (ADH), according to Leksell[45].

PG effects on food intake are equally unsettled. Several PGs are reported to affect the food intake of rats[46]. Baile et al. found in the rat that when PGE_1 was injected into certain hypothalamic sites, food intake was inhibited; this was not true when injection was into other hypothalamic sites[47]. Further studies by this group have also shown that PGE_1 suppressed or stimulated food intake in sheep depending upon the CNS injection site[48].

Since the psychological symptoms of PMS – irritability, depression and lability – are similar to those associated with affective disorders, it is useful to review Horrobin's studies of PG levels in these disorders. He has reported that both a deficiency and an excess of PGE_1 may be associated with affective disorders[49]. He also noted that a change in the ratio of this PG to dopamine may be related to schizophrenia[50].

PMS-type psychological symptoms have been induced by injection of prolactin. However, most women with PMS have normal levels of this hormone. Horrobin suggests that low PGE_1 levels may make PMS sufferers more prone to the effects of prolactin[51]. Data on PGE_1 levels in PMS, however, are not available. In the future, it is possible that treatments for depression and PMS might begin with attention to substances that affect PGE_1 synthesis, such as essential fatty acids, pyridoxine, zinc and vitamin C.

Migraine attacks are part of the symptomatology of some PMS sufferers. PGE_2 has been associated with this painful condition. Since it acts to constrict intracranial and dilate extracranial blood vessels, PGE_2 may be involved in more than one phase of migraine headache[52].

Women with epilepsy may have more seizures during the premenstruum than at other times. Studies of $PGF_{2\alpha}$ levels have shown that epileptics have significantly higher levels of this PG in their cerebrospinal fluid than do non-epileptics. However, levels vary widely between patients and in the same patient at different times[53].

Kidneys

It is well established that PGs are synthesized in the cortex as well as the medulla of human kidneys[54-57], with the medulla responsible for the bulk of production[58]. Though the precise role of renal PGs is still unclear, there is sufficient evidence at present to conclude that they affect the normal functioning of this organ. PGE_2 and $PGF_{2\alpha}$ are the chief products of the renal microsomal arachidonic acid cascade, but PGD_2 and TxA_2 are also produced in the kidney[59]. The rate and amount of renal PG production is affected by numerous factors, such as local enzyme concentrations, which, in turn, depend upon circulating catecholamines and hormones[59]. Renal disorders, such as ureteral obstruction[60], hypertension[61], Bartter's syndrome[62] and hypokalaemic nephropathy[63], may also affect the synthesis of PGs.

A wide variety of animal experiments and clinical studies provide clues to how PGs affect renal function and how their biosynthesis may vary. Three broad groups of drugs have been used to investigate these questions:

(1)   glucocorticoids and mepacrine, which are antagonists of phospholipase A_2, the catalyst for the release of arachidonic acid from membrane phospholipids;

(2)   non-steroidal anti-inflammatory drugs, which inhibit cyclo-oxygenase, the enzyme catalysing production of intermediate PGG_2 and PGH_2 from arachidonic acid;

(3)   imidazole and the hydroperoxy derivatives of indole, which inhibit synthesis of thromboxane or prostacyclin.

Experimental and clinical studies utilizing these drugs have shown that PGs are involved in several renal activities.

(1)   They influence renal blood flow, possibly through the vasodilator actions of PGI_2 and the vasoconstrictor functions[64] of TxA_2.

(2)   They also may mediate the action of loop diuretics by indirectly stimulating PG synthesis and/or inhibiting PG metabolism[6]. PGE_2 and PGI_2 may also be responsible for renin release[65].

(3)   The kidneys' water permeability response to vasopressin may also be modulated by PGs[66]. Several studies with various mammals suggest that a feedback loop is involved. Vasopressin increases production of PGE_2 and stimulates adenylate cyclase activity and water flow. The PGE_2 then inhibits the adenylate cyclase activity, thus modulating vasopressin's water resorption actions.

(4)   Renal PGs maintain the glomerular filtration rate at normal levels through feedback-type interactions with the kidneys' renin-angiotensin system. Circulating angiotensin II (AII) stimulates synthesis of vasodilator PGs. These substances oppose the vasoconstricting effects of AII on renal blood vessels[59].

(5)   The PGI_2 induced by AII increases the release of renin, which then leads to further synthesis of AII.

(6)   PGs may play a role in urinary electrolyte excretion, but precisely how this occurs is still unknown. Some evidence suggests that PGs increase the renal excretion of sodium chloride, though this remains a matter of controversy[67]. If sodium excretion is promoted by PGs, it may be mediated by PG effects on either renal vascular tone or on the transport properties of nephron segments.

(7)   PGs impede the renal concentrating actions of vasopressin[68].

Beyond these seven functions in the normal kidney, PGs – on their own or as mediators of other agents – influence many biochemical actions in various renal disorders.

Because PGs influence water and electrolyte balance, they may contribute to such PMS symptoms as oedema, abdominal bloating, increased thirst and craving for salty foods[10]. These actions may be affected by circulating hormones. If PGs do play a role in causing these symptoms, the PG inhibitors should provide relief.

Gastrointestinal tract

PGs occur in the GI tract of many mammals, including man[69-72]. PGE_2 and PGI_2 appear to have important effects throughout the GI tract[69], while PGA and PGF have been found in the gastric mucosa[73]. PGs perform a number of physiological functions in the gut. They control gastric mucosal microcirculation[74,75], inhibit gastric acid secretion[73], induce bicarbonate secretion in the dog stomach[76], and regulate fluid and ion transport in the human small intestine[71].

Robert *et al.* have suggested that, in rats, PGs are also responsible for 'cytoprotection'[73]. This conclusion resulted from a study in which rats were administered oral doses of boiling water or concentrated acidic, basic or salt solutions. The treatments caused extensive gastric erosions. However, when the tests were repeated with PG-pretreated rats, no lesions developed. Robert and co-workers have suggested that PGs may exert this cellular protection by causing a sudden secretion of mucus, by maintaining a sodium pump in the stomach or by improving resistance of mucosal cells to injury through cyclic AMP stimulation.

In view of these findings, it is not surprising that non-steroidal anti-inflammatory drugs, which inhibit PG synthesis, may cause gastric irritation, bleeding and even ulceration[77,78]. Rainsford *et al.* found a direct relationship between lowered PG levels in the stomach and gastric damage in the pig[79]. In rats, gastric mucosal blood flow was slowed by indomethacin[75]. The same phenomenon occurred with aspirin treatment. Aspirin also allowed the back diffusion of acid into the mucosal tissue, an action which is normally prevented by the mucosa[80].

Paradoxically, there are also numerous reports suggesting that PGs may cause inflammatory changes and damage throughout the gut. In one study, opossums were protected from radiation-induced oesophagitis by PG inhibitors, indomethacin or aspirin. Other animals, however, given PGE_2 before radiation developed worse oesophagitis[81] than non-treated animals. In mice, indomethacin has also reduced intestinal motility caused by radiation[82]. In addition, women irradiated for uterine cancer had significantly less diarrhoea, abdominal pain and flatulence after administration of a PG inhibitor. The authors of that study concluded that the symptoms were probably caused by excess PGs[83].

Other investigations have shown that the diarrhoea, nausea, vomiting and abdominal pain of some GI disorders may be related to PGs. Endogenous PGs appear related to the abdominal pain and vomiting associated with food intolerance, since prostaglandin biosynthesis inhibitors can prevent the symptoms[84,85]. These agents can also minimize the nausea, diarrhoea and vomiting normally associated with PG administration for abortion or induction of labour. They appear to act by inhibiting PG synthesis in the upper GI tract[86].

PGs may also mediate the inflammatory response in ulcerative colitis. Compared to normals, significantly higher PG levels have been found in rectal biopsies[87] and stools[88] of patients with this illness. This suggests that some of sulphasalazine's therapeutic effect in this disease may be due to its inhibition of PG synthesis.

The various E type PGs have also been associated with diarrhoeal diseases such as cholera[89,90] and with diarrhoea resulting from carcinoid tumours[91] or endotoxin[92]. Bismuth subsalicylate, a PG inhibitor, can effectively abate the diarrhoea, vomiting, cramps, nausea and anorexia of viral gastroenteritis[93].

It remains unclear, however, precisely how PGE causes diarrhoea, creates loss of fluid into the intestine, inhibits sodium absorption and stimulates chloride secretion. One theory holds that PGE interacts with cyclic AMP to affect small intestine secretion[90]. Another view states that PGE may mediate the actions of the cholera exotoxin[89]. Proponents of both views suggest that PG inhibitors may be useful in cholera treatment.

These studies suggest that PGs in the GI tract are responsible for many gastrointestinal complaints, such as nausea, vomiting, abdominal cramping and diarrhoea. Since these symptoms usually respond to PG inhibitors, these drugs might also relieve similar PMS-related symptoms.

PMS THERAPY

Numerous therapeutic approaches have been advocated to alleviate the symptoms of PMS. Hormones, neurotransmitters and their receptor agonists, vasodilators, diuretics, oral contraceptives, vitamins, an essential fatty acid precusor and PG inhibitors themselves have been studied.

Dalton in 1964 was the first to administer natural progesterone in open studies for relief of PMS symptoms[94]. Later, researchers using double-blind study methods duplicated her open studies. They found that progesterone was no more effective than placebo[10,11]. Other practitioners have reported positive results with progesterone. High-dose progesterone suppositories have been shown to be effective in reducing anxiety and producing sedation in a Boston PMS clinic[95]. Dydroges-terone and norethindrone, synthetic progestogens, are also reported to lessen mental PMS symptoms and fluid retention[11]. To date, there is no agreement on the effectiveness of progesterone and progestogens in the treatment of PMS. In addition, progesterone, a natriuretic, has not proved consistently effective in reducing fluid retention[7].

The effects of angiotensin II, aldosterone and the catecholamines – norepinephrine, epinephrine and dopamine – on fluid retention have been investigated with uneven results. Dietary changes like salt reduction and xanthine restriction provide relief from bloatedness[7]. Bromocriptine, a powerful prolactin antagonist, also appears to reduce oedema and breast swelling. Although a double-blind cross-over bromocriptine trial by Benedek-Jazmann showed the drug more effective than placebo[96], other controlled trials have not[97]. Results of trials with pyridoxine (vitamin B_6) for migraine and fluid retention have been erratic and contradictory[98]. Studies with other agents also showed mixed results. For example, the vasodilator, Bellergal, reduced psychological and breast symptoms of PMS in a double-blind study[5]; diuretics like chlorthalidone, thiazide, triamterene, furosemide and metolazone have been used with some success in treating bloating. Oral contraceptives, though useful in the treatment of dysmenorrhoea, do not reduce PMS symptoms consistently. However, they are effective in 15–25% of patients. Danazol, a synthetic androgen, appears to relieve breast tenderness in fibrocystic disease, but has unpleasant side-effects, such as acne, oedema, deepening of the voice, etc.[99].

Vitamin E, a known antioxidant, has also been tested in the treatment of PMS. Because lipogenase and cyclo-oxygenase pathways of arachidonic acid metabolites involved free-radical mediated reactions, it is possible that vitamin E may inhibit the chemical cascade leading to the formation of PGs, thromboxane and HETE [(S)-12-hydroxy-5,8,14-cis, 10-trans-eicosatetraenoic acid (Samuelssons' HETE)] London et al. recently found that vitamin E was significantly better than placebo in improving most PMS symptoms[100]. It did not improve symptoms of the PMT-H class, however, as defined by Abraham. These symptoms – breast tenderness, swelling, weight gain and abdominal bloating – increased significantly after vitamin E administration[100].

Prostaglandin biosynthesis inhibitors

There is ample evidence that PG inhibitors are of value in the treatment of several gynaecologic disorders. They provide analgesic relief of dysmenorrhoea[25,26,101] and reduce excessive blood loss in patients with menorrhagia[30]. Several studies suggest they may also be effective in alleviating the symptoms of PMS.

One of the earliest trials was a double-blind single-cross-over study of mefenamic acid vs. placebo, conducted by Budoff[7]. 43 women with a variety of PMS complaints were enrolled in the 8-month trial. Because the participants were being treated for dysmenorrhoea as well as PMS, and because in the United States mefenamic acid may only be given for 7 days, the women were directed not to begin medication until 4 days before the expected onset of menstrual flow. The dosage was 250 mg, four times a day. Although patients actually began therapy on the average 2·8 days before menses, reductions in breast tenderness, abdominal bloating, ankle swelling, menstrual pain and associated nausea were statistically significant. Mefenamic acid was no more effective than placebo in treating tension, lethargy and depression, however. There were improvements, though not statistically significant, in premenstrual back pain, weakness/dizziness and nervousness/irritability. It is possible that dosage and/or length of treatment were inadequate to affect some of these parameters.

Because there is no 7-day limit on the use of mefenamic acid in Australia, however, 37 patients in a trial conducted by Wood and Jakubowicz began medication at the onset of PMS symptoms and continued at least until menses[102]. In this randomized double-blind cross-over trial, patients served as their own controls for one cycle and then were given either mefenamic acid or placebo during the luteal phase of the next two cycles[102]. The dosage was 500 mg, three times a day. This drug significantly alleviated tension, irritability, depression, pain, anger and headaches. In contrast to Budoff's data, it was not effective in treating fluid retention or breast symptoms. The Australian authors reported that the prostaglandin inhibitors did not appear effective in a few women with premenstrual symptoms referable to the breast only. They suggest that mefenamic acid may relieve symptoms by reducing synthesis of endometrial PGs, by antagonizing PG tissue receptors, or by influencing hormone production in the corpus luteum. The study also demonstrated that most women (34 of 37 patients) with premenstrual symptoms also have menstrual symptoms. In a second, longer study with 80 patients, Jakubowicz et al. reported that PMS symptoms were substantially improved or completely eliminated in 86% of patients after administration of mefenamic acid[103]. Dosage was 500 mg three times daily. The mean duration of treatment was 13 months. In 15% of patients, however, four 500 mg doses per day were necessary for optimal relief. 19 of the 80 patients also took part in a randomized, double-blind, controlled trial spanning three menstrual cycles. Patients received either no treatment, placebo or drug during each cycle. Among this group, 68% reported their symptoms were best relieved by mefenamic acid. Because many women have PMS symptoms for 7–14 days, the 7-day limit on use of mefenamic acid in the US makes it unsuitable for the treatment of PMS in that country.

Another PG inhibitor, naproxen sodium (NS), was recently evaluated by Budoff[104]. 42 patients suffering from PMS were treated in this randomized, double-blind cross-over study spanning six cycles. Of this group, the 21 patients who finished at least five cycles were evaluated. These patients were also evaluated for improvements in fibrocystic breast disease symptoms. Participants were randomized to receive NS for three cycles followed by three cycles of placebo or vice versa. Therapy, 500 mg of NS every 12 h or matching placebo, started 10 days before the expected onset of menses and ceased when flow began. Each evening during treatment, patients recorded the maximum severity of 70 potential PMS symptoms on a scale of 0 (no symptom) to 99 (severe symptom). In addition, fibrocystic breast disease was evaluated on the basis of symptoms, ultrasonography and by physical examination at admission, cross-over and completion of the trial. At the end of menstrual flow patients completed a Menstrual Log sheet, on which they reported the characteristics of their menstrual flow. Standard laboratory data were obtained at the visit before cross-over and at the final visit.

Regarding the statistical analysis, the basic measure of efficacy was mean severity of PMS symptoms. For any given PMS symptom, analysis was limited to those patients who had reported that symptom prior to this study. Because of the non-normal distributions of the parameters, non-parametric statistical analysis was appropriate. Cross-over analyses were conducted using the Wilcoxon/Mann–Whitney test (two-tailed) based on the Grizzle model[105], which assumes equal residual treatment effects at cross-over.

In addition, a global measure of symptom severity was constructed by obtaining for each patient a mean of the individual assessments of mean severity for each symptom that (1) was reported by that patient as being characteristic of her PMS, and (2) was reported by at least 75% of the patients. These global mean severity scores were then analysed using the same statistical methods described above.

Of the 70 symptoms investigated, 46 were identified by ten or more patients as characteristic of the PMS. Cross-over analysis of the mean severities of these 46 symptoms showed 37 with treatment effects favouring NS and nine favouring placebo. Of these 46 symptoms, four exhibited statistically significantly less mean severity during NS cycles than in placebo cycles. These four symptoms were jittery feeling ($p = 0.02$), headache ($p = 0.02$), lack of self-control ($p = 0.03$) and tension ($p = 0.03$). Figure 18.1 presents the results of treatment for ten major categories encompassing those PMS symptoms (of the 70 investigated) that were reported by at least 16 of the 21 patients prior to

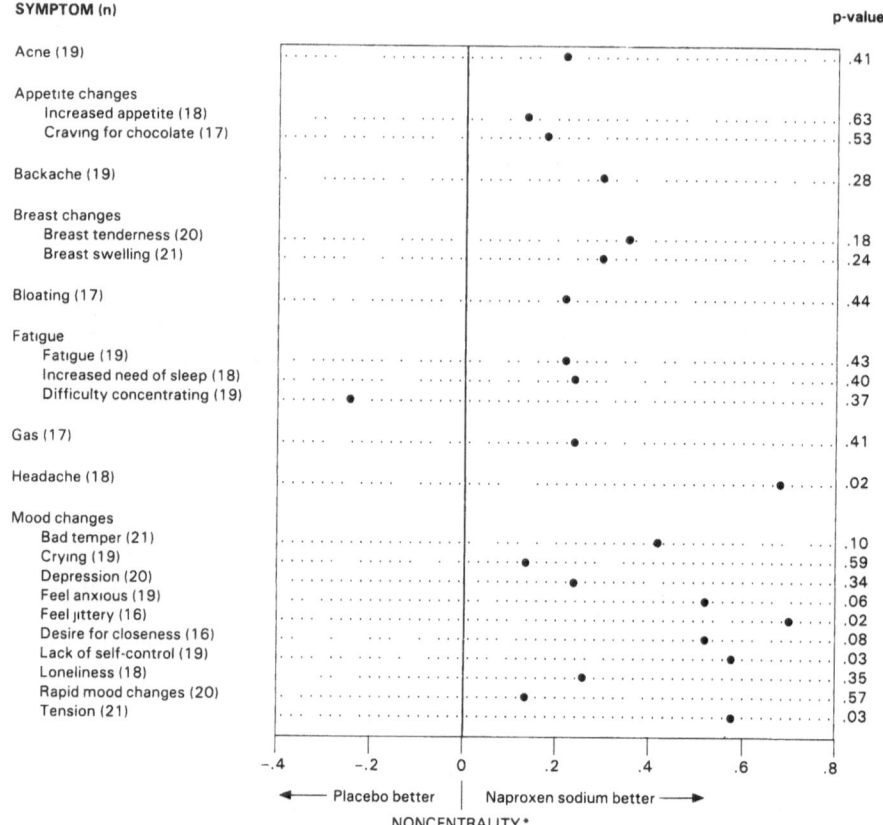

Figure 18.1   Mean severity, at baseline and during treatment, of those PMS symptoms reported by at least 16 of the 21 patients who were valid for efficacy analysis. *Noncentrality is a direct measure of the trend of the data. A noncentrality value of 0 indicates no difference between treatments. Minimum and maximum values are −1 and +1. Noncentrality values are calculated as 2U/mn −1 where U is the Mann-Whitney U statistic and m and n are the respective sample sizes. U/mn is the probability that a random patient on NS treatment would have less severity than a random patient on placebo treatment

treatment. NS was significantly more effective than placebo in relieving symptoms falling within two of these categories (mood changes and headache).

The means for the global measure of symptom severity during each treatment period are plotted in Figure 18.2. Overall mean severity was significantly less during naproxen treatment than during placebo treatment (U/mn = 0·78; $p = 0·03$).

Cross-over analysis of menstrual characteristics following premen-

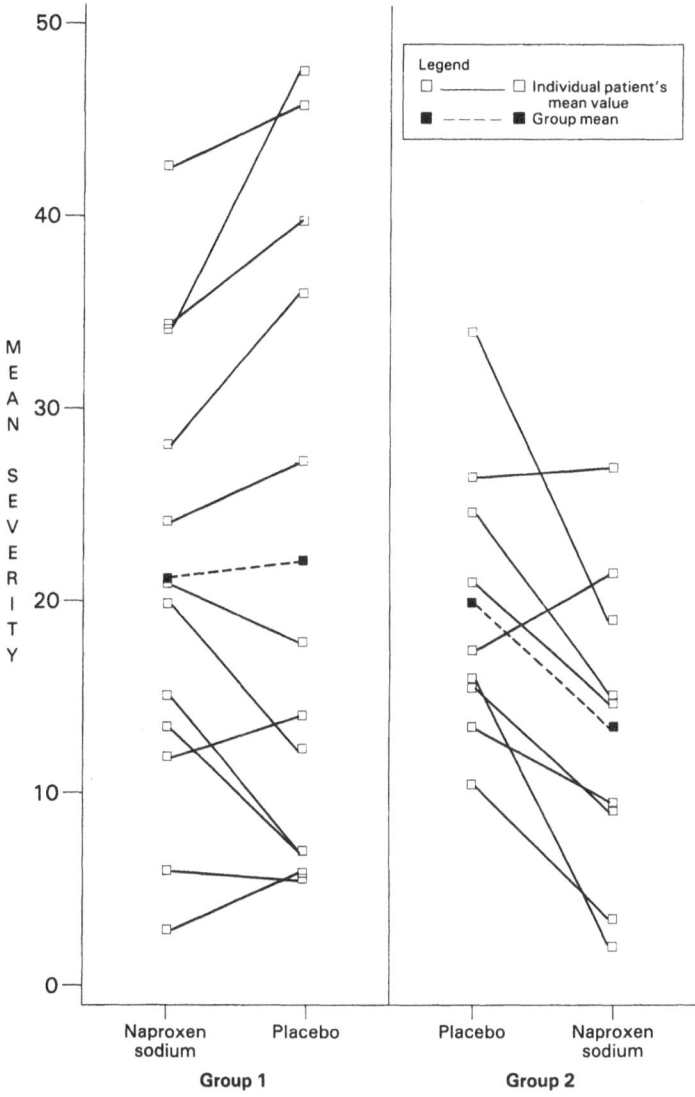

Figure 18.2 Overall means for symptom severity for the symptoms reported by at least 16 of the 21 patients who were valid for efficacy analysis. Each patient's overall mean represents only those symptoms for which she had a baseline mean severity score greater than zero

strual treatment (i.e. timing of cycle, days of menstruation, amount of flow, duration of flow and menstrual pain) indicated no significant differences between NS and placebo treatments.

Breast examination, symptoms reports and ultrasound all showed

NS lessened fibrocystic symptoms better than placebo, although none of the differences between these regimes were statistically significant.

NS provided significantly greater overall symptom relief than did placebo. For the many symptoms and the many organ systems that cause symptoms premenstrually, the ubiquitous PGs may best explain the syndrome, recognizing that production of PGs and sex hormones are intertwined and are dependent upon one another.

THE TREATMENT OF PMS TODAY

Because there are still many questions to be answered about the aetiology of PMS, I favour treating women with this disorder in a very conservative, step-wise fashion. They begin by permanently forgoing all methylxanthine products, including coffee, tea, cola and chocolate. 7–10 days before the expected onset of menses, they start to consume less sodium. They are also instructed to eat three well-balanced meals a day and three high-protein snacks during the premenstruum. A daily multivitamin/mineral formulation is added at this point. If, after 2 months, this regimen is not effective, the patient takes a potassium-sparing diuretic, once a day for the 7 days prior to menses. Nearly 75% of patients have subjective relief of symptoms within 3–4 months. Those who were heavy coffee drinkers can almost be guaranteed that they will feel less tense and have less breast tenderness and swelling.

Patients who still suffer from their PMS symptoms despite this regimen may be placed on naproxen sodium, given 2–4 times a day beginning at onset of symptoms, or vaginal or rectal progesterone suppositories, or 100 mg of natural progesterone in oil administered intramuscularly 7–10 days before menses. Small doses of an antianxiety or antidepressive agent may be added for women with primarily psychological symptoms, or an oral contraceptive might be tried (Figure 18.3).

PMS is an age-old problem. Yet only in recent years has it become the focus of research attention. Far more study is needed, especially to define PMS symptoms more uniformly and precisely, to better understand which PGs may be related to which PMS symptoms, to develop more accurate methods for measuring the presence of PGs in body fluids and tissues and to clarify patient subtypes responding to different therapies. If, as Abraham suggests, there are several types of PMS, it is possible that each would have its own aetiology and its own therapy.

Given our present state of knowledge, the safest treatment for PMS should be utilized first, whenever possible. Simple dietary methods can help large numbers of patients. Drugs and hormones should not be employed unless other efforts have been unsuccessful. Of great impor-

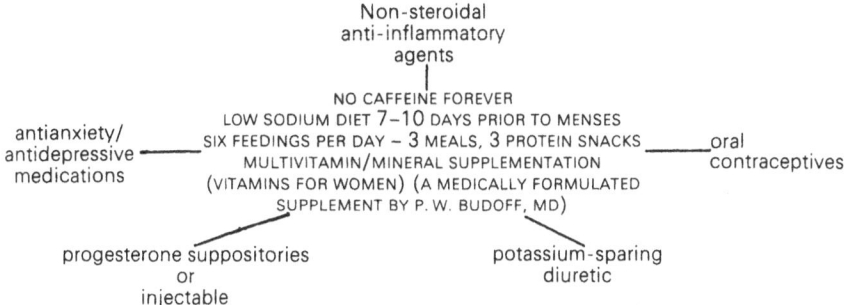

Most women with PMS tend to have poor eating habits. Nutritional counselling therefore is most helpful. A multivitamin/mineral supplement is therefore included in therapy.

Figure 18.3   PMS Therapy flowchart

tance, women with PMS should not be treated with drug regimens that have more side-effects than their disorder. Bromocriptine and danazol, for example, which may adversely affect health and mental well-being, should not become chronic therapy for healthy young patients.

ACKNOWLEDGEMENT

Much of the material for this chapter was derived from Budoff, P. W. (1983) The use of prostaglandin inhibitors for the premenstrual syndrome. *J. Reprod. Med.*, **28**, 469.

References

1 Frank, R. T. (1931). The hormonal causes of premenstrual tension. *Arch. Neurol. Psychiatr.*, **26**, 1053

2 Gray, L. A. (1941). Use of progesterone in nervous tension states. *South. Med. J.* **34**, 1004

3 Dalton, K. (1979). *Once a Month.* (Claremont, California: Hunter House)

4 Abraham, G. E. (1983). Nutritional factors in the etiology of the premenstrual tension syndromes. *J. Reprod. Med.*, **28**, 446

5 Robinson, K., Humtington, R. M. and Wallace, M. G. (1977). Treatment of the premenstrual syndrome. *Br. J. Obstet. Gynaecol.*, **87**, 784

6 Sutherland, H. and Stewart, I. (1965). A critical analysis of the premenstrual syndrome. *Lancet*, **1**, 1180

7 Budoff, P. W. (1980). *No More Menstrual Cramps and Other Good News.* (New York; G. P. Putnam's Sons). Paperback Edition: (New York: Penguin Books) 1981

8 Rosseinsky, D. R. and Hall, P. G. (1974). An evolutionary theory of premenstrual tension. *Lancet*, **2**, 1024

9 Hargrove, J. T. and Abraham, G. E. (193). The ubiquitousness of premenstrual tension in gynecologic practice. *J. Reprod. Med.*, **28**, 435

10 Reid, R. L. and Yen, S. S. C. (1981). Premenstrual syndrome. *Am. J. Obstet. Gynecol.*, **139**, 85

11 Sharma, V. (1982). Premenstrual syndrome. *Practitioner*, **226**, 1091

12 Labrum, A. H. (1983). Hypothalamic, pineal and pituitary factors in the
   premenstrual syndrome. *J. Reprod. Med.*, **28**, 438

13 Goldberg, V. J. and Ramwell, P. W. (1975). The role of prostaglandins in
   reproduction. *Physiol. Rev.*, **55**, 325

14 Schwartz, A., Zor, U., Lindner, H. R. *et al.* (1974). Primary dysmenorrhea: alleviation by an inhibitor of prostaglandin synthesis and action.
   Obstet. Gynecol., **44**, 709

15 Lindner, H. R., Zor, U., Kohen, F. *et al.* (1980). Significance of prostaglandins in the regulation of cyclic events in the ovary and uterus. In
   Samuelsson, B., Ramwell, P. W. and Paoletti, R. (eds.) *Advances in Prostaglandin and Thromboxane Research.* Vol. 8, pp. 1371-90. (New York:
   Raven Press)

16 Mitchell, M. D., Biddy, J. G., Hicks, B. R. *et al.* (1978). Possible role for
   prostacyclin in human parturition. *Prostaglandins*, **16**, 931

17 Chobsieng, P., Naor, Z., Koch Y. *et al.* (1975). Stimulatory effect of
   prostaglandin E₂ on LH release in the rat: evidence for hypothalamic site
   of action. *Neuroendocrinology*, **17**, 12

18 Lindner, H. R., Tsafriri, A. and Lieberman, M. E. (1974). Gonadotropin
   action on cultured Graafian follicles: induction of maturation division of
   the mammalian oocyte and differentiation of the luteal cell. *Recent Prog.
   Horm. Res.*, **30**, 79

19 Armstrong, D. T. and Grinwich, D. L. (1972). Blockade of spontaneous
   and LH-induced ovulation in rats by indomethacin, an inhibitor of prostaglandin biosynthesis. *Prostaglandins*, **1**, 21

20 Grinwich, D. L., Kennedy, T. G. and Armstrong, D. T. (1972). Dissocia-
   tion of ovulatory and steroidogenic actions of luteinizing hormone in
   rabbits with indomethacin, an inhibitor of prostaglandin biosynthesis.
   Prostaglandins, **1**, 89

21 Caspo, A. I. (1980). A rationale for the treatment of dysmenorrhea. *J.
   Reprod. Med. (Suppl.)*, **4**, 213

22 Ylikorkala, O. and Dawood, M. Y. (1978). New concepts in dysmenorrhea. *Am. J. Obstet. Gynecol.*, **130**, 833

23 Budoff, P. W. (1981). Antiprostaglandins for primary dysmenorrhea. *J.
   Am. Med. Assoc.*, **246**, 2576

24 Budoff, P. W. (1977). Treatment of dysmenorrhea. *Am. J. Obstet. Gynecol.*, **129**, 232

25 Budoff, P. W. (1979). Use of mefenamic acid in the treatment of primary
   dysmenorrhea. *J. Am. Med. Assoc.*, **241**, 2713

26 Henzl, M. R., Massey, S., Hanson, F. W. *et al.* (1980). Primary dysmenorrhea: the therapeutic challenge. *J. Reprod. Med. (Suppl.)*, **25**,
   226

27 Budoff, P. W. (1982). Zomepirac sodium in the treatment of primary dysmenorrhea syndrome. *N. Engl. J. Med.*, **307**, 714

28 Buttram, V., Izu, A. and Henzl, M. R. (1978). Naproxen sodium in uterine pain following intrauterine contraceptive device insertion. *Am. J. Obstet. Gynecol.*, **134**, 575

29 Davies, A. J., Anderson, A. B. M. and Turnbull, A. C. (1981). Reduction by naproxen of excessive menstrual bleeding in women using intrauterine devices. *Obstet. Gynecol.*, **57**,

30 Jakubowicz, D. L. and Wood, C. (1978). The use of the prostaglandin synthetase inhibitor mefenamic acid in the treatment of menorrhagia. *Aust. N. Z. J. Obstet. Gynaecol.*, **18**, 135

31 Massey, S. E., Varady, J. C. and Henzl, M. R. (1974). Pain relief with naproxen following insertion of an intrauterine device. *J. Reprod. Med.*, **13**, 226

32 Wood, C., Larsen, L. and Williams, R. (1979). Menstrual characteristics of 2,343 women attending the Shepherd Foundation. *Aust. N.Z. J. Obstet. Gynaecol.*, **19**, 107

33 Horrobin, D. F. (1979). Cellular basis of prolactin action: involvement of cyclic nucleotides, polyamines, prostaglandins, steroids, thyroid hormones, Na/K ATPases and calcium: relevance to breast cancer and the menstrual cycle. *Med. Hypotheses*, **5**, 599

34 Bennett, A., McDonald, A. M., Simpson, J. S. *et al.* (1975). Breast cancer, prostaglandins and bone metastases. *Lancet*, **1**, 1218

35 To, D., Smith, F. L. and Carpenter, M. P. (1980). Mammary gland prostaglandin synthesis: effect of dietary lipid and propyl gallate. In Samuelsson, B., Ramwell, P. W. and Paoletti, R. (eds.) *Advances in Prostaglandin and Thromboxane Research*. Vol. 1, pp. 1807–15. (New York: Raven Press)

36 Rolland, P. H., Martin, P. M., Rolland, A. M. *et al.* (1979). Benign breast disease: studies of prostaglandin E_2, steroids, and thermographic effects of inhibitors of prostaglandin biosynthesis. *Obstet. Gynecol.*, **54**, 715

37 Ferreira, S. H. and Nakamura, M. (1980). Pathogenesis and phamacology of pain. In Weissmann, G., Paoletti, R. and Samuelsson, B. (eds.) *Advances in Inflammation Research*. Vol. 1, pp. 317–39. (New York: Raven Press)

38 Weissman, G., Smolen, J. E. and Korchak, H. (1970). Prostaglandins and inflammation: receptor/cyclase coupling as an explanation of why PGEs and PGI_2 inhibit functions of inflammatory cells. In Samuelsson, B., Ramwell, P. W. and Paoletti, R. (eds.), *Advances in Prostaglandin and Thromboxane Research*. Vol. 8, pp. 1637–53. (New York: Raven Press)

39 Hamilton, T. and Rankin, M. E. (1975) Changes in volume of the breast during the menstrual cycle. *Br. J. Surg.* **62**, 660

40 Preece, P. E., Hughes, L. E., Baum, M. *et al.* (1977). Studies on breast pain. *Br. J. Surg.*, **61**, 322

41 Preece, P. E., Hughes, L. E., Mansel, R. E. *et al.* (1976). Clinical syndromes of mastalgia. *Lancet*, **2**, 670

42 Wolfe, L. S. and Coceani, F. (1979). The role of prostaglandins in the central nervous system. *Ann. Rev. Physiol.*, **41**, 699

43 Epstein, A. N. (1980). The neuroendocrinology of thirst and salt appetite. In Ganong W. F., and Martini, L. (eds.) *Frontiers in Neuroendocrinology.* Vol. 5, pp. 101–31. (New York: Raven Press)

44 Andersson, B. and Leksell, L. G. (1975). Effects on fluid balance of intraventricular infusions of prostaglandin E_1. *Acta Physiol. Scand.,* **93**, 286

45 Leksell, L. G. (1976). Influence of prostaglandin E_1 on cerebral mechanisms involved in the control of fluid balance. *Acta Physiol. Scand.,* **98**, 85

46 Scaramuzzi, O. E., Baile, C. A. and Mayer, J. (1971). Prostaglandins and food intake of rats. *Experientia,* **27**, 256

47 Baile, C. A., Simpson, C. W., Bean, S. M. *et al.* (1973). Prostaglandins and food intake of rats: A component of energy balance regulation? *Physiol. Behav.,* **10**, 1077

48 Baile, C. A., Martin, F. H., Forbes, J. M. *et al.* (1974). Intrahypothalamic injections of prostaglandins and prostaglandin antagonists and feeding in sheep. *J. Dairy Sci.,* **57**, 81

49 Horrobin, D. F. and Manku, M. S. (1980). Possible role of prostaglandin E_1 in the affective disorders and in alcoholism. *Br. Med. J.,* **280**, 1363

50 Horrobin, D. F. (1979). Schizophrenia: reconciliation of the dopamine, prostaglandin and opioid concepts and the role of the pineal. *Lancet,* **1**, 529

51 Horrobin, D. F. (1983). The role of essential fatty acids and prostaglandins in the premenstrual syndrome. *J. Reprod. Med.,* **28**, 465

52 Pickard, J. D., MacDonell, L. A., MacKenzie, E. T. *et al.* (1977). Prostaglandin-induced effects of the primate cerebral circulation. *Eur. J. Pharmacol.,* **43**, 343

53 Wolfe, L. S. and Mamer, O. A. (1975). Measurement of prostaglandin $F_{2\alpha}$ levels in human cerebrospinal fluid in normal and pathological conditions. *Prostaglandins,* **9**, 183

54 Friesinger, G. C., Oelz, O., Sweetman, B. J. *et al.* (1978). Prostaglandin D_2, another renal prostaglandin? *Prostaglandins,* **15**, 969

55 Grenier, F. C. and Smith, W. L. (1978). Formation of a 6-keto-prostaglandin F_1 by collecting duct cells isolated from rabbit renal papilla. *Prostaglandins,* **16**, 759

56 Lee, J. B., Covino, B. G., Takman, B. H. *et al.* (1965). Renal medullary vasodepressor substance, medullin: isolation, chemical characterization and physiological properties. *Circ. Res.,* **17**, 57

57 Lee, J. B., Crowshaw, K., Takman, B. H. *et al.* (1967). The identification of prostaglandins E_2, $F_{2\alpha}$ and A_2 from rabbit kidney medulla. *Biochem. J.,* **105**, 1251

58 Hassid, A. and Dunn, M. J. (1980). Microsomal prostaglandin biosynthesis of human kidneys. *J. Biol. Chem.,* **255**, 2472

59 Levenson, D. J., Simmons, C. E. and Brenner, B. M. (1982). Arachidonic acid metabolism, prostaglandins and the kidney. *Am. J. Med.,* **72**, 354

60 Cadnapaphornchai, P. and Aisenbrey, G. (1978). Prostaglandin-mediated hyperemia and renin-mediated hypertension during acute ureteral obstruction. *Prostaglandins,* **16**, 965

61 Levy, J. V. (1977). Changes in systolic arterial blood pressure in normal

and spontaneously hypertensive rats produced by acute administration of inhibitors of prostaglandin biosynthesis. *Prostaglandins*, **13**, 153

62 Bowden, R.E., Gill, J.R., Radfar, N. *et al.* (1978). Prostaglandin synthetase inhibitors in Bartter's syndrome. *J. Am. Med. Assoc.*, **239**, 117

63 Linas, S.L. and Dickman, D. (1981). Mechanism of the decrease in renal blood flow in the potassium depleted conscious rat. *Clin. Res.*, **29**, 73

64 Gerber, J.G. and Nies, A.S. (1979). The hemodynamic effects of prostaglandins in the rat: Evidence for important species variation in renovascular responses. *Circ. Res.*, **44**, 406

65 Lee, J.B. (1980). Prostaglandins and the renin–angiotensin axis. *Clin. Nephrol.*, **14**, 159

66 Zusman, R.M. (1981). Prostaglandins, vasopressin and renal water absorption. *Med. Clin. N. Am.*, **65**, 915

67 Tannenbaum, J., Splawinski, J.A., Oates, J.A. *et al.* (1975). Enhanced renal prostaglandin production in the dog: I. Effects on renal function. *Circ. Res.*, **36**, 197

68 Jamison, R.L. (1981). Urine concentrations and dilution: the role of antidiuretic hormone and urea. In Brenner, B.M. and Rector, F.C. (eds.) *The Kidney*. 2nd Edn., pp. 495–550. (Philadelphia: W.B. Saunders)

69 Moncada, S., Salmon, J.A., Vane, J.R. *et al.* (1978). Formation of prostacyclin and its product 6-*oxo*-PGF$_{1\alpha}$ by the gastric mucosa of several species. *J. Physiol.*, **275**, 4

70 Peskar, B.M., Seyberth, H.W. and Peskar, B.A. (1980). Synthesis and metabolism of endogenous prostaglandins by human gastric mucosa. In Samuelsson, B., Ramwell, P.W. and Paoletti, R. (eds.) *Advances in Prostaglandin and Thromboxane Research*. Vol. 8, pp. 1511–14. (New York: Raven Press)

71 Rask-Madsen, J. and Bukhave, K. (1980). Saturation kinetics applied to the inhibition by PGE$_2$ of ion transport in the isolated human jejunum. In Samuelsson, B., Ramwell, P.W. and Paoletti, R. (eds.) *Advances in Prostaglandin and Thromboxane Research*. Vol. 8, pp. 1595–1601. (New York: Raven Press)

72 Whittle, B.J.R. (1981). Temporal relationship between cyclooxygenase inhibition, as measured by prostacyclin biosynthesis, and the gastrointestinal damage induced by indomethacin in the rat. *Gastroenterology*, **80**, 94

73 Robert, A., Nezamis, J.E., Lancaster, C. *et al.* (1979). Cytoprotection by prostaglandins in rats. *Gastroenterology*, **77**, 433

74 Main, I.H.M. and Whittle, B.J.R. (1975). Investigation of the vasodilator and antisecretory role of prostaglandins in the rat gastric mucosa by use of non-steroidal anti-inflammatory drugs. *Br. J. Pharmacol.*, **53**, 217

75 Whittle, B.J.R. (1977). Mechanisms underlying gastric mucosal damage induced by indomethacin and bile-salts, and the actions of prostaglandins. *Br. J. Pharmacol.*, **60**, 455

76 Allen, A. and Garner, A. (1980). Mucus and bicarbonate secretion in the stomach and their possible role in mucosal protection. *Gut*, **21**, 249

77 Lanza, F.L., Royer, G.L., Nelson, R.S. *et al.* (1979). The effects of

ibuprofen, indomethacin, aspirin, naproxen and placebo on the gastric mucosa of normal volunteers. *Dig. Dis. Sci.*, **24**, 823

78 Pemberton, R. E. and Strand, L. J. (1979). A review of upper-gastro-intestinal effects of the newer nonsteroidal antiinflammatory agents. *Dig. Dis. Sci.*, **24**, 53

79 Rainsford, K. D., Peskar, B. M. and Brune, K. (1981). Relationship between inhibition of prostaglandin production and gastric mucosal damage induced by antiinflammatory drugs may depend on type of drugs and species. *J. Pharm. Pharmacol.*, **33**, 127

80 Whittle, B. J. R., Higgs, G. A., Eakins, K. E. *et al* (1980). Selective inhibition of prostaglandin production in inflammatory exudates and gastric mucosa. *Nature (London)*, **284**, 271

81 Northway, M. G., Bennett, A., Carroll, M. *et al.* (1980). Comparative effects of anti-inflammatory agents and radiotherapy on normal esophagus and tumors in animals. *Gastroenterology*, **78**, 1229

82 Borowska, A., Sierakowski, S., Mackowiak, J. *et al.* (1979). A prostaglandin-like activity in small intestine and postirradiation gastrointestinal syndrome. *Experientia*, **35**, 1368

83 Mennie, A. T., Dalley, V. M., Dineen, L. C. *et al.* (1975). Treatment of radiation-induced gastrointestinal distress with acetylsalicylate. *Lancet*, **2**, 942

84 Buisseret, P. (1980). Drug treatment of allergic gastroenteritis. *Am. J. Clin. Nutr.*, **33**, 865

85 Lieb, J. (1978). Prostaglandin-synthesis inhibitors in prophylaxis of food intolerance. *Lancet*, **2**, 157

86 Walker, S. L. (1973). Prostaglandins and the gastrointestinal tract. *Gut*, **14**, 402

87 Sharon, P., Ligumsky, M., Rachmilewitz, D. *et al* (1978). Role of prostaglandins in ulcerative colitis. *Gastroenterology*, **75**, 638

88 Goudl, S. R., Brash, A. R. and Conolly, M. E. (1977). Increased prostaglandin production in ulcerative colitis. *Lancet*, **2**, 98

89 Bennett, A. (1971). Cholera and prostaglandins. *Nature (London)*, **231**, 536

90 Jacoby, H. I. and Marshall, C. H. (1972). Antagonism of cholera enterotoxin by anti-inflammatory agents in the rat. *Nature (London)*, **235**, 163

91 Williams, E. D., Karim, S. M. M. and Sandler, M. (1968). Prostaglandin secretion by medullary carcinoma of the thyroid. *Lancet*, **1**, 22

92 Bennett, A. (1980). Prostaglandins as factors in diseases of the alimentary tract. In Samuelsson, B., Paoletti, R. and Ramwell, P. W. (eds.) *Advances in Prostaglandin and Thromboxane Research.* Vol. 2, pp. 547-55. (New York: Raven Press)

93 Steinhoff, M. C., Douglas, R. G., Greenberg, H. B. *et al.* (1980). Bismuth subsalicylate therapy of viral gastroenteritis. *Gastroenterology*, **78**, 1495

94 Dalton, K. (1977). *Premenstrual Syndrome and Progesterone Therapy.* (New York: James H. Heinemann)

95 Jakubowicz, D. L. (1982). Personal communication

96 Benedek-Jazmann, L. V. and Hearn-Sturtevant, M. D. (1976). Premen-

strual tension and functional infertility: aetiology and treatment. *Lancet*, **1**, 1095

97 Ghose, K. and Coppen, A. (1977). Bromocriptine and premenstrual syndrome: controlled study. *Br. Med. J.*, **1**, 147

98 O'Brien, P. M. S. (1982). The premenstrual syndrome: a review of the present status of therapy. *Drugs*, **24**, 140

99 Day, J. (1979). Danazol and the premenstrual syndrome. *Postgrad. Med. J.* (suppl. 5), **55**, 87

100 London, R. S. *et al.* (1983). Effects of alpha tocopherol on premenstrual symptomatology: a double-blind study. *J. Am. Col. Nutr.*, **2**, 115

101 Anderson, A. B. M., Haynes, P. J., Fraser, I. S. *et al.* (1978). Trial of prostaglandin-synthetase inhibitors in primary dysmenorrhoea. *Lancet*, **1**, 345

102 Wood, C. and Jakubowicz, D. L. (1980). The treatment of premenstrual symptoms with mefenamic acid. *Br. J. Obstet. Gynaecol.*, **87**, 627

103 Jakubowicz, D. L., Godard, E. and Dewhurst, J. (1984). The treatment of premenstrual tension with mefenamic acid: analysis of prostaglandin concentrations. *Br. J. Obstet. Gynaecol.*, **91**, 78

104 Budoff, P. W. (1985). Naproxen sodium vs. placebo in the treatment of premenstrual syndrome and fibrocystic breast disease. In Press

105 Grizzle, J. E. (1965). The two-period change-over design and its use in clinical trials. *Biometrics*, **21**, 467

19
Endometriosis

E. L. MARUT

INTRODUCTION

Endometriosis is classically defined as 'the presence of ectopic tissue which possesses the histological structure and function of the uterine mucosa. It also includes the abnormal conditions which may result not only from the invasion of organs and other structures by this tissue, but also from its reaction to menstruation'[1].

The presence of endometrium in an ectopic location is associated with pelvic pain, dysmenorrhoea, abnormal bleeding, infertility, spontaneous abortion, and obstruction or invasion of other organs. While endometriosis has been identified in such distant sites as the nose and knee, the thrust of this chapter is pelvic endometriosis. It is not our purpose to review the history, clinical presentation or pathology of endometriosis[2-7], but rather to consider the endocrinological implications of endometriosis and its manifestations on pelvic pain and reproductive dysfunction. More specifically, this chapter will consider the evidence for the involvement of prostaglandins in the pathogenesis of endometriosis.

The incidence of endometriosis varies; in series of women undergoing laparotomies, it is as high as 50%[2,8-10], but these are selected groups of women and do not represent the general female population. Since there is only a weak correlation between the presence of endometriosis and the severity of its symptoms, a substantial degree of underdiagnosis must exist. The apparent increase in diagnoses observed in recent years results from the more liberal use of laparoscopy than from the recognition of so-called risk factors (race, career, personality)[11-13]. Unless symptoms such as pain, infertility and other organ involvement lead the woman to seek medical attention, she may continue throughout her reproductive life in a relatively fertile, painless state, albeit in the presence of mild, moderate and occasionally severe endometriosis. Simply stated, endometriosis is a common condition; a high index of suspicion must precede the making of this diagnosis, and direct visualization or histological biopsy is required to substantiate its presence.

Whereas discovery of any degree of endometriosis was once followed by immediate therapy, temporization has been suggested more recently in cases of mild endometriosis[14]. The role of endometriosis in infertility is not completely clear[15], since pregnancy often occurs in its presence. A comparison of endometriosis with normal (eutopic) endometrium needs to be made in order to explain the abnormalities found when the endometrium grows outside the uterine cavity. It is easy to understand how the cyclic exposure of endometriosis to the proliferative, secretory and ultimately sloughing effects of consecutive ovulatory cycles results in extension and progression of the initial disease process. However, the finding of endometriosis in euoestrogenic, anovulatory women, as well as adolescents in early post-pubertal years, is more difficult to explain, since the same sequence does not occur.

AETIOLOGY: THEORIES PAST AND PRESENT

The origin of endometriosis has been ascribed to several mechanisms; most common is the theory of tubal regurgitation in which menstrual endometrium flows retrograde into the pelvis[1,16-18]. From an anatomic point of view, this process explains the most common sites of endometriosis, i.e. the ovary, the structure closest to the tubal ostium, and the cul de sac, the most dependent portion of the pelvis. By way of providing additional clinical corroboration of these observations, women with Muellerian anomalies and obstruction to menses have a high rate of endometriosis[11,19]. Nonetheless, retrograde menstruation has been observed in many women without endometriosis. Thus, at best retrograde menstruation is only a contributing factor, while the true aetiology must lie in the ability to accept the ectopic growth of endometrial tissue rather than treating it as a foreign substance and degrading and removing it. Other theories which require transportation and implantation of normal endometrium to an abnormal site involve vascular or lymphatic spread[20-23] tend to explain distant endometriosis, as well as direct implantation such as in laparotomy and episiotomy scars[24]. Direct implantation may also explain intrapelvic non-contiguous spread of disease.

In contrast, the concept of coelomic metaplasia holds that tissues differentiating from the embryonic coelomic epithelium retain their capacity for metaplasia into tissues of similar origin. The capacity of epithelial ovarian tumours to mimic other Muellerian tissues is also thought to be related to this mechanism. When one considers that the 'less differentiated' tissues of the peritoneum and ovary are most likely to be affected by endometriosis, this concept seems tenable. The remote existence of endometriosis can also be explained by the embryonic migration and trapping of coelomic cells which then become differentiated into endometriosis under the influence of specific modulatory regulators. As yet, the genetic message for this metaplastic process has not been determined, nor is there evidence that specific stimuli cause metaplasia solely to endometrium. Once again, the underlying process which permits or nurtures the development of ectopic endometrium appears to represent the real answer, and this is not at hand. There must be some truth to the concept of coelomic metaplasia, however, since reports of endometriosis in a female without a uterus and in a man on oestrogen therapy have been published[25].

The more recently described induction theory combines elements of retrograde menstruation (along with vascular or lymphatic spread and direct implantation) with coelomic metaplasia[23,26,27]. Accordingly, viable endometrium does not grow in the place where it arrives or is transported, but rather provides an indirect effect on the coelomic tissue with which it makes contact and thus leads to metaplasia to endometrial tissue. Although tissue-free extracts of endometrium have been shown to cause development of endometrial cells from connective tissue[28], no similar studies have proven the association of a local or circulating substance in the *in vivo* development of endometriosis[26]. Women with severe pelvic inflammatory disease or pelvic cancer do not have an equal incidence of endometriosis compared to women without these diseases. Experiments in primates have shown that endometriosis occurs in the face of intraperitoneal menstruation[29], as well as when direct implantation of endometrium is performed under various stimulation regimens[30]. Taken together with the *in vitro* evidence that menstrual endometrium is capable of growth in culture[31-33], the overwhelming evidence favours retrograde menstrual flow as the primary event in initiating pelvic endometriosis. The question still remains, however, why does every woman with retrograde menstruation not develop endometriosis, and why does every woman who undergoes uterine surgery not develop either peritoneal or abdominal wall endometriosis? The missing factor is that of susceptibility: something which allows the endometrium to grow and behave as if it were still in the uterine cavity. The apparent multifactorial inheritance of endometriosis has been reported, suggesting a genetic predisposition to development of the disease, irrespective of all other circumstances[34,35].

In this regard, the problem seems to be that of recognition or non-recognition of tissues. The cellular immune response of women with endometriosis seems to be attenuated and similar to the response seen in monkeys with experimental endometriosis[36]. This would suggest a permissiveness of the individual enzyme system, at least at the cellular levels, that allows implantation and subsequent growth of ectopic endometrium. Since the studies on this point have been retrospective, i.e. the endometriosis was already established, it remains possible that the blunted immune response may be a consequence, not a prerequisite of the development. Moreover, a large antigenic 'load' of endometrium may render the woman anergic to the stimulus so that endometriosis may develop successfully. The reports of increased antibody formation to endometrium as well as ovarian tissue in women with endometriosis[37] are confusing in this regard. That author attempts to explain reproductive dysfunction by the presence of these antibodies, but the antibodies do not explain the presence of the endometriosis. Other workers have reported the presence of IgG and IgA in the endometrium of patients with endometriosis, but without correlation with infertility[38]. Badawy et al. reported precipitation reactions between endometrial homogenates and serum of peritoneal fluid in some women with endometriosis[39], again suggesting an autoimmune phenomenon.

If an autoimmune situation does exist, the tissue itself should be obliterated by the exaggerated immune response and this is not the case. One explanation for this discrepancy would be a divergence between cell-mediated and humoral immunity; whereas the former is deficient and permits the endometriosis to grow, the latter reflects the B-cell recognition of 'foreign' tissue and is merely an appropriate host response.

The question remains, however, how can this immunological process be responsible for the reproductive dysfunction? An inflammatory element is obvious in endometriosis; histological sections reveal both acute and chronic changes, and a high number of macrophages in the peritoneal fluid is consistently reported in patients with endometriosis compared to other patients[40]. If, on the other hand, total immunological quiescence were possible, endometriosis would not have its devastating effects on the pelvic structures. Inflammation is a critical factor in the development of adhesions, fibrosis and possibly humoral effects. The significance of increased macrophages in the peritoneal fluid in endometriosis is unclear, however: this observation suggests a recognition of the endometriosis by the immune system, coupled with a response insufficient to attack and remove the abnormal tissue. Its relevance is possibly diagnostic, suggesting microscopic endometriosis if high macrophage counts are found without visible disease[41]. On the other hand, the macrophages may affect sperm function, either in the cul de sac or in the tube, since tubal secretions and peritoneal fluid seem to form a common pool[42]. The macrophages may affect oocyte viability as well.

An alternative hypothesis proposed by Weed et al. is that an auto-immune response results from the endometriosis rather than causing it[43]. The antigenic agents are typical endometrial proteins which are not expelled from the uterus as is the menstrual effluent, but are processed as foreign material by the immune system. The autoimmune reaction varies from patient to patient; thus, an immune response to eutopic endometrium which could influence implantation may or may not result in infertility. The severity of endometriosis would correlate well with infertility, since a larger antigenic load may evoke a greater response. Thus, one could postulate that suppression of endometriotic implants would reduce the immune response; however, clinical use of immunosuppressive agents has not been reported in endometriosis.

An interesting concept in the aetiology of endometriosis arises from reports of associated luteal deficiency and luteinized unruptured follicles. While the presence of endometriosis may affect ovulation and corpus luteum function, it has been hypothesized that the opposite relationship may also exist. That is, the woman who luteinizes but does not rupture the preovulatory follicle permits endometriosis to grow by failure to provide high levels of peritoneal fluid steroids. These high steroid levels are thought to locally impair cyclic endo-metriotic growth[44]. While there may be some merit to this theory, inconsistencies have also been noted. First of all, fertile women without endometriosis have been shown to have presumptive luteinized unrup-tured follicles[45]. Also, the rate of endometriosis in anovulatory women who do not rupture follicles is significantly lower than in normally ovulatory women[46]. Thus, the absence or presence of follicular fluid in the peritoneal cavity alone cannot explain the development of endo-metriosis. This hypothesis will be discussed below.

In summary, the pathogenesis and maintenance of endometriosis remains speculative after more than five decades of consideration with this exception; it is now recognized that the underlying factor seems to be an individual response to the presence of ectopic endometrium. It is also now well accepted that the fibrosis, adhesions, and direct infiltration of ovarian and tubal tissue have a mechanical effect on tubo-ovarian function that impairs fertility. However, the effect of a mild or moderate endometriosis without mechanical distortion is much more uncertain. In fact, the current treatment of minimal or mild endometriosis thought to be causing infertility is expectant[14]; that is, a period of time is allowed to elapse following laparoscopic diagnosis before conservative surgery or suppressive therapy is undertaken. Thus, the diagnosis of mild (stage I by AFS classification[47]) endometriosis does not preclude pregnancy when untreated for a period of 6–12 months. Nonetheless, there is an increased association between all stages of endometriosis (possibly even microscopic endometriosis[48]) with infertility. In the absence of direct mechanical involvement, the likely aetiology becomes humoral; the most likely and amply studied humoral agents are the prostaglandins. This will be the main focus of the final section.

CONVENTIONAL THERAPY

The traditional treatment of endometriosis has been suppressive, extirpative, or both. The first approach began with the use of continuous diethylstilboestrol in an attempt to overstimulate and exhaust the disease[49]. This treatment was later replaced by that of 'pseudopregnancy' which utilized continuous oestrogen–progestin (oral contraceptive) administration or progestin alone to cause stimulation, decidualization and finally atrophy of the endometriosis[50,51]. The observation that functional hypogonadal states such as lactational amenorrhoea[52] prevented recurrence or exacerbation of endometriosis following pregnancy led to the concept of providing a 'pseudomenopause' or state of minimal ovarian function. Most commonly this is achieved by the use of danazol, the attenuated androgen which affects hypothalamic–pituitary activity as well as ovarian and adrenal steroid synthesis[53–56]. Most recently, the use of GnRH agonists has been advocated in the treatment of endometriosis[57,58]. GnRH agonists are apparently devoid of drug-related side-effects, and desensitize the pituitary to endogenous GnRH, resulting in a 'medical oophorectomy' or more properly stated, a 'selective medical hypophysectomy'.

Compared to medical suppressive therapy, conservative surgery has virtually equal success rates in terms of relief of symptoms and achieved pregnancy rates. Surgical approaches include operative laparoscopy with lysis of adhesions and electrocauterization or laser ablation of endometriotic implants[59–61]. Alternatively, conservative laparotomy compares favourably to medical therapy for equivalent staging[62,63], and obviously laparotomy remains the treatment of choice with those patients with extensive adhesive disease, endometriomata and anatomic distortion[64–68]. It is also preferred in those patients who do not wait the 6 or 9 months required for suppression. Data on results for various modes of therapy may be found in reviews[6,7]. The use of combined suppressive and surgical therapy has changed from the provision of surgery followed by suppression to suppression followed by surgery[69,70]. Recurrence of disease (about 15% in the first year) is similar following either therapy[71–73]. Early pregnancy should be attempted to avoid a return of disease and the initial infertile state.

Modern treatment of endometriosis is aimed at removal or regression of endometriotic implants and surgical lysis of adhesions. Unfortunately, all treatment modalities have side-effects and also carry with them the risk of incomplete removal or regression and the inability to recognize the presence of microscopic endometriosis.

This returns us to the theme raised previously: what is the role of prostaglandins in endometriosis-associated infertility and, more importantly, is there a role for prostaglandin synthetase inhibitors in treatment of endometriosis?

LUTEINIZED UNRUPTURED FOLLICLES

The two hypotheses currently espoused in regard to endometriosis are the luteinized unruptured follicle theory and the prostaglandin theory. A full discussion on the former is necessary because of its contrast to the latter, as well as its hypothesized dual cause and effect role in endometriosis. The issue of prostaglandins and endometriosis needs to include a full discussion of the luteinized unruptured follicle hypothesis, since interactions between these two have been proposed.

The association between luteal abnormalities and endometriosis has been made on several occasions. Grant[74] found 45% of patients with endometriosis had corpus luteum dysfunction as judged by several criteria, and Kistner[75] noted a 25% incidence of luteal defects in patients after conservative surgery for endometriosis. Brosens et al.[76] first suggested a relationship between corpora lutea lacking ovulatory stigmata and endometriosis; 21% of patients with endometriosis were found to have a luteal stigma as compared to 94% of controls. Koninckx[44,47] extended this observation by noting that 50% of patients with or without endometriosis had viable endometrial cells in the cul de sac at laparoscopy; at the same time, those with ovulatory stigmata had peritoneal fluid concentrations of progesterone and oestradiol 5–20 times greater than those with luteinized unruptured follicles. This finding, taken with evidence that peritoneal fluid was ovarian in origin, led Koninckx et al.[78] to propose the hypothesis that the higher peritoneal fluid steroid concentration in normal ovulatory women precluded the ability of endometrial cells to implant and grow; conversely, those women with luteinized unruptured follicles were more likely to have a peritoneal environment conducive to the development of endometrosis. Later studies have failed to demonstrate this association consistently, however. Dmowski et al.[79] found no difference in the incidence of endometriosis among infertile patients with luteinized unruptured follicles, while Vanrell noted 50% of fertile women to have no visible stigma[45], and Portuondo et al. reported conception in cycles where no stigmata had been seen at laparoscopy[80]. Lesorgen et al.[81] used peritoneal fluid concentrations of oestradiol and progesterone as well as visual detection of corpora lutea without stigmata to diagnose luteinized unruptured follicles. They found an increased incidence of unruptured follicles in infertile patients but no significant excess in endometriosis patients compared to those with inflammatory disease unless biochemical criteria were used. Peripheral steroid levels and endometrial dating were not different among all patients. On the other hand, using laparoscopy Dhont et al.[82] noted a stigma in about half of patients at different times of the luteal phase, irrespective of fertility status, presence or absence of endometriosis, or time of the luteal phase. Oestradiol and progesterone levels were highest early in the luteal phase whether or not a stigma was present. Most importantly, no differences in steroid levels were noted between patients with or without endometriosis, or those with or without ovulatory stigmata. Crain and Luciano[83] noted no difference in luteal phase peritoneal fluid volumes or steroid concentration between patients with or without endometriosis. In summary, sufficient diversity of findings, observations and opinions exists to cast doubt on a direct cause or effect relationship between endometriosis and luteinized unruptured follicles.

PROSTAGLANDINS

The search for the mechanism by which endometriosis causes infertility has concentrated on the potential role of prostaglandins, and the remainder of this discussion will explore the evidence relating to the role of prostaglandins in the pathophysiology of the disease. Pickles et al.[84] have shown that prostaglandins are produced by the endometrium of both normal and dysmenorrhoeic patients. Since endometriosis is histologically similar to normal endometrium, it is reasonable to consider that humoral effects of endometriosis are also mediated by prostaglandins. Prostaglandins are known to participate in numerous reproductive processes, and the events which have been targeted as possibly being moderated by the action of prostaglandins are luteal function and tubal function. The controversy over the association between the luteinized unruptured follicle syndrome and endometriosis has already been discussed. Because prostaglandins increase in the preovulatory rabbit follicle[85] and are thought to be important in the release of the ovum, excess prostaglandins would not be likely to impair ovulation. In fact, the human preovulatory follicle $PGF_{2\alpha}$ content is a marker of oocyte maturity[86]. Nonetheless, Ylikorkala and Tenhunen[87] noted no difference in follicular fluid prostaglandins among patients with endometriosis or other tubal diseases. While subhuman primate ovulation may be blocked by prostaglandin synthetase inhibitors[88], no similar situation exists in humans[89]. The possibility of luteal insufficiency being caused by excess prostaglandins in the human is also unlikely[90,91] since only one study infusing pharmacological amounts of prostaglandin impaired normal luteal function[92]. However, prostaglandin-induced luteolysis in the monkey has been readily demonstrated[93].

The more likely effect of prostaglandins is on tubal motility; studies have shown that $PGF_{2\alpha}$ enhances tubal contractility and ovum transport in rabbits[94,95] and women[96], although ovum transport is not necessarily affected in the latter[97]. PGE_2 has an opposite effect on tubal contractility[98], as does PGI_2 which has been shown to relax tubal musculature in humans[99]. In addition, the administration of an esterified $PGF_{2\alpha}$ to rhesus monkeys affected ovum transport[100]. The question still remains whether the local production of prostaglandins by endometriotic implants results in increased levels of these compounds in the peritoneal fluid. Normal endometrium produces the classic prostaglandins as well as PGI_2 and TxA_2[101]. PGE_2 and $PGF_{2\alpha}$ are secreted in greater amounts by the eutopic endometrium of patients with endometriosis, and these patients also excrete more prostaglandins than controls[102]. The production of the stable metabolites of PGI_2 and TxA_2 (6-keto-$PGF_{1\alpha}$ and TxB_2) is similar in endometriotic implants as it is in eutopic endometrium, and the peritoneal endometriosis secretion of these prostaglandins is greater than that produced by ovarian endometriosis[103]. Since the *in vitro* as well as the *in vivo* release of $PGF_{2\alpha}$ by ovarian endometriosis is greater than that of the surrounding normal ovarian tissue, the possibility of an as yet undefined deleterious effect on ovarian function still exists[104,105].

Conflicting results have been obtained when peritoneal fluid has been assayed for the presence of prostaglandins. In a preliminary study, Meldrum et al.[106] found increased $PGF_{2\alpha}$ in cul de sac fluid obtained from infertile patients. Drake et al.[107] demonstrated an increase of 6-keto-$PGF_{1\alpha}$ and TxB_2 in the peritoneal fluid patients with endometriosis, as well as an increase of fluid volume[41,107]. Interestingly, patients with unexplained infertility also had elevated levels of these compounds, leading these authors to suggest microscopic endometriosis as the source of infertility. Cycle day was not reported in this series, although it becomes a point of contention in other studies. However, in patients with unexplained infertility matched for cycle day with controls, the same increase in peritoneal fluid 6-keto-$PGF_{1\alpha}$ and TxB_2 was noted[108]. On the other hand, Rock et al.[109] studied patients with and without endometriosis in the late follicular phase and found no difference between patients and controls with respect to peritoneal fluid volume or its prostaglandin content or any correlation of these parameters with the severity of endometriosis. This study did not consider differences in the luteal phase when one could expect the endometriosis to be at its peak: Singh et al.[110] showed $PGF_{2\alpha}$ to be lowest in normal proliferative endometrium and highest in secretory endometrium. Badawy et al.[111] found no correlation between peritonal fluid levels of the $PGF_{2\alpha}$ metabolite 13,14-dihydro-15-keto $PGF_{2\alpha}$, PGFM, or PGE_2 and the stage of endometriosis. However, PGFM content (but not PGE_2) was greater in patients with endometriosis or unexplained infertility than controls. This finding suggests the presence of microscopic endometriosis in unexplained infertility patients to account for their increased PGFM. There were wide individual variations in all groups, and no association between any prostaglandin levels and cycle day was found. Sgarlata et al.[112] also observed no association between endometriosis and prostaglandins. These workers found no difference in peritoneal fluid volumes or peritoneal plasma or prostaglandins in women undergoing proliferative phase laparoscopy with or without endometriosis[112]. Similarly, Dawood et al.[113] only noted increased peritoneal fluid concentrations (but not total content) of 6-keto-$PGF_{1\alpha}$ in patients with endometriosis.

Another possibility explaining the presence of prostaglandins in peritoneal fluid of patients with endometriosis comes from the study of Haney et al.[40], who noted increased macrophages as well as peritoneal fluid volume in patients with endometriosis. Macrophages are known[114] to secrete TxA_2 as well as to release lysosomal phospholipases which convert arachidonic acid to prostaglandins[115]. Macrophages also can phagocytize sperm[42] independent of any effect on prostaglandins, and thus may play a dual role in the initiation of the process of infertility. Halme et al.[116] found no difference in peritoneal fluid prostaglandins, but did note an increase in activated macrophages in patients with endometriosis. Interestingly, the peritoneum itself (in rabbits) is capable of prostaglandin synthesis which may be induced by irritant stimuli[117].

Badaway et al.[118] have noted increased macrophages, activated macrophages, $PGF_{2\alpha}$, PGE_2 and complement components in the peritoneal fluid of patients with endometriosis, thus associating the cellular, prostaglandin and immunological aspects of endometriosis for the first time.

Animal models have provided additional insight into the pathophysiology of endometriosis. DiZerega et al.[30] demonstrated the dependence of experimental implants on steroid hormone support in monkeys. Schenken and Asch[119] induced endometriosis in rabbits and found increased peritoneal fluid $PGF_{2\alpha}$ but not PGE_2. They also found that an ovulatory defect was important in the decreased fertility in this species. In further studies in monkeys, Schenken et al.[120] found increased peritoneal fluid $PGF_{2\alpha}$ in the luteal phase of monkeys who underwent surgical induction of endometriosis compared to those whose surgery consisted of peritoneal adipose grafts. Compared to preoperative levels, peritoneal fluid $PGF_{2\alpha}$ was elevated only in moderate or severe cases of endometriosis. Abnormalities in the infertile monkeys included luteinized unruptured follicles, luteal phase defects, and adhesive disease[121]. The levels of $PGF_{2\alpha}$ were not higher in infertile cycles, nor was there an increased incidence in spontaneous abortion. Microscopic disease was found in nearly one third of the animals. These authors concluded that the contradictory observations in women were due to the fact that studies were conducted during the proliferative phase of the menstrual cycle as well as to possible differences in functional and cellular aspects of endometriosis between the two species[120].

This last suggestion seems to have been described very well by Vasquez et al.[122] who performed scanning electron microscopy on biopsies of endometriosis. Three types of endometriosis were detected; differences may explain inconsistencies of peritoneal fluid measurements and effects on reproductive function. Some intraperitoneal endometriosis had no gland openings but was associated with deeper glands and stroma; lesions such as these or even retroperitoneal endometriosis might be expected to make little contribution to peritoneal fluid, while intraperitoneal endometriosis with surface epithelium and stroma is far more likely to release increased secretory products. In addition, standard histology of ectopic endometrium does not show cyclic changes as seen in eutopic endometrium. This may be explained by different steroid receptors in endometriosis compared to normal endometrium[123], and thus a different cellular response to the circulating steroid hormones.

PROSTAGLANDIN SYNTHETASE INHIBITORS AND ENDOMETRIOSIS

The massive amount of apparently contradictory data regarding prostaglandins, macrophages, fluid volume and effects of endometriosis can be explained by diverse host responses and variations in functional and morphological qualities of endometriosis. Although it is reasonable to conclude that prostaglandins are involved in the pathophysiology of endometriosis, the individual importance of these compounds cannot be easily predicted. Only two studies to date have considered therapy of endometriosis with prostaglandin synthetase inhibitors, and the results of these efforts are contradictory. The first, by Kauppila *et al.*[124] used a placebo-controlled double-blind study to show improvement of symptoms of secondary dysmenorrhoea with tolfenamic acid (which also blocks prostaglandins at the target organ) but not with aspirin or indomethacin. The other, by Ylikorkala[103], found no benefit from any of these same three drugs.

To fully consider the possibility of therapy of endometriosis using prostaglandin synthetase inhibitors when adhesive disease is not a component would require comparing individual patients. Serving as their own controls, peritoneal fluid parameters would have to be measured before and after prostaglandin synthetase inhibitor therapy, with care to control for cycle day and to identify secretory function and cellular morphology. Animal studies could be similarly constructed, since the diversity of findings in human studies also seem to exist in primate models. Of course, symptomatology as well as fertility would also have to be monitored longitudinally in order to confirm or refute an effect of these drugs.

Treatment of ovulatory dysfunction would logically take place in the periovulatory interval, while that of tubal dysfunction may require exposure of the early conceptus to pharmacological therapy. The wide-ranging differences in mechanistic hypotheses and the controversy surrounding detection of prostaglandins in endometriosis make further studies mandatory before any unifying theory may be advanced.

ACKNOWLEDGEMENT

The author wishes to express his sincere appreciation to Mrs Mary Coppolillo for her expert assistance in the preparation of the manuscript for this chapter.

References

1 Sampson, J. A. (1940). The development of the implantation theory for the origin of peritoneal endometriosis. *Am. J. Obstet. Gynecol.*, **40**, 549

2 Kitchin, J. D. (1982) Endometriosis. In Sciarra, J. J. (ed.) *Gynecology and Obstetrics*. Vol. 1, pp. 1–12 (Philadelphia: Harper & Row)

3 Novak, E. R. and Woodruff, J. D. (1979). *Novak's Gynecologic and Obstetric Pathology*. 8th edn., p. 561. (Philadelphia: W. B. Saunders)

4 Ranney, B. (1980). Endometriosis: pathogenesis, symptoms, and findings. *Clin. Obstet. Gynecol.*, **23**, 865

5 Ranney, B. (1980). Etiology, prevention, and inhibition of endometriosis. *Clin. Obstet. Gynecol.*, **23**, 875

6 Batt, R. E. and Naples, J. D. (1982). Conservative surgery for endometriosis in the infertile couple. In Leventhal, J. M. (ed.) *Current Problems*

in Obstetrics and Gynecology. Vol. 6, No. 1, pp. 11–39. (Chicago: Year-book Medical Publishers Inc.)

7 Polan, M. L. (1984). Endometriosis. *Sem. Reprod. Endocrinol.*, **2**, 186

8 Kistner, R. W. (1979). Endometriosis and infertility. *Clin. Obstet. Gynecol.*, **22**, 101

9 Williams, T. J. and Pratt, J. H. (1977). Endometriosis in 1000 consecutive celiotomies: incidence and management. *Am. J. Obstet. Gynecol.*, **129**, 245

10 Ranney, B. (1978). Endometriosis. *Obstet. Gynecol. Annu.*, **7**, 219

11 Schifrin, B. S., Erez, S. and Moore, J. G. (1973). Teenage endometriosis. *Am. J. Obstet. Gynecol.*, **116**, 973

12 Goldstein, D. P., deCholnoky, C., Emans, S. J. and Leventhal, J. M. (1980). Laparoscopy in the diagnosis and management of pelvic pain in adolescents. *J. Reprod. Med.*, **24**, 251

13 Chatman, D. L. (1976). Endometriosis and the black woman. *J. Reprod. Med.*, **16**, 303

14 Schenken, R. S. and Malinak, L. R. (1982). Conservative surgery versus expectant management for the infertile patient with mild endometriosis. *Fertil. Steril.*, **37**, 183

15 Muse, K. N. and Wilson, E. A. (1982). How does mild endometriosis cause infertility? *Fertil. Steril.*, **38**, 142

16 Sampson, J. A. (1921). Perforating haemorrhagic (chocolate) cysts of the ovary: their importance and especially their relation to pelvic adenomas of endometrial type. *Arch. Surg.*, **3**, 245

17 Sampson, J. A. (1927). Peritoneal endometriosis due to menstrual dissemination of endometrial tissue into peritoneal cavity. *Am. J. Obstet. Gynecol.*, **14**, 422

18 Sampson, J. A. (1925). Heterotopic or misplaced endometrial tissue. *Am. J. Obstet. Gynecol.*, **10**, 649

19 Nunley, W. C., Jr. and Kitchin, J. D. III. (1980). Congenital atresia of the uterine cervix with pelvic endometriosis. *Arch. Surg.*, **115**, 757

20 Sampson, J. A. (1925). Endometrial carcinoma of the ovary arising in endometrial tissue in that organ. *Arch. Surg.*, **10**, 1

21 Halban, J. (1924). Metastatic hysteroadenosis. *Wien. Klin. Wochenschr.*, **37**, 1205

22 Javert, C. T. (1949). Pathogenesis of endometriosis based on endometrial homeoplasia, direct extension, exfoliation and implantation, lymphatic and hematogenous metastasis. *Cancer*, **2**, 399

23 Scott, R. B., Novak, R. J. and Tindale, R. M. (1958). Umbilical endometriosis and Cullen's sign: study of lymphatic transport from pelvis to umbilicus in monkeys. *Obstet. Gynecol.*, **11**, 556

24 Ridley, J. H. (1968). The histogenesis of endometriosis. *Obstet. Gynecol. Surv.*, **23**, 1

25 Pinkert, T. C., Catlow, C. E. and Straus, R. (1979). Endometriosis of the urinary bladder in a man with prostatic carcinoma. *Cancer*, **43**, 1562

26 Scott, R. B., TeLinde, R. W. and Wharton, L. R. Jr. (1953). Further studies on experimental endometriosis. *Am. J. Obstet. Gynecol.*, **66**, 1082

27 Scott, R. B. and TeLinde, R. W. (1950). External endometriosis: the scourge of the private patient. *Ann. Surg.*, **131**, 697

28 Merrill, J. A. (1966). Endometrial induction of endometriosis across Millipore filters. *Am. J. Obstet. Gynecol.*, **94**, 780

29 TeLinde, R. W. and Scott, R. B. (1950). Experimental endometriosis. *Am. J. Obstet. Gynecol.*, **60**, 1147

30 diZerega, G. S., Barber, D. L. and Hodgen, G. D. (1980). Endometriosis: role of ovarian steroids in initiation, maintenance, and suppression. *Fertil. Steril.*, **33**, 649

31 Keetel, W. C. and Stein, R. J. (1951). The viability of the cast-off menstrual endometrium. *Am. J. Obstet. Gynecol.*, **61**, 440

32 Geist, S. H. (1933). The viability of fragments of menstrual endometrium. *Am. J. Obstet. Gynecol.*, **25**, 751

33 Cron, R. S. and Gey, G. (1927) The viability of cast-off menstrual endometrium. *Am. J. Obstet. Gynecol.*, **43**, 645

34 Simpson, J. L., Elias, S., Malinak, L. R. and Buttram, V. C. Jr. (1980). Heritable aspects of endometriosis. I. Genetic studies. *Am. J. Obstet. Gynecol.*, **137**, 327

35 Malinak, L. R., Buttram, V. C., Jr., Elias, S. and Simpson, J. L. (1980). Heritable aspects of endometriosis. II. Clinical characteristics of familial endometriosis. *Am. J. Obstet. Gynecol.*, **137**, 332

36 Dmowski, W. P., Steele, R. W. and Baker, G. F. (1981). Deficient cellular immunity in endometriosis. *Am. J. Obstet. Gynecol.*, **141**, 377

37 Mathur, S., Peress, M. R., Williamson, H. O., Youmans, C. D., Maney, S. A. Garvin, A. J., Rust, P. F. and Fudenberg, H. H. (1982). Autoimmunity to endometrium and ovary in endometriosis. *Clin. Exp. Immunol.*, **50**, 259

38 Saifuddin, A., Buckley, C. H. and Fox, H. (1983). Immunoglobulin content of the endometrium in women with endometriosis. *Int. J. Gynecol. Pathol.*, **2**, 255

39 Badawy, S. Z. A., Cuenca, C., Stitzel, A., Jacobs, R. and Tomar, R. (1984). Autoimmune phenomena in infertile patients with endometriosis. *Obstet. Gynecol.*, **63**, 271

40 Haney, A. F., Muscato, J. J. and Weinberg, J. B. (1981). Peritoneal fluid cell populations in infertility patients. *Fertil. Steril.*, **35**, 696

41 Drake, T. S., O'Brien, W. F., Ramwell, P. W. and Metz, S. A. (1981). Peritoneal fluid thromboxane B_2 and 6-keto-prostaglandin $F_{1\alpha}$ in endometriosis. *Am. J. Obstet. Gynecol.*, **140**, 401

42 Muscato, J. J., Haney, A. F. and Weinberg, J. B. (1982). Sperm phagocytosis by human peritoneal macrophages: a possible cause of infertility in endometriosis. *Am. J. Obstet. Gynecol.*, **144**, 503

43 Weed, J. C. and Arquembourg, P. C. (1980). Endometriosis: can it produce an autoimmune response resulting in infertility? *Clin. Obstet. Gynecol.*, **23**, 885

44 Koninckx, P. R., Ide, P., Vandenbroucke, W. and Brosens, I. A. (1980). New aspects of the pathophysiology of endometriosis and associated infertility. *J. Reprod. Med.*, **24**, 257

45 Vanrell, J. A., Balasch, J., Fuster, J. S. and Fuster, R. (1982). Ovulation stigma in fertile women. *Fertil. Steril.*, **37**, 712

46 Soules, M. R., Malinak, L. R., Bury, R. and Poindexter, A. (1976). En-

dometriosis and anovulation: a coexisting problem in the infertile female. *Am. J. Obstet. Gynecol.*, **125**, 412

47 The American Fertility Society. (1979). Classification of endometriosis. *Fertil. Steril.*, **32**, 633

48 Dmowski, W. P. and Scommegna A. (1976). A rationale for treatment of endometriosis with danazol. In Greenblatt, R. (ed.) *Recent Advances in Endometriosis.* pp. 87–99. (Princeton: Excerpta Medica)

49 Karnaky, K. J. (1948). The use of stilbesterol for endometriosis. *South Med. J.*, **41**, 1109

50 Kistner, R. W. (1958). The use of newer progestins in the treatment of endometriosis. *Am. J. Obstet. Gynecol.*, **75**, 264

51 Moghissi, K. S. and Boyce, C. R. (1976). Management of endometriosis with oral medroxyprogesterone acetate. *Obstet. Gynecol.* **47**, 265

52 Meigs, J. V. (1922). Endometrial hematomas of the ovary. *Boston Med. Surg. J.*, **187**, 1

53 Greenblatt, R. B., Dmowski, W. P., Mahesh, V. B. and Scholer, H. F. L. (1971). Clinical studies with an antigonadotropin: danazol. *Fertil. Steril.*, **22**, 102

54 Barbieri, R. L. and Ryan, K. J. (1981). Danazol: endocrine pharmacology and therapeutic applications. *Am. J. Obstet. Gynecol.*, **141**, 543

55 Madanes, A. E. and Farber, M. (1982). Danazol. *Ann. Intern. Med.*, **96**, 625

56 Barbieri, R. L., Evans, S. and Kistner, R. W. (1982). Danazol in the treatment of endometriosis: analysis of 100 cases with a 4-year follow-up. *Fertil. Steril.*, **37**, 737

57 Meldrum, D. R., Chang, R. J., Lu, J., Vale, W., Rivier, J. and Judd, H. L. (1982). 'Medical oophorectomy' using a long-acting GNRH agonist – a possible new approach to the treatment of endometriosis. *J. Clin. Endocrinol. Metab.*, **54**, 1081

58 Lemay, A., Maheux, R., Faure, N., Jean, C. and Fazekas, A. T. A. (1984). Reversible hypogonadism induced by a luteinizing hormone-releasing hormone (LH-RH) agonist (Buserelin) as a new therapeutic approach for endometriosis. *Fertil. Steril.*, **41**, 863

59 Cohen, M. R. (1980). Laparoscopic diagnosis and pseudomenopause treatment of endometriosis with danazol. *Clin. Obstet. Gynecol.*, **23**, 901

60 Daniell, J. F. and Christianson, C. (1981). Combined laparoscopic surgery and danazol therapy for pelvic endometriosis. *Fertil. Steril.*, **35**, 521

61 Hasson, H. M. (1979). Electrocoagulation of pelvic endometriotic lesions with laparoscopic control. *Am. J. Obstet. Gynecol.*, **135**, 115

62 Puleo, J. G. and Hammond, C. B. (1983). Conservative treatment of endometriosis external: the effects of danazol therapy. *Fertil. Steril.*, **40**, 164

63 Guzick, D. A. and Rock, J. A. (1983). A comparison of danazol and conservative surgery for the treatment of infertility due to mild or moderate endometriosis. *Fertil. Steril.*, **40**, 580

64 Sadigh, H., Naples, J. D. and Batt, R. E. (1977). Conservative surgery for endometriosis in the infertile couple. *Obstet. Gynecol.*, **49**, 562

65 Brosens, I., Boeckx, W. and Gordts, S. (1978). Conservative surgery of ovarian endometriosis in infertility. *Eur. J. Obstet. Gynecol. Reprod. Biol.*, **8**, 277

66 Schenken, R. S. and Malinak, L. R. (1978). Reoperation after initial treatment of endometriosis with conservative surgery. *Am. J. Obstet. Gynecol.*, **131**, 416

67 Buttram, V. C., Jr. (1979). Surgical treatment of endometriosis in the infertile female: a modified approach. *Fertil. Steril.*, **32**, 635

68 Rock, J. A., Guzick, D. A., Sengos, C., Schweditsch, M., Sap, K. C. and Jones, H. W. Jr. (1981). The conservative surgical treatment of endometriosis: evaluation of pregnancy success with respect to the extent of disease as categorized using contemporary classification systems. *Fertil. Steril.*, **35**, 131

69 Buttram, V. C. Jr., Belue, J. B. and Reiter, R. (1982). Interim report of a study of danazol for the treatment of endometriosis. *Fertil. Steril.*, **37**, 478

70 Wheeler, J. M. and Malinak, L. R. (1981). Postoperative danazol therapy in infertility patients with severe endometriosis. *Fertil. Steril.*, **36**, 460

71 Dmowski, W. P. and Cohen, M. R. (1978). Antigonadotropin (danazol) in the treatment of endometriosis. Evaluation of post-treatment: fertility and three-year follow-up data. *Am. J. Obstet. Gynecol.*, **130**, 41

72 Punnonen, R., Klemi, P. and Nikkanen, V. (1980). Recurrent endometriosis. *Gynecol. Obstet. Invest.*, **11**, 307

73 Andrews, W. C. and Larsen, G. D. (1974). Endometriosis: treatment with hormonal pseudopregnancy and/or operation. *Am. J. Obstet. Gynecol.*, **118**, 643

74 Grant, A. (1966). Additional sterility factors in endometriosis. *Fertil. Steril.*, **17**, 514

75 Kistner, R. W. (1975). Management of endometriosis in the infertile patient. *Fertil. Steril.*, **26**, 1151

76 Brosens, I. A., Koninckx, P. R. and Corveleyn, P. A. (1978). A study of plasma progesterone, oestradiol-17β, prolactin and LH levels, and the luteal phase appearance of the ovaries in patients with endometriosis and infertility. *Br. J. Obstet. Gynaecol.*, **85**, 246

77 Koninckx, P. R., DeMoore, P. and Brosens, I. A. (1980). Diagnosis of the luteinized unruptured follicle syndrome by steroid hormone assays on peritoneal fluid. *Br. J. Obstet. Gynaecol.*, **87**, 929

78 Koninckx, P. R., Renair, M. and Brosens, I. A. (1980). Origin of peritoneal fluid in women: an ovarian exudation product. *Br. J. Obstet. Gynaecol.*, **87**, 177

79 Dmowski, W. P., Rao, R. and Scommegna, A. (1980). The luteinized unruptured follicle syndrome and endometriosis. *Fertil. Steril.*, **33**, 30

80 Portuondo, J. A., Pena, J., Otaola, C. and Echanojaurequi, A. D. (1983). Absence of ovulation stigma in the conception cycle. *Int. J. Fertil.*, **28**, 52

81 Lesorgen, P. R., Wu, C. H., Green, P. J., Gocial, B. and Lerner, L. J. (1984). Peritoneal fluid and serum steroids in infertility patients. *Fertil. Steril.*, **42**, 237

82 Dhont, M., Serreyn, R., Duvivier, P., Vanluchene, E., De Boever, J. and

Vandekerckhove, D. (1984). Ovulation stigma and concentration of progesterone and estradiol in peritoneal fluid: relation with fertility and
endometriosis. *Fertil. Steril.*, **41**, 872

83 Crain, J. L. and Luciano, A. A. (1983). Peritoneal fluid evaluation in
infertility. *Obstet. Gynecol.*, **61**, 159

84 Pickles, V. R., Hall, W. S., Best, R. A. and Smith, G. N. (1965). Prostaglandins in endometrium and menstrual fluid from normal and dysmenorrhoeic subjects. *J. Obstet. Gynaecol. Br. Commonw.*, **72**, 185

85 LeMaire, W. J., Leinder, R. and Marsh, J. M. (1973). Pre and post ovulatory changes in the concentration of prostaglandins in rabbit graafian
follicles. *Prostaglandins*, **3**, 367

86 Seibel, M. M., Swartz, S. L., Smith, D., Levesque, L. and Taymour, M. L.
(1984) *In vivo* prostaglandin concentrations in human preovulatory follicles. *Fertil. Steril.*, **42**, 482

87 Ylikorkala, O. and Tenhunen, A. (1984). Follicular fluid prostaglandins
in endometriosis and ovarian hyperstimulation. *Fertil. Steril.*, **41**, 66

88 Maia, H., Barbosa, I. and Coutinho, E. M. (1978). Inhibition of ovulation
in marmoset monkeys by indomethacin. *Fertil. Steril.*, **29**, 565

89 Chaudhuri, G. and Elder, M. G. (1976). Lack of evidence of inhibition of
ovulation by aspirin in women. *Prostaglandins*, **11**, 727

90 Jones, G. S. and Wentz, A. C. (1972). The effect of $PGF_{2\alpha}$ infusion on
corpus luteum function. *Am. J. Obstet. Gynecol.*, **114**, 393

91 Jewelewicz, R., Cantor, B. and Dyrenfurth, J. (1972). Intravenous infusion of prostaglandin $F_{2\alpha}$ in the midluteal phase of the normal human
menstrual cycle. *Prostaglandins*, **1**, 443

92 Wentz, A. C. and Jones, G. S. (1973). Transient luteolytic effect of $PGF_{2\alpha}$
in the human. *Obstet. Gynecol.*, **42**, 172

93 Sotrel, G., Helvacioglu, A., Dowers, S., Scommegna, A. and Auletta, F. J.
(1981). Mechanism of luteolysis: effect of estradiol and $PGF_{2\alpha}$ on corpus
luteum LH/HCG receptors and cyclic nucleotides in the rhesus monkey.
Am. J. Obstet. Gynecol., **139**, 134

94 Chang, M. C., Hunt, D. M. and Polge, C. (1973). Effects of prostaglandins (PGs) on sperm and egg transport in the rabbit. *Adv. Biosci.*, **9**, 805

95 Aref, I. and Hafez, E. S. (1976). Effects of prostaglandins on oviductal
contractility and egg transport in rabbits. In Harper, M. J. K., Pauerstein,
C. J., Adams, C. E., Coutinho, E. M., Croxatto, H. B. and Paton, D. M.
(eds.) *Ovum Transportation and Fertility Regulation.* p. 320. (Copenhagen: Scriptor)

96 Coutinho, E. M. and Maia, H. S. (1971). The contractile response of the
human uterus, Fallopian tubes and ovary to prostaglandins *in vivo*. *Fertil.
Steril.*, **22**, 539

97 Croxatto, H. B., Oritz, M.-E., Guiloff, E., Ibarra, A., Salvatierra, A.-M.,
Croxatto, H.-D. and Spilman, C. H. (1978). Effect of 15(S)-15-methyl
prostaglandin $F_{2\alpha}$ on human oviductal motility and ovum transport. *Fertil. Steril.*, **30**, 408

98 Lindblom, B., Hamberger, L. and Wiqvist, N. (1978). Differentiated contractile effects of prostaglandins E and F on the smooth muscle of the
human oviduct. *Fertil. Steril.*, **30**, 553

99 Omini, C., Pasargiklian, R., Folco, G. C., Fano, M. and Berti, F. (1978). Pharmacological activity of PGI_2 and its metabolite 6-oxo-$PGF_{1\alpha}$ on human uterus and fallopian tubes. *Prostaglandins*, **15**, 1045

100 Eddy, C. A. (1980). Ovum transport in the rhesus monkey following postovulatory intravaginal 15(S)-15-methyl prostaglandin $F_{2\alpha}$-methyl/ester administration. *Am. J. Obstet. Gynecol.*, **137**, 966

101 Abel, M. H. and Kelly, R. W. (1979). Differential production of prostaglandins within the human uterus. *Prostaglandins*, **18**, 821

102 Willman, E. A., Collins, W. P. and Clayton, S. G. (1976). Studies in the involvement of prostaglandins in uterine symptomatology and pathology. *Br. J. Obstet. Gynaecol.*, **83**, 337

103 Ylikorkala, O. and Viinikka, L. (1983). Prostaglandins and endometriosis. *Acta Obstet. Gynecol. Scand. (Suppl.)*, **113**, 105

104 Moon, Y. S., Leung, P. C., Yuen, B. H. and Gomel, V. (1981). Prostaglandin F in human endometriotic tissue. *Am. J. Obstet. Gynecol.*, **141**, 344

105 Moon, Y. S., Gomel, V., Yuen, B. H. and Nickerson, K. G. (1983). The role of prostaglandin F in the symptoms of endometriosis. *Can. Med. Assoc. J.*, **129**, 458

106 Meldrum, D. R., Shamonki, I. M., Clark, K. E., Rubinstein, L. M. and Lebherz, T. B. (1977). Prostaglandin content of ascitic fluid in endometriosis: A preliminary report. Presented at the *Twenty-Fifth Annual Meeting of the Pacific Coast Fertility Society*, October, Palm Springs, C A

107 Drake, T. S., Metz, S. A., Grunert, G. M. and O'Brien, W. F. (1980). Peritoneal fluid volume in endometriosis. *Fertil. Steril.*, **34**, 280

108 Drake, T. S., O'Brien, W. F. and Ramwell, P. W. (1983). Peritoneal fluid prostanoids in unexplained infertility. *Am. J. Obstet. Gynecol.*, **147**, 63

109 Rock, J. A., Dubin, N. H., Ghodgaonkar, R. B., Bergquist, C. A., Erozan, Y. S. and Kimball, A. W. Jr. (1982). Cul-de-sac fluid in women with endometriosis: fluid volume and prostanoid concentration during the proliferative phase of the cycle – days 8 to 12. *Fertil. Steril.*, **37**, 747

110 Singh, E. J., Baccarini, I. M. and Zuspan, F. P. (1975). Levels of prostaglandins $F_{2\alpha}$ and E_2 in human endometrium during the menstrual cycle. *Am. J. Obstet. Gynecol.*, **121**, 1003

111 Badawy, S. Z., Marshall, L., Gabal, A. A. and Nusbaum, M. L. (1982). The concentration of 13,14-dihydro-15-keto prostaglandin $F_{2\alpha}$ and prostaglandin E_2 in peritoneal fluid of infertile patients with and without endometriosis. *Fertil. Steril.*, **38**, 166

112 Sgarlata, C. S., Hertelendy, F. and Mikhail, G. (1983). The prostanoid content in peritoneal fluid and plasma of women with endometriosis. *Am. J. Obstet. Gynecol.*, **147**, 563

113 Dawood, M. Y., Khan-Dawood, F. S. and Wilson, L. Jr. (1984). Peritoneal fluid prostaglandins and prostanoids in women with endometriosis, chronic pelvic inflammatory disease, and pelvic pain. *Am. J. Obstet. Gynecol.*, **148** 391

114 Sun, F., Chapman, J. and McQuire, J. (1977). Metabolism of prostaglandin endoperoxide in animal tissues. *Prostaglandins*, **14**, 1055

115 Franson, R. and Waite, M. (1973). Lysosomal phospholipases A_1, and A_2 of normal and bacillus guerin-induced alveolar macrophages. *J. Cell. Biol.*, **56**, 621

116 Halme, J., Becker, S., Hammond, M. G., Raj, M. H. and Raj, S. (1983). Increased activation of pelvic macrophages in infertile women with mild endometriosis. *Am. J. Obstet. Gynecol.*, **145**, 333

117 Herman, A., Claeys, M., Moncada, S. and Vane, J. R. (1979). Biosynthesis of prostacyclin PGI_2 and 12-HETE by pericardium, pleura, peritoneum, and aorta of the rabbit. *Prostaglandins*, **18**, 439

118 Badawy, S. Z. A., Cuenca, V., Marshall, L., Munchbach, R., Rinas, A. C. and Coble, D. A. (1984). Cellular component in peritoneal fluid in infertile patients with and without endometriosis. *Fertil. Steril.*, **42**, 704

119 Schenken, R. S. and Asch, R. H. (1980). Surgical induction of endometriosis in the rabbit: effects on fertility and concentrations of peritoneal fluid prostaglandins. *Fertil. Steril.*, **34**, 581

120 Schenken, R. S., Asch, R. H., Williams, R. F. and Hodgen, G. D. (1984). Etiology of infertility in monkeys with endometriosis: measurement of peritoneal fluid prostaglandins. *Am. J. Obstet. Gynecol.*, **150**, 349

121 Schenken, R. S., Asch, R. H., Williams, R. F. and Hodgen, G. D. (1984). Etiology of infertility in monkeys with endometriosis: luteinized unruptured follicles, luteal phase defect, pelvic adhesions and spontaneous abortions. *Fertil. Steril.*, **41**, 122

122 Vasquez, G., Cornillie, F. and Brosens, I. A. (1984). Peritoneal endometriosis: scanning electron microscopy and histology of minimal pelvic endometriotic lesions. *Fertil. Steril.*, **42**, 696

123 Gould, S. F., Shannon, J. M. and Cunha, G. R. (1983). Nuclear binding sites in human endometriosis. *Fertil. Steril.*, **39**, 520

124 Kauppila, A., Puolakka, J. and Ylikorkala, O. (1979). Prostaglandin biosynthesis inhibitors and endometriosis. *Prostaglandins*, **18**, 655

20
Epilogue

G. S. BERGER, M. BYGDEMAN and L. G. KEITH

Key concepts regarding the structure, metabolism, and bioregulation of the prostaglandins were summarized in Chapter 1. In this final chapter we emphasize specific aspects of reproductive physiology and their clinical applications, with brief mention of current research interests and the possible future applications of prostaglandins and their inhibitors in clinical obstetrics and gynaecology.

In Chapter 4, Ulmsten describes the process of cervical ripening as pregnancy reaches term. For reasons that are not well understood, cervical ripening may begin either too early or too late and present a background for numerous problems of pregnancy, including premature delivery, spontaneous midtrimester abortion, and postmaturity.

Unlike the uterine fundus, the uterine cervix is composed not of muscle but primarily of collagen and connective tissue. The process of cervical ripening depends upon the biochemical and biophysical changes occurring within this connective tissue. Endogenous prostaglandins appear to be involved in ripening, and there is clear evidence that the exogenous administration of prostaglandin E_2 by any route produces physiological changes identical to those which occur naturally. The effect of prostaglandin E_2 administration is to increase collagenase activity which results in dissolution of the collagen fibrils. The mechanism by which PGE_2 activity increases to affect cervical ripening is unknown at present.

Despite the high degree of efficacy and safety of prostaglandin E_2 to effect cervical priming, its clinical use in most countries is restricted due to the commercial unavailability of pharmaceutical preparations. While individual physicians may be able to have a local pharmacist prepare prostaglandin E_2 in doses and vehicles suitable for use in cervical ripening, it will be necessary to have commercially produced products in the future if this effective clinical treatment is to become more widely available. Although prostaglandins are involved in the process of cervical ripening, the use of prostaglandin inhibitors to delay the onset of labour or to treat the problems of the incompetent cervix or habitual spontaneous abortion have received little research attention to date.

In Chapter 5, Lundström describes how prostaglandins modulate uterine contractility both during pregnancy and in the non-pregnant state. In sheep, the fetal pituitary initiates labour by the increased production of ACTH which results in an increase in adrenal steroid production, which then affects placental function to decrease the ratio of progesterone to oestrogen production. Progesterone acts as a stabilizer and oestrogen as a labilizer of membrane lysosomes. With the release of phospholipase A_2, phospholipids are enzymatically converted into arachidonic acid which then results in increased prostaglandin synthesis within the decidua which stimulates uterine contractions.

The modulation of uterine contractility varies for different endogenous prostaglandins. During pregnancy PGE_2 and $PGF_{2\alpha}$ cause increased uterine contractility. During the non-pregnant state $PGF_{2\alpha}$ always results in increased uterine contractility, whereas the effect of PGE_2 varies depending upon the particular stage of the menstrual cycle. During the proliferative and secretory phase, PGE_2 stimulates uterine contraction whereas at the time of ovulation and menstruation, PGE_2 acts as an inhibitor of contractility. Thus, the effect of the natural prostaglandins on the pattern of uterine contraction also varies depending upon the particular stage of the cycle. At the time of menstruation, strong contractions with decreased frequency are seen, whereas at ovulation there is a high frequency and low amplitude of contraction, and in early pregnancy, the uterus is almost silent in its contraction pattern. Interestingly, in dysmenorrhoea where an excess $PGF_{2\alpha}$ production has been noted, uterine contraction recordings have shown an increased amplitude, increased basic tone, and increased frequency of contraction which results in reduction in blood supply to the myometrium and ischaemic pain. Prostaglandin biosynthesis inhibitors do not stop uterine contractions in dysmenorrhoeic patients but rather change them back to the normal contractility pattern.

In Chapter 6, Lindblom describes the anatomy and physiology of the Fallopian tube. The Fallopian tube is a conduit for spermatozoa and for the egg. It has the dual functions of recovering the egg following ovulation and transporting it toward the uterus and assisting the migration of spermatozoa toward the ovarian end of the tube. Furthermore, the Fallopian tube must permit fertilization and early embryo development and actively hold the early fertilized egg within it until endometrial secretory activity is optimal for implantation, at which time the early embryo is discharged into the uterus.

Endogenous prostaglandins influence uterotubal junction activity to delay passage of the fertilized ovum until the optimal time for implantation; at this time the sphincter relaxes. This activity of the uterotubal junction is thought to be regulated by ovarian hormones and modulated by the prostaglandins.

Although the effects of prostaglandins on uterine muscle contraction and on the different layers of tubal musculature are known, the precise correlation between ovarian hormone production and tubal function is not yet entirely worked out. It seems likely that prostaglandin biosynthesis inhibitors may interfere with tubal function, but as yet there is inadequate information available on this subject.

In Chapter 7, Hamberger *et al.* describe the ovarian functions of hormonogenesis and ovulation. Prostaglandins appear to be involved in the initial recruitment of follicles and in follicular rupture. A definite increase in prostaglandin concentration in the follicular fluid is present immediately prior to rupture of the follicle, and the production of proteolytic enzymes appears to be influenced by the availability of prostaglandins. Prostaglandin $F_{2\alpha}$ results in the contraction of smooth muscle fibrils in the follicular wall which then results in expulsion of the ova. Furthermore, prostaglandins are involved in the regulation of the life span of the corpus luteum. An increase in $PGF_{2\alpha}$ production results in atrophy of the corpus luteum as demonstrated in a variety of animal species and possibly in humans as well. The source of $PGF_{2\alpha}$ production, however, may vary in different species since in animals, hysterectomy results in pseudopregnancy due to a prolonged life span of the corpus luteum, whereas in humans, hysterectomy does not necessarily lead to any changes in ovarian function. In animals, the administration of prostaglandin biosynthesis inhibitors prevents follicular rupture, although this phenomenon is not known to occur in humans. Whether prostaglandin biosynthesis inhibitors affect corpus luteum function in animals is not clear.

One of the major problems in assessing the relationship between prostaglandins and ovarian function is that the ovary, in contrast to the uterine cervix, is not easily accessible. In sheep a countercurrent mechanism exists in which a pulsatile release of prostaglandin $F_{2\alpha}$ from the endometrium can be traced in the uterine vein. If the uterine artery is separated from the vein, the corpus luteum will remain without luteolysis. Furthermore, radioactive labelled prostaglandin in the uterine vein can be followed into the uterine artery. In the sheep, an ovarian transplant to the neck with blood supplied by the carotid artery results in a persistent corpus luteum. The administration of $PGF_{2\alpha}$ into the carotid artery results in a normal hormonal ovarian cycle, and the amount of $PGF_{2\alpha}$ needed to cause corpus luteum regression of the transplanted ovary is equivalent to the amount found in the ovarian artery at the time of corpus luteum regression in the normal animal. While similar information has not been collected in the human, it is possible that a countercurrent mechanism may also exist, the disruption of which may account in part for the occurrence of the so-called 'poststerilization' syndrome.

In Chapter 8, Poyser describes the process of implantation in animals. Studies indicate that prostaglandin biosynthesis influences implantation, although the clinical implications of this observation are unclear. The administration of indomethacin inhibits implantation in animals and this effect can be partly overcome by giving prostaglandins at the same time. As a result of increasing interest in *in vitro* fertilization and embryo replacement, where the major attrition rate is associated with failure of implantation, there is corresponding clinical interest in the factors regulating implantation. Hopefully, this process will become better understood in the future and practical means of enhancing (or inhibiting) implantation will be developed.

Chapter 9, by Bygdeman, provides a notable contrast to the other chapters in this book because of its attention to the male and because of the dearth of information that has been accumulated on this subject over the past several decades. Since a number of prostaglandins are present in seminal fluid in vastly larger quantities than in any other tissue found in the body, it seems logical to believe that there must be a reason for this. One attractive hypothesis is that the seminal fluid prostaglandins exert their primary effects on female reproductive function following coitus at mid-cycle, but as yet there is no experimental evidence to support this. There are associations of sperm characteristics with seminal fluid prostaglandin content; sperm density is associated with the concentrations of PGE and 19-hydroxy-PGE in seminal fluid, and sperm motility is correlated with the concentrations of 19-hydroxy-PGE and 19-hydroxy-PGF in seminal fluid. It appears that an optimal prostaglandin concentration of the seminal fluid is a prerequisite for normal sperm characteristics and that prostaglandins are involved in the ejaculatory process, although which prostaglandins are involved is not known. In the short term, administration of prostaglandin biosynthesis inhibitors does not appear to influence sperm characteristics, although insufficient information has been accumulated to render a judgment about the possible effect of long-term prostaglandin inhibitors on male reproductive function. It is clear that in the future attention needs to be given to the role of prostaglandins in male reproductive function.

In Chapter 10, Calder discusses the problem of the gravida in whom obstetrical indications exist for delivery but who has an unripe cervix.

Although the cervix may ripen with further passage of time, the reasons requiring delivery often preclude lengthy delay. When the cervical score is 5 or above (by Bishop's criterion) and delivery is indicated, induction can usually be performed with oxytocin. When the cervical score is less than 5, however, prostaglandin E_2 can convert an unripe cervix to a ripe cervix within 6–12 h in most cases. This use of prostaglandin is likely to become one of the most important and widely applied applications for these compounds in clinical obstetrics. Unfortunately, the general unavailability of pharmaceutically prepared preparations suitable for cervical ripening has limited the application of this clinical treatment. Many such preparations are currently in the process of development in various European countries and these are likely to be available within the next 2 or 3 years. In the meantime, clinicians must depend on a local pharmacist preparing PGE_2 in Tylose gel or administer it via extra-amniotic infusion.

Chapter 11 by Lange describes the induction of labour with PGE_2 or $PGF_{2\alpha}$, both of which are equally effective as oxytocin infusion but which have different risks and side-effects. The main disadvantages with the use of intravenous prostaglandin are local irritation at the site of infusion, gastrointestinal side-effects, and a relatively narrow gap between effective dosage and toxicity. Overdosage with prostaglandins may result in overstimulation of uterine contractions with tetany. This also occurs with oxytocin overdosage, but the risk of water intoxication which accompanies oxytocin infusion does not accompany induction of labour with prostaglandin. Another difference between oxytocin induction and prostaglandin induction involves the dose and timing of administration of the pharmaceutical agent. Unlike oxytocin, which is generally given in increasing or doubling doses in order to achieve effective treatment, prostaglandin induction involves repetition of the same dose at shorter intervals. Recognition of these differences is required to ensure safe induction of labour with prostaglandins. Oxytocin and prostaglandins cannot be used in combination because they may have synergistic effects. Lange considers intravenous oxytocin infusion as the 'gold standard' for the induction of labour since it has been a recognized method in clinical use for over two decades, and he points out that any alternative method of treatment must be significantly better than this standard in order to replace it. It is doubtful that intravenous prostaglandin induction is significantly more effective than oxytocin induction. One particular advantage, however, accompanies the use of oral PGE_2 tablets in that they are more convenient to use than intravenous infusion of either prostaglandin or oxytocin. In combination with amniotomy, oral E_2 is as effective as i.v. oxytocin combined with amniotomy. While oral E_2 tablets are currently available in Great Britain for induction of labour, they have not replaced oxytocin induction but do provide an alternative method of treatment for low risk patients.

In Chapter 12, Thiery and Amy discuss the use of prostaglandin biosynthesis inhibitors for the inhibition of labour. They thoroughly review all clinically available methods of labour inhibition and their mechanisms of action, as well as their efficacy and safety. The use of β-mimetic agents, particularly of the β-2 type such as ritodrine, Fenoteral, and Salbutanol all appear comparable with respect to safety and efficacy. While the use of prostaglandin inhibitors such as indomethacin or naproxen is highly effective in inhibiting premature labour, the major problem limiting their use is the potential of detrimental effects to the fetus, in particular the risk of premature closure of the ductus arteriosus leading to pulmonary hypertension. Thiery and Amy point out that the risk of this occurrence is quite rare, however. Because of their high degree of effectiveness in inhibition of uterine contractions, prostaglandin biosynthesis inhibitors may be used when other means of inhibiting premature labour have proved ineffective. Perhaps the use of prostaglandin inhibitors will increase in the future if further evidence indicates a low probability of fetal risk, because they are so highly effective.

Chapter 13, by Toppozada, presents a thorough yet succinct review of the various causes of postpartum haemorrhage and of all the clinical methods of its treatment. When other causes of postpartum haemorrhage have been ruled out, such as retained placenta, uterine or cervical trauma, and coagulation defects, and when other available methods of treatment have failed, the administration of 15-methyl-$PGF_{2\alpha}$ intramuscularly in a dose of 0.25 mg every 1-1½ h can be life saving. In fact, this treatment is so effective that persistent bleeding following prostaglandin administration suggests a diagnosis other than atonic bleeding. Even in severe cases of uterine atony when the uterus is less likely to react to oxytocin or ergot alkaloids the administration of 15-methyl-$PGF_{2\alpha}$ is highly successful. The success rate in the two studies cited has been over 90%. This is remarkable considering that almost all of those patients were *in extremis* and the therapy itself consists of only intermittent intramuscular injections. In countries where 15-methyl-$PGF_{2\alpha}$ is not available, $PGF_{2\alpha}$ can be given as an injection directly into the cervix or myometrium.

In Chapter 14, Ratnam and Prasad review the available methods of treatment of abnormal uterine pregnancies including missed abortion, anencephaly, and fetal death *in utero*. They conclude that the use of intramuscular prostaglandin analogues represents the best available treatment for these conditions.

The patient with an abnormal intrauterine pregnancy represents a problem to the clinician; oxytocin induction may not be effective early in pregnancy, and surgical methods have increased risks the later they are applied in pregnancy. They may be especially risky if the patient has developed a coagulation disorder accompanying intrauterine fetal death.

The choice of treatment depends largely upon whether membranes are present or not. The molar pregnancy behaves differently because of the absence of membranes and thus the absence of phospholipase necessary for the production of prostaglandins. In Ratnam's opinion, the method of choice for the management of molar pregnancy and of missed abortion is intramuscular injection of prostaglandin analogues, two of which are currently commercially available, 15-methyl-PGF$_{2\alpha}$ and sulprostone. Both of these compounds have proven highly effective with short durations of labour, although gastrointestinal side-effects can occur. In countries where the risk of choriocarcinoma is high, vacuum aspiration of the uterus at any stage of pregnancy may be preferred over medical induction of labour. In countries where the risk of choriocarcinoma is low, on the other hand, the use of prostaglandins is preferred by some to vacuum aspiration, particularly in the second trimester of pregnancy.

Chapter 15 on induced abortion by Lauersen provides one of the most thorough reviews of the existing voluminous literature on this subject. In the first trimester of pregnancy, the method of choice for induced abortion is vacuum aspiration because of its high efficacy and low risk of complications. However, the most difficult part of this procedure may be mechanical dilatation of the cervix, especially in the primiparous patient. In this situation, preoperative treatment with prostaglandin to ripen and dilate the cervix facilitates the procedure and reduces postoperative complications. In the second trimester of pregnancy, intra-amniotic or extra-amniotic administration of natural prostaglandins (E$_2$ or F$_{2\alpha}$) has been widely used for induced abortion and has proven to be more effective and more rapid than the use of hypertonic saline, urea, and/or oxytocin infusion. On the other hand, the frequency of gastrointestinal side-effects and the risk of cervical laceration accompanying prostaglandin induction is higher than with hypertonic saline infusion. The major risk accompanying the intra-amniotic instillation of hypertonic saline, that is, serious coagulation defects or direct intravascular injection, may be avoided with the use of prostaglandins. The availability of prostaglandin analogues suitable for non-invasive intramuscular injection represents a great advantage over intra-amniotic or extra-amniotic instillation of the natural prostaglandins.

Although some authorities, particularly those in the United States, have argued that surgical dilatation and evacuation is the preferred method of induced abortion in the second trimester as well as in the first, in our opinion surgical termination of pregnancy in the second trimester should only be performed in centres where physicians have considerable experience in this technique, since risks to patients can be severe when performed by the inexperienced or occasional operator. Moreover, for second trimester abortion, a simple method of treatment such as intramuscular prostaglandin injection is advantageous because the outcome does not depend upon physician's skill to the same extent as does surgical termination.

While prostaglandins have the capacity to terminate very early pregnancies (within the first 2 weeks after a 'missed' menstrual flow and thus these compounds can be used for 'menstrual regulation'), the frequency of gastrointestinal side-effects and uterine pain limits their use to selected cases where surgical treatment may not be desired by the patient or the physician.

In Chapter 16, Russel and Owens discuss the use of prostaglandin biosynthesis inhibitors for the treatment of primary dysmenorrhoea. This clinical application represents the most common indication for PG inhibitors in gynaecology at the present time. The authors present a rational basis for the use of these agents based on the pathophysiological excess of prostaglandin production by the endometrium in the majority of patients with primary dysmenorrhoea. Reference to Figure 16.3 will lead the clinician to a simple and rational decision for the use of prostaglandin inhibitors in appropriate cases.

The fourth section of this book, Chapters 17–19, deals with ongoing clinical research and possible future applications for prostaglandins and their analogues or inhibitors in clinical obstetrics and gynaecology. In this section, the pathophysiology of toxaemia of pregnancy, premenstrual syndrome, and endometriosis are reviewed. Unfortunately, the pathophysiological mechanisms of these conditions are incompletely understood and, although prostaglandins may be involved in various ways, possible clinical application of prostaglandins or prostaglandin synthesis inhibitors has only a limited rationale at present. Further research will be required before specific prostaglandin analogues or more specific prostaglandin biosynthesis inhibitors may be shown to have clinical application. The same can be said for the problems of infertility, particularly that of male infertility where the role of prostaglandins has been almost entirely neglected in basic research.

The decision to use prostaglandins and prostaglandin biosynthesis inhibitors in the clinical treatment of obstetric and gynaecologic disorders rests with the individual clinician, and of necessity remains within the limits of pharmaceutical availability and governmental regulations regarding the use of pharmaceutical products. We believe that this volume accurately summarizes the use of natural prostaglandins, synthetic prostaglandin analogues, and prostaglandin biosynthesis inhibitors for clinical use in the 1980s, and we thank this international panel of experts who have so willingly provided this useful reference for the interested clinician.

Index